AF556024

Food Adulterants and Contaminants

NIPA® GENX ELECTRONIC RESOURCES & SOLUTIONS P. LTD.
New Delhi-110 034

About the Editors

Dr. Gopal Kumar Sharma, completed his MSc (1981) and PhD in Chemistry (1986) from Lucknow University. He started his career as Scientist 'B' in DFRL in the year 1987 and rendered more than 3 decades of service in the field of food technology and nutritional sciences to developed instant and ready-to-eat processed foods and their induction in Armed forces operational rations i.e. Emergency flying ration, Meals-Ready-to-Eat (MRE) ration, Main Battle Tank ration and submarine crew ration which are based on quick cooking dhals, kadhi, stuffed parotha, chapathies, edible rolls, quick reconstituting jiffy products, energy dense bars such as protein bar, omega 3 rich bar flax oat bar.

Dr GK Sharma designed low cost processing equipments such as Infra Red (IR), Microwave and automatic chapati making machine for up scaling the food processing technologies. Shelf life of cereals, legumes and their milled products were enhanced using IR and Micro wave based processing without usage of preservatives.

He is also the inventor of patented anti-oxygenic salt for enhancing shelf life of deep fried products, cold shock dehydration technology for instantisation of cereal/pulses/legumes and vegetables which reconstitute in 2 to 5 minutes, combination processing for Jiffy products which reconstitute within 1 to 2 minutes by addition of hot water, stack encapsulation technology for sugar/salt to keep them free flowing in high humid areas. Dr GK Sharma also standardised use of natural antioxidants to retard oxidative spoilage of oil and processed foods.

Dr GK Sharma has 115 national and international publications, 16 patents, 41 popular articles, 78 products, 13 technologies, 120 technology transfer,29 National and DRDO Awards to his credit. He also guided 3 students for Ph.D, 23 M.Sc/M.Tech/B.E/PGD for their dissertation. His research accomplishments are well documented in reputed international journals with H index of 18, i10 index of 34 and citation index of 1139. His efforts in up scaling and commercialization of processes and technologies helped in large scale production and supply of processed foods to Indian population and reduced the post-harvest losses in the primary agricultural produce.

1. Fellow, Association of Food Scientist & Technologist of India (AFSTI)
2. NN Mohan Memorial Award (2015)-AIFPA
3. Kejriwal Award (2014)- Studies on the development of Panner stuffed paraothas by thermal processing- AIFPA
4. NN Mohan Memorial Award (2006)- AIFPA
5. National Technology Day Award (2011)-DRDO
6. National Science Day Oration Award (2009)-DRDO

7. Technology Group Award (2003) Development of freeze-thaw dehydration of Instantisation of whole legumes and pulses-DRDO
8. Technology Group Award (2008) Development of shelf stable thermally processed chapaties with better sensory attributes to cater to the requirement of Armed Forces. -DRDO
9. Technology Group Award (2010) Development of microwave based technology for control of infestation and elimination of enzymatic browning in food grains.-DRDO
10. Technology Group Award (2012) Development of various value added products from VCM and evaluation of storage stability of VCO under different packaging conditions as well as anti-diabetic and hypocholesterimic effects of VCO in experimental rats.-DRDO
11. Technology Group Award (2014) by Defence Institute of Physiology and Allied Science (DIPAS) for developing specific technology for defence. -DRDO
12. Technology Group Award (2016) Development of Nutritious bars , ready to eat and ready to reconstitute idlis , instant sambhar mix and chutneys and atomization low cost bar making as well as idli making machine. -DRDO
13. Laboratory Scientist of The Year Award 2010.-DFRL, DRDO
14. All Rounder Scientist Award (1999) For making significant contribution in Food Science , supply of ration during Kargil and Welfare activities in -DFRL- DFRL, DRDO
15. DRDO Cash Award (1990) Investigation on the interaction of plastic packaging materials with food stuff.- DFRL,DRDO
16. DRDO Cash Award (1995) Development of instant foods and mixes and methodology for separation of carbonyls- DFRL,DRDO
17. Best Research Paper Award (1996) Development of instant Kadhi Mix-DFRL, DRDO
18. Best Research Paper Award (1996) Flavour degradation in dehydrated convenience foods - changes in carbonyl in quick cooking rice and Bengal gram dhal- DFRL, DRDO 1
19. Best Research Paper Award (1997) Suitability of Antioxigenic salts for stabilization of fried snacks- DFRL, DRDO
20. Best Research Paper Award (2007) Leaching of aluminum from utensils during cooking of foods¬ DFRL, DRDO
21. Best Research Paper Award (2008) Antioxidant potential of synthetic and natural antioxidants and its effect on WOF in different species of meat- DFRL, DRDO

22. Best Research Paper Award (2008) Star-anise and black caraway as natural antioxidants- DFRL,DRDO
23. Best Research Paper Award (2009) Soybean – A unique legume for improving the health and nutritional status of Indian population- DFRL,DRDO
24. Best Research Paper Award (2010) Lipid oxidation and its control in meat and meat products¬ DFRL,DRDO
25. Best Research Paper Award (2011) Storage stability of omega-3-rich bar-DFRL,DRDO
26. Best Research Paper Award (2012) Development and evaluation of long shelf-life ambient stable chapaties without the use of chemical preservatives-DFRL,DRDO
27. Best Research Paper Award (2015). Studies on the optimization and stability of instant wheat porridge mix- DFRL,DRDO
28. Patra-Patrika Protsahan Yojna Award (1997)-Ministry of Defence
29. Hindi Essay Writing Competition Award (1987,1988, 1990)- DFRL,DRDO
30. Hindi Elocution Award (1991)- DFRL,DRDO

1. Editorial Board Member: J Food Science and Technology (2003-04)
2. Editorial Board Member : Indian Food Industry (1999-2000)
3. Member: Bureau of Indian Standards (1999-Till Date)
4. Member: Board of Studies in Food Science and Technology, University of Mysore (1999-2000)
5. Member: Steering Committee, Disaster Management in the field of Food Science and Nutrition
6. Member; Research Advisory Committee, Indian Institute of Crop Processing (Ministry of Food Processing Industries, Govt. of India), Thanjavur (2007-2013)

Janifer Raj Xavier, Scientist 'E' completed her M.Sc. in Agricultural Biotechnology in 2006 at Center for Plant Molecular Biology, Tamil Nadu Agricultural Research University (TNAU), Coimbatore. She served in Defence Institute of High-Altitude Research (DIHAR), Leh, Ladakh from 2008-2012 and at Defence Food Research Laboratory (DFRL), Mysore since 2012. Her significant scientific contributions include: Development of light weight biodegradable cutlery and films, scale-up of bacterial cellulose production and preparation of oligosaccharides from food processing by-products, using in-house microbial enzymes for prebiotic applications. Designed and modernized clean kitchen for safe and clean production as well as packaging of foods for Indian Space programme 'Gaganyaan'. She has been conferred with awards such as Young Scientist award of Mycological Society of India, 2014 for developing a DNA based detection system for mycotoxins, Defence technology spin-off award 2017 of DRDO for developing and commercializing edible cutlery and edible films, Technology group award 2017 and 2021 of DFRL for the best team in developing technologies for defence forces during Chemical, Biological, Radiation and Nuclear warfare (CBRN) scenario and DFRL laboratory scientist of the year award (2019). She has a granted patent on development of biodegradable containers (2021), 02 edited books and more than 50 research publications in international and national journals to her credit. She is an also an expert evaluator for BIRAC Innovation fellowships for Igniting new Market-led Ideas and review editor for the journal frontiers for young minds.

Dr. Ajay Singh, has about 8+ years of teaching and research experience. He has published around 40 articles, more than 30 book chapters in publishing houses of international repute and edited 07 books (03 CRC Press, 01 NIPA, and 01 Objective Food Science with Brillion Publishers). Dr. Singh presented his talks and demonstrated workshops at various national and international venues. He is the potential reviewer for many international journals (Springer, Elsevier, Wiley, and Taylor & Francis). He serves as an Academic Editor(s) of the Journal of Food Processing and Preservation and the Journal of Food Biochemistry, both published by Wiley-Hindawi. He is the lead guest issue editor for the International Journal of Bioinformatics (PubMed & SCOPUS) and Journal of Food Processing and Preservation (SCI). Dr. Singh's area of research and teaching is composite blended foods, gluten-free baked optimization, and functional approaches among eatables. He handled a sponsored project entitled "Snacks- Technology, Process & Automation" funded from RONDO-Switzerland. He also served as BOS member among various academic bodies for curriculum development, paper setters and potential reviewer for lists of journals of immense height among food sciences domain.

Food Adulterants and Contaminants

Gopal Kumar Sharma
Defence Food Research Laboratory
Defence Research and Development Organization
(Ministry of Defence, Government of India)
Myosre, Karnataka 570 011, India

Janifer Raj Xavier
Defence Food Research Laboratory
Defence Research and Development Organization
(Ministry of Defence, Government of India)
Myosre, Karnataka 570 011, India

Ajay Singh
Department of Food Technology
Mata Gujri College,
Fatehgarh Sahib, Punjab-140406, India

NIPA® GENX ELECTRONIC RESOURCES & SOLUTIONS P. LTD.
New Delhi-110 034

NIPA® GENX ELECTRONIC RESOURCES & SOLUTIONS P. LTD.

101,103, Vikas Surya Plaza, CU Block
L.S.C.Market, Pitam Pura, New Delhi-110 034
Ph : +91 11 27341616, 27341717, 27341718
E-mail: newindiapublishingagency@gmail.com
www: www.nipabooks.com

For customer assistance, please contact
Phone: + 91-11-27 34 17 17
Fax: + 91-11- 27 34 16 16
E-Mail: feedbacks@nipabooks.com

Print ISBN: 978-81-19235-04-9

ebook ISBN: 978-81-19235-02-5

Composed and Designed by NIPA®.

GSTN : 03 AACHM 9649 M 1 ZV FSSAI : 12118721000032
CIN No. L 15311 CH 2011PLC 033393

MEGASTAR FOODS LTD.

Registered Office: Plot No. 807 Industrial Area Phase-2 Chandigarh - 160 002
Works: Kurali-Ropar Road, Village Solkhian-140 108, Distt. Roopnagar, Punjab
Ph.: + *91 1881 240 401 (6 Lines) E-mail: info@megastarfood.com

Dr. Bhupendar Singh Khatkar
Ph.D. (University of Reading, UK)
General Manager (Quality, R&D)

Foreword

In recent years, there has been a growing concern due to ill effects of food adulteration on health and wellbeing. Food Adulteration and Contamination are ever-present challenges that threaten our health as well as the very essence of our food heritage. It brings together leading experts and scholars from diverse fields to dissect the multifaceted challenges of food safety and authenticity. This book is a compendium of the collaborative efforts of experts from diverse fields who have come together to provide insights into this multifaceted issue. With each turn of the page, readers will gain a deeper understanding of the challenges we face and the innovative solutions that are being offered to address them. From the fundamental principles of food toxicology to the intricacies of detecting adulterants in products as diverse as coffee, spices, and dietary supplements, this book provides a comprehensive and up-to-date overview of the field.

The threat of food bioterrorism looms large in our modern world, and its implications for human physiology are dissected. The book takes us on a journey through food processing, additives, and packaging materials, all of which can be sources of adulterants and contaminants. The field of foodomics provides a powerful tool for the detection of adulterants, while nanotechnology offers promising avenues for enhancing food quality and safety. Chapters dedicated to specific food categories, such as cereals, oils, fruits, and meats, shed light on the unique challenges each faces in terms of adulteration. Moreover, it does not stop at detection but also delves into prevention strategies, tracing the evolving landscape of food fraud prevention and the dynamic relationship between authenticity, traceability, branding, and consumer preference. They also culminate in a discussion of strategies to prevent food fraud, addressing the ever- evolving challenges in this domain.

This book is more than just a compilation of scientific knowledge; it is a call to action. It challenges us to be vigilant consumers, informed policymakers, and proactive researchers. It urges us to prioritize food safety and authenticity in a world where the global food supply chain is more interconnected than ever before. I commend the dedication and expertise of the editors and authors who have contributed to this book and am confident that their work will inspire further research, innovation, and, most importantly, a renewed commitment to safeguarding the purity and trustworthiness of our food. May this book serve as a beacon of knowledge and a catalyst for positive change in the world of food safety and authenticity.

Bhupendar Singh Khatkar

Preface

Food adulteration involves the deliberate act of diminishing the quality of food products for the purpose of financial gain. This unethical practice can take various forms, including mixing or substituting with subpar ingredients or removing valuable components. Food adulteration is a pervasive issue that affects various aspects of our daily food consumption. In recent times, the field of food science has witnessed remarkable technological advancements, such as the development of nutraceuticals, functional foods, probiotics, and composite blends, all aimed at enhancing the joy of eating and promoting healthier living. However, the flip side of this progress is the adoption of adulteration techniques by unscrupulous individuals solely for monetary gain. This poses significant health risks to consumers and extends beyond the realm of food, encompassing sectors like pharmaceuticals, cosmetics, fuels, and chemicals. Today, we witness a distressingly long list of malpractices among food industries, wherein adulteration of raw and processed cereals, spices, fruits, vegetables, dairy products, and other additives is generally practiced. To combat these fraudulent practices effectively, amendments in labeling policies and their rigorous enforcement are imperative. Government initiatives, such as penalties and revisions in food laws, coupled with the widespread dissemination of home-based detective measures, represent the most viable means of addressing this issue. Initiatives like "DART: Detect Adulteration Rapid Test, Pink book," led by FSSAI (Food Safety and Standards Authority of India), exemplify such efforts. Governments must intensify their endeavors to ensure that policies are effectively communicated, with the overarching goal of ensuring safe and healthy food for all citizens, thus contributing to the development of a healthier nation. Given that food adulteration is a global concern, various aspects, including its detection, have gained paramount importance in the current scenario. However, the steps involved in sample preparation for the analysis of adulterants in food, utilizing high-end technologies, have made the process cumbersome and time-consuming. Consequently, there is a growing interest in developing rapid methods for detecting food adulteration, especially those that enable real-time, non-destructive assessment of food quality, which is highly desirable in the present scenario. The need of the hour is to create composite in silico tools and computer-vision systems that require minimal analytical technology for rapid, non-destructive, highly efficient, and cost-effective detection of food adulteration. These tools should be accessible for use in the field or at the point of consumption by individuals with limited training, ensuring reproducible results.

The primary aim of this book is to provide a comprehensive understanding of food adulteration and toxicology, food safety standards for quality assurance, food bioterrorism, the adverse effects of food contact packaging materials, nanotechnology, and omics technologies in the detection of food adulteration. It delves into the adulteration of various food categories, such as cereals, pulses, oils, fats, fruits, vegetables, milk and milk-based products, meat and meat-based products, and nutraceuticals, while highlighting recent trends in detection methods. Additionally, the book sheds light on global trends in food authenticity, traceability, branding, consumer preferences, and the challenges faced in preventing and detecting food adulteration and food fraud. This book serves as an illuminating guide to the latest advancements in the detection of adulteration in foods and beverages, as well as the diverse range of raw materials used in their production. It is an invaluable resource for anyone with an interest in ensuring the safety and authenticity of food on a global scale. The comprehensive collection of insights into adulteration, detection methods, food science, food culture, and the rapid techniques employed to identify food adulterants places a strong emphasis on both classical and contemporary approaches to detecting food adulteration in a wide array of food categories, making this book unique. It covers cereals, legumes, fruits, vegetables, dairy products, fish, and meat, and highlights emerging trends and utilization of point-of-care devices, as well as cutting-edge omic technologies. Beyond the laboratory, it also explores global trends in food authenticity, traceability, branding, and consumer preferences. It also covers the strategic measures taken to prevent food fraud and addresses the multifaceted challenges faced in the ongoing battle against adulteration and food fraud. In essence, this book offers a holistic perspective on the critical issue of food adulteration, making it an indispensable reference for researchers, food professionals, policymakers, and anyone passionate about preserving the integrity of our food supply in an ever-evolving world.

Editors

Contents

List of Contributors

Aastha Deewan, Department of Food Technology, Guru Jambheshwar University of Science and Technology, Hisar, Harayana, India

Abhishek B.S., Defence Food Research Laboratory, Defence Research and Development Organization, Mysore, Karnataka-570011, India

A.D. Semwal, Defence Food Research Laboratory, Defence Research and Development Organization, Mysore, Karnataka-570011, India

Ajay Singh, Department of Food Technology, Mata Gujri College, Fatehgarh Sahib Punjab-140406, India

Alan Jerish J., Defence Food Research Laboratory, Defence Research and Development Organization, Mysore, Karnataka-570011, India

B.S. Khatkar, Department of Food Technology, Guru Jambheshwar University of Science and Technology, Hisar, Harayana, India

Balakrishna K., Department of Food Microbiology, Defence Food Research Laboratory (DFRL) Siddhartha Nagar, Mysore-570011, India

Bibha Kumari, Department of Zoology, Magadh Mahila College, Patna University, Patna, India

C. Sarathambal, ICAR-Indian Institute of Spices Research Marikunnu (PO), Kozhikode-673012

E. Jayashree, ICAR-Indian Institute of Spices Research Marikunnu (PO), Kozhikode-673012

Elizabeth Thomas, Dairy Technology Division, ICAR-National Dairy Research Institute Karnal-132001, Haryana, India

Gopal Kumar Sharma, Defence Food Research Laboratory, Defence Research and Development Organization, Mysore, Karnataka-570011, India

Heena Sharma, ICAR-National Dairy Research Institute, Karnal-132001, Haryana, Indiaa

Janifer Raj Xavier, Defence Food Research Laboratory, Defence Research and Development Organization, Mysore, Karnataka-570011, India

Jyoti Rani, Department of Nutrition and Dietetics, Chandigarh University, Mohali-140413

K. Anees, ICAR-Indian Institute of Spices Research Marikunnu (PO), Kozhikode-673012, India

K.R. Anilakumar, Defence Food Research Laboratory, Defence Research and Development, Organization, Mysore, Karnataka-570011, India

K.G. Nelum P. Piyasena, Biochemistry Division, Tea Research Institute of Sri Lanka, Talawakelle, Sri Lanka

Kavanashree N., Defence Food Research Laboratory, Defence Research and Development Organization, Mysore, Karnataka-570011, India

Mahasen A.B. Ranatunga, Plant Breeding Division, Tea Research Institute of Sri Lanka, Talawakelle, Sri Lanka

Mrithula Mahalakshmi Madhan Kumar, Department of Food Process Engineering, University of New South Wales, Sydney, Australia

Mukul Sain, ICAR-National Dairy Research Institute, Karnal, Haryana, India

Narender Raju Panjagari, Senior Scientist, Food Packaging Laboratory, Dairy Technology Division, ICAR-National Dairy Research Institute Karnal, Haryana-132001, India ICAR-National Dairy Research Institute, Karnal-132001, Haryana, India

Neha Chaubey, Research Scholar; Dairy Technology Division, ICAR-National Dairy Research Institute, Karnal Haryana, India

Neha Thakur, Department of Livestock Products Technology, Lala Lajpat Rai University of Veterinary and Animal Sciences (LUVAS), Hisar-125001, Haryana, India

Pal Murugan M., Defence Food Research Laboratory, Defence Research and Development Organization, Mysore, Karnataka-570011, India

Parveen Kumari, Assistant Professor, Department of Food Technology, Guru Jambheshwar, University of Science and Technology, Hisar, Harayana, India

Poornima, ICAR-National Dairy Research Institute, Karnal, Haryana, India

Pradyuman Kumar, Department of Food Engineering and Technology, Sant Longowal Institute of Engineering & Technology, Longowal, Sangrur-148106, Punjab, India

Priyanka Kajla, Assistant Professor, Department of Food Technology, Guru Jambheshwar, University of Science and Technology, Hisar, Harayana, India

Priyanka Minhas, College of Veterinary Science, GADVASU, Ludhiana, Punjab, India

Rajan Sharma, ICAR-National Dairy Research Institute, Karnal, Haryana, India

Richa Singh, ICAR-National Dairy Research Institute, Karnal 132 001, Haryana, India

Rosy Bansal, GSSDGS Khalsa College Patiala-147001, Punjab, India

Ruby Siwach, Assistant Professor, College of Dairy Science and Technology, Lala Lajpat Rai, University of Veterinary and Animal Sciences, Hisar, Haryana, India

Sangita Ganguly, Scientist (Senior Scale), Functional Ingredients & Applications Laboratory, Dairy Technology Division, ICAR-National Dairy Research Institute Karnal, Haryana-132001. Haryana, India

Sanjay Yadav, Department of Livestock Products Technology, Lala Lajpat Rai University of Veterinary and Animal Sciences (LUVAS), Hisar-125001, Haryana, India

Santosh Pal, Defence Food Research Laboratory, Defence Research and Development Organization, Mysore, Karnataka, India-570011, India

Semwal A.D., Defence Food Research Laboratory, Defence Research and Development Organization, Mysore, Karnataka, India-570011, India

Sheetal Thakur, University Centre for Research and Development, Department of Biotechnology, Chandigarh University, Gharuan-Mohali, Punjab, 140413, India

Shylaja R., Department of Food Microbiology, Defence Food Research Laboratory (DFRL) Siddhartha Nagar, Mysore-570011, India

Snigdha Mohandas, ICAR-Indian Institute of Spices Research Marikunnu (PO), Kozhikode-673012, India

Suresh Kumar, National Institute of Biologicals (NIB), Ministry of Health & Family Welfare, Govt. of India, Sector-62, Noida,Uttar Pradesh-201309, India

Swati Haldar, Department of Genetics, Albert Einstein College of Medicine, Bronx, New York, 10461, USA

Twinkle Suri, Department of Food Engineering and Technology, Sant Longowal Institute of Engineering & Technology, Longowal, Sangrur-148106, Punjab, India

Vandana Chaudhary, Assistant Professor, College of Dairy Science and Technology, Lala Lajpat Rai, University of Veterinary and Animal Sciences, Hisar, Haryana, India

Vijay K. Bharti, DRDO-Defence Institute of High Altitude Research (DIHAR), Leh, UT Ladakh, India

Viney Kumar, Department of Biotechnology, Indian Institute of Technology Roorkee, Uttarakhand 247667, India and Department of Pathology, Albert Einstein College of Medicine, Bronx New York, 10461, USA

Wadikar, D.D., Defence Food Research Laboratory, Defence Research and Development Organization, Mysore, Karnataka-570011, India

1

Food Adulteration and Toxicology Impact of Raw Materials/Environment in Food Safety

Parveen Kumari[1], Priyanka Kajla[1], Aastha Deewan[1] and B.S. Khatkar[2]

[1]Department of Food Technology, Guru Jambheshwar University of Science and Technology, Hisar, Harayana, India
[2]Department of Food Technology, Guru Jambheshwar University of Science and Technology, Hisar, Harayana, India

Introduction

In the ancient era, adulteration was practiced as economical fraud like the addition of water to milk and stone to grains and pulses but nowadays it has become a global concern due to adverse health effects of adulterants on a human being. Words like substitution and admixture can help define adulteration. In general, adulteration is defined as either removal of valuable constituents from food or intentionally/unintentionally addition of substandard quality material to degrade food quality. Food adulteration not only negatively affects the quality of food products but also results in numerous adverse health effects along with significant economic problems (Cheng *et al.*, 2015). Toxicants/adulterants are found in traces in food products that are being added intentionally (to enhance aesthetic quality) or unintentionally (due to processing or mishandling) and it is almost impossible to visually detect these contaminants and judge the wholesomeness of the food products (Sharma *et al.*, 2017). Therefore, it is the demand of time for authentic testing of food products for quality assurance and consumer protection from being deceived. The Indian government has established a national-level authority -Food Standards and Safety Authority of India (FSSAI) which looks after the concerns about adulteration in food products thereby providing quality assured products to consumers (Bansal *et al.*, 2017; Babu and Shenolikar, 1995).

According to FSSA (Food Safety Standard Act of India), adulterant is defined as any particle or material that could be added to make food unsafe for consumption or substandard or it may be extraneous matter. According to FFDCA (Federal Food, Drug and Cosmetic Act) and FDA (Food and Drug Administration) food is considered adulterated if:

- a substance added is injurious to health
- a cheaper or inferior quality ingredient added to food
- any valuable constituent is extracted from the main food article
- quality of food is below the standards
- any substance has been added to increase bulk or weight
- to make it appear more valuable

Adulterated foods that contain toxins and have ill effects on health or lack nutrients required for good health or that may cause toxic reactions including allergies are considered dangerous (Azuara *et al.*, 2016).

Adulteration can be categorized into two groups: intentional and unintentional. Intentional includes the addition of cheaper material possessing quality similar to products hence it becomes quite difficult to detect or identify. Adulterants used in this category can be physical or chemical. Some of the examples include removal of cream from milk, replacement of milk solids from products, the addition of unnecessary material/powders to spices, etc. The unintentional category of adulteration includes the inclusion of unnecessary material to food items due to the absence of hygiene and proper condition during food processing (Bansal *et al.*, 2017). This can be of an attained type such as food contamination by microflora including fungi/ bacteria/ yeast, toxic and harmful chemicals residues from packing material, due to presence of stones and dust, food spoilage by rodents, etc. or it can be intrinsic adulteration like the presence of naturally occurring chemicals in some foods for examples toxin containing varieties of mushrooms, pulses, green vegetables, fish and some other seafood. PFA (Prevention of adulteration) rules, 1955 (presently under Food Safety Standard Act, 2006) has covered one more adulteration crop contaminant under the unintentional category which gets incorporated into food during production like harvesting, processing, packaging, transport, storage in an infected environment. Metallic contamination includes the inclusion of toxic metals like lead, mercury, arsenic, cadmium, iron, etc. Out of which first four are highly toxic and may lead to severe health issues. Contamination by microorganisms like fungi, bacteria may at various steps (harvesting, processing, transport, storage) from different sources. Unintentional adulteration in raw material due to packaging is one of the major concerns regarding food safety.

Major foodborne health-associated outbreaks are linked with the adulteration of food with chemical adulterants. There are numerous sources of chemical adulterants like soil, environment, industrial waste, disinfectant, packaging materials, personal care items, pesticide residues, and water (Faille *et al.*, 2017). Due to the excess use of plastics, disinfectants, pesticides, and other chemicals, these days even consumption of water and food has become risky (Robertson *et al.*, 2014). Raw food can be contaminated with soil, sewage water, live animals, organs of dead animals, poultry, outer surfaces one of the excluded reasons from diseased animals (Marriott and Gravani, 2006). Consumption of food contaminated with heavy metals and other chemical residues may lead to severe health issues related to gastro-intestine. In Nigeria, approx. 400-500 kids died of severe lead poisoning because of the consumption of lead-contaminated food. (Song *et al.*, 2017). (Tirima *et al.*, 2020).

Unintentional contaminants mostly include contaminants during food processing (baking, canning, fermentation, roasting, heating, germination), environmental (in water, soil, air, or human originating contaminates), packing material migrants, adulterants and food additives (legally unapproved) (Rather *et al.*, 2017). The migration of hazardous chemicals from packaging material to raw or processed food can lead to severe chemical adulteration.

Based on nature, toxic chemicals are divided into two groups: intrinsic and extrinsic. Intrinsic factors include naturally existing chemicals like mushroom toxins, some of the seafood toxins, phytotoxins, and extrinsic examples include microbial toxins, marine biotoxins, and mycotoxins (Zhang *et al.*, 2014). Some of the important examples of toxin containing natural resources include mycotoxins/ aflatoxins in nuts, cereals, and fruits, ciguatoxins, and mercury and its salts in seafood, a higher concentration of selenium in grains, polycyclic aromatic hydrocarbons from smoked meat, and beta-thujone from some of the essential oils (Yeung *et al.*, 2020). One of the more complicated issues is the existence of multiple toxins in the same commodity that may alter their toxicity by giving synergistic effects to make it more hazardous (Lee and Ryu, 2017). These severe toxins present in food can cause severe health complications including acute poisoning, cancer, complications in pregnancy, transgenerational effects on fetal development (Patial *et al.*, 2018).

Role of Different Raw Materials as Adulterants

Packaging Materials

Diverse kinds of food packaging materials are utilized in the food industry for food protection as well as to maintain the supply chain otherwise it will be very expensive and inefficient to handle food (Robertson, 2006; Brody *et al.*, 2008). Recently, there is an increasing demand for intelligent packaging, active packaging due to changes in consumer perceptions as a packaging material is an initial standard to purchase packaged food. The ultimate objective of packaging is to protect food from various kinds of possible hazards (physical, chemical, microbiological, or miscellaneous) affecting the quality and safety of food products (Lee, 2010). Design of package, selection of packaging material is the major criteria for food purchase. The present competitive environment is endorsing smart and innovative packaging like package material's interaction with the environment to release desired substance or absorption of unwanted material from package headspace to enhance the shelf life of the product (Robertson, 2006). The most common packaging material includes plastics, glass, metals, and papers, and out of these plastics are available in rigid as well as flexible forms. Recently, laminates are gaining popularity due to their numerous fundamental properties that improve the functionality of final packaging material. Different kinds of chemicals such as preservatives, flavoring, colorants, micronutrients, mycotoxins pesticides, and insecticides are reported in food products along with some of the additives as plasticizers, monomers, oligomers are part of packaging materials. These chemicals could transfer from packaging material to food due to contact or during processing and storage and this process of transferring chemicals from package to food is termed as migration (Arvanitoyannis and Bosnea, 2004). Migration may alter the food quality and safety; aroma and color could change due to absorption and movement of unwanted material to food from the package. Concerning polymer-based synthetic packaging material, it seems quite difficult to estimate migration mechanism and detrimental effect on food quality. Although the interaction between package and food products is not detrimental, it might result in some positive or desirable outcomes for example oxygen scavenging packaging film helpful in the prevention of microbial growth and absorbing oxygen along with undesirable flavors (Hotchkiss, 1997). The mass transfer has been diversely explained as a physical interaction in which chemical migration happens at package – food interface, chemical interaction that majorly results from the corrosive reaction of food constituents on the metallic pack, and microbial adulteration occurring due to contact with contaminated packages (Lee *et al.*, 2008). A judicious selection is required for particular packaging material to avoid detrimental effects on the quality, safety, and shelf life of processed food as the interaction between food and packaging material is dependent on various factors. Design of appropriate packaging material, method of preparations, polymer to food ratio, and type of polymer are important criteria to assess migration of chemicals into food products (Hotchkiss, 1997). Product criteria include color alteration, flavor sensitivity, chemical constituents, and microbial sensitivity along with processing and storage conditions.

Di-(ethylhexy) adipate (DEHA), a plasticizer in polyvinyl chloride, also can migrate from package into fat rich foods and TDI (total daily intake) is limited < 0.3mg/kg body weight. Phthalates or its ester's salt are having main application as plasticizers. These compounds might have migrated from packaging material into food are reported to have severe effects on animal's reproductive hormones and are supposed to have similar effect on human hormonal system. Nitrosamines, a compound present in waxed papers and cardboard was also reported in some food products mainly beverages. Amines might have migrated in food through surface contact of package during processing and storage of food for longer duration. These amines are reported to be potentially genotoxic and carcinogenic (Robertson, 2006).PET (polyethylene terephthalate) is generally characterised by the presence of lesser concentration of LMW (low-molecular weight) oligomers of cyclic dimers to pentamers. Lesser migration was reported of the available unreacted oligomer in PET as compared to high amount of migration of polyamides oligomers (Kajla and Sharma, 2018). Temperature along with time of exposure are major factors influencing the migration rate of PET oligomers. Migration of these polymers was reported lesser in microwave heating compared to oven heating that might be due to less exposure time in the former than the latter. Recently, Bisphenol A has gaining attention by the food scientists around the world, as this compound poses severe health issues. BPA act as an endocrine disruptor, which may cause detrimental effects on brain development as well as impair the overall development. As it is a matter of great concern different studies have been carried out to investigate the level of BPA release from linings of can and polycarbonate bottles (Kubwabo *et al.*, 2009; Nam *et al.*, 2010). Some of the solvents such as esters, ethers, ketones and alcohols used in preparation of food packaging materials are also found in food products that might have migrated due to direct contact with food or after release into package headspace. Isocyanates, ingredients of polyurethanes and some of food packaging materials along with aromatic amines are reported in food products migrated from epoxy polymers and azo dyes. These compounds are highly toxic to human health (Lau and Wong *et al.*, 2000). Tin and lead are commonly observed contaminant in beverages (carbonated and non-carbonated) and in food products packed in metals. Higher concentration of tin in food can lead to severe gastro-intestinal disorders. Lead toxicity could affect human organs specially central nervous system (Alamri *et al.*, 2021). Chemical compounds like monomers, oligomers, adipate and phthalates migrater faster into food from package at elevated temperature (freezer to cooking). Plasticizers have low molecular weight thus can easily migrate from packaging material to wrapped food (Guo *et al.*, 2010; Fasano *et al.*, 2012). Simoneau *et al.*, (2011) studied phthalate migration from baby bottles kept under hot filled condition (2 hrs at 70^0C) and observed migration level of dibutyl phthalate and diisobutyl phthalate that varied from 50 to 150 micro gram per kg. Styrene, residuals monomers were reported in polyethylene packed breads, yoghurt and cheese. Fatty acids amides (erucamide, oleamides, oleyl palmitamide, stearamide and stearyl erucamide) most common ingredients in plastics are reported in olive oil as migrants from packaged material (Fasano *et al.*, 2012; Bhunia *et al.*, 2013). There is a great need of time to thoroughly monitor health hazard due to packaging material and chemicals used in food packing. Compounds causing severe health issues like cancer, central nervous disorder, impaired fertility and other chronological disease should be eliminated from food packaging material (Claudio, 2012).

Biological Adulterants

Biological adulterants can comprise visible agents i.e. living organisms like weevils, flies, cockroaches, worms, caterpillars, etc as well as invisible agents i.e. microbiological organisms like bacteria, yeast, protozoa, virus, fungi. Accidental adulterants are pests such as rodents and insects that trespass on food to a great degree and produce contamination in the form of excreta, bodily fluids, and degradation through microorganisms (Choudhary *et al.*, 2020). Growth and contamination of foodborne pathogens can cause food poisoning and food infections. Biologically contaminated and adulterated food is hazardous since it can be poisonous and have a detrimental effect on one's health. It can also result in lacking one of the essential nutrients needed to stay healthy, and it can cause intoxication or allergies to certain foods. Food safety concerns are influenced by a series of chemical, biological, and physical risks. Bacterial contaminants, as well as environmental pollutants such as pesticide residues, mycotoxins, and adulterants, have been blamed for both large-scale outbreaks and single episodes of food poisoning.

Hazardous foods containing infectious microbes and lethal synthetics are linked to more than 200 ailments, ranging from minor to life-threatening. Food and waterborne diarrheal illnesses claim the lives of an estimated 2 million people each year around the world, including adolescent boys and girls. According to studies, a 10% reduction in foodborne infections would prevent 5 million people from being ill (Padhye, 2019).

Escherichia coli O157, also known as enterohemorrhagic *E. coli*, creates verotoxins, which are primarily transmitted to humans by the consumption of contaminated foods, such as raw and half-cooked meats and raw milk, as cattle appear to be the primary reservoir for microscopic organisms. It's a major cause of both bloody and non-bloody bowel movements. *Listeria monocytogenes* is a type of bacteria that can be found in a variety of raw and processed foods, including milk and cheddar, beef, seafood, and fish. Pregnant women, newborns, and the elderly are the people who are most vulnerable to microorganisms. This bacterium is mostly responsible for meningitis, encephalitis, and septicemia. When pregnant women are tainted, they may have fetal removal, stillbirth, or early birth. *Salmonella enteritidis* serotype is mostly found in poultry and eggs, and it's the main cause of human salmonellosis, which manifests as obtrusive disease and responsive joint inflammation (Padhye, 2019).

Mycotoxins are hazardous byproducts of particular parasites that can sometimes be found on or in plant and creature root meals. Several mycotoxins have been identified in over 200 different types of growths. The identification of pathogenic organisms necessitates the use of research center offices and trained personnel; nevertheless, a huge number of the harmful synthetic concoctions commonly found in foods as contaminants can be identified with a simple technique and even the use of test units. Deceptive traders and brokers use a variety of chemical compounds (adulterants) that are not allowed to be placed in foods on purpose to make quick money. These compounds, on the other hand, are sometimes mistakenly mixed up with foods. Milk and milk products, atta, edible oil, and cereals are all routinely debased foods. (Padhye, 2019).

Food poisoning is a widespread problem in the food industry, impacting both the formal and informal segments. *Enterobacter* spp., *Escherichia* spp., *Staphylococcus* spp., and *Pseudomonas* spp. were the most common bacteria isolated in starchy foods, accounting for 20, 16, 12, and 11 percent, respectively. Soup, stew, macaroni, salad, as well as other food samples, were the most contaminated (Yeleliere *et al.*, 2017). Various "food poisonings" have been recorded in Indian newspapers from time to time, as well as epidemics of Lathyrism, pandemic dropsy, Veno-occlusive illness, various mycotoxicoses, and foodborne disease associated with chemical toxins (Alauddin, 2012). Although not every food events are harmful to people's health, they do weaken end-users trust in food safety and costs businesses and government money.

Natural and unintentional contamination can occur in a variety of food matrices as these provide excellent conditions for the growth of fungus, viruses, and bacteria. Natural and accidental contamination can range from mild to severe from product losses to major human health problems, depending on the agents involved and the food chain. The food business has targeted testing schemes on the most likely dangers and essential control points (HACCP) due to the intrinsic costs of food analysis in terms of investment, workforce, and analysis time. As a result, as indicated by various national and international food safety legislation, food safety testing has advanced considerably in the food business. There are also other more stringent private security plans in place. Humidity, pH, and the amount of water in the system are all factors to consider. Pesticides, heavy metal analysis, collagen, and protein content (meat) are all indicators. As the pollutants are foreign to the food chain in issue, deliberate contamination can be difficult to detect using conventionally generated safety testing for chemicals (Pedersen and colleagues, 2016)

Chemical Adulterants

Chemical adulterants can be defined as the occurrence of chemicals in an unwanted place or exceeded amount than a safer limit. Invasion of chemicals in dangerous concentrations in the food we eat and beverages we drink can be incidental or deliberate, is a terrible act that has several negative consequences for human health (Rather *et al.*, 2017).

Chemical food adulterants that are released during the production and processing of food, can hurt human health. Cancer, cardiovascular disease, renal and liver dysfunction, hormone imbalance, reproductive abnormalities, inflammatory diseases, mental health difficulties, and other illnesses are associated with these factors. Several chemicals exploited as adulterants, such as formalin, calcium carbide, melamine, and histamine, can have a detrimental effect on health (Mridha, 2011).

Chemical food contaminants are compounds that aren't naturally present in the raw materials used in food production and aren't added during the manufacturing process. Environmental pollutants or toxins resulting from agricultural production of crops or livestock, as well as insufficient food product manufacture, are examples. The classification of compounds created during routine manufacturing, such as products of heat operations, including flavoring chemicals, is more complex. Compounds known for their negative consequences, such as acrylamide, are commonly referred to as contaminants in these circumstances, but components that contribute to the food's distinctive flavor, such as Maillard products created during roasting, baking, and so on, are not. This distinction is not always apparent from a toxicological standpoint. Due to background concentrations of these chemicals in our environment, metals like mercury and lead, as well as persistent organic pollutants like polychlorinated biphenyls (PCBs) and related pollutants, have been detected in several types of food. Furthermore, natural poisons produced by microbes or plants, as well as chemicals formed during food thermal processing, are of great interest (Schrenk, 2019).

Several chemicals and dyes added to fruit and vegetable products have been observed to affect the digestive system, eyes, and liver, causing vomiting and diarrhea in youngsters, as well as renal failure. as well as the cause of the majority of ailments. Calcium carbide, which is found in mangoes and bananas, copper sulfate, which is employed to speed up the ripening of fruits, and oxytocin, which is used to hasten the growth of pumpkin, watermelon, brinjal, gourds, and cucumber, can all damage the brain. Wax gives sparkle to apples and pears. To give bitter gourd and leafy vegetables a new color, cheap green colors including synthetic dyes and harmful chemicals such as metallic lead are added. In the cultivation of fruits and vegetables, pesticides and herbicides are employed excessively (Gupta and Panchal 2019).

Calcium carbide being commonly used for the ripening of fruits results in dizziness, irritation in the mouth and nose, weakness, vomiting, skin ulcer due to the presence of arsenic and phosphorous. Ethylene glycol's direct consumption might cause renal failure, as it is commercially used in the acceleration of fruit ripening (Rehman *et al.*, 2008; Mamun *et al.*, 2021).

There is a myriad of approaches to adulterate food, including formalin treatment for preservation, the addition of calcium carbide to speed up the ripening process, and the inclusion of DDT powder in ski, among others (Asif, 2012, Bhuyian *et al.*, 2009). Some surprising symptoms of infestation include fishermen neglecting to properly dry fish owing to weight loss, which is why they often apply a mixture of organochlorine (DDT and heptachlor) insecticides (Bhuyian *et al.*, 2009) to prevent dried fish from infestation. Farmers who are exposed to DDT have a higher rate of nonallergic asthma (Brow, 2007). Organochlorine chemicals have been linked to diabetes in the past (Jones *et al.*, 2008). A study of malaria workers who work with these Organochlorine substances discovered a significant incidence of liver and biliary tract tumors (Rogan and Chen, 2005). Puberty causes also raise the chance of breast cancer (Clapp *et al.*, 2008). Furthermore, there is no special regulation in place to regulate it, and the use of dangerous industrial chemicals is widespread (Mamun *et al.*, 2021).

Mustard oil is widely used as an edible oil in northern and eastern regions of India, where it is commonly adulterated by argemone seed oil which has toxic effects by causing hypertension, dropsy, diarrhea, vomiting, and anemia for its non-edible portion. Argemone seed oil contains two alkaloids sanguinarine and dihydro-sanguinarine which are carcinogenic (Mamun *et al.*, 2021).

Polycyclic aromatic hydrocarbons (PAHs) are produced predominantly through incomplete combustion or pyrolysis of organic matter, as well as industrial activities. Food contamination can arise due to the environment, industry, or domestic cooking (on using biomass fuels). These substances emerge to be carcinogenic as well as genotoxic in nature. Oil leaks from container ships in the water are all too prevalent, polluting seafood. In addition to petroleum-related PAH compounds, chemical diffusions are routinely employed to minimize the impact of oil in the water. Following the 2010 BP Deepwater Horizon oil leak in Louisiana, the US government acknowledged the concerns about seafood safety by establishing sampling and analytical methodologies for determining food safety. Food contamination with polychlorinated biphenyls (PCBs) can cause harm and lifelong problems at considerably lower concentrations than those that can affect adult brain function (Rather *et al.*, 2017).

Acrylamide is found in a variety of foods and is produced for commercial and industrial purposes. It's frequently associated with high-temperature cooking methods (for example, in bread and baked or fried potatoes). The commercial application of acrylamide is in paper and dye production as well as in wastewater treatment. International Agency for Research on Cancer (IARC) categorized acrylamide as a potential human carcinogen.

Foodborne diseases cause around 48 million illnesses in the United States each year. (Gould and colleagues, 2013) Chemically contaminated food has a significant negative impact on people's health. The negative repercussions range from minor gastrointestinal problems to serious health problems. Chemical pollutants have major repercussions, a loss of personal control, and long-term ramifications (Kher *et al.*, 2011). Food is the most prevalent way for humans to be exposed to metals. Cadmium and lead are two metals that easily pass through the food chain. Heavy metals cause prenatal development retardation by depleting specific nutrients in the body, decreasing immune defenses, altering psychosocial processes, and lowering immunological defenses. Heavy metal consumption in the form of metallic contamination has been linked to malnutrition and an increased risk of digestive issues (Khan *et al.*, 2008).

Toxins in drinking water and food can have serious health repercussions for humans, as well as marine life and other living organisms consuming that contaminated water. All potential causes of contamination include industrial and municipal discharges, natural geological formations, urban and rural run-off, drinking water treatment procedures, and water supply materials (Calderon, 2000). Pollutants found in drinking water include heavy metals like arsenic, aluminum, lead, fluoride along with other industrial waste like disinfection by-products, radon, and pesticides. These pollutants result in cancer, cardiovascular disease, poor reproductive system, neurological diseases, etc. Currie *et al.* (2013) discovered that moms who drink chemically polluted water have a significant impact on their infants' gestation and birth weight, particularly those who are less informed. Arsenic, PCBs, and lead are all industrial contaminants. Other potential causes of contamination include industrial and municipal discharges, natural geological formations, urban and rural run-off, drinking water treatment procedures, and water supply materials (Calderon, 2000). Pollutants found in drinking water include arsenic, aluminum, lead, fluoride, disinfection by-products, radon, and pesticides. They cause cancer, cardiovascular disease, poor reproductive outcomes, and neurological diseases, among other things. Currie *et al.* (2013) also discovered that moms who drink chemically polluted water have a significant impact on their infants' gestation and birth weight, particularly those who are less informed. Neurological impairments such as cerebral palsy, attention deficit disorder, autism, and mental retardation can be caused by industrial pollutants such as arsenic, PCBs, and lead in both food and water. At much lower levels than those previously mentioned Exposure to such drugs during the embryonic phases of development can cause brain damage and long-term deficits, which can change adult brain function (Grandjean and Landrigan, 2006).

Foodborne exposure to polychlorinated biphenyls (PCBs) can harm children's brain development and immune response (Schantz *et al.*, 2004). (Schantz *et al*, 2004). Pesticides in food have major health repercussions as contaminants. Excessive amounts of these compounds in food can induce neurological and renal impairment, as well as birth abnormalities and reproductive difficulties, in addition to being carcinogenic (Bassil *et al.*, 2007). Pesticide buildup in the body's tissues has the potential to cause metabolic malfunction (Androutsopoulos *et al.*, 2013). Industrial pollutants such as arsenic, PCBs, and lead in both food and water have been related to neurodevelopment illnesses such as attention deficit disorder, autism, cerebral palsy, and mental retardation. Such drugs may induce brain damage if they are exposed during the embryonic stages of development. Simple and quick methods are required to determine the concentration of adulterants in food products. The nanotechnology-based assay has proven to have potential scope in analytical chemistry and has been exploited to detect adulterants like melamine, detergent, urea using nanogold particles (Kumar *et al.*, 2014; Kumar *et al.*, 2016).

Several compounds found in food are governed by laws that limit their quantities. Laws should be passed prohibiting the use of adulterants and hazardous additives. Inclusion of chemical hazards can be avoided from entering in the food chain and causing suffering to the public by robust surveillance and response mechanisms. The FDA defines minimum chemical levels that can be present in food, such as pesticide residues that must not exceed a set limit (Bajwa and Sandhu, 2011). Problems may arise if the concentration and criteria are not followed precisely. In many developing and underdeveloped countries, particularly in the case of developing and undeveloped countries, the administration of the concentration of harmful chemical adulterants in food is still insufficient. Pesticides escape into the environment because some countries are so reliant on agriculture. Both food and water are poisoned by groundwater (Rather *et al.*, 2017). Individual consumer concerns are vital because they have the potential to impact how persons monitor their health (Liang and Scammon, 2016). Because of the prominence and ubiquitous use of the internet, customers can also gather answers and reduce the health risks related to food contamination scenarios. Most importantly, the food sector must acknowledge the need to be more open and accountable in the future when it comes to developing safe commercial food items and protecting the public from foodborne illness (Rather *et al.*, 2017).

However, to make a stronger impact, WHO launched a strategic plan in 2010 to address priority concerns in the areas of food safety and foodborne zoonoses from 2013 to 2022. The plan's scope includes food safety in all of its manifestations, including the farm-to-table approach and foodborne zoonotic diseases. It was created in consultation with food safety specialists at the international, regional, and national levels (of WHO).

The strategic plan's main components are:

- Science-based decision-making, which provides the scientific foundation for measures along the entire food chain to reduce foodborne health risks
- Cross-sectorial collaboration, which aims to improve international and national cross-sectorial collaborations as well as communication and advocacy, and leadership, and technical assistance, which aims to provide leadership and technical assistance.
- Leadership and technical assistance to give leadership and aid in the building and improvement of risk-based, integrated national food safety systems (Siruguri, 2018)

In most cases, societal attitudes on food safety and contamination are shaped by history and habit. Regulatory organizations retain control over food safety, even though customers choose their diets based on social and financial concerns. These groups can keep a lookout for pollution while also enforcing rules. Food contamination has political repercussions as well. Food safety requirements, as well as food and water contamination monitoring and efforts to minimize and remove environmental pollution exposure, are all essential, as previously indicated. Following an occurrence involving environmental pollution, a government must establish public health, legal, and ethical frameworks as soon as possible. Society should be educated on the need for safe crop production and livestock husbandry. It was also emphasized the significance of eating a well-balanced diet and cooking with safer methods.

Governments should work to decrease urban inequities in environmental exposures on a national scale. Although some contaminants have localized impacts, the majority are distributed worldwide. As a result, an international approach to food safety is required, to reduce environmental and food contamination while also ensuring global trade in safer food products (Thompson and Darwish, 2019).

1.3 Impact of Environmental Contaminants on Food Safety and Toxicology

The environment and food safety have a significant impact on human health (Fung *et al.*, 2018; Gallo *et al.*, 2020). Contaminants are migrating from the environment to food, and then from food to humans, according to reports. Heavy metals can migrate into fishes by biomagnification in the aquatic environment (Korkmaz *et al.*, 2019; Arisekar *et al.*, 2020), and heavy metals can enter human bodies through the ingestion of these species. (Saha *et al.*, 2016; Gholamhosseini *et al.*, 2021). Although heavy metal concentrations in fish are frequently lower than maximum legal limits (Velusamy *et al.*, 2014; Safiur Rahman *et al.*, 2019), human health and safety must be considered because humans are at the top of the food chain. In other words, because people consume a variety of foods, heavy metals accumulate and accumulate to a level that poses major health hazards, such as cancer (Badamasi *et al.*, 2019; Yu *et al.*, 2020). Pesticides, viruses, particulate matter, and other developing types of contaminants such as microplastics, personal care items, medications, and so on are all substantial contaminants in addition to heavy metals. (Aghilinasrollahabadi *et al.*, 2020; Zhang *et al.*, 2020).

Pesticides, fertilizers, feed additives, and livestock medication are used in intensive agricultural production practices. As an example, honeybees are overfed with honey, antibiotics, and industrial sugars during the primary nectar period to retrieve more honey from hives (Wu *et al.*, 2017). Fruit, cereals, vegetables, meat, and dairy products all contain the residues from the above-mentioned chemicals that intrude the food chain. Furthermore, bio-accumulative chemicals viz., heavy metals, and persistent organic pollutants often found in fish, meat, and dairy products are also added to the list (Thompson and Darwish, 2019). Unsafe livestock feeding methods, by which contaminants enter the food chain unintentionally, for example, dioxins from contaminated poultry feed can make their way up the food chain, and feeding livestock feed contaminated with sick/dead animal remains might induce "mad cow disease" (BSE), reported to cause Creuzfeldt-Jacobs disease in consumers/humans. These contaminants pose serious health hazards to the consumers which have raised the attention of researchers towards safety concerns of these unintentional as environmental contaminants.

Through several anthropogenic actions, a pollutant/contaminant becomes a component of the environmental medium - air, water, and soil. Contaminants migrate into food sources via treatment plants of textiles, canneries, and pharmaceuticals industries as well as dumping of toxic/solid waste and wastewater in water bodies and on land as well (Amin *et al*., 2011; Chakraborty *et al*., 2013), discharge of untreated effluent and toxic heavy metals from sewage and industry and agricultural runoff of pesticides (Baki *et al*., 2018).

In addition to environmental contaminants, different foods including fish, meat milk, raw and processed fruits & vegetables get contaminated through human malpractice/adulteration of preservatives, additives, stabilizers, texture improvers, etc., to enhance shelf stability as well overall aesthetic quality (Labu *et al*., 2011; Ali, 2013a). Contaminants enter the food chain from a variety of sources and accumulate in the human body, reaching dangerous concentration levels depending on the rate of consumption and absorption. As a result, the harmful compounds have an impact on food consumers and spread across food chains, resulting in a phenomenon known as bio-magnification. The bio-magnification process describes how a contaminant accumulates in tissue and concentrations rise when it progresses through two or more trophic levels in the food chain (Nowell *et al*., 1999). It occurs as a result of contamination from heavy metals or other persistent contaminants that cannot be destroyed by natural processes (Thompson and Darwish, 2019).

According to Zenker *et al*. (2014), calculations derived from extrapolations and applications of uncertainty factors reported as chemical risk assessment methods for contaminants or chemical residues found in food and potable water were established as acceptable daily intake (ADI) or tolerable daily intake (TDI) based on toxicological and epidemiological databases. In case the ingestion of contaminated food containing chemical residue accumulates in the human body and its level lies within the allowable daily intake value of the corresponding contaminant/residue as per WHO specifications, then there is the possibility that contaminant will pose zero/negligible harmful effect on human health. But on other hand, if the contaminant/residue accumulates higher than the allowable daily intake limits and will get absorbed in the body will cause serious health effects (Thompson and Darwish, 2019).

Mercury and its related compounds are well-known bio-magnifiable contaminants that are hazardous even at very low concentration levels both at aquatic as well as terrestrial ecosystems. Since mercury is a long-lasting chemical, it can accumulate or bio-accumulate in living beings. Mercury is transformed into methyl mercury in the environment, which is an extremely poisonous chemical compound. Methyl mercury is produced in the aquatic environment by bacterial activity resulting from anthropogenic activities and natural sources of elemental mercury. Anthropogenic activities include coal combustion, chloralkali process, industrial effluent discharge, waste, and untreated water disposal which disrupts and contaminate the ecosystem and hence food chain (Yadav *et al.*, 2015). Methylmercury biomagnifies through the food chain and gets absorbed in body tissues from food sources. Biomagnification of this hazardous compound keeps on increasing with the step of the food chain viz., plants to animals to another animal to humans and eventually reaches and accumulate in consumers to such higher levels which have toxic and further life-threatening effects. Mercury compounds have an abrogating impact on the functioning of the nervous system leading to the development of deadly brain disorders. Due to the accumulation of methyl mercury, it has been reported that consumers might get paralyzed and severe brain disorders often resulted in blindness and hearing impairment. DDT is another harmful environmental contaminant that is discovered in each food web and could have caused bio-magnification from the lowest to the highest trophic levels (Siriwong *et al.*, 2009). However, DDT as a pesticide is banned around the globe.

Polychlorinated biphenyls, polynuclear aromatic hydrocarbons, petroleum products, cyanide, selenium, and heavy metals viz., copper, cadmium, chromium, lead, zinc, nickel, and tin, could be biomagnified in addition to DDT and mercury. In terms of food resources, the aquatic source is crucial in terms of high risk to human beings (Majed,2016). Because humans are at the top position of the food chain, therefore, they consume fish at higher trophic levels. Fish or aquatic food contains the highest quantities of pollutants/contaminants that can bio-magnify through the consumption of other fish and related phytoplanktons/zooplanktons. Many popular dishes are a good source of essential fats and oils, which can store a lot of pollutants. It can be concluded that toxic chemical exposure can be reduced by feeding lower trophic level animals and most importantly avoiding aquatic food from polluted sources. Lead has been discovered to be a harmful compound, and its contamination of food can produce severe toxic symptoms. Lead causes harmful abnormalities in the kidneys and hepatic arteries when turmeric is coated with lead chromate by ignorant manufacturers in India. Nausea, abdominal pain, anemia, sleeplessness, muscular paralysis, and brain damage are all common symptoms of lead poisoning. (Gautam and Singh, 2013).

The Indian Council of Medical Research conducted a survey that discovered high amounts of pesticide residues in bovine milk, as well as metals arsenic, cadmium, and leads in infant foods, canned products, and turmeric (Srilakshmi, 2001). The unethical substitution of brown rice syrup with high fructose corn syrup in some organic foods has sparked researchers' qualms owing to its high arsenic level (Siddiqui *et al.* 2017). According to a study led by Lia, *et al.*, (2015) baby formulae containing organic brown rice syrup have an elevated arsenic level over the drinking water norms, and there is no law to manage this specific instance. Long-term ingestion of HFCS causes renal failure in rats, according to Li *et al.* (2015). Chronic consumption/intake of any sugar syrup, such as sucrose or fructose leads to elevated levels of cholesterol, triglycerides, as well as insulin leading to diabetes (Fakhlaei *et al.*, 2020)

Contaminants such as metals may pose additional concerns in urban farms and gardens. Furthermore, polluted drinking water is a possibility. Even rainwater has been found to contain xenoestrogen chemicals (Kohrman and Chamberlain, 2014). Pollution of the marine biota may occur as a result of water contamination, compromising the suitability of seafood for consumption (Enault *et al.*, 2015). As an outcome, those who eat a lot of seafood will have higher amounts of these pollutants in their bodies. There is no such thing as a perfectly safe dose level for most pollutants (Nelms *et al*., 2018). Acceptable limits, on the other hand, have been estimated for many—levels below which no evidence of toxicity should be visible. The evident toxicological effects are dependent on the type of pollutant, the dose, and the individual. Many pollutants, for example, have been linked to an increased risk of cancer. Long-term exposure to arsenic-contaminated drinking water causes stomach cancer, liver, and gastrointestinal cancer due to lead poisoning, and skin cancer from mercury-contaminated grains (Jarup, 2003; Jonker *et al*., 2004).

Another key route of chemical exposure is through food consumption. Selenium enters the food chain when inorganic selenium is converted to organically bound forms by plants and microorganisms (Sors *et al*. 2005). Selenium poisoning has been reported on a wide scale in seleniferous regions of China as a result of the increased consumption of selenium-containing foods (Yadav *et al*., 2015). Humans are exposed to mercury in specific occupational sectors as a result of eating fish. The discharge of untreated methyl mercury into the environment in Japan caused Minamata disease. Exposure occurs when methyl mercury is absorbed into the food chain, advancing up the food chain when each predator consumes the smaller, less fortunate species. Methyl mercury is concentrated near the top of the food chain in fish such as bonito (*Sarda* spp), halibut (*Hippoglossus* spp), mackerel (*Scomberomorus* spp), marlin (*Makaira* spp*)*, shark (all species), swordfish (*Xiphias gladius*), and bluefin tuna (*Xiphias gladius)* (Hutt *et al*. 2007).

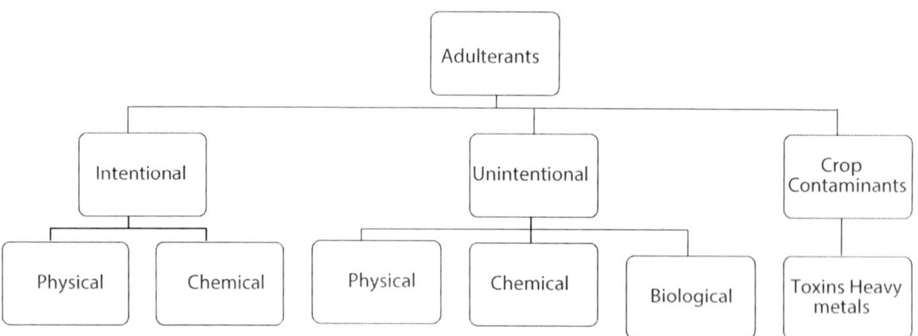

Fig. 1: Types of adulterants

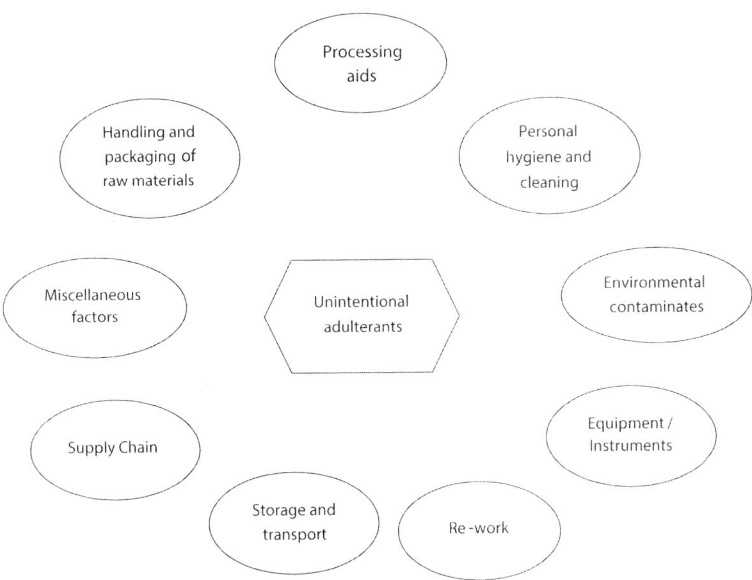

Fig. 2: Sources of Unintentional Adulterants

1.4 Conclusions

Food adulteration is a crime against humanity. Chemicals are used in every step of the food processing chain i.e. right from the farm to table. Pesticides, fertilizers are used while producing different food crops, antibiotics and hormones while rearing animals. Some of the contaminants intrude on food products inadvertently as a result of mishandling during storage, transportation. The residue of these chemicals enters the human body and accumulates reaching levels that cause toxicity and different health disorders. These contaminants persist in the environment and affect human bodies for decades to come. Different other toxicants also enter the food chain via environment viz, heavy metals, synthetic dyes, etc. due to improper treatment of commercial waste into water bodies thus imbalancing aquatic life. When the contaminated/residues containing aquatic food are consumed, it leads to severe health risks. Therefore, in the end, it can be concluded that consumption of such foods should be avoided which are produced in polluted areas. There is a need to implement such policies that promote as well as ensure safe and sustainable agriculture & livestock production practices. Food adulteration done advertently or inadvertently needs to be stopped and addressed at the national as well as international level to secure beautiful lives.

References

Aghilinasrollahabadi, K., Salehi, M., Fujiwara, T. (2020). Investigate the influence of microplastics weathering on their heavy metals uptake in stormwater. J Hazard Mater. 408:124439.

Alamri, M.S., Qasem A. A. A., Mohamed, A.A., Hussain, S., Ibraheem, M. A., Shamlan, G., Hesham, H. A. and Qasha, A. A. S. (2021). Food packaging's materials: A food safety perspective. Saudi Journal of Biological Sciences. 28 (8), 4490-4499.

Alauddin, S. (2012). Food adulteration and society. Global Research Analysis, 1(7), 3-5.
Amin, M.N., Begum, A., and Mondal, M.G.K. (2011). Trace element concentrations are present in five species of freshwater fish of Bangladesh. Bangladesh J. Sci. Ind. Res. 46, 27–32.
Arisekar, U., Shakila, R.J., Shalini, R. and Jeyasekaran, G. (2020) Human health risk assessment of heavy metals in aquatic sediments and freshwater fish caught from Thamirabarani River, the Western Ghats of South Tamil Nadu. Mar Pollut Bull. 159:111496.
Arvanitoyannis I and Bosnea L. (2004). Migration of substances from food packaging materials to foods. Critc Rev Food Sci Nutr. 44(2):63–76
Asif, M. (2012). Physico-chemical properties and toxic effect of fruit-ripening agent calcium carbide. Annals of Tropical Medicine & Public Health, 5(3).
Azuara, G., José, L., Tornos, J. and Luis, S. (2012). Improving RFID traceability systems with verifiable quality. Industrial Management & Data Systems. 112(3): 340 – 359.
Babu, S. and Shenolikar, I.S. (1995). Health and nutritional implications of food colours. Indian J. of Medical Research. 102: 245-249
Badamasi, I., Odong, R. and Masembe, C. (2019) Implications of increasing pollution levels on commercially important fishes in Lake Victoria. J Great Lakes Res. 45:1274–1289.
Bajwa, U., and Sandhu, K. S. (2014). Effect of handling and processing on pesticide residues in food-a review. Journal of food science and technology. 51(2), 201-220.
Baki, M.A., Hossain, M.M. and Akter, J., (2018) Concentration of heavy metals in seafood (fishes, shrimp, lobster, and crabs) and human health assessment in Saint Martin Island, Bangladesh. Ecotoxicol Environ Saf. 159:153–163.
Bansal, S., Singh, A., Mangal, M., Mangal, A.K. and Kumar, S. (2017). Food adulteration: Sources, health risks, and detection methods. Crit Rev Food Sci Nutr. 13;57(6):1174-1189.
Beaglehole, R. and Bonita, R. (2008). Global public health: a scorecard. The Lancet, 372(9654), 1988-1996.
Beaglehole, R., Bonita, R., Horton, R., Adams, O. and McKee, M. (2004). Public health in the new era: improving health through collective action. The Lancet. 363(9426), 2084-2086.
Bhuiyan, M. N. H., Dawlatana, M., Bhuiyan, H. R., Lucky, F. A., Saha, B. K. and Bhuiyan, M. N. I. (2009). Quality assessment of dry fishes of Bangladesh with the special emphasis on harmful additives. Bangladesh Journal of Scientific and Industrial Research. 44(3), 311-318.
Bhunia, K., Sablani, S., Tang, J. and Rasco, B. (2013). Migration of Chemical Compounds from Packaging Polymers during Microwave, Conventional Heat Treatment, and Storage. Comprehensive Reviews in Food Science and Food Safety. 12 (5): 523-545.
Brody, A.L., Bugusu, B., Han, J.H., Sand, C.K. and McHugh, T.H. (2008). Innovative food packaging solutions. J. Food Sci. 73 (8), 107–116.
Brow, A. J. (2007). Pesticide exposure is linked to asthma. Sci Am. 162, 890-97
Calderon, R. L. (2000). The epidemiology of chemical contaminants of drinking water. Food and chemical toxicology. 38, S13-S20.
Chakraborty, C., Huq, M. M., Ahmed, S., Tabassum, T., and Miah, M. R. (2013). Analysis of the causes and impacts of water pollution of buriganga river: a critical study. Int. J. Sci. Technol. Res. 2, 245–252.
Cheng, J., Sun, D.W., Zeng, X.A. and Liu, D. (2015). Recent advances in methods and techniques for freshness quality determination and evaluation of fish and fish fillets: a review. Crit. Rev. Food Sci. Nutr. 55 (7): 1012–1225.
Choudhary, A., Gupta, N., Hameed, F. and Choton, S. (2020). An overview of food adulteration: Concept, sources, impact, challenges, and detection. International Journal of Chemical Studies. 8(1), 2564-2573.

Clapp, R. W., Jacobs, M. M. and Loechler, E. L. (2008). Environmental and occupational causes of cancer: new evidence 2005-2007. Reviews on environmental health. 23(1), 1-38.

Claudio, L. (2012). Our food: Packaging & public health. Environ. Health Perspect. 120 (6), A232–7.

Currie, J., Graff Zivin, J., Meckel, K., Neidell, M. and Schlenker, W. (2013). Something in the water: Contaminated drinking water and infant health. Canadian Journal of Economics/ Revue canadienne d'économique. 46(3), 791-810.

Da Araujo, C. F., Lopes, M. V., Vasquez, M. R. (2016). Cadmium ′ and lead in seafood from the Aratu Bay, Brazil, and the human health risk assessment. Environmental Modeling & Assessment. 188 (4); 259.

Enault, J., Robert, S., Schlosser, O. de Te, C., and Loret, J.F. (2015). Drinking water, diet, indoor air: Comparison of the contribution to environmental micropollutants exposure. International Journal of Hygiene and Environmental Health. 218(8): 723–730

Faille, C., Cunault, C., Dubois T. and Bénézech, T. (2017). Hygienic design of food processing lines to mitigate the risk of bacterial food contamination with respect to environmental concerns. Innovat. Food Sci. Emerg. Technol. 46:65-73.

Fakhlaei, R., Selamat, J., Khatib, A., Razis, A. F. A., Sukor, R., Ahmad, S. and Babadi, A. A. (2020). The toxic impact of honey adulteration: A review. Foods. 9(11), 1538.

Fasano, E., Bobo-Blay, F., Cirillo, T., Montuori, P. and Lacorte, S. (2012). Migration of phthalates, alkylphenols, bisphenol A and di(2-ethylhexyl) adipate from food packaging. Food Control. 27:132–8.

FSSAI. (2008). Advisory no. 2. Advisory to the State Health Authorities on Monitoring Melamine Contamination of Foods. FSSAI, India. Available at http://www.fssai.gov.in/ portals/0/Pdf/AdvisoryMelamineinFoods[1].pdf)

Fung, F., Wang, H.S., Menon, S. (2018). Food safety in the 21st century. Biom J. 41:88–95.

Gallo, M., Ferrara, L. and Calogero, A. (2020) Relationships between food and diseases: what to know to ensure food safety. Food Res Int. 137: 109414.

Gautam, A. and Singh, N. (2013). Hazards of New Technology in Promoting Food Adulteration. IOSR Journal of Environmental Science, Toxicology and Food Technology. 5(1): 08-10.

Gholamhosseini, A., Shiry, N., Soltanian, S. and Banaee, M. (2021). Bioaccumulation of metals in marine fish species captured from the northern shores of the Gulf of Oman, Iran. Reg Stud Mar Sci. 41:101599.

Gould, L. H., Walsh, K. A., Vieira, A. R., Herman, K., Williams, I. T., Hall, A. J. and Cole, D (2013). Surveillance for foodborne disease outbreaks—the United States, 1998–2008. Morbidity and Mortality Weekly Report: Surveillance Summaries, 62(2), 1-34.

Grandjean, P. and Landrigan, P. J. (2006). Developmental neurotoxicity of industrial chemicals. The Lancet. 368(9553), 2167-2178.

Guo, Z., Wang, S., Wei, D., Wang, M., Zhang, H., Gai, P. and Duan, J. (2010). Development and application of a method for analysis of phthalates in ham sausages by solid-phase extraction and gas chromatography–mass spectrometry. Meat Sci. 84:484–90.

Gupta, N. and Panchal, P. (2009). Extent of awareness and food adulteration detection in selected food items purchased by homemakers. Pakistan journal of nutrition. 8(5), 660-667.

Hotchkiss, J.H. (1997). Food-packaging interactions influencing quality and safety. Food Addit. Contam. 14(6–7), 601–607.

Hutt, P.B., Merrill, R.A. and Grossman, L.W. (2007). Food and Drug Law, 3rd ed.; Foundation Press: New York, NY, USA, pp 369.

Jarup, L. (2003). Hazards of heavy metal contamination. British Medical Bulletin. 68(1):167–182

Jones, O., Maguire, M. and Griffin, J. (2008). Environmental pollution and diabetes: a neglected association. The Lancet. 371(9609), 287-288.

Jonker, D., Freidig, A. and Groten, J. (2004). Safety Evaluation of Chemical Mixtures and Combinations of Chemical and NonChemical Stressors. Reviews on Environmental Health. 19(2):83–140.

Kajla, Priyanka and Sharm, Alka, (2018). Polymer migration. Food Frontiers. Edi (1). Pg 189-200.

Khan, S., Cao, Q., Zheng, Y. M., Huang, Y. Z. and Zhu, Y. G. (2008). Health risks of heavy metals in contaminated soils and food crops irrigated with wastewater in Beijing, China. Environmental pollution. 152(3), 686-692.

Kher, S. V., De Jonge, J., Wentholt, M. T., Deliza, R., de Andrade, J. C., Cnossen, H. J. and Frewer, L. J. (2013). Consumer perceptions of risks of chemical and microbiological contaminants associated with food chains: a cross-national study. International Journal of Consumer Studies, 37(1), 73-83.

Kohrman, H. and Chamberlain, C. P. (2014). Heavy metals in produce from urban farms in the San Francisco Bay Area. Food Additives & Contaminants: Part B. Surveillance. 7 (2):127–134,

Korkmaz, C. and Ay Ö Ersoysal, Y. (2019) Heavy metal levels in muscle tissues of some fish species caught from the north-east Mediterranean: evaluation of their effects on human health. J Food Compos Anal 81:1–9

Kubwabo, C., Kosarac, I., Stewart, B., Gauthier, B. R., Lalonde, K. and Lalonde, P. J. (2009). Migration of bisphenol A from plastic baby bottles, baby bottle liners and reusable polycarbonate drinking bottles. Food Addit Contam. 26(6):928–37.

Kumar, N., Kumar, H., Mann, B. and Seth, R. (2016). Colorimetric determination of melamine in milk using unmodified silver nanoparticles. Spectrochimica Acta Part A: Molecular and Biomolecular Spectroscopy. 156, 89-97.

Kumar, N., Seth, R. and Kumar, H. (2014). Colorimetric detection of melamine in milk by citrate-stabilized gold nanoparticles. Analytical biochemistry. 456, 43-49.

Labu, Z. K., Mohiuddin, A. K., and Bake, M. A. (2011). Food Contamination: Bangladesh Perspective. Available online at: http://www.pharmatutor.org/ articles/food-contaminants-bangladesh-perspective

Lai, T.N.H., André, C., Rogez, H., Mignolet, E., Nguyen, T.B.T. and Larondelle, Y. (2015). Nutritional composition and antioxidant properties of the sim fruit (Rhodomyrtus tomentosa). Food Chem. 168, 410–416.

Lau, O. and Wong, S. (2000). Contamination in food from packaging material. J. Chromatogr. A. 882: 255–270.

Lee, D., Yam, K. and Piergiovanni, L. (2008). Food Packaging Science and Technology. CRC Press, Boca Raton.

Lee, H. J. and Ryu, D. (2017). Worldwide Occurrence of Mycotoxins in Cereals and Cereal-Derived Food Products: Public Health Perspectives of Their Co-occurrence. J. Agric. Food Chem. 65(33):7034-7051.

Lee, K. T. (2010). Quality and safety aspects of meat products as affected by various physical manipulations of packaging materials. Meat Science. 86(1):138-150,

Li, L., Zhao, Z., Xia, J., Xin, L., Chen, Y., Yang, S. and Li, K. (2015). A long-term high-fat/ high-sucrose diet promotes kidney lipid deposition and causes apoptosis and glomerular hypertrophy in bama minipigs. PLoS ONE 10, e0142884

Liang, B. and Scammon, D. L. (2016). Food contamination incidents: what do consumers seek online? Who cares?. International Journal of Nonprofit and Voluntary Sector Marketing. 21(4), 227-241.

Majed, N., Real, M.I.H., Akter, M. and Azam, H.M. (2016). Food Adulteration and Bio-Magnification of Environmental Contaminants: A Comprehensive Risk Framework for Bangladesh. Front. Environ. Sci. 4:34

Marriott N. and Gravani R. (2006). Food Contamination Sources. 76–82.

Md. Abdullah, Al Mamun, Biplob Kumar Biswas, Sumita, Tabassum Tamanna and Md. Bashirul Islam (2021). International Journal of Scientific and Research Publications. 11(5), 780-796.

Mridha, Z. U. (2011). Food safety-A global concern. Bangladesh Journal of Medical Biochemistry, 4(2), 3-5.

Multari, R. A., Cremers, D. A., Dupre, J. A. M. and Gustafson, J. E. (2013). Detection of biological contaminants on foods and food surfaces using laser-induced breakdown spectroscopy (LIBS). Journal of agricultural and food chemistry. 61(36), 8687-8694.

Nam, S. H, Seo, Y. M. and Kim, M.G. (2010). Bisphenol A migration from polycarbonate baby bottle with repeated use. Chemosphere 79: 949–52.

Nelms, S. E., Galloway, T. S. Godley, B. J. Jarvis, D.S. and Lindeque, P. K. (2018). Investigating microplastic trophic transfer in marine top predators. Environmental Pollution.

Padhye, R. R. (2019). Public Health, Nutrition and Food Safety. Think India Journal, 22(13), 639-646.

Patial, V., Asrani, R. K. and Thakur, M. (2018). Food-borne mycotoxicoses: pathologies and public health impact. In: Holban AM, Grumezescu AM, editors. Foodborne Diseases. Cambridge, USA: Academic Press; p. 239–274.

Pedersen, B., Gorzkowska-Sobas, A. A., Gerevini, M., Prugger, R., Belenguer, J., Maletti, M. and Davidson, R. K. (2016). Protecting our food: Can standard food safety analysis detect adulteration of food products with selected chemical agents?. TrAC Trends in Analytical Chemistry. 85, 42-46.

Pitini, E., Baccolini, V., Rosso, A., Massimi, A., De Vito, C., Marzuillo, C. and Villari, P. (2021). How public health professionals view mandatory vaccination in italy—A cross-sectional survey. Vaccines. 9(6), 580.

Rather, I. A., Koh, W. Y., Paek, W. K. and Lim, J. (2017). The sources of chemical contaminants in food and their health implications. Frontiers in pharmacology. 8, 830.

Robertson, G.L. (2006). Safety and legislative aspects of packaging. Food packaging: principles and practice, vol. 3. CRC Press, US, pp. 473–502.

Robertson, L., Sprong, H., Ortega, Y., van der Giessen, J. and Fayer, R. (2014). Impacts of globalisation on foodborne parasites. Trends Parasitol. 30, 37–52.

Rogan, W. J. and Chen, A. (2005). Health risks and benefits of bis (4-chlorophenyl)-1, 1, 1-trichloroethane (DDT). The Lancet. 366(9487), 763-773.

Safiur Rahman, M., Solaiman Hossain, M. and Ahmed, M.K. (2019). Assessment of heavy metals contamination in selected tropical marine fish species in Bangladesh and their impact on human health. Environ Nanotechnol Monit Manag 11:100210.

Saha, N., Mollah, M Z.I., Alam, M.F. and Safiur Rahman, M. (2016) Seasonal investigation of heavy metals in marine fishes captured from the Bay of Bengal and the Implications for human health risk assessment. Food Control. 70:110–118.

Sharma, A., Batra, N. Garg, A. and Saxena, A. (2017). International Journal for Research in Applied Science & Engineering Technology. Food Adulteration: A Review. 5(3):686.

Siddiqui, A.J. Musharraf, S.G. and Choudhary, M.I. (2017). Application of analytical methods in authentication and adulteration of honey. Food Chem. 217, 687–698.

Siriwong, W., Thirakhupt, K., Sitticharoenchai, D., Rohitrattana, J., Thongkongowm, P. and Borjan, M. (2009). DDT and Derivatives in Indicator Species of the aquatic food web of rangsit agricultural area, Central Thailand. Ecol. Indic. 9, 878–882.

Siruguri, V. (2018). The journey from food toxins to food safety: Transition over a century in service of the nation. The Indian Journal of Medical Research. 148(5), 488.

Song, Q., Zheng, Y. J., Xue Y., Sheng, W. G. and Zhao M. R. (2017). An evolutionary deep neural network for predicting morbidity of gastrointestinal infections by food contamination. Neurocomputing. 226, 16–22.

Sors, T.G., Ellis, D.R. and D.E. Salt (2005). Selenium uptake, translocation, assimilation, and metabolic fate in plants. Photosynth. Res. 86: 373-389.

Srilakshmi, B. (2001). Food Adulteration, Food science, New Age International(P)Limited Publisher, pp.304.

Thompson, L. A. and Darwish, W. S. (2019). Environmental chemical contaminants in food: a review of a global problem. Journal of Toxicology. Pp 14.

Tirima, S., Bartrem, C., Lindern, I., von Braun, M., von Lind, D. and Anka, S. M. (2020). Food contamination as a pathway for lead exposure in children during the 2010–2013 lead poisoning epidemic in Zamfara, Nigeria. J. Environ. Sci. 10 (26).

Tompson, L. A., Ikenaka, Y. and Yohannes, Y. B. (2018). Human Health Risk from Consumption of Marine Fish Contaminated with DDT and Its Metabolites in Maputo Bay, Mozambique. Bulletin of Environmental Contamination and Toxicology. pp. 1– 5,

Ur-Rahman, A., Chowdhury, F. R., & Alam, M. B. (2008). Artificial ripening: what we are eating. Journal of Medicine, 9(1), 42-44.

Vejarano, R., Siche, R. and Tesfaye, W. (2017). Evaluation of biological contaminants in foods by hyperspectral imaging: A review. International Journal of Food Properties. 20; 1264-1297.

Velusamy, A., Satheesh Kumar, P., Ram, A. and Chinnadurai, S. (2014). Bioaccumulation of heavy metals in commercially important marine fishes from Mumbai Harbor, India. Mar Pollut Bull. 81:218–224.

Wu, L., Du, B., Vander Heyden, Y., Chen, L., Zhao, L., Wang, M. and Xue, X. (2017). Recent advancements in detecting sugar-based adulterants in honey—A challenge. TrAC Trends Anal. Chem. 86, 25–38.

Yadav, A., Kumar, R. and Pandey, A. K. (2015). Some common food toxins and their effect on human health. Journal of the Kalash Science. 3(3) (Special Volume): 1-6

Yeleliere, E., Cobbina, S. J. and Abubakari, Z. I. (2017). Review of microbial food contamination and food hygiene in selected capital cities of Ghana. Cogent Food & Agriculture. 3(1), 1395102.

Yeung, A. W. K., Tzvetkov, N.T., Jóźwik, A., Horbanczuk, O.K., Polgar, T., Pieczynska, M. D., Sampino, S., Nicoletti, F., Berindan-Neagoe, I., Battino, M. and Atanasov, A. G. (2020) Food toxicology: quantitative analysis of the research field literature. Int J Food Sci Nutr. 71(1):13-21.

Yu, B., Wang, X. and Dong, K.F. (2020). Heavy metal concentrations in aquatic organisms (fishes, shrimp, and crabs) and health risk assessment in China. Mar Pollut Bull. 159:111505.

Zenker, A., Cicero, M. R., Prestinaci, F., Bottoni, P. and Carere, M. (2014). Bioaccumulation and biomagnification potential of pharmaceuticals with a focus to the aquatic environment. J. Environ. Manage. 133, 378–387.

Zhang, S., Wang, J. and Yan, P. (2020). Non-biodegradable microplastics in soils: a brief review and challenge. J Hazard Mater. 409:124525.

Zhang, Z., Zhao, C., Ma, Y. and Li, G. (2014). Rapid analysis of trace volatile formaldehyde in aquatic products by derivatization reaction-based surface enhanced Raman spectroscopy. Analyst. 139 (14), 3614–3621.

2

Food Bioterrorism Chemistry, Toxicology, Evaluation of Contaminants and Impact on Human Physiology

Shylaja R. and Balakrishna K.

Department of Food Microbiology, Defence Food Research Laboratory (DFRL) Siddhartha Nagar, Mysore, Karnataka, India

Introduction

Terrorism has been experienced throughout human history by unlawful violence against innocent population to achieve financial or religious or political objective (Rakesh Gupta, 1998). Terrorism has no geographic or geopolitical boundaries and is adapting to changes in the socio-political environment. Progress in the area of science and technology have the potential to be used for both beneficial and harmful purposes leading to dual-use dilemma. Of late, more scenarios are witnessed where the applications of well-intentioned scientific research are being used for vicious purposes. Earlier, terrorism has been associated primarily with physical acts of violence and crime. Eventually the global militant and terrorism strategies evolved to chemical, nuclear, radioactive explosive agents to biological terrorism. Recently, cyber terrorism controlled by computers has seen a new form of criminal activities (www.hannover-re.com).

Bioterrorism is a global threat today. With advances in biotechnology, potential of creating more hazardous biological warfare agents are on the increase. Bioterrorism can be defined as terrorism by deliberate spread of biologic agents (bacteria, viruses, or toxins); naturally occurring or in a human-altered form. The ready availability of both technology and trained personnel to operate is leading to multiple intentional outbreaks. The greatest challenge about bioterrorism is the possibilities of bio agents prevailing in food, water, air and the human being himself. Biowarfare (BW) agents are comparatively economical to generate, stockpile, and transfer across the world. Bioagents can be toxic, contagious, even fatal and are present abundantly in nature, but can be transformed to upsurge their capability to cause illness, and make them resistant to existing drugs, or to enhance their capacity to proliferate into the environment (Magdy Hefnawy and Dina Solodoukhina, 2011). Although toxins are not self-replicating, their toxic power is high which can be further enhanced by biotechnology, and they are extremely lethal even in small quantities. These agents are produced by biotechnology are of dual use, thus they are not easy to prohibit. Moreover, safety measures of these BW agents is very meager in number of facilities. Inadequate border jurisdiction policy makes possible illegal trafficking of these dangerous materials, or weapons of mass destruction by antisocial people (Robert, 2003).

History of Bioterrorism

The practice of bio-war using biologic agents is not a new concept. Even bible describes about Plague outbreak by Israelites over Philistines (King James version of The Bible). In 400 BC, nomadic archers in Scythia used arrows soaked in decomposed cadavers and blood with infectious agents. Romans were well recognized to throw feces on their rivals as a practice of biologic terrorism. In the middle ages, smallpox infected animals were thrown on the walls of forts (Bellal *et al.*, 2013). At 600 BC itself, the use human cadavers and animal carcasses to destruct the enemy were in practice (Eitzen *et al.*, 1997). Biological war was waged more than 600 years to the siege of Caffa in 1346 (Wheelis, 2002). In the fifth century BC, Athens, water supply was infected with the plague bacteria by the Spartans. The Romans had contaminated water sources with the dead corpses. Poisoning water sources of the opposing army was a frequently followed strategy that persisted through many European, the American Civil War, and even continued in the 20th century. Plague pandemic of 14th and 15th century was the reason for the death of more than 25 million Europeans. Spaniards had spiked French wine with leprosy patient's blood in 1495. Water supply poisoning occurred even in the American Civil War, the Boer War, and, allegedly, in Turkish Kurdistan as recently as 1997. Use of smallpox infected people to spread disease was even mentioned in a letter from General Alexander Leslie to General Cornwallis in 1781 during the American War of Independence (http://www.cbwinfo.com/History/History.html/> accessed 30 September 2003). At Fort Pitt, Ohio River Valley, the British troops intentionally dispersed smallpox among native Indian population by presenting them with blankets and linens used by smallpox victims (Wheelis, 1991). There is reference in history of spreading smallpox via infected British soldiers and their clothing during the American Revolutionary War (1776–1781) and during the American Civil War (1980–1864). The incidents of the use of biological agents during the past 2000 years mentioned in historical recordings is as depicted in the Table 1.

Bioterrorism as a Burning Global Issue

Bioterrorism has the prospective to destroy millions and incur losses in billions and create political instability. It was estimated by the chief of US Army's BW laboratories that one jet of bioweapon spray would perhaps be more effective in killing human than a ten-megaton hydrogen bomb (Preston, 1998; Miller *et al.*, 2002. The National Academies, 2004; Jack Alderson, 2012). The jeopardy of catastrophic bio-threat is extended by globalization, increased tourism, travels across the globe and rapid advances in technology leading to the manipulation of pathogens with high pandemic potential like COVID 19 virus. The need of the hour is to strengthen biosecurity, decrease biological risks posed by new dual research. More stringent and new modified surveillance approaches are required continuously to identify bio-threat outbreaks. Identification of the weakness and filling gaps to strengthen global health security capabilities are uninterruptedly required to combat with the advent of new bio-risk agents created by the vicious antisocial elements. Regardless of these challenges, more importance is not given to biosecurity issues. The stockpile of more than a dozen BW agents that includes smallpox virus, anthrax, Ebola and typhus, along with an unspecified number of genetically engineered organisms and maintained by a couple of countries may well be an ever threat to the mankind (Debajit Sarkar, 2020). The ongoing Covid-19 crisis has compelled the international society to rethink and analyze about present and forthcoming biological threats (Ophélie Guillouet-Lamy, 2020).

Historical Recordings of Intentional Food Outbreak Incidents with Biological Agents

Table 1: Historical Recordings of Intentional Food Outbreak Incidents with Biological Agents

Time	Event
600 BC	Solon used the hellebore plant, a laxative during the siege of Krissa
1155	Royal leader Barbarossa poisoned water wells with human bodies in Tortona, Italy
1346	Tartar forces catapult cadavers of plague over the city walls of Caffa, Crimean Peninsula (now Feodosia, Ukraine)
1495	Spanish mix wine with blood of leprosy patients to sell to their French foes in Naples, Italy
1675	German and French military agreed not to use "poisonous bullets"
1710	Russian troops threw bodies of plague victims in Swedish cities
1763	British issued blankets of smallpox patients to Americans
1797	Napoleon spreads malaria in Mantua and Italy
1863	Yellow fever and smallpox patients clothes were issued to troops in the US Civil War
World War I	German and French used glanders and anthrax to incapacitate enemies
World War II	Japan and other countries experimented with plague, anthrax, and other disease causing agents and developed BW programs
1980–1988	Iraq employed mustard gas, sarin, and tabun against Iran and ethnic groups inside Iraq during the Persian Gulf War
1984	Rajneeshee religious followers attacks, The Dalles, Oregon contaminated restaurant salad bars with *Salmonella* Typhimurium to incapacitate the population to win the elections resulted in 751 illnesses. Initially CDC declared event as natural outbreak. One of the followers was arrested on unrelated charge confessed involvement with the event.
1995	Aum Shinrikyo used sarin gas in subway of Tokyo.
1987-1990	David J. Acer, Florida dentist, a HIVpositive infected 6 patients with virus.

Time	Event
1998	Richard J. Schmidt, a gastroenterologist in Louisiana convicted of attempt to second degree murder for injecting nurse Janice Allen with AIDS patient blood.
1964-1966	Dr. Mitsuru Suzuki, a physician in.Japan to take revenge on prevailing seniority system, tainted sponge cake with *Shigella dysenteriae.* and *Salmonella typhi.* After official investigation by Ministry of Health and Welfare, he was found guilty. The incident implicated in 200 – 400 illnesses and 4 deaths.
1996	Diane Thompson, a clinical laboratory technician, Dallas, TX infected co-workers with *Shigella dysenteriae* Type 2 stolen from hospital's collection and contaminated pastries. Infected 12 of her coworkers, she was arrested, convicted, & sentenced to 20 years in prison.
2001 Amerithrax"	Letters containing anthrax spores were mailed to media offices and senators. Alleged culprit was a US DOD scientist. 22 were infected in the incident and 5 deaths happened.

Source: Stefan Riedel (2004); https://en.wikipedia.org/wiki/Bioterrorism

Initiatives Taken Across the World to Prevent Bioterrorism

The first initiative was taken during 1925 Geneva Protocol, wherein international restrictions were imposed on biological warfare but allowed the possession or development of chemical and biological weapons (Text of Geneva protocol). However, the US program on biological weapons was initiated in 1943 and by end of World War II, they had spent over $ 400 million on biological weapons (Guillemin and Jeanne 2005). Britain, France, and the Soviet Union affirmed that they obey to the 1925 Geneva protocol as long as their enemies did not use chemical or biological weapons. The American Biological Defense Policies and Programs were dismissed in 1969 by then President Nixon.

Biological and Toxin Weapons Convention (BTWC): was the first multilateral disarmament treaty made categorically to ban the development, manufacturing, accumulation/stockpiling, or relocation of biological agents and toxins of "types and quantities" that have no justification for defensive or peaceful use. Draft of the BWC, submitted by the British was signed on 10 April 1972 and entered into power by 26 March 1975 (http://www.bbc.co.uk/history/worldwars/coldwar/pox_weapon_01.shtml). The BWC established a strong global standard against biological weapons (Cross *et al.*, 2020). The convention was of unlimited period. This standard reflected in the treaty's foreword, which states that the use of biological weapons would be "repugnant to the conscience of mankind". This norm is demonstrated by the fact that not a single state today declares to possess or seek biological weapons, or asserts that their use in war is legitimate (Feakes, 2017). However, the convention's effectiveness was inadequate as it lacked any official verification system to monitor compliance (Tucker, 2001). The unintentional dissemination of anthrax from a military testing unit in the former Soviet Union in 1979 and possession of anthrax, botulinum toxin and aflatoxin in Iraq in 1995 called attention to do research and development of these agents in spite of the 1972 Biological Weapons Convention (Thavaselvam, 2010). As of July 2021, 183 states have become part of the treaty.

Characteristics of a Biowarfare Agent

All pathogens and the toxins produced by them are not used as potential BW agents. To be suitable as a biowarfare agent, certain properties and characteristics are to be qualified. Following are the key criteria for the microbial agents/toxins to qualify as bioterrorist.

- Easily producible
- Pathogenicity of the organism
- Able to infect the large population
- Sufficient stability to disseminate rapidly and to survive in the atmosphere until it reaches its target
- Low production cost
- Flexibility to attack humans, plants or animals
- Agents can be either incapacitating or lethal
- Low infection dose and high toxicity, with minimum incubation time
- Challenging to detect, identify and diagnose at an initial stage
- High morbidity or high infectivity, high transmissibility and/or contagiousness
- Stable in diverse environment
- Lack of availability of protective mechanism
- Easy mode of transmission and host range
- Accessibility of effective preventive measures (e.g., vaccines or medical treatment)
- Availability of effective treatment (e.g., antibiotics)
- Scientific expertise to handle
- Able to produce in mass and deliver in the proper size for dissemination as aerosol.

The Routes of Entry of Bio-agents

The most effective route of transmission of biowarfare agents in order to incapacitate large population is through the aerosol form. Dispersal of aerosolizable agents may possibly be through bomblets or through spray guns mounted on aircrafts/ tall buildings. The entry of aerosolizable bio agents into human body mainly occur through inhalation and skin contact. The most severe infections occur through aerosol/droplet infections caused by the natural process of inhalation, from person to person through direct contact, by vectors and or by other methods. Anthrax, plague, brucellosis, smallpox, viral encephalitides and haemorrhagic fever causing virus can be and dispersed over large topographical areas in aerosolized form (Hayoun, 2021).

Additional easily accessible important routes include food and water which mainly affects gastrointestinal tract once the agent is ingested. The deliberate contamination can happen at any stage from farm to fork. Diverse targets of bioterrorism outbreaks in the farm-to-table food gamut include soil, air, crops, water used for irrigation, livestock, food products and personnel involved in the processing and distribution chain, wholesale and retail outlets, storage facilities, transportation, food and agriculture research laboratories. Other methods include delivery by post or by deliberate infiltration of infected animals, vectors and pests through the international border.

Potential Biological Warfare agents and Its Categorization

More than 180 pathogens and their toxins have been employed as bio threat agents, including anthrax, tularemia, brucellosis, plague, Legionnaire's disease, Q fever, glanders, melioidosis, smallpox, viral hemorrhagic fevers, influenza, ricin, botulinum toxin, staphylococcal enterotoxin B, coccidiosis, rice blast, and wheat rust (Hayoun, 2021). Centers for Disease Control and Prevention (CDC) demarcated and published three distinct groups of potential bioterrorism biological agents in the category list based on assessment of the threat level of biological agents which are as mentioned in detail in the Fig 1.

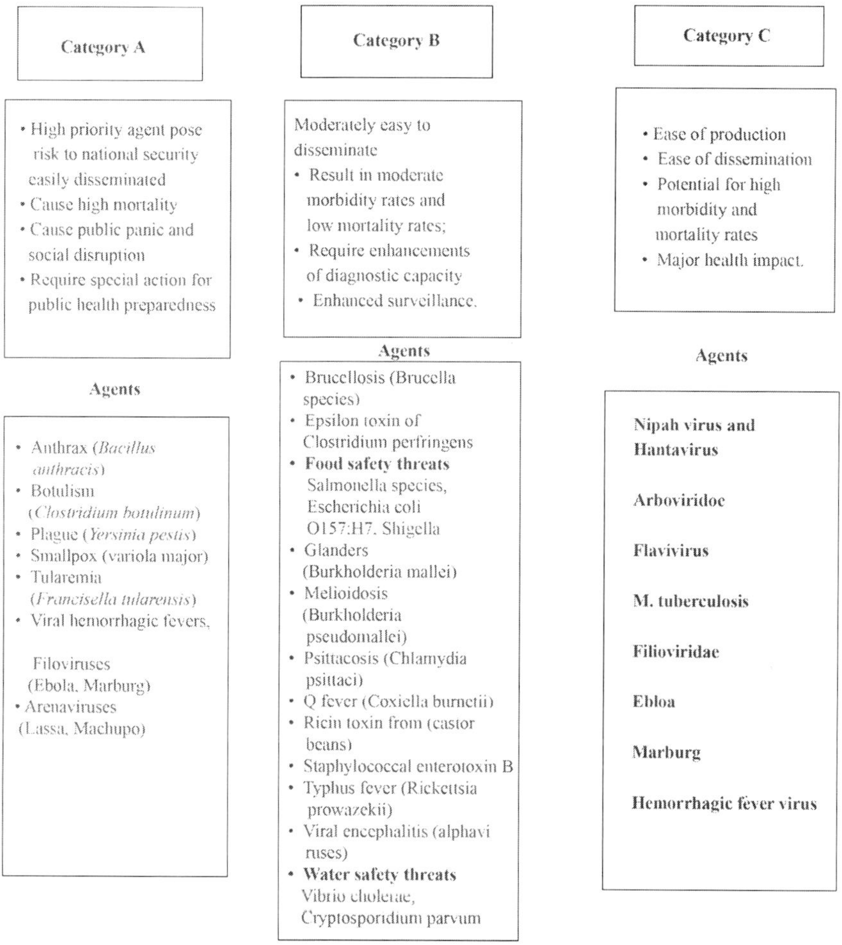

Category A

- High priority agent pose risk to national security easily disseminated
- Cause high mortality
- Cause public panic and social disruption
- Require special action for public health preparedness

Agents

- Anthrax (*Bacillus anthracis*)
- Botulism (*Clostridium botulinum*)
- Plague (*Yersinia pestis*)
- Smallpox (variola major)
- Tularemia (*Francisella tularensis*)
- Viral hemorrhagic fevers,

  Filoviruses (Ebola, Marburg)
- Arenaviruses (Lassa, Machupo)

Category B

Moderately easy to disseminate
- Result in moderate morbidity rates and low mortality rates;
- Require enhancements of diagnostic capacity
- Enhanced surveillance.

Agents

- Brucellosis (Brucella species)
- Epsilon toxin of Clostridium perfringens
- **Food safety threats** Salmonella species, Escherichia coli O157:H7, Shigella
- Glanders (Burkholderia mallei)
- Melioidosis (Burkholderia pseudomallei)
- Psittacosis (Chlamydia psittaci)
- Q fever (Coxiella burnetii)
- Ricin toxin from (castor beans)
- Staphylococcal enterotoxin B
- Typhus fever (Rickettsia prowazekii)
- Viral encephalitis (alphavi ruses)
- **Water safety threats** Vibrio cholerae, Cryptosporidium parvum

Category C

- Ease of production
- Ease of dissemination
- Potential for high morbidity and mortality rates
- Major health impact.

Agents

Nipah virus and Hantavirus

Arboviridoe

Flavivirus

M. tuberculosis

Filioviridae

Ebloa

Marburg

Hemorrhagic fever virus

Fig. 1: Categorization of Biowarfare agents: The CDC classifies biological agents into three categories

Food Bioterrorism Threats

COVID 19 changed the societal, political and economic condition in the world, which forced us to consider the newly emerging bio threats to the health and lives of human kind. These threats can enter through any mode of transmission. Food, the dire essential of mankind is easily accessible to terrorists (Just *et al.*, 2009). Intentional contamination of food for terrorism purposes at one place could have worldwide public health implications due to global food trade. As compared to other transmission modes, such as dissemination through air or liquids, food is an ideal means of distributing contaminants to humans, since it can reach multiple targets in many different ways, and the control measures can be easily circumvented, either intentionally or accidentally. Biological agents can be introduced at multiple susceptible points along the food supply continuum. Contaminating food is simple, yet effective, means to incapacitate large populations. Ensuring the safety of food is a long-standing and critical challenge of public health agencies.

Food is an extremely effective vehicle for distribution of pathogens, such as anthrax, *Salmonella, E. coli* O157:H7, *Shigella* and other bacterial BW agents (Osterholm, 2005), chemical agents including biological toxins (e.g., botulinum toxin, aflatoxin, T2 toxin) and poisons (cyanide, dioxin) directly to humans including non-infectious microorganisms (Khan *et al*., 2001).

An estimated 600 million – almost 1 in 10 people in the world – fall ill after eating contaminated food and 420 000 die every year (WHO 2020). Unsafe food creates cycle of illness and malnourishment, affecting infants, young children, elderly and the sick in particular. Foodborne diseases hinder socioeconomic growth of a nation by overburdening health care systems, and harming national economies, tourism and trade. If an unintentional contamination can affect millions, a determined, deliberate attack on food could be devastating. Intentional food contamination is of greatest concern worldwide that can paralyze world.

Few of the Category A and majority of the Category B agents represent the food and water safety threats as per CDC (Fig. 1). These include *Clostridium botulinum* (Botulism), Shiga-toxin producing *E. coli O157* (STEC), *Salmonella typhi* (Typhoid fever), *Shigella dysentaeriae,* Staphylococcal enterotoxin B, Trichothecene, Aflatoxin and other bacterial pathogens, such as *Listeria monocytogenes* and *Campylobacter jejuni* are foodborne agents. Amongst the water borne are *Vibrio cholera* (cholera) and *Cryptosporidium parvum* are listed. Though, Category A biothreats more intensely capture the community and scientific attention, in the age of globalization of food supply system, the terror by highly potent and virulent foodborne pathogens should not be undervalued.

The list of potential biowarfare agents along with food bioterrorism agents, their mode of transmission, their morbidity and mortality rate, effect of these agents on human physiology, clinical syndromes caused and their treatment are as described in the Table 2 :

Table 2: Potential biowarfare agents along with food bioterrorism agents, their mode of transmission, chemical nature, virulence factors, the clinical syndromes treatment

Agent / Disease	Chemical nature / Virulence factors	Transmission	Morbidity/ Mortality rate	Changes in Human Physiology, Clinical syndrome and Treatment
Agent: ***Bacillus anthracis*** (Gram + ve spore forming bacteria) **Disease:** Anthrax **Food threat agent**	**Proteins:** Protective antigen (PA), Lethal factor (LF) and Edema factor (EF); Capsule (poly-D-glutamate)	By inhalation, direct contact and ingestion	High/high	**Physiological changes:** LT and ET increase permeability of vasculature leading to hemorrhage. PA enables transfer of LF and EF to the cytosol. LT inactivates mitogen-activated protein and activates the release of sepsis-related cytokines TNF- alpha and IL1β. EF, a calmodulin (CaM)-dependent adenylate cyclase, upsurges cyclic AMP levels, damages neutrophil function and disturbs water balance resulting in massive tissue edema. Altogether, the toxins effectively inhibit host innate and adaptive immune responses. Systemic anthrax affects the cardiovascular system leading to cyanosis, decreases blood pressure. Mediastinal widening in association with pleural effusion and edema (Klempner *et al.*, 2010). **Clinical syndrome:** Subcutaneous, Gastrointestinal, and Inhalational Anthrax Fever, respiratory distress, chest pain. Widened mediastinum, malaise, **Treatment:** Ciproflox / Doxycycline
Agent: *Clostridium botulinum* (Gram + ve spore forming bacteria) **Disease:** Botulinum	Neurotoxin Eight antigenically different exotoxins (A, B, C1, C2, D, E, F and G). Type A is the most potent toxin.	Ingestion	High/high	**Physiological changes:** Neurotoxin produced by *Clostridium botulinum* involves the release of neurotransmitters from the enteric nervous system. Alters the action of intestine smooth muscle. Inhibits release of presynaptic acetylcholine from peripheral neurons, leading to flaccid paralysis. **Clinical syndrome:** Acute bilateral descending flaccid paralysis with cranial nerve palsies. **Treatment:** Supportive treatments like rehydration, Equine antitoxin.

Agent / Disease	Chemical nature / Virulence factors	Transmission	Morbidity/ Mortality rate	Changes in Human Physiology, Clinical syndrome and Treatment
Agent: *Yersinia pestis* (Gram – ve bacillus) **Disease:** Plague	Type 3 secretory system proteins (TTSS) encoded by pCD1 plasmid viz.,YopE, YopJ/ YopP, YopM, YopH, YopT, and YpkA/ YopO along with adhesins	Inhalation and vector	High/high	**Physiological changes:** Bubonic plague, infection of the lymph nodes situated in the groin, armpit or neck; Pneumonic plague, an infection of the lungs; Septicemic plague, an infection of the blood. *Y. pestis* invades macrophages and expresses virulence factors to inhibit subsequent phagocytosis. Systemic infection takes place 12–24 h subsequent to lymph node colonization. Lung colonization/secondary pneumonic plague by systemic spread, spleen colonization followed by death if left untreated (Cathelyn *et al.*, 2006). **Clinical syndrome:** Acute pneumonia, shock, hemoptysis, cyanosis, gastrointestinal symptoms. **Treatment:** Streptomycin intramuscular or intravenous or Doxycycline / Chloramphenicol.
Food and water **Agent:** *Fransciella tularensis* **(Gram-ve bacillus)** **Disease:** Tularemia	Capsule; Lipo polysaccharide layer; Type IV pili; Outer membrane; Periplasm; Inner membrane and secretion systems	By inhalation, direct contact, ingestion and vector	High/high	**Physiological changes:** After cutaneous bacterial infection and multiplication, spread of bacteria to regional lymph nodes, liver, spleen, and lungs is observed with bacteraemia and intense inflammatory response is stimulated. PMN invasion, necrosis and accumulation of mononuclear cells, macrophages and T-lymphocytes leading to formation of granuloma. In A type Tularemia, excessive tissue necrosis and death is observed due to the formation of granuloma and lobular pneumonia (Rowe 2015; Tärnvik and Berglund., 2003). **Clinical syndrome:** Typhoidal, pneumonic fever, chills, rigors, headache, myalgias, sore throat; uneasiness below the sternum, arid cough. Chest radiograph shows infiltrate/effusion. **Treatment:** Streptomycin.

Agent / Disease	Chemical nature / Virulence factors	Transmission	Morbidity/ Mortality rate	Changes in Human Physiology, Clinical syndrome and Treatment
Agent: *Variola major* (Virus) **Disease:** Smallpox	Smallpox inhibitor of complement enzymes (SPICE); complement control protein; Chemokine binding protein type II	Direct contact	High/high	**Physiological changes:** Virus attacks oropharyngeal mucosa and Viremia in observed in local lymph nodes due to viral replication. Localizes in small blood vessels of the dermis and the oropharyngeal mucosa. Rarely the central nervous system is involved with encephalitis (Pittmann *et al*.,, 2019). **Clinical syndrome:** Fever with papular rashes on the face and uniformly progresses to vesicles and pustules; **Treatment:** Supportive.
Agent: *Filo virus* Marburg virus & Ebolavirus **Disease:** Viral hemorrhagic fever	Viral capsid proteins such as VP24, VP30, VP 35 etc and RNA the nucleic acid	Direct contact and vector	High/high	**Physiological changes:** Lysis of monocytes/macrophages, dendritic cells, and hepatocytes that suppresses innate immune response. Cytolysis and massive release of IL1, IL6, interferons, cytokines, chemokines results in severe illness. Lymphocyte apoptosis takes place contributing to immunosuppression by weakening adaptive immune responses. Disseminated intravascular coagulation (DIC) also takes place (Mahanty, 2004; Yamaoka, 2021). **Clinical syndrome:** Viral haemorrhagic fever. Mucous membrane bleeding, Petechiae, Thrombocytopenia and low blood pressure. **Treatment**: Supportive treatment and Ribavirin.

Agent / Disease	Chemical nature / Virulence factors	Transmission	Morbidity/ Mortality rate	Changes in Human Physiology, Clinical syndrome and Treatment
Bacteria **Agent:** *Brucella spp* **Disease :** Brucellosis	Hsp60 (heat shock protein); LPS, lipo polysaccharide; Surface protein (41-kDa); (T4SS); Brucella virulence factorA (BvfA); BvrR/BvrS	By direct contact, inhalation and ingestion	High/low	**Physiological changes:** WBCs carries pathogen after phagocytosis through hematologic/lymphatic route to different organs especially reticuloendothelial system. Endotoxic LPS suppresses phagosome-lysosome fusion internalizing bacteria into ER. T4SS (VirB) and T3SS, controls intracellular survival and transferring. BvrS/BvrR codes for histidine kinase, which controls cell invasion. Hemolysins induce cell necrosis (Pappas, *et al.*, 2005). **Clinical syndrome:** Fever, shivering, sick feeling, headache, pleuritic chest pain. **Treatment:** Intramuscular Streptomycin with Doxycycline.
Food threat agent:*Clostridium perfrigens* (Gram +ve bacillus) **Disease:** Enteritis	*Clostridium Perfringens Enterotoxins* (CPE): **CPE is a polypeptide of 35.5 kDa.**	Ingestion	High/low	**Physiological changes**: CPE enterotoxin is fatal, cytotoxic and enterotoxic activity. Activates the adenylyl cyclase letting the increase of cAMP in epithelial cells,ultimately leading to diarrhea. **Clinical syndrome** Enteritis in humans **Treatment**: No specific treatment for Clostridium perfringens toxins. Supportive treatment (IV fluids, controlling fever and pain). Epsilon toxin (ETX) causes enterotoxemia, a fatal disease particularly in sheep (domestic ruminants).

Agent / Disease	Chemical nature / Virulence factors	Transmission	Morbidity/ Mortality rate	Changes in Human Physiology, Clinical syndrome and Treatment
Food threat agent Agent: *Salmonella spp* (Gram negative bacteria) **Disease:** Salmonellosis	High cell concentration of Bacteria, Somatic antigen (O) of Endotoxins and Type III secretion system Flagellar antigens, Virulence plasmid etc	Ingestion and oral-faecal route	Moderate/ Low	**Physiological changes**: Gastric acid barrier survived pathogenic *Salmonellae* attack the mucosa of the small and large intestine and produce toxins. Release of proinflammatory cytokines by invading epithelial cells to induce diarrhea by an acute inflammatory reaction. This causes ulceration and damage of the mucosa. Bacteria becomes systemic, infects liver, spleen, gallbladder, bones, meninges, and other organs (Giannella. 1996). Typhoidal salmonella causes enteric fever. Non-typhoidal salmonella causes transient intestinal infections. **Clinical syndrome**: Salmonella gastroenteritis with blood-tinged diarrhea, abdominal cramps, fever, and occasionally nausea and vomiting. **Treatment:** Rehydration and intravenous fluids. Antibiotics are recommended in complicated cases of systemic infection. Ciprofloxacin or Ofloxacin.
Food threat agent: *E. coli* O157:H7 (bacteria) **Disease:** GE, HC, HUS and TTP	Somatic O antigen 157 and flagellar H antigen, Virulence plasmid. hemolysins, Shiga toxins	Ingestion and oral-faecal route	Moderate/ Low	**Physiological changes**: Enterohemorrhagic *E. coli* (EHEC) damages the lining of kidneys and intestines. Infection can lead to fatal hemolytic uremic syndrome (HUS), toxin destroys red blood cells, kidney failure, neurologic damage, and death. 5 – 10% of HUS cases are fatal. **Clinical syndrome:** Severe abdominal cramps, diarrhea become bloody diarrhea (hemorrhagic colitis) within 24 hours with little or no fever. Followed by hemolytic uremic syndrome (HUS), in which RBCs are destroyed (hemolytic anemia), thrombocytopenia, kidney failure and death. **Treatment:** Antibiotics are **not recommended** for *E. coli O157* infections. Rehydration and intravenous fluids.

Agent / Disease	Chemical nature / Virulence factors	Transmission	Morbidity/ Mortality rate	Changes in Human Physiology, Clinical syndrome and Treatment
Food threat agent: *Shigella* (bacteria) **Disease:** *Shigellosis*	**Shiga toxins,** invasive plasmid antigens Lipopolysaccharide, T3SS system, Outer shigella protein etc.,	Ingestion and oral-faecal route	Moderate/ moderate	**Physiological changes**: Death of cells by inhibiting protein synthesis. Shiga toxin influences the severity of bacillary dysentery by inducing colonic vascular damage, bloody stools, intestinal ischemia, and increase of a PMN intestinal compartment. Inflammatory vasculitis of the peritoneal mesothelium, and major efflux to the intestinal lumen (Fontaine *et al*., 1988). **Clinical syndrome:** Bloody diarrhea, fever, stomach pain, urge to pass stool even with empty bowels. Rarely, HUS occurs when bacteria produce shigatoxin that destroys red blood cells. Bloodstream infections due to damage of the intestinal linings by the organism. High fever with abnormal blood electrolyte levels. **Treatment:** Ciprofloxacin and azithromycin are two recommended oral antibiotics.
Water and Food threat agent: *Vibrio cholera* (bacteria) **Disease:** Cholera	Cholera toxin (CT), which is encoded by a lysogenic bacteriophage (CTXΦ), and toxin-co-regulated pilus (TCP)	Ingestion and oral-faecal route	High/low	**Physiological changes**: Adenyl cyclase gets activated; inflates the levels of cAMP promoting fluid and electrolytes secretion in the intestinal epithelium, causing diarrhea. **Clinical syndrome:** Acute watery diarrheal resembling rice-water stools, emesis, dehydration, leg cramps, restlessness, rapid heart rate, low BP, dryness of skin and mucous membranes. Kidney failure. Untreated patient can experience shock, can enter coma and subsequently death. **Treatment:** Supportive treatment. Tetracycline or doxycycline. Even Ofloxacin, trimethoprim-sulfamethoxazole (TMP-SMX), and ciprofloxacin are effective.

Agent / Disease	Chemical nature / Virulence factors	Transmission	Morbidity/ Mortality rate	Changes in Human Physiology, Clinical syndrome and Treatment
Agent: *Burkholderia mallei* (bacteria) **Disease:** Glanders	Capsular polysaccharide, lipopolysaccharide, and type III and VI, V secretion systems.	Inhalation and direct contact	High/low	**Physiological changes**: Critical infection with subcutaneous tissue ulceration, lymph nodes, and salivary glands. Pneumonia due to lung infection, abscesses in lungs, and pleural effusions. Septicemia, abscesses in skin, intramuscular, limbs and cutaneous abscesses. Liver, spleen, skin are also affected. **Clinical syndrome:** Fever, chills, excessive sweat, muscle ache and tightness, chest pain, headache, nasal discharge. Pulmonary infection with pneumonia, pulmonary abscesses. Without treatment, septicemia is fatal. **Treatment:** Sulfadiazine has been found to be effective in the treatment of glanders. In addition, Tetracyclines, Ciprofloxacin, Streptomycin, Novobiocin, Gentamicin, Imipenem, Ceftazidime,
Agent: *Burkholderia pseudomallei* (bacteria) **Disease:** Melidiosis	Burkholderia lethal factor; Cytotoxin capsular polysaccharide I; the cluster I; T4SS, T5SS, hemagglutinins and invasins	Inhalation and direct contact	High/low	**Physiological changes**: *B. pseudomallei* invades lungs (pulmonary), bloodstream, causes local, and dispersed infections. In pulmonary melidiosis bronchitis and severe pneumonia is observed and blood stream melidiosis leads to septic shock. Pulmonary infection can progress to septicemia and is the most serious form of melioidosis. **Clinical syndrome:** If local infection, pain, inflammation, rise in temperature, ulceration, abscess; in case of pulmonary infection, headache, high fever, cough, chest pain, anorexia, abdominal discomfort, joint pain, central nervous system/brain infection, seizures. People with comorbidities are at greater risk. **Treatment:** Intravenous with Ceftazidime /Oral antimicrobial therapy: Trimethoprim-sulfamethoxazole.

Agent / Disease	Chemical nature / Virulence factors	Transmission	Morbidity/ Mortality rate	Changes in Human Physiology, Clinical syndrome and Treatment
Food threat agent: *Coxiella burnetti* (bacteria) ***Disease:*** Q fever	Lipo polysachharide	Inhalation and direct contact, Zoonotic		**Physiological changes**: TIVBSS translocates over 100 effector proteins into host. These effectors modulate many pathways of host cells such as blocking cell death, inhibiting immune reactions, and altering vesicle trafficking (Clemente *et al*., 2018). **Clinical syndrome:** Fever, shivering or sweat, fatigue, headache, muscle aches, nausea, emesis, diarrhea, chest pain, stomach pain, weight loss, dry cough. Abortion in human and animals. **Treatment:** Doxycycline
Agent: *Chlamydia psittaci* (bacteria) **Disease: *Psittacosis***	Chlamydia-specific protein CT135; protease and other effector proteins.	Inhalation and Zoonotic especially through birds	High/low	**Physiological changes**: Pathogen invades host cells from epithelial cells to macrophages, and multiplies rapidly, translocate to mitochondria and the Golgi apparatus. Associates with energy-providing cell compartments, as well as destabilizes the regulation of host cell and expresses effectors of the T3SS and suppresses host defense mechanism (Michael, 2015). **Clinical syndrome:** Fever and chill, headache, muscle aches and dry cough. **Treatment:** *Chlamydia psittaci* treated with tetracyclines.

Agent / Disease	Chemical nature / Virulence factors	Transmission	Morbidity/ Mortality rate	Changes in Human Physiology, Clinical syndrome and Treatment
Food threat agent: ***Staphyloccus aureus*- SEB** (bacterial toxin) **Disease:** Enteritis, Toxic shock	Food poisoning by Heat stable protein Enterotoxins	Ingestion	Moderate/ Low	**Physiological changes:** Enterotoxins are thermostable and superantigenic in nature that causes massive non-specific activation of the immune system enhanced production of T cells including lymphocytes and macrophages leading to multiple organ failure. Secretion in intestinal segments. Stimulation in the short circuit current (Isc) and potential difference in the chamber; this result in the secretion of electrogenic anions. In addition, SEB can impair NaCl absorption, resulting in ion secretion. Superantigenic action causes inflammation of the vascular system, fever followed by multi-organ failure and toxic shock. **Clinical syndrome:** High fever, shivering, headache, dry cough and body ache. Profuse watery diarrhoea, if ingested. Toxic shock due to multiple organ failure. **Treatment:** Supportive therapy.

Agent / Disease	Chemical nature / Virulence factors	Transmission	Morbidity/ Mortality rate	Changes in Human Physiology, Clinical syndrome and Treatment
Water borne **Agent:** *Cryptosporidium parvum* (Protozoa) **Disease:** Cryptosporidiasis	Invasins; Adhesins,; Serine protease; Hemolysins; Glycoproteins; HSPs; Proteases	**Water borne**	Moderate/ High	**Physiological changes**: Parasite invades intestinal surface epithelial cells and signals the submucosal intestinal compartment to yield an intestinal secretory response (Bouzid *et al*., 2013). **Clinical syndrome:** Gastroenteritis: profuse watery diarrhea, abdominal pain, fever, emesis, Rarely muscle pain, weakness, headache and loss of appetite. Cryptosporidiosis can be fatal to immuno-compromised patients. Management of diarrhea by rehydration and other supportive therapy. Treatment: Nitazoxanide has been FDA-approved treatment for Cryptosporidiosis.
Agent: Ricinus communis (Plant) Rarely transmitted through food	Ricin poisoning by ricin toxin a glycoprotein (toxalbumin) in nature	Inhalation Air borne	High/high	**Physiological changes**: Ricin kills host cells by inhibiting protein synthesis by inhibiting RNA translation and transcription (Laurence Knott, 2011). **Clinical syndrome: Ricin Aerosolized:** Fever, chest pain, tightness in the chest, nausea, cough, acute respiratory distress and hypoxemia, heavy sweating, pulmonary edema, skin might turn blue, pulmonary edema; **Ingestion:** Emesis, diarrhea that might perhaps develop bloody. Extensive dehydration, low BP, seizures, blood in the urine and multiple organ failure and death; **Skin and eye exposure:** Conjunctivitis in eyes, painful skin. **Treatment:** Supportive.

Agent / Disease	Chemical nature / Virulence factors	Transmission	Morbidity/ Mortality rate	Changes in Human Physiology, Clinical syndrome and Treatment
Agent: Niphavirus **Disease:** Encephalitis	**Capsid proteins Phosphoprotein (P), glycoprotein (F), matrix protein, N, P, V, W and C protein**	Aerosol, Zoonotic Fruit bats	High/high	**Physiological changes**: G and F protein helps in binding to the host cell receptor and damages DNA pathway and inhibition of nucleolar protein and virus multiplication takes place. Viral P gene plays a vital role in pathogensis. V protein increases the level of UBXN1 and suppresses induction of innate interferon (Singh RK 2019). **Clinical syndrome:** Fever, boy pain, emesis, headaches and painful throat, Giddiness, lethargy and neurological disorders of acute encephalitis. Atypical pneumonia, Critical respiratory disorder, encephalitis and seizures in extreme cases, leading to coma.
Agent: Hantavirus **Disease:** Hemorrhagic fever with renal syndrome (HFRS) and pulmonary disease (HPS)	**Capsid**	Airborne Zoonotic especially through Mice	High/high	**Physiological changes**: Hantavirus replicates in vascular endothelial cells and macrophages particularly in the lung and the kidney. Hantavirus impair the integrity of endothelial barrier due to extreme innate immune response that is major act played in the pathogenesis and is a characteristic of hantavirus disease (Naveed munir 2020). **Clinical syndrome:** Tiredness, rise in temperature, myalgia. Headache, giddiness, shivering, and abdominal cramps, nausea, emesis, diarrhea. Cough and fluid accumulation in the lungs leading to shortness of breath. Treatment: No specific treatment. However, Ribavirin is administered.

Centers for Disease Control and Prevention, National Center for Emerging and Zoonotic Infectious Diseases (NCEZID) , Division of High-Consequence Pathogens and Pathology (DHCPP)

The major mission of a nation should be safeguarding the food supply chain. In order to mitigate, the unintentional/intentional contamination of food many government agencies like WHO, FAO, USFDA, FSIS, GFSI. EFSA, EU, FSSAI and many more agencies of various countries are operational across the world. The main aim of these agencies is to safeguard public health by assuring the food safety and from food-related risks including food bioterrorism. Food related risks are dealt with different headings such as food security, food safety, and food fraud and food defense. All are interrelated assessment concepts with intense influence on quality of human life and are as explained in Fig.2.

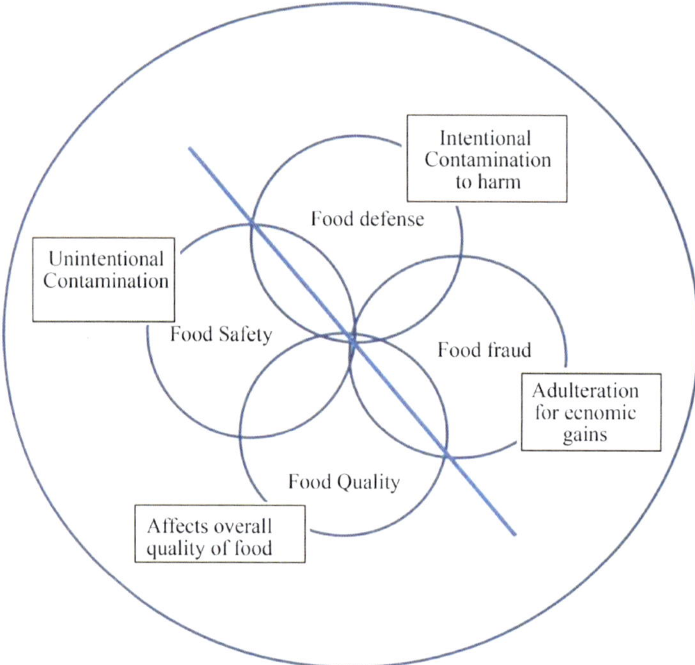

Fig. 2: Food Risk Types

Food Security: Food security is ensuring that all population has access to sufficient and nutritious foods. Food security has been well defined by the Food and Agriculture Organization (FAO) of the United Nations (UN) as; "Food security exists when all people, at all times, have physical, social and economic access to sufficient, safe and nutritious food which meets their dietary needs and food preferences for an active and healthy life. Household food security is the application of this concept to the family level, with individuals within households as the focus of concern" (FAO 2003).

Food Safety

Food safety is taking steps to inhibit the unintentional adulteration of food. The globalized food chain has created substantial difficulties to agencies ensuring food safety. Food safety focuses on the unintentional contamination of food products at different stages of processing starting from farm to fork including handling, preparation, storage and transportation. Food safety umbrella covers many facets of hazards mainly microbes, chemicals and foreign objects. Allergens and poisonous chemicals are of top priority as these are life threatening to sensitive individuals. Particles such as glass, metal saw dust etc comprise physical hazards whereas, pathogenic bacteria, viruses and toxins produced by microorganisms comes under biological hazards. Products can become contaminated due to lack of attention that can occur during any stage of food processing from farm to fork. Foods supply chain currently cross various national borders. Proper collaboration between governments, producers and consumers are required to guarantee food safety. Food tampering at various points must be dealt internationally to meet regulatory requirements and to ensure consumer safety. Creating food safety systems in resource-poor nations are confronting, as they are deficient in the basic infrastructure facilities required. Yet the accomplishment of this objective is need of the hour to prevent the introduction and circulation of newly emerging foodborne threats across the industrialized world. WHO, in collaboration with the Food and Agriculture Organization (FAO) of the United Nations and other federal agencies are taking action to this requirement by performing assessments of risks associated with foodborne illness and providing training and capacity building for nations venturing to create and meet risk-based standards for food safety (WHO, 2002a,b). This attempt targets to facilitate developing countries build competent and suitable food safety systems that meets the standards of successful systems in developed countries and to ensure food safety (Schlundt 2014).

"Intentional food contamination is of greatest concern world wide and food defense is guarding against the intentional contamination or adulteration of food.

Food Defense

Unlike food safety, food defence is the efforts to inhibit intentional contamination of food products by biological, chemical, physical, or radiological agents in the food supply by human intervention with malicious intention. Intentional acts normally occur rarely, thus challenging to detect, and are tough to predict. These agents are materials that are not normally present in food and thus not routinely tested for in food products. This has all the prospects to cause major health consequences and fear amongst public. Public fear leads to loss of confidence in the efficiency of government, disruption of food trade and decrease in economy. Countries across the world have started developing `food defence' strategies to protect food supply chain, to prevent and to mitigate any malevolent manipulation of food by terrorists, criminals and fantasists, and to counteract the adverse effects.

Food defense is setting preventive measures in place that decreases the probabilities of intentional contamination. A Food Defense Program is based on four major activities - prevention, preparedness, mitigation response and recovery. Food Defense Plan (FDP) is a set of documented protocols formulated on food defense principles encompassing vulnerability assessment, mitigation strategies, and define food defense monitoring, counteractive measures, and verification protocols to be followed. FDP augment vigilance and preparedness and this plan shall be in place always. During emergencies, when trauma is high and response time is crucial, a documented set of protocols enhances capacity to respond swiftly. Food defense guidelines must be executed at every food processing unit and at national levels. Federal agencies, such as the FDA, FAO, FSIS, GFSI and have adopted risk assessment tool the CARVER+ Shock tool to assess the vulnerabilities in a system or in an infrastructure. CARVER stands for: Criticality - a measure of impacts of an attack on public health and economy; Accessibility- the capability to acquire physical access and egress from target; Recuperability the ability of food system to recover from an attack; Vulnerability—the ease of accomplishing the attack; Effect—the direct economical loss from an attack; Recognizability—the ease of identifying the target (Manning, 2016). Vulnerability assessment (VA) tool shall be established to execute at the every food processing facility as well. Support is rendered by the federal agencies to food business operators to improve personalized FDP by incorporating existing FDA tools and guidance into single application (FDA 2015) to address the risk of deliberate food contamination.

Food protection is integration of all of food security, food safety and food defense.

Agroterrorism

Agroterrorism is a part of bioterrorism, which can be defined as the deliberate introduction of plant or animal pathogens and pests with the aim of agricultural sabotage creating panic, affecting economy and to disturb social stability and loss of confidence in government. It is relatively easy to obtain agroterrorism agents straightaway from the environment with comparatively less efforts and minimum cost. Agricultural lands are usually dispersed as open fields and pastures throughout the countryside without security and hence diseases can be spread easily. Livestock often graze in confined locations and transported in herds. Under such circumstances, spreading of the disease to more animals is very easy and quick. Many animal diseases that are eradicated in one country might still remain as endemic in foreign countries which might get spread easily by globalization of food the trade. Contaminated plant/animal food easily reaches the food chain harming human health.

Food Recalls Studies

Of the 365 cases recorded across the globe since 1958 till date, the food terrorism incidents which occurred with biological agents are listed in the Table 3. Rest includes chemical and radiological agents. All these events happened at consumer points. The incident of Salmonella outbreak in United States by Rajneeshee group is frequently held as the example of the potential biological threats.

Detection Systems

A perfect detection system for a BW agent should be capable of detect and confirm biological risk factors at very low concentration, directly from diverse matrices without showing false positive and false negative results. These should be workable at field conditions, shall be portable, easy to use and capable of detecting multiple agents simultaneously. List of the platforms of detection methods which are available is as provided in the following table 4 and 5.

Table 3: Intentional contamination of the food supply chain with bio threat agents

Date	Agent	Place	Deaths	Injury	Description /reference
1964	Typhoid fever Salmonella and Shigella	Japan	4	200-400	Dr. Mitsuru Suzuki contaminated cake with *Shigella* and *Salmonella* with malicious intention of infecting his colleagues. More number of outbreaks of typhoid and dysentery were linked Suzuki by Police investigation during 1964-1966 (Carus 2001).
1970	Ascaris suum Parasite	canada	0	4	Eric Kranz contaminated his roommate with parasite (Carus 2001)
1979	Anthrax	Rhodesia	182	10,000	South African special forces were allegedly responsible for poisoning the foodstuffs of villagers living in Cochemane, Tete province (John yeld 1998).
1980s	Anthrax	Zambia	1	2	Dr. Schaik Van Reesburg, Director at Rodeplaat Research Laboratories in South Africa alleged his company supplied anthrx for attempted assassination of three Russian advisors to Zambia (Carus, 2001).
1984	*Salmonella* typhimurium	USA	0	751	In an effort to affect the local elections and in order to win the elections of Wasco country, the restaurant salad bars were contaminated with *Salmonella* by Osho group which lead to 751 illness (Carus 2001).
1984	*Salmonella* Typhimurium	USA	0	2	Members of the Rajneeshees cult poisoned drinking water (Carus 2001)
1990	*Bacillus anthracis* *Clostridium botulinum*	Tokyo, Japan			*Bacillus anthracis, Clostridium botulinum* and the *Ebola* virus were used by Aum Shinrikyo in Japan to to fulfil the apocalyptic prophecy.
1995	*Yersinia pestis* *Shigella dysenteriae*	Las vegas			*Yersinia pestis* and *Shigella dysenteriae* were employed by Larry Wayne Harris to infect her co-workers.
1995	Ricin	Virginia	0	1	Dr. Ray W. Mettetal a neurologist found with ricin with the intension of killing his boss.
1995	Ricin	USA	0	1	Dr. Dobora Green was accused of poisoning her husband with ricin.
1996	*Shigella*	USA	0	13	Diane Thompson poisoned her co-workers with Shigella tainted pastries (Dalziel, 2009)
1999	*Oleander*	Australia	0	1	Annmarie Hughes poisoned her husband with oleander in tea (Dalziel, 2009).

Table 4: Detection platforms for various BW agents

S.No.	Product	Company	Assay	Pathogens/Toxins detected	Total run time
1	Zephyr	PathSensors, Inc.	CANARY® Cell based biosensor technology	B. anthracis, Y. pestis, ricin, F. tularensis, orthopoxviruses, Salmonella sp.	15 min
2	NIDS handheld biothreat assay and handheld reader	ANP technologies	Lateralflow/handheld immunoassay	Bacillus anthracis, vaccinia virus, Brucellae, Venezuelan equine encephalitis virus, Listeria, SEB, Francisella tularensis, botulinum toxin A, Vibrio cholerae, Escherichia coli 0157, ricin, Coxiella burnetii, Yersinia pestis, Salmonella sp.	15 min
3	PrimeAlert	GenPrime, Inc.	Lateral flow/ handheld immunoassay	Ricin, botulinum toxin, SEB, F. tularensis, Y. pestis	15 min
4	ENVI Assay System Gold	Environics Oy	Immunoassay	Ricin, botulinum toxin, SEB, orthopoxviruses, B. anthracis, Y. pestis and F. tularensis	<20 min
5	RAMP 200 Biowarfare Detection System	Response Biomedical Corporation	Lateral flow/ handheld immunoassay	B. anthracis, ricin, botulinum toxin, variola virus	>30 min
6	Aegis 1000	BioDetection Instruments (BDI)	Immunoassay	Foodborne pathogens, toxins, infectious agents, protein biomarkers, waterborne pathogens	<30 min
7	PRO STRIPS 5 Agent biowarfare threat detection kit	ADVNT Biotechnology	Lateral flow/ handheld immunoassay	B. anthracis, ricin, botulinum toxin A and B, Y. pestis and SEB	15 min

S.No.	Product	Company	Assay	Pathogens/Toxins detected	Total run time
8	BioFire Defense, LLC: FilmArray	BioFire Diagnostics, Inc.; Biomerieux, France	Multiplex PCR	*Bacillus anthracis, Francisella tularensis, Yersinia pestis, Clostridium botulinum, Brucella species, Burkholderia mallei/pseudomallei, Ebola virus, MarBurg virus, Lassa virus, Coxiella burnetii, Ricin toxin, Variola virus, Rickettsia prowazekii, Venezuelan equine encephalitis virus, Eastern equine encephalitis virus, Western equine encephalitis virus*	60 min
9	RAZOR EX	BioFire Defense, LLC, USA	RT-PCR	*Bacillus anthracis, Brucella* species, *Coxiella burnetii, E. coli 157:H7, Francisella tularensis, Ebola virus, MarBurg virus, Lassa virus, Yersinia pestis.*	30 min
10	one3	Biomeme, USA	PCR	*Bacillus anthracis, Ricin toxin, Abrin toxin (Abrus precatorius), Yersinia pestis, Staphylococcal Enterotoxin B (SEB), Francisella tularensis, Brucella species, Coxiella burnetii, Venezuelan equine encephalitis virus, Variola virus, Ebola virus, MarBurg virus, Lassa virus, Epsilon Toxin (Clostridium perfringens), Burkholderia mallei/ pseudomallei, Chlamydophila psittaci, Rickettsia prowazekii, Vibrio cholerae , Cryptosporidium parvum*	60 min
11	Bio-Seeq PLUS	BioFire Diagnostics, Inc.; Biomerieux, France	LATE-PCR (Linear-After-The-Exponential- PCR)	*Bacillus anthracis, Brucella species, Coxiella burnetii, E. coli 157:H7, Francisella tularensis, Salmonella, Yersinia pestis, Small Pox, Ricin toxin*	60 min

S.No.	Product	Company	Assay	Pathogens/Toxins detected	Total run time
12	POCKIT Micro Nucleic Acid Analyzer	GeneReach Biotechnology Corporation, Taiwan	Convection PCR	*Bacillus anthracis, Salmonella, Brucella species, Ricin toxin*	30 min
13	POCKIT	GeneReach Biotechnology Corporation, Taiwan	PCR	*Bacillus anthracis, Brucella species, Salmonella, Ricin toxin, Dengue virus, Middle east respiratory syndrome coronavirus, Rift valley fever virus.*	60 min
14	FilmArray	BioFire, Salt Lake City, UT	Nested Multiplex PCR	*Rotavirus, Campylobacter, Clostridium difficile, Norovirus, Salmonella*	60 min
15	T-COR 8	Tetracore, Inc. Rockville, USA	Multiplexed PCR	*Abrin toxin (Abrus precatorius), Ricin toxin, Bacillus anthracis, Variola virus, Plague, Burkholderia mallei/pseudomallei, Staphylococcal Enterotoxin B (SEB), Brucella species, Francisella tularensis, Yersinia pestis*	20–45 min

Table 5: Various types of biosensors used for the detection of biothreat agents

S.No.	Transducer	Biothreat agents	Limit of detection	Readout time	Samples tested	References
1	Electrochemical biosensor with bimetallic gold and palladium nanoparticles or boron-nitrite nanosheets	B. anthracis surface array proteins	5 pg/ml to 100 ng/ml	1 h	Cell culture	Sharma *et al.*, (2016)
2	Electrochemical biosensor with gold-nanoparticle	Botulinum neurotoxin type E	10 pg/ml to 10 ng/ml	65 min	Orange juice and milk	Narayanan *et al.*, (2015)
3	Surface plasmon resonance and electrochemical impedance spectroscopy	Brucella abortus	0.05 pM	10 min	Cell culture	Gupta *et al.*, (2009)
4	Surface plasmon resonance coupled with electrochemical impedance spectroscopy	Botulinum neurotoxin A	0.045 fM		Cell culture	Tomar *et al.*, (2016)
5	Impedometric biosensor with gold nanoparticle	Brucella melitensis	4×10^5 CFU/ml	1.5 h	Milk	Wu *et al* (2013)
6	Piezoelectric Immunosensor coupled with quartz crystal microbalance (QCM) and covalently immobilized F. tularensis antigen	F. tularensis	5×10^6 cells/ml	35 min	Cell culture	Pohanka *et al.*, (2007)
7	Surface plasmon resonance with antibody against F1 antigen	Y. pestis	10^6 CFU/ml	1 h	Environmental samples	Huynh *et al.*, (2015)
8	Quartz crystal microbalance	Staphylococcal enterotoxin A (SEA)	0.02 mg/L	25 min	Milk	Salmain *et al.*, (2012)

Prevention of Food Bioterrorism

Prevention of food related risks of intentional food bioterrorism and unintentional threats to preserve food integrity is a key matter of concern in society and an indispensable feature of global trade. As seen earlier, food manufacturing chains are highly susceptible to planned acts of terrorists at multiple points. Effective surveillance mechanism, traceability methods combined with analytical detection and fingerprinting of food components can be employed efficiently to inhibit and mitigate the consequences of these dangers. State-of-the-art research in making of strategies and micro devices for analysis of damage of food chain starting from farm, production, processing, distribution and preparation till fork offers new confidence to address the issues of food bioterrorism. World Health Organization (WHO) plays a major role along with FAO of the UN and various international organizations related to food threat. To response to food terrorism, expansion and strengthening of international systems related to emergency response and disease surveillance systems in compliment with other guidelines and advices developed by WHO is need of the hour. A complete integrated system which gives micro details about the history and movement of the entire steps of the food chain, quality and safety actions at critical points would undoubtedly upsurge readiness towards intentional and accidental threats (Opara, 2003). Prevention is best achieved through a cooperative effort between government, industry and public for minimizing food risks lie with the food industry.

Surveillance

The major necessity of rapid detection of a food terrorism outbreak is a sensitive surveillance system at local and national levels to identify and to differentiate intentional or unintentional contamination and its expected frequency. Active communication with clinicians, laboratories and public health professionals is also important.

Detection of Suspected Food Biothreat Agents

The ability to detect the reason of a disease as an intentional food terrorist act is determined by the availability of the analytical procedures. Quick response depends on effective acquaintances to laboratories with the ability to identify different food threat agents, including uncommon ones with a suitable expert and analytical systems to detect all agents in food samples.

Epidemiological investigations: Further strengthening of existing epidemiological investigation, detection of the agent responsible for the disease, the transmission vehicle is also vital aspect of the investigation for timely treatment and to prevent further dissemination of the agent. Tracing the source and the identification of the product flow along the supply chain is of supreme importance in an effective recall (Wang *et al*., 2006; Wang *et al*., 2009).

Preparedness

Principles: The efficacy of a response depends on preparedness plans that are developed and implemented well in advance of any event. Different mechanisms of preparedness that empower an operative emergency response include:

- Sensitive Surveillance systems;
- Preparedness Planning principle's execution;
- Effectiveness testing of Preparedness Plan;
- Assessment of vulnerability to the specific threat or incident;
- Ability to investigate and verify a threat or incident;
- Linking of government and other agencies to manage the consequences of public health.

The Indian Government must be prepared to fight bioterrorism by developing preparedness goals such as strengthening medical countermeasures such as keeping stock of drugs, medical devices, vaccines and other biological products required against biological agents which are required during emergencies. Enhancing and strengthening of existing health care and other medical facilities to respond efficiently to of both natural and man-made emergencies and biological threats. Half of the battle is won if everything is prepared and organized to face the adversity.

References

Bouzid, M., Hunter, P. R., Chalmers, R.M., & Tyler, K. M. (2013). Cryptosporidium pathogenicity and virulence. Clinical Microbiology Reviews, 26(1), 115-134.

Carus, W. S. (2001). Bioterrorism and biocrimes: the illicit use of biological agents since 1900. National Defense Univ Washington DC.

Cathelyn, J. S., Ellison, D. W., Hinchliffe, S. J., Wren, B. W., & Miller, V. L. (2007). The RovA regulons of Yersinia enterocolitica and Yersinia pestis are distinct: evidence that many RovA-regulated genes were acquired more recently than the core genome. Molecular Microbiology, 66(1), 189-205.

Centers for Disease Control and Prevention, National Center for Emerging and Zoonotic Infectious Diseases (NCEZID), Division of High-Consequence Pathogens and Pathology (DHCPP).

Clemente, T. M., Mulye, M., Justis, A. V., Nallandhighal, S., Tran, T. M., & Gilk, S. D. (2018). Coxiella burnetii blocks intracellular interleukin-17 signaling in macrophages. Infection and Immunity, 86(10), e00532-18.

Cross, G., & Klotz, L. (2020). Twenty-first century perspectives on the Biological Weapon Convention: Continued relevance or toothless paper tiger. Bulletin of the Atomic Scientists, 76(4), 185-191.

Dalziel, G. R. (2009). Food defense incidents 1950–2008. A chronology and analysis of incidents involving the maliciuos contamination of the food supply chain, Centre of Excellence For National Security (CENS), S. Rajaratnam School of International Studies, Nanyang Technology University, Singapore.

Eitzen, E. M., & Takafuji, E. T. (1997). Historical overview of biological warfare. Medical aspects of chemical and biological warfare, 415-423.

Feakes, D. (2017). The biological weapons convention. Revue scientifique et technique (International Office of Epizootics), 36(2), 621-628.

Fontaine, A., Arondel, J., & Sansonetti, P. J. (1988). Role of Shiga toxin in the pathogenesis of bacillary dysentery, studied by using a Tox-mutant of Shigella dysenteriae 1. Infection and immunity, 56(12), 3099-3109.

Giannella, R.A. (1996) Chapter 21 Salmonella. In: Baron, S., Ed., Medical Microbiology, 4th Edition, University of Texas Medical Branch at Galveston, Galveston, 1-7.

Guillemin, J. (2005). Biological Weapons: From the Invention of State-Sponsored Programs to Contemporary Bioterrorism. Columbia University Press. http://www.jstor.org/stable/10.7312/guil12942.

Gupta, G., Kumar, A., Boopathi, M., Thavaselvam, D., Singh, B., & Vijayaraghavan, R. (2011). Rapid and quantitative determination of biological warfare agent Brucella abortus CSP-31 using surface plasmon resonance. Anal. Bioanal. Electrochem, 3(1), 26-37.

Hayoun, M. A., & King, K. C. (2021). Biologic Warfare Agent Toxicity. StatPearls. https://www.ncbi.nlm.nih.gov/books/NBK441942/

https://en.wikipedia.org/wiki/Bioterrorism.

https://www.financialexpress.com/defence/beyond-coronavirus-why-india-needs-to-be-prepared-for-bioterrorism/1899820/

https://www.hannover-re.com/1133518/political-violence-and-terrorism-2021.pdf

https://www.who.int/news-room/fact-sheets/detail/food-safety

https://www.hannover-re.com

Huynh, H. T., Gotthard, G., Terras, J., Aboudharam, G., Drancourt, M., & Chabrière, E. (2015). Surface plasmon resonance imaging of pathogens: the Yersinia pestis paradigm. BMC Research Notes, 8(1), 1-8.

Bellal, J., Brown, C. V., Diven, C., Bui, E., Aziz, H., & Rhee, P. (2013). Current concepts in the management of biologic and chemical warfare causalities. Journal of Trauma and Acute Care Surgery, 75(4), 582-589.

Just, D. R., Wansink, B., & Turvey, C. G. (2009). Biosecurity, terrorism, and food consumption behavior: using experimental psychology to analyze choices involving fear. Journal of Agricultural and Resource Economics, 91-108.

Khan, A. S., Swerdlow, D. L., & Juranek, D. D. (2001). Precautions against biological and chemical terrorism directed at food and water supplies. Public Health Reports, 116(1), 3.

King James Version. Encyclopedia Britannica. ttps://www.britannica.com/topic/King-James-Version.

Klempner, M. S., Talbot, E. A., Lee, S. I., Zaki, S., & Ferraro, M. J. (2010). Case 25-2010: A 24-Year-Old Woman with Abdominal Pain and Shock. New England Journal of Medicine, 363(8), 766-777.

Knittler, M. R., & Sachse, K. (2015). Chlamydia psittaci: update on an underestimated zoonotic agent. Pathogens and Disease, 73(1), 1-15.

Laurence Knott, 2011. https://www.google.com/url?q=https://patient.info/doctor/ricin-poisoning.htm&sa=U&ved=2ahUKEwjd4djsl5rzAhUcxDgGHTH2B5AQFnoECAkQAg&usg=AOvVaw0-saIPz4IcXZwE9MfBUZCQ.

Lentzos, F. (2011). Hard to prove: The verification quandary of the Biological Weapons Convention. The Nonproliferation Review, 18(3), 571-582.

Lentzos, Filippa (2019). "Compliance and Enforcement in the Biological Weapons Regime". WMDCE Series, 4: 7–8. doi:10.37559/WMD/19/WMDCE4.

Lim, D. V., Simpson, J. M., Kearns, E. A., & Kramer, M. F. (2005). Current and developing technologies for monitoring agents of bioterrorism and biowarfare. Clinical microbiology reviews, 18(4), 583-607.

Magdy Hefnawy and Dina Solodoukhina (2011). Food Safety and Bioterrorism from Public Health Perspective. Advances in Food Protection. 17–25. doi: 10.1007/978-94-007-1100-6_2.

Mahanty, S., & Bray, M. (2004). Pathogenesis of filoviral haemorrhagic fevers. The Lancet Infectious Diseases, 4(8), 487-498.

Manning, L., & Soon, J. M. (2016). Food safety, food fraud, and food defense: a fast evolving literature. Journal of Food Science, 81(4), R823-R834.

Michael R. Knittler, Konrad Sachse, Chlamydia psittaci: update on an underestimated zoonotic agent, Pathogens and Disease, Volume 73, Issue 1, February 2015, Pages 1–15, https://doi.org/10.1093/femspd/ftu007.

Miller, J., Engelberg, S., Broad, W. J., (2002). Germs: Biological Weapons and America's Secret War. Simon and Schuster. ISBN 9780684871592.

Narayanan, J., Sharma, M. K., Ponmariappan, S., Shaik, M., & Upadhyay, S. (2015). Electrochemical immunosensor for botulinum neurotoxin type-E using covalently ordered graphene nanosheets modified electrodes and gold nanoparticles-enzyme conjugate. Biosensors and Bioelectronics, 69, 249-256.

Naveed Munir, N., Jahangeer, M., Hussain, S., Mahmood, Z., Ashiq, M., Ehsan, F., ... & Sana, A. (2021). Hantavirus diseases pathophysiology, their diagnostic strategies and therapeutic approaches: A review. Clinical and Experimental Pharmacology and Physiology, 48(1), 20-34.

Notes for Project SHAD presentation by Jack Alderson given to Institute of Medicine on April 19, 2012 for SHAD II study.

Opara, L. U. (2003). Traceability in agriculture and food supply chain: a review of basic concepts, technological implications and future prospects. Food, Agriculture & Environment, 1(1), 101e106.

Ophélie Guillouet-Lamy. COVID-19, a Biological Weapon? A Guide to Biological Weapons to Answer that Question. nct-magazine.com/nct-magazine-May-2020.

Osterholm M. (2005) Session II: The Food Supply "Threat Spectrum." Presentation at the Forum on Microbial Threats Workshop Meeting, Foodborne Threats to Health: The Policies and Practices of Surveillance, Prevention, Outbreak Investigations, and International Coordination, Washington, D.C., Institute of Medicine, Forum on Microbial Threats.

Pappas, G., Akritidis, N., Bosilkovski, M., & Tsianos, E. (2005). "Brucellosis". The New England Journal of Medicine, 352 (22), 2325-2336.

Perry, R. D., & Fetherston, J. D. (1997). Yersinia pestis--etiologic agent of plague. Clinical Microbiology Reviews, 10(1), 35-66.

Pittman, P. R., Hahn, M., Lee, H. S., Koca, C., Samy, N., Schmidt, D., ... & Chaplin, P. (2019). Phase 3 efficacy trial of modified vaccinia Ankara as a vaccine against smallpox. New England Journal of Medicine, 381(20), 1897-1908.

Pohanka, M., Pavliš, O., & Skládal, P. (2007). Diagnosis of tularemia using piezoelectric biosensor technology. Talanta, 71(2), 981-985.

Preston, R. (1998) The Bioweaponeers: As a Soviet scientist, Ken Alibek engineered an arsenal of mass destruction. New Yorker-New Yorker Magazine Incorporated, 52-65.

Rakesh Gupta (1998). A comparative perspective on the causes of terrorism. International Studies, 35(1), 23-53.

Rimmington, Anthony (2018). Stalin's Secret Weapon: The Origins of Soviet Biological Warfare. Oxford University Press.

Rissanen, J. (2003). The Biological Weapons Convention. Nuclear Threat Initiative-Ten Years of Building a Safer World.

Robert. J. Einhorn,., & Flournoy, M. A. (2003). Protecting Against the Spread of Nuclear, Biological, and Chemical Weapons: Agenda for action (1). CSIS; 27–28.

Rowe, H. M., & Huntley, J. F. (2015). From the outside-in: the Francisella tularensis envelope and virulence. Frontiers in Cellular and Infection Microbiology, 5, 94.

Salmain, M., Ghasemi, M., Boujday, S., & Pradier, C. M. (2012). Elaboration of a reusable immunosensor for the detection of staphylococcal enterotoxin A (SEA) in milk with a quartz crystal microbalance. Sensors and Actuators B: Chemical, 173, 148-156.

Schlundt, J. (2014). Institutions Involved in Food Safety: World Health Organization (WHO). Encyclopedia of Food Safety, 359.

Sharma, M. K., Narayanan, J., Pardasani, D., Srivastava, D. N., Upadhyay, S., & Goel, A. K. (2016). Ultrasensitive electrochemical immunoassay for surface array protein, a Bacillus anthracis biomarker using Au–Pd nanocrystals loaded on boron-nitride nanosheets as catalytic labels. Biosensors and Bioelectronics, 80, 442-449.

Singh, R. K., Dhama, K., Chakraborty, S., Tiwari, R., Natesan, S., Khandia, R., ... & Mourya, D. T. (2019). Nipah virus: epidemiology, pathology, immunobiology and advances in diagnosis, vaccine designing and control strategies–a comprehensive review. Veterinary Quarterly, 39(1), 26-55.

Stefan Riedel, S. (2004). Biological warfare and bioterrorism: a historical review. In Baylor University Medical Center Proceedings 17(4). 400-406.

Tärnvik, A., & Berglund, L. (2003). Tularaemia. European Respiratory Journal, 21(2), 361-373. Text of the 1925 Geneva Protocol. United Nations Office for Disarmament Affairs. Retrieved 16 February 2021.
(http://cns.miis.edu/pubs/reports/ricin_chron.html)

Thavaselvam, D., & Vijayaraghavan, R. (2010). Biological warfare agents. Journal of Pharmacy and Bioallied Sciences, 2(3), 179-188.

The Bible (King James version): Exodus chapters 8–12.

The National Academies; The Center for Research Information, Inc. (2004). Health Effects Of Project Shad Biological Agent: Bacillus Globigii, (Bacillus licheniformis), (Bacillus subtilis var. niger), (Bacillus atrophaeus) (PDF) (Report). Prepared for the National Academies. Contract No. IOM-2794-04-001.

Tomar, A., Gupta, G., Singh, M. K., Boopathi, M., Singh, B., & Dhaked, R. K. (2016). Surface plasmon resonance sensing of biological warfare agent botulinum neurotoxin A. J Bioterror Biodef, 7(2), 142-154.

Tucker, J. (2001). The Biological Weapons Convention (BWC) Compliance Protocol. Nuclear Threat Initiative. August, 1.

Wang, X., & Li, D. (2006). Value added on food traceability: a supply chain management approach. In 2006 IEEE International Conference on Service Operations and Logistics, and Informatics (pp. 493-498). IEEE.

Wang, X., Li, D., & O'Brien, C. (2009). Optimisation of traceability and operations planning: an integrated model for perishable food production. International Journal of Production Research, 1e22.

Wheelis M. Biological warfare before 1914. Biological and toxin weapons: Research, development, and use from the middle ages to 1945. (1991). Vol. 1. Stockholm International Peace Research Institute. 8–34.

Wheelis, M. (2002). Biological warfare at the 1346 siege of Caffa. Emerging infectious diseases, 8(9), 971-975. doi:10.3201/eid0809.010536.

Wu, H., Zuo, Y., Cui, C., Yang, W., Ma, H., & Wang, X. (2013). Rapid quantitative detection of brucella melitensis by a label-free impedance immunosensor based on a gold nanoparticle-modified screen-printed carbon electrode. Sensors, 13(7), 8551-8563.

Yamaoka, S., & Ebihara, H. (2021). Pathogenicity and Virulence of Ebolaviruses with Species- and Variant-specificity. Virulence, 12(1), 885-901. Download citation https://doi.org/10.1080/21505594.2021.1898169.

3

Undeclared (Frozen/Double Dead Meat) /Non Labeled (Irradiation) Food Processing: Preservation or Poisoning

Swati Haldar[1], *Ajay Singh*[2], *Jyoti Rani*[3] *and Viney Kumar*[4,5*]

[1]*Department of Genetics, Albert Einstein College of Medicine, Bronx New York, 10461, USA*
[2]*Department of Food Technology, Mata Gujri College, Fatehgarh Sahib Punjab, India*
[3]*Department of Nutrition and Dietetics, Chandigarh University Mohali, India*
[4*]*Department of Biotechnology, Indian Institute of Technology Roorkee Uttarakhand, India*
[5*]*Department of Pathology, Albert Einstein College of Medicine, Bronx New York, USA*

Introduction

Organic substances consumed for nutritional purposes can be called as "food". Food contains biomolecules, like, protein, lipid and carbohydrate, inorganic minerals as salts, and organic substances, such as, vitamins, emulsifiers, acids, antioxidants, pigments, polyphenols and flavour-producing compounds. They can be of plant or animal origin. Depending on the durability in terms of suitability of consumption, foods can be perishable, semi-perishable and non-perishable. Food spoilage occurs due to chemical, microbial, or physical actions. eventually affecting its nutritional values, colour, edibility and texture (M. Rahman, 2007). Therefore, in order to improve the shelf-life food needs preservation. So, the food preservation may be defined as processes or techniques employed to contain external and internal factors responsible for spoilage of food. The preservation method is mainly based on the types of food that need to be preserved. Food preservation aims to increase life of the preserved food while retaining its original nutritional values, colour, texture, and flavour. The history of 'Food Preservation' dates back to ancient civilizations when nomadic hunter-gatherer humans first felt the necessity for preserving a big hunt or surfeit gathering, that could not be consumed all at a time. Therefore, undoubtedly, inventing the food preservation techniques, played a key role in further development of human civilizations as more and more time could be dedicated towards inventive activities, instead of engaging in the basic need of food provision. Interestingly, although the ancient human civilizations were spatially and temporally separated, the food preservation techniques noted in these different cultures were quite similar (Nummer, 2002). These primitive preservation techniques included drying, freezing, and preservation through salting or dipping in sugar syrup. The evolution of food preservation has made a long journey from basic drying-freezing-salting to irradiation, hyperbaric and hurdle technologies. The journey still continues with scientific advancements towards improving the existing food preservation techniques and inventing newer ones (R. Rahman, 2014). The current trends in food preservation in a highly integrative one bringing multi-disciplinary approaches together to preserve food at different stages of its generation, like, growing/manufacturing, harvesting, processing, packaging, and distribution. This integrated approach, is more effective. The 7 trillion dollar global food preservation market is rapidly expanding its value due to seamless globalization and aggressive industrialization (*World Food Market Overview Marketing Essay*, 2013). An interesting analysis of the UNIDO Industrial Statistics Database (2005) shows that the food processing industries in developing countries contribute significantly towards their national GDP rises (Kumar Kar, 2014; Wilkinson & Roch, 2008). So, a timely accounting of food preservation techniques for a better understanding of their principles and food-type-linked requirements, is pertinent for a broader readership of food industry, particularly, when this sector has the potential to directly affect the national GDP. In line with the above-mentioned justification, this chapter represents the food items classification and discusses different chemical, physical, and biological factors of food spoilage, to begin with. Subsequently, basic and advanced food preservation techniques, effective against food spoilage and helpful for prolonging shelf life, are discussed in detail elaborating on their mechanisms, application conditions, advantages, and disadvantages. The food preservation techniques based on the principles of inactivation, inhibition, and avoidance of recontamination have been discussed in detail. Following the motto of United Nations World Commission on Environment and Development "meeting the needs of the present generation without compromising the ability of future generations to meet their own needs.", the aim and scopes of incorporating innovation, sustainability, and safety in food processing industry has also been discussed. A sustainable approach offers a competitive edge with a positive public image (French, 2004). The element of sustainability comes with demands of finer modulations with varying degrees of executional challenges. In turn, this calls for improved understanding of the technical matrix of the food preservation industry. With all the aspects mentioned above, that the readers would be coming across, this chapter aims to provide a holistic account of the concept of food preservation, classical and modern, alike, with an intention for improved insight into the changing demands of food quality, convenience, and safety, and how they can be met. Fig. 1 provides an overview of the content mentioned in this chapter.

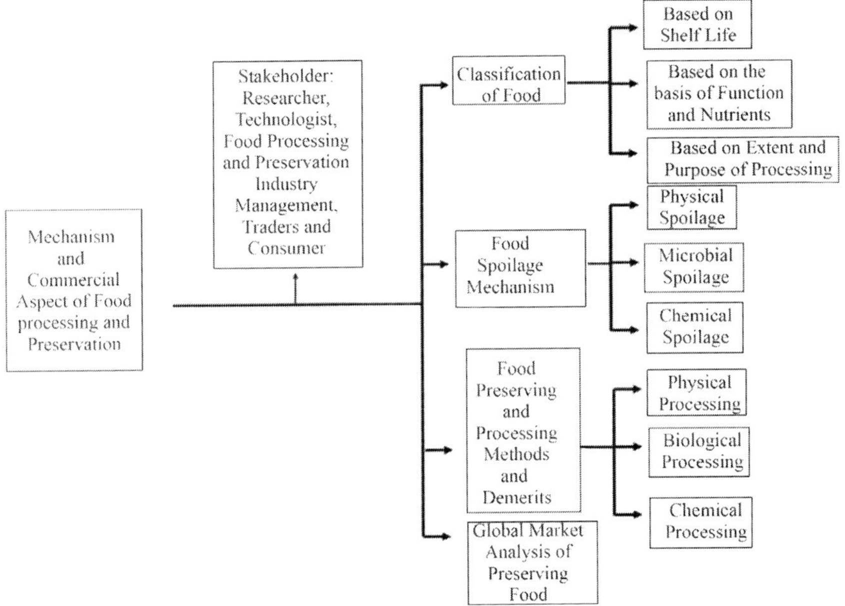

Fig. 1: An overview of the content summarized in this chapter.

Classification of Foods

Foods are classified based on their functions, shelf life, processing mechanisms and nutrient value (Fig. 2). Different food categories are summarized and briefly discussed in the following sections.

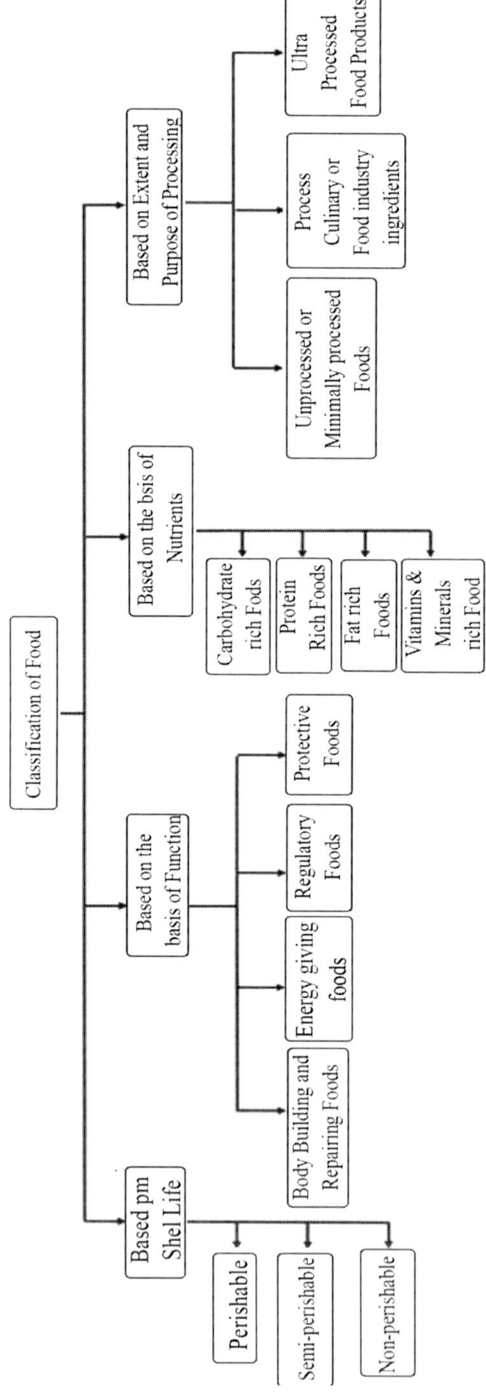

Fig. 2. Classification of food

Recreated from (Amit *et al.*, 2017) Under Creative Commons Attribution 4.0 International License (http://creativecommons.org/licenses/by/4.0/)

Food Categories Based on Shelf-life

Shelf-life of food is dependent on the how long it can resist spoilage. Food spoilage is a natural process leading to loss of colour, texture, flavour, nutritional qualities, and edibility of the food. Spoiled food consumption cause illness and in the extreme case can be fatal as well (Steele, 2004). Based on shelf-life, food items can be semi-perishable, perishable or non-perishable (Sperber & Doyle, 2009).

Perishable foods that have shelf life from days to about three weeks after preservation, which, if left unpreserved spoil right away in matter of hours. Milk and dairy products, poultry, meats, seafood and eggs come under this category and require efficient preservation to extend their shelf-life from hours to days and weeks (Sperber & Doyle, 2009).

Semi-perishable food without preservation can last for days, but, if subjected to specific preservation techniques, their shelf-life can be extended considerably, to months (about six months). Semi-perishable food items are like fruits, vegetables, potatoes and cheese.

Non-perishable foods by nature have long shelf-lives which can be further prolonged indefinitely through preservation. Such foods stay edible and retain nutritive values for several years or longer. Packaged foods with years of shelf-lives also come under this category. Few examples of non-perishable foods are nuts, dry beans, four, sugar, mayonnaise, canned fruits, and peanut butter.

Food Categories Based on Functions and Nutrients

Foods serve a diverse range of purposes for the body. Depending on these functions performed, foods can be energy-giving, repairing and body building, protective and regulatory. All these requirements met by the foods in turn depend on the constitution of the food that determines its nutritive value. Based on the nutritive value, foods can be protein-rich, carbohydrate-rich, fat-rich, and mineral/vitamin-rich. Classification of foods based on their functions and nutritive values are summarized in Table 1.

Table 1: Based on functions and nutrients food classification

Functions	**Source**
Repairing and body building foods	Milk, fish, meat, pulses, nuts and vegetables
Energy-giving foods	Oil, sugar, butter, dry fruits, cereals, and starch foods
Regulatory foods	Water, citrus fruits, raw vegetables, and beverages
Protective foods	Milk, meat, whole grain cereals, fruits and vegetables
Nutrients	
Carbohydrate-rich foods	Rice, starchy vegetables and wheat
Protein-rich foods	Milk, fish, meat, nuts and eggs
Fat-rich foods	Oils, egg yolk and butter
Vitamin and minerals rich foods	Vegetables and fruits

Adapted from (Amit *et al.*, 2017) Under Creative Commons Attribution 4.0 International License (http://creativecommons.org/licenses/by/4.0/)

Food Categories Based on the Extent and Purpose of Processing

The techniques for food processing and extent to which foods require to be processed are the aspects that requires careful consideration in the food processing industry. The ultimate decisions regarding the techniques and extent of its application during food processing is determined based on the intended outcome with respect to the usage of the food item. Accordingly, these factors are also known to divide foods into different categories, like, unprocessed or minimally processed foods, processed culinary or food industry ingredients, and ultra-processed food products. This classification is summarized in Table 2.

Table 2: Food Classification based on the extent and purpose of processing

Food group	Extent and purpose of processing	Example
Minimally or unprocessed foods	No processing or primarily physical techniques are employed to increase the availability, accessibility, palatability, or safety of one complete food.	Fresh, frozen, chilled, vacuum-packed fruits, cereals, vegetables, dried beans and fresh frozen and other pulses; unsalted nuts, dried fruits, and seeds; dried, fresh, chilled, poultry, frozen meats, and fish; pasteurized and fresh milk, fresh yoghurt, tea, eggs, and coffee
Processed culinary or food industry ingredients	Extraction and purification of food components that produce ingredients used in the cooking and preparation of dishes or in the creation of ultra-processed foods.	Vegetables, milk cream, butter, sweeteners, noodles and raw pastas; food industry ingredients, like preservatives, corn syrup with high-fructose, and cosmetic additives
Ultra-processed food products	Processing a mixture of food components that are used in the food business or in the culinary arts to create readily available, appetizing, ready-to-eat, or ready-to-heat food items with a longer shelf life	Breads, cakes, biscuits, and pastries; chocolates, ice-cream, chips, sugared fruits, cereal bars, soft drinks and milk drinks; poultry, pre-pared meat, vegetable, fish; meat including hot dogs, chicken nuggets, burgers; pickled, salted, cured or smoked, fish and meat; fish canned in oil and vegetables bottled or canned in brine

Adapted from (Amit *et al*., 2017)

3. Food Spoilage: Mechanism

Food spoilage is the process in which food consuming quality is compromised, therefore, it is directly related to food safety (Steele, 2004). Change in colour, smell, flavour, and/or texture of food are the primary indications of spoilage. There are several reasons of food spoilage, which could be physical, chemical or biological/ microbial and may affect food independently or in tandem, stimulating one another.

Different factors that affect food spoilage are depicted in Fig. 3 and briefly discussed in the following sections.

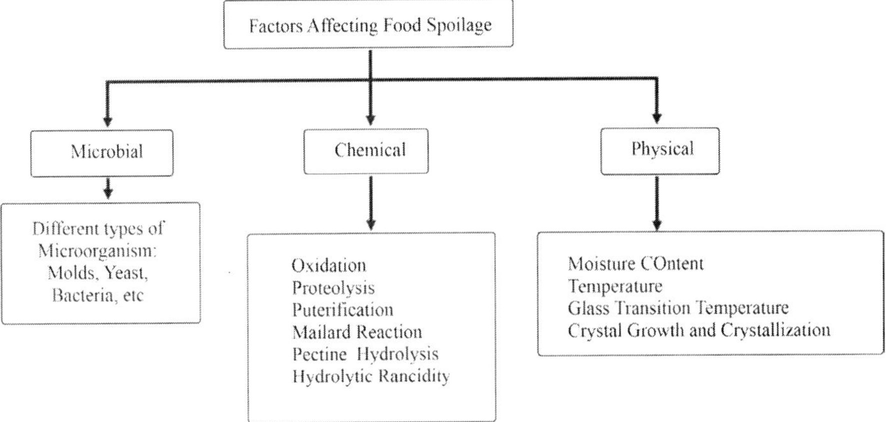

Fig. 3. Different factors that affect food spoilage

Recreated from (Amit *et al.*, 2017)

Physical Spoilage

Food spoilage due to changes in physical factors is called physical spoilage. These include but not restrict to loss or gain of moisture, change in temperature and change in physical properties like crystallinity (Barbosa-Cánovas *et al.*, 2005; Kader *et al.*, 1989; Karmas *et al.*, 1992; M. S. Rahman, 1995; Steele, 2004; White & Cakebread, 1966). These factors are briefly discussed in the following paragraphs.

Moisture Content

Change in moisture content of the food item is a very common cause behind spoilage, and could both due to water loss or gain, or even migration of water from one constituent to another within the food (Fabunmi *et al.*, 2015). This transfer of moisture within the food depends on the thermodynamic property, water activity (w_a) of the food item (Balasubramanian & Viswanathan, 2010; Steele, 2004). Water activity is the ratio of the vapour pressure of water in a system to the vapour pressure of pure water at the same temperature (Barnwal *et al.*, 2010; M. S. Rahman, 1995). Another way of perceiving water activity is equilibrium relative humidity at a given temperature. In general, water activity of foods at normal temperature is 1.0, whereas at −20 and −40 °C temperatures the water activities are 0.82 and 0.68, respectively (Barbosa-Cánovas *et al.*, 2005; Kader *et al.*, 1989). Water activity in food products reduces with temperature, plausibly that is why, lowering the temperature helps in food preservation, of course only applicable to food items not susceptible to chilling injury.

Temperature

Temperature is the most common physical factor behind vegetable and fruit spoilage. Using an optimum temperature and humidity range, technically known as modified atmospheres (MA), slow ripening of fruits and vegetables are facilitated to enhance post-harvest life. Temperatures fluctuations, both high and low, make it difficult to attain desired MA (Kader *et al.*, 1989). In fact, low temperature are detrimental to food prone to chilling injury, like, most tropical vegetables and fruits. Injuries caused by chilling happen just before the food product begins to freeze, when the temperature is between 5 and 15 °C., just before the food starts freezing and damages the cells. The frozen food items with chilling injury therefore, are pre-inflicted with maladies (Steele, 2004).

Glass Transition Temperature

Food items contain solids in crystalline or metastable amorphous forms acting as plasticizers, thereby, making glass transition temperature (Tg) an important determinant of its shelf-life (Harry Levine & Slade, 1986). Food items with amorphous solids are subject to more degrees of freedom as amorphous matrices are usually in a dynamic state with co-existence viscous glass or liquid rubbery forms (Karmas *et al.*, 1992). Temperature and relative humidity are the other two factors that play in-concert with the solid composition of food to affect the glass transition (White & Cakebread, 1966). For example, dry food products kept in highly humid conditions, become physically unstable due to glass transition phenomena (Steele, 2004).

Crystal Growth and Crystallization

As mentioned earlier, lowering temperature is not always a good choice for preserving food, because not only freezing causes chilling injury but also facilitate growth of ice crystals Particularly, rapid freezing is more conducive for formation of ice crystals with food cells (Reid, 1990). Adding emulsifiers to bind water molecules can reduce crystal growth during freezing cycles (Levine, 1988). Besides ice crystal, food items can also get endowed with sugar crystals due to moisture accumulation or rise in temperature. Staling of sugar cookies, graininess in candies, and ice creams are the results of sugar crystallization (Steele, 2004).

Microbial Spoilage

When microorganisms spoil food due to various microbial metabolic activities, it is called microbial spoilage. Microbial spoilage is a serious issue with perishable foods and is a major cause of food poisoning. Reducing storage temperature, pH, water activity, addition of preservatives, and ensuring proper packaging against downstream microbial contamination are the common procedures for preventing microbial food spoilage (Tianli *et al.*, 2014).

Microorganisms Involved in Food Spoilage

Microorganisms involved in food spoilage can be divided into three major categories, namely, molds, yeasts, and bacteria. Table 3 presents the active conditions of different microorganisms that affect foods.

Factors Affecting Microbial Spoilage

Microbial spoilage is an outcome of various intrinsic and extrinsic factors, besides, microbial contamination of the food (Jay *et al.*, 2000). However, the intrinsic and extrinsic factors work in co-ordination. For example, the intrinsic factor, availability of microbial enzymes depends on the microbial growth, which in turn is linked to conducive growth conditions determined by the innate temperature, pH, nutrient-richness-linked substrate availability, humidity/water activity and redox potential of the food (in't Veld, 1996; Jay *et al.*, 2000; Sperber & Doyle, 2009; Steele, 2004).

Table 3: Active conditions of different microorganisms and affected foods

Microorganisms	Active condition				Affected food
	pH	**Temperature**	**Water activity**	**Heat sensitivity**	
Molds	3.0-8.0	Grow across a wide range of temperature	0.62-1.0	Heat sensitivity	Bottled minerals water, fermented foods
Yeasts	Grow around a broad range of acidic pH	Grow across a wide range of temperature, but prefer natural ambient temperature	Above 0.9	Heat-resistant and able to sustain intense sunshine	Fermented foods
Bacteria	Broad pH range	Prefer growth at high temperature ($\geq$55°C)	Above 0.9 for gram positive and above 0.98 for gram negative	Mostly thermophiles	Fresh meats, poultry, sea foods, eggs, and heat-treated foods

Adapted from (Amit *et al*., 2017)

Chemical spoilage

Chemical spoilage is usually physical manifestation of the underlying microbial actions the food undergoes, resulting in unpleasant and undesirable alterations. Changes at an elementary level due to various biochemical and /chemical reactions led to major changes, such as, fluctuations in pH, accumulation of toxic compounds, lipid oxidation, *etc.* that become detectable as unpleasant deviations from expected smell/aroma, colour, texture and flavour (in't Veld, 1996; Van Boekel, 2008). Major underlying causes of chemical spoilage of food are elaborated below.

Oxidation

Biomolecules, like, amino acids and lipids present in refrigerated fresh meat and fish undergo oxidation into (Jay *et al.*, 2000). Amino acids oxidize into organic acids and ammonia.

$$\text{Amino acid} + \text{O2} \xrightarrow{\text{deaminase}} NH_3 + \text{Organic acid}$$

In case of lipid oxidation, unsaturated fatty acids react with oxygen and undergo 'rancidification' resulting in changes in colour and flavour due to formation of toxic substances, like, carbonyl compounds. These compounds are mainly responsible for imparting rancid taste to the food upon lipid oxidation. Rancidification is catalysed by metal oxides and exposure to light enhances its rate (Enfors S-O., 2008; Steele, 2004).

Hydrolysis

Free fatty acids undergo hydrolysis in the presence of water by the action of lipolytic enzymes. This process in called hydrolytic rancidification, in which, the free fatty acids are cleaved into triglycerides, making the spoiled food malodorous and rancid in taste (Steele, 2004). Hydrolytic rancidification is most noticeable in butter (Rodriguez & Mesler, 1984).

Pectin are the other biomolecules prone to hydrolysis. These are a complex combination of polysaccharides constituting nearly one-third of the cell wall of dicotyledons and some monocotyledons (Hoff & Castro, 1969; Walter & Taylor, 1991). Inherent pectin hydrolysis is required during fruit ripening. But pectinase activation due to mechanical reasons initiate microbial infestation (Enfors S-O., 2008).

De-esterification

Besides hydrolysis, pectin are prone to de-esterification by pectin methyl esterase. Pectin de-esterification gets seeded at sites of tissue damage that involves calcium mediated intercellular cohesion. During de-esterification, heat-labile fruit pigments, consisting of pectin, getting catalytically decomposed by metal ions. That is why the fruit jams and jellies discolour upon de-esterification (Walter & Taylor, 1991). Thus, glass containers are used for preserving fruit jams and jellied instead of metallic jars.

3.3.4 Proteolysis

Peptide and iso-peptide bonds between the amino acids in proteins are cleaved by the actions of various protease enzymes. This process in called proteolysis and is frequently found to occur in foods containing nitrogen compounds (Igarashi *et al.*, 2007; Rogers & Overall, 2013). The following reaction presents proteolysis mechanism:

$$\text{Proteins} \xrightarrow{\text{proteinase}} \text{Peptides} \xrightarrow{\text{peptidase}} \text{Amino Acids}$$

Due to proteolysis, the food developers unpleasant taste (Enfors S-O., 2008). Table 4 presents the taste of various amino acids.

Table 4: Taste of different amino acids

Taste	**Amino acids**
No taste/barely perceptible taste	D-alanine, L-arginine, D-arginine, D-aspartate, D-lysine, L-aspartate, L-lysine L- histidine, , D-glutamate, D-isoleucine, D-proline, L-isoleucine, L-proline, L-serine, D-serine, L-threonine, L-valine
Sweet taste	D-Tryptophan, D-phenylalanine, D-histidine, D-tyrosine, L-alanine, D-leucine, glycine
Bitter taste	L-Tryptophan, L-tyrosine, L-phenylalanine, L-leucine
Sulfurous taste	D-Cysteine, D-methionine, L-cysteine, L-methionine
Unique taste	L-Glutamic acid

Adapted from (Amit *et al.*, 2017)

Putrefaction

Meat and protein-rich foods, at temperatures above 15°C, provide extremely conducive environment for anaerobic bacteria. These anaerobic bacteria facilitate the biochemical process of putrefaction. This process converts amino acids into a mixture of amines, organic acids, indole and phenol and foul-smelling ammonia and sulphur compounds, like, hydrogen sulphide and mercaptans (Enfors S-O., 2008; Panda, 2003).

Maillard Reaction

Maillard reaction, the non-enzymatic browning due to the presence amino acids, is an important cause of spoilage in stored dry milk, breakfast cereals and dry whole eggs. This reaction makes proteins present in the food stuff unstable reducing the nutritional availability of the amino acids. Maillard reaction also compromises the aesthetic appeal of the food as the colour darkens and bitter flavour developers (Desrosier & Singh, 2018).

Food Preservations

Efficient preservation strategies require customization with respect to the food being preserved. A thorough analysis based on the knowledge of different processes involved from growing/generating that food item to its intended shelf-life is required to design the optimum preservation strategy. Since, we have already seen that food spoilage is an outcome of co-operative actions of physical, chemical and microbiological processes, therefore, preservation approach also needs to be of integrative nature. It is not just prolonging the shelf-life that the preservation strategies are geared at. It is equally important to prioritize the food properties that require to be preserved based on the end-user expectations in terms of appearance/texture, aroma and taste. Drying fruits collapse and developing pores are undesirable. But the same pore formation is one of the most desired changes in case of breakfast cereals. Likewise, crust formation is encouraged in breakfast cereals. Expected final features of food strongly drive the employed preservation techniques. For example, the apple juice is expected to be clear but orange juice cloudy. All these requirements are addressed by the employed food preservation technique/s.

Why Preservation?

The main reason of food preservation is to prolong the shelf life of food items. In doing so, a substantial amount of food wastage is avoided. Agriculture has undergone revolutionary changes across the world to meet the growing demands of the food requirements of the expanding world population. Global agricultural sector is now producing surplus food, which, if not preserved will get spoiled. Similarly, dairy and meat industries have also undergone tremendous modernization resulting in surfeit production in case of these food items also. Again, the excess portion of the produce, which is not consumed immediately, requires processing and preservation. Besides surplus production, farm, dairy and meat produces also require transportation, which means, they need to be preserved on the go until they reach the consumers and beyond that until they are consumed. Moreover, lifestyles have changes and have become very fast-paced. Therefore, gourmet cooking has become more of a hobby unlike erstwhile days when home-cooked square meals were the norms. In today's world, it is more of processed or semi-processed food that just require 'fixing', like, milk heating, microwaving, or mixing, before they are ready to be eaten. So, our fast-paced lifestyle has made the convenient foods more popular. Therefore, unprecedented improvement in food production, expansion in the food distribution patterns and changes in the day-to-day lifestyles of the people have triggered the requirement for food preservation, giving it the shape of a full-fledged industry. Additionally, food preservation also adds variety to the diet (M. S. Rahman, 1999).

Food preservation facilitates long transportation of the food stuff without the risk of spoilage. This helps in meeting the food crisis in famine-affected underdeveloped countries. Besides, it also helps in adding variety to the diets of underdeveloped countries. Preservation quite often enhances the nutritive value of the food, suggesting that, besides introducing variety to the diet, preservation also ensures nutrient supplementation. Therefore, the following aspects serve as critical guidelines for food preservation:

- The desired level of quality
- The preservation length (shelf-life post preservation)
- The group for whom the products are preserved

Food preservation needs to take into consideration several parameters. Therefore, one particular technique of preservation might not suffice the desired outcome. Hence, collection of techniques required to be optimized for a particular type of food to ensure safety of its consumption. Preservation techniques might partially compromise certain desirable properties of the otherwise unpreserved food. But, a trade-off between retaining the desirable food qualities at the cost of their ready perishing and preservation with an extended shelf-life needs to be decided (R. P. Singh, 1994).

Since, lack of preservation can result in hazardous changes in the food items over a short period of time, therefore, the above-mentioned trade-off almost always provide a win-win situation when it comes to the overall consumable quality of the food (Gould, 1989).

How Long to Preserve?

Preservation of foods prolongs the shelf-life, but over time, the preserved food does reach an undesirable stage, that marks the end of its shelf life. So, based on extensive research on every food item being preserved, the shelf-life post preservation is determined and as per regulatory requirements clearly mentioned on the food packing by the manufacturers. The suggested shelf-life is conditionally to storage conditions, that is, it is applicable only if the preserved food item is stored at the recommended conditions. Therefore, packaged/processed food product are labelled with the production date as "pack date" and "best-before date." Alternatively, food items may come with an expiry data (Bengtsson, 2003). Shelf-lives of food items are decided by monitoring the undesired changes occurring in it with respect to fixed quality parameters (R. P. Singh, 1994). These parameters could be appearance, texture, aroma, colouration, edibility, nutritive values and chemical compositions and characteristics of the preserved food (Sebranek, 1996). Quality loss of food due to preservation is modulatable at every processing step. Therefore, an optimized preservation matrix with respect to each food item is very essential. The major quality-loss mechanisms are shown in Table 5 (Amit *et al.*, 2017).

Table 5: Major Quality-Loss Mechanisms

Microbiological	Enzymatic	Chemical	Physical	Mechanical
Microorganism growth	Browning	Colour loss	Collapse	Bruising due to vibration
Off-flavour	Colour change	Flavour loss	Controlled release	Cracking
Toxin production	Off-flavour	Nonenzymatic browning, Nutrient loss, Oxidation-reduction, Rancidity	Crystallization Flavour encapsulation Phase changes Recrystallization Shrinkage Transport of component	Damage due to pressure

Adapted from (Amit *et al.*, 2017).

Post preservation shelf life of food actually depends on the specific type of consumer it is intended to cater. For example, packed meals do not require long shelf-life whereas, foods to be taken on long journeys, as extreme exemplar being foods for space travellers or soldiers at the front. Nutritive requirements of the consumers affect the opted preservation techniques for a particular preserved food.

5. Food Preservation Methods

Processes and techniques employed to prevent spoilage of food stuff and increase their shelf-life are collectively constitute food preservation (Fig. 4) (Karel & Lund, 2003; Ohlsson & Bengtsson, 2002; M. Rahman, 2007; Rodriguez-Gonzalez *et al.*, 2015).

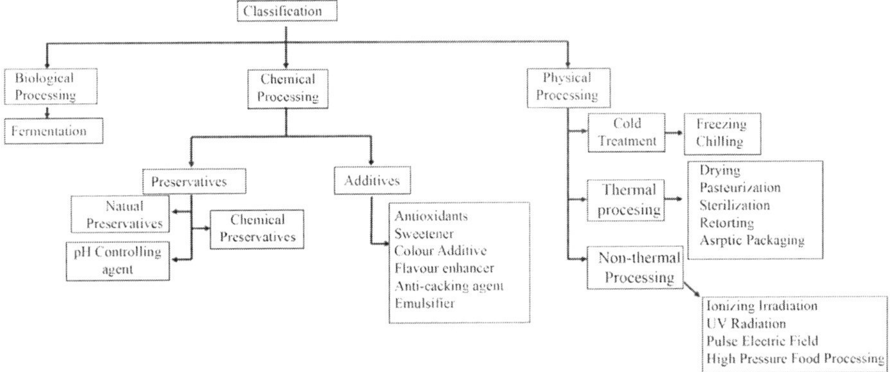

Fig. 4: Processes and techniques employed to prevent spoilage and used for food preservation. *Adapted from (Amit et al., 2017). Under Creative Commons Attribution 4.0 International License (http://creativecommons.org/licenses/by/4.0/)*

Physical Processing

Thermal Processing

Drying

Removing the moisture content of food through evaporation as a method to preserve food is called drying or dehydration. Drying is one of the classical methods of food preservation and results in a solid product with sufficiently low water content (Berk, 2013). The extent of availability of the moisture content of the food for microbial growth is called water activity. Most microorganisms can thrive at water activity above 0.95. Bacteria are inactive at water activity below 0.9. Most of the microorganisms cannot grow at water activity below 0.88 (Leniger & Beverloo, 1975; Syamaladevi *et al.*, 2016). Drying lowers the moisture content of the food to such an extent that it is no more able to facilitate microbial growth and/or enzymatic activities responsible for food spoilage. (Jay *et al.*, 2000; Rayaguru & Routray, 2010).

Besides, weight and volume reduction, drying facilitates ease of food storage, packaging, and transportation. With all these advantages, drying appears to be the least expensive way to preserve food. (Agrahar-Murugkar & Jha, 2010). However, this process also has limitations, such as, it might in some cases lead to significant loss of flavour and aroma of the food. Some functional compounds like vitamin C, thiamine, protein, and lipid are also lost because of drying (Jangam *et al.*, 2010; Kutz, 2008; Salvato *et al.*, 2003).

Classification of drying

Depending on the movement pattern of the hot air being used in drying, it can be classified as convective, conductive, and radiative, out of which, Convective drying is the most popular method used to obtain over 90% dehydrated foods. Dryers can operate in batch or continuous modes; batch being preferred for smaller scale and shorter time. Continuous drying is preferred when long periodic operations are required (Baker, 1997).

Drying of different foods

Food items preserved by drying are fruits, vegetables, meats, and fishes. Other food items preserved through drying are instant coffee and tea (Bhat & Alias, 2012; Sagar & Suresh Kumar, 2010). Processing temperature and drying time of different food items are presented in Table 6.

Table 6: Processing temperature and time for different food products

Foods to be dried	Processing temperature and time
Fruits	
Cherries	70°C for 2-3 h; 55°C until dry
Coconuts	45°C until dry
Pineapples	70°C for 1-2 h; 55°C until dry
Persimmons	60°C for 1-2 h; 55°C until dry
Pears (Asian)	60°C
Vegetables	
Asparagus	60°C for 2-3 h; 55°C until dry
Beans, green	60°C for 2 h; 55°C until dry
Mushrooms	25-30°C for 2-3 h; increase to 50°C until dry
Onions	70°C for 1-2 h; 55°C until dry
Parsley	30°C to 50°C; may be room dried
Fish	
Carp	4°C under high pressure for 15-20 min
Prawn	70°C for 30 min
Meat	
sAll	80°C for 2h

Adapted from (Amit *et al.*, 2017)

Pasteurization

In pasteurization, food is heated up to a specific temperature which destroys spoilage-causing microorganisms and enzymes (Baker *et al.*, 1997; Shenga *et al.*, 2010). Pasteurization efficiently gets rid of pathogenic bacteria, yeasts, and molds resulting in increased shelf-life of the food (Cavazos-Garduño *et al.*, 2016; Laudan, 2009) . This process was invented by the French microbiologist, Louis Pasteur (1822–1895), and was rightfully named after him. He experimented in the applicability of this process in preserving wine and beer in 1862 (Brown, 2007). Table 7 presents the applications of pasteurization process to preserve different food items.

Table 7: Pasteurization of different foods

Food type	**Main purpose**	**Sub-purpose**	**Typical temperature-time combination used**
Fruit juice (pH<4.5)	Inactivation of enzymes (pectin esterase, poly-galacturonase)	Destruction of spoilage-causing microorganisms (*Salmonella enterica, Cryptosporidium parvum*)	65°C for 30 min,77°C for 1 min, or 88°C for 15 s
Beer (pH<4.5)	Destruction of spoilage-causing microorganisms (wild yeasts, *Lactobacillus species*)	Destruction of spoilage-causing microorganisms	65 to 68°C for 20 min (in bottle) or 72-74°C for 1-4 min at 900-1000 kPa
Milk (pH>4.5)	Destruction of pathogens *(Brucellaabortis, Mycobacterium tuberculosis)*	Destruction of spoilage-causing microorganisms (*Streptococcus laptis, Streptococcus cremoris*) and enzyme	63°C for 30 min or 71.5°C for 15 s
Liquid egg	Destruction of pathogens (*Salmonella seftenburd*)	Destruction of spoilage-causing microorganisms	64.4°C for 2.5 min or 60°C for 3.5 min

Adapted from (Amit *et al.*, 2017)

Pasteurization techniques

The temperature–time combination critically affects the efficiency of pasteurization. This combination is optimized from the thermal death-time studies of heat-resisting microorganisms (Kutz, 2008). Based on the temperature–time combination, pasteurization can be done through high temperature short time (HTST) exposure, and ultra-high temperature (UHT); both HTST and UHT are continuous processes (Arcand & Boye, 2012; Barbosa-Cánovas *et al.*, 2005). The third type of pasteurization is called vat pasteurizer and is suitable for handling small-scale capacities of 100–500 gallons (Salvato *et al.*, 2003) Vat and HTST pasteurization processes effectively eliminate pathogenic microorganisms, but heat –resistant spores persist. UHT pasteurization is more effective than vat and UHST pasteurization processes in eliminating heat-resistant spores. Therefore, food items preserved through UHT pasteurization have longer shelf life than those processed through other types of pasteurization. Table 8 presents the comparisons between the three pasteurization methods (Kutz, 2008). Following pasteurization, food items are packed aseptically (Clark *et al.*, 2008). Despite treatment at high temperatures during various pasteurization processes, physical, chemical and biological alterations of the food items are minimum and the overall quality of the food remains uncompromised (Tamime, 2009). However, for certain food items, particularly, those rich in vitamins, minerals and beneficial bacteria, pasteurization should not be the preferred choice of preservation technique. Extreme thermal exposure during this process, damage these thermos-labile constituents of the food. For example, pasteurization can reduce vitamin C content by 20 %, soluble calcium and phosphorus by 5 % and thiamine and vitamin B12 by 10 %. Therefore, pasteurization cannot be used to preserve fruit juices, whose nutritive value and edibility depend on ascorbic acid, vitamin C, and carotene. However, these losses can be considered minor from nutritional point of view, in case of food items for which nutritive quality is not determined by these thermos-labile constituents (Fellows, 2009; M. Rahman, 2007).

Table 8: Comparison between different pasteurization techniques

Criteria	**Vat**	**HTST**	**UHT**
Process type	Batch	Continuous	Continuous
Typical temperature-time combination	65°C for 30 min	72°C for 15-30 s	135-150°C for a few sec
Food preserved	Butter milk and sour cream	Milk, fruit juices, frozen dessert mixes etc.	Milk
Shelf-life increase(milk)	Several days when refrigerated	2-3 week when refrigerated	6-9 months when aseptically packaged
Type of microbes destroyed	Vegetative pathogens	Vegetative pathogens	All bacteria and spores

Adapted from (Amit *et al.*, 2017)

5.1.1.3 Sterilization

Thermal sterilization is a heat treatment that destroys all viable microorganisms (yeasts, molds, vegetative bacteria, and spore formers) resulting in a longer shelf life (M. Rahman, 2007). Retorting and aseptic processing are two types of thermal sterilization (Knechtges, 2012; M. Rahman, 2007). Thermal sterilization is different from pasteurization, and a comparison of these two techniques is given in Table 9.

Table 9: Comparison between pasteurization and sterilization

Criteria	Pasteurization	Sterilization
Temperature level	Mild heat treatment process. Temperature level 65-75°C (exception UHT)	Severe heat treatment process. 135-140°C and up to 150°C are applied
Status of heat-resisting microorganism	Many heat-resisting microorganisms, viruses, and spores may remain alive	Bacteria species, spores, and thermophiles
Change in nutrition capacity and profile	Negligible	Fats, protein, and sugar may decompose; calcium, minerals, and vitamins may escape
Storage	Refrigerated conditions	Ambient temperature
Product parameter (pH)	3.5<pH<4.6	pH>4.6
Shelf-life extension	For few days to weeks	For months

Adapted from (Amit *et al.*, 2017)

Retorting

Retorting means sterilization of food after packaging it in a container (Knechtges, 2012). Retorting is suitable for foods with pH above 4.5 that require more than 100 °C as sterilization temperature. Such high temperature can be attained in batch or continuous retorts, however, batch retorts are being gradually replaced with continuous ones (Kirk-Othmer, 2007). The most common continuous systems employed in food industries are hydrostatic retorts and rotary cookers (G.S. Tucker, 2007). Different requirements of batch and continuous retorts are summarized in Table 10.

Table 10: Comparison between batch and continuous retort

Criteria	Batch retort	Continuous retort
Capital investment	Low capital investment and higher flexibility	Initial investment is high
Throughputs	Lower	Higher
Energy	Energy and labour intensive	Scope for energy saving
Time of heating to sterilization temperature	2-6 h	0.5-3 min
Sterilization time	20-60 min	15-90 s
Cooling time after sterilization	4-10 h	5-10 min

Adapted from (Amit *et al.*, 2017)

Aseptic Packaging

In aseptic packaging, packaging and food items are sterilized separately and then the food is put in the sterile packaging in an aseptic environment. Sterilization is done by heat and/or chemical treatments (Potter & Hotchkiss, 1999). Aseptic packaging is the popular choice of preservation technique for juices, dairy products, tomato paste, and fruit slices (Kirk-Othmer, 2007). Aseptic packaging can prolong the shelf-life of food significantly. In fact, aseptic packaging when combined with UHT pasteurization increases the shelf life of liquid mild from 19-90 days (in case of UHT pasteurization alone) to more than 6 months. Although, paper and plastic materials are used for conventional aseptic packaging, other materials, like metal cans, foils, which can withstand high temperatures and pressures are also used (Miller *et al.*, 2007). Availability of a wide variety of packaging materials to choose from has kept the overall cost of aseptic packaging technique competitive.

In direct aseptic packaging is done through steam injunction and steam infusion. In contrast, in case of indirect aseptic packaging, plate, scrapped surface and/or tubular heat exchangers are used (Ohlsson & Bengtsson, 2002). Steam injection, besides being one of the fastest methods of heating, also removes volatile substances from food products. On other hand, steam infusion offers better control over the preservation process so that risks of overheating are minimized (Ohlsson & Bengtsson, 2002). In case of indirect aseptic packaging, plate and scrapped surface heat exchangers are more amenable to changes in production capacities and can better accommodate to varying viscosities of the liquid foods being process. However, their demand for frequent cleaning has reduced their popularity in food industries. On the other hand tubular heat exchangers are adapted for high pressure and flow rates, but, they are suitable for low viscous liquids and not flexible to production capacity changes (Ohlsson & Bengtsson, 2002).

Cold Processing

Freezing

Freezing of food items converts the water present within them to ice crystals at a temperature below freezing point. This reduces the water content of the food thereby, decreasing its water activity and consequently, inhibiting the growth of microbes (Brennan, 2006; George, 2008; M. S. Rahman & Velez-Ruiz, 1999). During freezing, first the nucleation of ice crystal formation occurs followed by gradual growth in the size of the crystals. Efficiency of freezing depends on the freezing time, that is, the time required to reduce the initial temperature of a food item to a given temperature at its thermal centre. Slow freezing leads to formation of non-uniformly, mostly surface distributed, large ice crystals, whereas, rapid freezing forms uniformly distributed small crystals embedded within the food item, and thus, cause less damage to the cell structure and consequently, to the texture of the food (Ramaswamy & Tung, 1984). Through shorter freezing periods diffusion of salts and decomposition of foods during freezing can be prevented. Solid food items like cut vegetables, shrimps, fish and meat can preserved through Individual Quick Freezing (IQF), a type of rapid freezing in which a collection small food items are frozen simultaneously. IQF is suitable for solid small sized food items while the rapid freezing technique employed for liquid or semi-liquid or pulpy food items like, fruit juices and pulps, is simply referred to as quick freezing. IQF can be scaled up to commercial levels with eventual reduction in cost, although, requires initial capital investment (Pruthi, 1999). Quick freezing can be achieved through contact plate, air-blast and cryogenically. The comparison between different quick-freezing techniques for fishery products is presented in Table 11.

Table 11: Various quick-freezing techniques (fishery products)

Criteria	Contact plate freezing	Air-blast freezing	Cryogenic freezing
Capital cost	Low capital investment	Economic to construct and operate	Higher capital cost
Operation cost	Low operation cost	Higher operation cost	Higher operation cost
Heat transfer	Controlled heat transfer	Efficient heat transfer	Efficient heat transfer
Product line	Generally bulk freezing	Flexible	Flexible
Required floor space	Large	Large	Small
Refrigeration plant	Required	Required	Not required
Maintenance cost	Low	Low	Minimum
Dehydration loss	High	High	Minimum
Product quality	Reasonably good product quality	Good product quality	Superior product quality
Cooling time after sterilization	4-10 h	5-10 min	5-10 min

Adapted from (Amit *et al.*, 2017). Under Creative Commons Attribution 4.0 International License (http://creativecommons.org/licenses/by/4.0/)

Depending on the food to be preserved, freezing equipment being used, lattice structure of the food item, initial and aimed final temperatures, temperature of the refrigerating medium, and thermal properties of the food item, such as, surface heat transfer co-efficient, change in enthalpy and thermal conductivity, the International Institute of Refrigeration (1986) has defined various freezing time for different food products and to be used with different freezing equipments (Barbosa-Cánovas *et al.*, 2005).

Chilling

Chilling storage is best suited for short-term preservation of food as it can slow down microbial growth and metabolic and enzymatic activities in harvested plant and slaughtered animal tissues, prevent detrimental chemical reactions, like, lipid oxidation (Karel & Lund, 2003). Chilled foods are maintained between −1 and 8 °C. During chilling, the initial temperature of the food item is reduced to a certain temperature at which the it is maintained for over an extended period of time (Saravacos & Kostaropoulous, 2002). Chilling retards biochemical and microbiological changes that eventually extend the shelf-life of the preserved food (Indira & Sudheer, 2007). Sometimes when freezing is carried out at a temperature >15 °C, it is referred as chilling (Lund & Karel, 2005). In order to reduce formation of ice crystals during freezing of fresh food items, they are partially frozen, the process being referred to as super chilling (Magnussen *et al.*, 2008).

Chilling can be achieved through different types of coolers, like, continuous air cooler and ice bank cooler, or through heat exchanges, such as, plate heat exchanger and jacketed heat exchanger, or through other facilities, like, ice implementation system, vacuum attribution system, and cryogenic chambers (James, 2008). Thermal conductivity, initial temperature of the food/s to be preserved, lattice structure, moisture content, all affect the chilling rate (Light & Walker, 1990). Table 12 describes various methods for chilling solid and liquid food items.

Table 12: Chilling methods of solid and liquid foods

Solid foods	Batch air chillers	Warm food items are feed into large refrigerated room, widely used in industry
	Moving air	This cost-effective, hygienic, and widely used method incurs little damage to equipment. Surface dehydration of food is major disadvantages of this process
	Ice/ice water chilling	Food items are packed in boxes and then they are placed between layers of crushed ice. Melting ice assist to maintain the temperature at 0°C. However, this method is labour efficient and consume much time comparing to other processes
	Cryogenic cooling	This method involves the use of liquid nitrogen to freeze the products. Thermal shock conformation of food items makes this process vulnerable
	Immersion cooling/ hydrocooling	A cost-effective cooling method is suitable for small products. This technique involves immersing or spraying the product in cool water at 0°C. Hydrocooling moisturizes food items, which can be detrimental to some extents
Liquid foods	Batch cooling of liquids	A jacketed stainless-steel vessel of varying capacity with agitator inside usually used for this type of chilling. The coolant may circulate through the jacket of the vessel or through a coil placed in the liquid food stuff, or both while the agitator incurs uniform heat transfer
	Continuous cooling of liquids	The continuous cooling of liquids can involve multi-plates and tubes, aeration, and double pipe coolers. The most widespread piece of equipment is the multi-plate cooler, which has the best efficiency, high surface area for exchanging heat, easy cleaning opportunity, and less material requirement than others

Adapted from (Amit *et al.*, 2017)

Non-Thermal Processing

Ionizing Irradiation

Preservation through irradiation involves exposing the food item/s to a fixed doses of natural or artificial ionizing radiation (IR) (Arvanitoyannis, 2010). X-rays, gamma rays, and high-energy ultraviolet (UV) radiation are the natural IRs, whereas, accelerated electrons and induced secondary radiation are artificially generated IRs (Moniruzzaman *et al.*, 2016; Sommers, 2010). Food preservation through IR is the choice of preservation technique for over 60 different foods across 40 different countries across the globe (Arvanitoyannis, 2010). IR effectively disinfests grains, fruits, and vegetables, inhibits sprouting or decelerate maturation and senescence and thereby prolongs the shelf lives of fruits and vegetables and prevents microbial spoilage and maintains edible safety by destroying or inactivating foodborne pathogens (Heldman & Moraru, 2010; Kanatt *et al.*, 2006). Different factors of food irradiation techniques are listed in Table 13.

Table 13: Food Irradiation Technologies

Factors	Electron beam	X-Ray	Gamma ray
Source	Accelerated electrons, typically 5-10 MeV	Include by impingement of electron beam onto a metal plate. Conversion efficiency is 5-10%	Radioactive decay of Co-60 (2.5 MeV) or Cs-137 (0.51 MeV)
Processing time	Seconds	Seconds	Minutes
Penetration	6-8 cm, suitable for relatively thin or low-density products	30-40 cm, suitable for all products	30-40 cm, suitable for all products
Shielding for operator	≥2m concrete or 0.7m steel/iron/lead	>2m concrete or ~0.7 m steel/iron/lead	>5m water or >2m concrete or 0.7m stell/iron/lead

Adapted from (Amit *et al.*, 2017)

Regulatory limits of irradiation

IRs are delivered at fixed doses set by recognized regulatory bodies, and are designated in kilo grays (kGy). The amount of radiation absorbed by 1 g of irradiated material is 1 gray (Sommers, 2010). Table 14 presents different regulatory limits for food irradiation applications.

Table 14. Regulatory limits for food irradiation applications

Type of dose	Benefits	Dose	Products
Low dose (up to 1 kGy)	Inhibition of sprouting	0.05-0.15	Potatoes, onions, garlic, root ginger, yam, etc.
	Insect disinfestation and parasite disinfection	0.15-0.5	Cereals and pulses, fresh and dried fruits, dried fish and meat, fresh pork, etc.
	Delay of physiological process(ripening)	0.25-1.0	Fresh fruits and vegetables
Medium dose (1-10 kGy)	Extension of shelf life	1.0-3.0	Fresh fish, strawberries, mushrooms, etc.
	Elimination of spoilage and pathogenic microorganism	1.0-7.0	Fresh and frozen seafood, raw or frozen poultry and meat, etc.
	Effect on food properties	2.0-7.0	Grapes (increasing juice yield), dehydrated vegetables (reduced cooking time), etc.
High dose (10-50 kGy)	Industrial sterilization (in combination with mild heat)	30-50	Meat, poultry, seafood, prepared foods, sterilized hospital diets
	Decontamination of certain food additives and ingredients	10-50	Spices, enzyme preparations, natural gum, etc.

Adapted from (Amit *et al.*, 2017)

Effects of Irradiation

Although, lipids, carbohydrates, proteins and minerals, are unaffected by even high doses of IR, levels of certain nutrients, such as, vitamins A, B1, C, and E are reduced. According to FDA, IR affects nutritive values of the preserved food in a manner similar to conventional food processing techniques (J. S. Smith & Pillai, 2004).

Ultraviolet (UV) Radiation

UV radiation (treatment) is used to decontaminate food, improving both the shelf-life and safety of foodstuffs (Miller *et al.*, 2007; Steele, 2004). UV radiation is a form of energy considered to be non-ionizing in nature with general germicidal properties at wavelengths in the range of 200–280 nm (usually termed UV-C). Based on different ranges of wavelength, UV light can be categorized into UV-A (315–400 nm), UV-B (280–315 nm), UV-C (200–280 nm) and UV- Vacuum (100–200 nm) (Ohlsson & Bengtsson, 2002). UV radiation destroys the genetic constituent of the pathogen thereby, preventing division, multiplication and eventually, propagation (George, 2008; Steele, 2004). Different UV doses (measured in mJ/cm2) are required for different kinds of pathogens. For instance bacteria can be eradicated at 1–10 mJ/cm2, yeast at 2–8 mJ/cm2, 20–200 mJ/cm2 of UV irradiation is required for fungus, 100–15 for 0 mJ/cm2protozoa and 300–400 mJ/cm2 for algae (Ohlsson & Bengtsson, 2002;

M. S. Rahman & Velez-Ruiz, 1999). Besides, the type of microorganism being dealt with, the efficacy of UV irradiation is affected UV source and its physical alignment with respect to the food item being treated (Brennan, 2006).

High-pressure Food Preservation

In high hydrostatic pressure or Ultra High-Pressure processing (HPP) technology, microorganisms present in food items are killed using pressure upto 900 MPa. Naturally, this process effectively prevents chemical and enzymatic reaction driven food spoilage while retaining the physical and physiochemical characteristics. HHP is safe for vitamins also, besides, leaving flavours, and colour molecules unaffected (Bhat & Alias, 2012; Dunne, 2007; Koutchma *et al.*, 2016). Additionally, it imparts freshness and improved taste with high nutritional value to the foods which make it the best preservation technique so far. It is environment friendly, with low energy consumption and minimal effluent discharge. The only drawback of this technique is the one-time high capital cost. However, limited information on this technique has made scepticism shroud it, limiting its widespread application (Bhat & Alias, 2012; Nielsen *et al.*, 2009; Yeung & Huang, 2016).

HPP functions on Le-Chatelier's and isostatic principles (Bhat & Alias, 2012). A minimum 40% moisture level is critical for HPP to work (Ohlsson & Bengtsson, 2002). HPP can be run as container processing and bulk processing. Table 15 presents the advantages and limitations of in-container and bulk processing of foods under high pressure.

Table 15: Advantages and disadvantages of in-container processing and bulk processing

Type of processing	Advantages	Disadvantages
In-container processing	Applicable to all solid and liquid food Minimal risk of post-processing contamination Major development needed for high-pressure processing Easier cleaning	Complex materials handling Little flexibility in choice of container Greater dead time in use of pressure vessel
Bulk processing	Simple material handling Greater flexibility in choice of container Maximum efficiency in use of high-pressure vessel volume Minimum vessel dead time (no opening/closing of vessel needed, faster loading/unloading)	Only suitable for pumpable foods Aseptic filling of containers required potential post-processing contamination All pressure components in contact with food must have aseptic food design and suitable for cleaning in place and sterilizing in place

Adapted from (Amit *et al.*, 2017)

Pulsed Electric Field

Pulsed electric field (PEF) is a food processing technique of getting rid of microbial load by exposing the food transiently for less than a second to a field of high potential difference (voltage) (20–40 kV/cm) between two electrodes (Brennan, 2006). The efficiency of PEF depends on the strength of the electric field and the number of pulses (Ahmed & Rahman, 2012). This preservation technique due to low processing time retains the overall quality of the food maximally, and more importantly, the technique is non-toxic (Mohammed *et al.*, 2016; Sun, 2014). Cell deaths caused by PEF processing are dye to cell membrane disruption and electroporation. PEF treatment is applicable to only liquid food items, such as, fruit or vegetable juices, milk, liquid egg, and nutrient broth, because such a high electric field would be destructive to the lattice structure of the food constituents (Mohammed *et al.*, 2016). PEF processing is more effective against gram-negative bacteria than gram-positive bacteria. Similarly, vegetative cells are susceptible to this treatment than spores. (Jay *et al.*, 2000). Enzymes and spores are resilient to PEF treatment. and it is unsuitable for conductive materials. Moreover, this technique is energy intensive posing environmental risks (Fellows, 2009; Mathavi *et al.*, 2013). Different types of foods are processed using PEF process. Processing parameters of different PEF-treated foods are listed in Table 16.

Table 16: Processing parameters of PEF-treated food products

Products	**Processing Condition**				
	Electric field Strength (KV/cm)	**Number of Pulses**	**Duration of Pulses (μs)**	**Temperature of Product (°C)**	**Log Reduction (D)**
Orange juice	6.7	5	20	45-50	5
Milk	22	20	20	45-50	4.6
Skim milk	45	64	1.8-6	35	2
Yoghurt	23-28	20	100	63	2
Liquid egg	25.8	100	4	37	6
Pea soup	25-33	10-30	2	53-55	4.4
Fluid food	12-25	25	1-100	45-55	Shelf life extended from 3-7 days

Adapted from (Amit *et al.*, 2017). Under Creative Commons Attribution 4.0 International License (http://creativecommons.org/licenses/by/4.0/)

Biological processing

Fermentation

Microbial activities cause food spoilage. On an interesting note, microbial activity can also preserve food by decomposing carbohydrates through various enzymatic activities and the process is called fermentation (Shivasankar, 2002). Such helpful microbes can be bacteria, yeasts, and molds that can ferment a different variety of food items, like dairy products, cereal-based foods, and meat products (Battcock & Azam-Ali, 1998; Katz, 2001). Besides food preservation, fermentation improves nutritional value, and digestibility of foods, making it a healthy alternative to chemical preservatives (Lewin, 2012).

Fermentation involves different types of chemical reactions. For example, alcohol is produced by the action of yeast on 'hexose', a simple sugar, and emitting carbon dioxide in the process. For alcohol production fermentation takes place under anaerobic conditions. The following equation illustrates alcohol fermentation by conversion of hexose (Dagoon, 1993).

$C_6H_{12}O_6 + O_2 = C_2H_5OH + CO_2$ Hexose + Oxygen
= Ethanol + Carbon dioxide

Vinegar is produced due to aerobic fermentation when Acetobacter converts alcohol to acetic acid in the presence of excess oxygen. This process is used to preserve foods as pickles and relishes. Vinegar fermentation results in acetic acid and water by oxidation of alcohol (Battcock & Azam-Ali, 1998; Koutchma *et al.*, 2016).

$C_2H_5OH\ O_2 = CH_3COOH + H_2O$ Ethanol + Oxygen
= Acetic acid + Water

Lactic acid is produced by the fermentation by homo- and heterofermenters. Homofermenters follow the glycolytic (Embden–Meyerhof pathway) pathway to produce lactic acid, whereas heterofermenters work through the 6-phosphogluconate/phosphoketolase pathway to generate lactic acid along with ethanol, acetate, and carbon dioxide. Homofermentation produces lactic acid more efficiently than heterofermentation, as 1 mole of glucose yields 2 moles of lactic acid along the former pathway and 1 mole along the latter one (Battcock & Azam-Ali, 1998).

$C_6H_{12}O_6$(Glucose) ⟶ $2CH_3CHOHCOOH$
(Lactic acid)

$C_6H_{12}O_6 = CH_3CHOHCOOH + C_2H_5OH + CO_2$Glucose
= Lactic acid + Ethanol + Carbon dioxide

Different kinds of microorganisms are used in the fermentation process, exclusively to produce flavour in foods, and these are presented in Table 17 (Shivasankar, 2002).

Table 17. Microorganism used in food processing and flavour compounds produced

Food items	**Microorganisms**	**Flavour compounds produced**
Buttermilk	*Streptococcus lactis* *Streptococcus cremoris* *Lactobacillus bulgaricus*	Lactic acid, diacetyl, small amounts of acetaldehyde
Yoghurt	*Streptococcus thermophilus* *Lactobacillus bulgaricus*	Acetaldehyde and diacetyl acetoin
Alcoholic fermented milk	*Saccharomyces sp.* *Lactobacillus sp.*	Ethanol acetoin and diacetyl
Sauerkraut	Mixed cultures of *Lactobacillus brevis* *Leuconostoc mesenteroides* *Lactobacillus plantarum*	Acetate and small amounts of short chain fatty acids

Food items	Microorganisms	Flavour compounds produced
Soyabean milk	*Lactobacillus sp.* *Saccharomyces thermophiles*	Aldehydes including pentanal
Soya sauce	*Aspergillus oryzae* *Lactobacillus sp.* *Saccharomyces rouxii*	Organic acids, alkyl phenols, and pyrazines
Tempeh	*Rhizopus sp.*	Fatty acid
Bread	*Saccharomyces cerevisiae*	Ethanol
Swiss cheese	*Propionibacterium shermanii*	Propionic acid
Coca	*Saccharomyces sp.* *Lactobacillus sp.* *Acetobacter sp.*	Fatty acids and aromatic acids

Adapted from (Amit *et al.*, 2017)

Chemical Processes

Preservatives are usually added to easily perishable natural foods to increase their shelf lives but, may have some harmful effects on human health. These preservatives sometimes add flavour to the food and are called additives. These substances have anti-bacterial and anti-fungal properties that help prevent food from microbial spoilage (Blum, 2012).

Food additives can be natural or artificial, direct or indirect. The detrimental effects of food additives may be immediate or manifest themselves slowly over time.

Preservative

Natural Preservatives

Salt, alcohol and vinegar are natural food preservatives used traditionally at home while making pickles, jams and juices. Sometimes sugar combined with alcohol is used for preservation of luxury products such as fruits in brandy and other asperities (Table 18) (Friis, 2012; Garg *et al.*, 2010; Meyer *et al.*, 2002; M. Rahman, 2007; J. Smith, 2004).

Table 18: Some type of natural preservatives

Natural preservative	Example of food items	Functions
Salts	Salted fish	Salt and sugar draw the water out of microorganisms and retard the growth of microorganism
Sugar	Jam	
Vinegar	Pickled mango	Vinegar provides an acidic condition which creates an unfavourable condition for microorganism
Rosemary extract	Mayonnaise, margarine, oils and fats, etc.	Rosemary extracts work as antioxidant

Adapted from (Amit *et al.*, 2017)

Chemical Preservatives

Chemical substances added to food to prevent spoilage are called artificial preservatives (Abusaloua *et al.*, 2019; Arora *et al.*, 2010; Dar *et al.*, 2017; Inetianbor *et al.*, 2015; Sharma, 2015; Silva & Lidon, 2016) .

pH Controlling Agents

The pH of the food represents its acidity or alkalinity and the deciding factor behind the flavour of the food. Therefore, pH fluctuations can be a cause for spoilage. So, chemical additives are included as preservatives to prevent pH fluctuations. Natural acidulants, like, acetic acid or vinegar and citric acid from citrus, malic acid and tartaric acid help in maintaining the acidic pH of the originally acidic foods. Likewise, Potassium citrate, Calcium carbonate, Calcium acetate, Sodium bicarbonate and Sodium lactates are also used to neutralize excess acids in non-acidic foods. Some of the preservatives and their classification are listed in Table 19 and 20.

Table 19: Some important Chemical preservative and pH controlling agents

Chemical Preservative		**pH Controlling Agents**	
Chemical	Code	Chemical	Code
Acetic acid	E260	Acetic acid	E260
Benzoic acid	E210	Fumaric acid	E2976
Propionic acid	E280	Phosphoric acid	E338
Sorbic acid	E200	Malic acid	E296
Sulphur dioxide	E220	Lactic acid	E270
Citric acid	E330		

Table 20. Classification of artificial preservatives

	Anti-microbial agents	**Anti-oxidant agents**	**Anti-enzymatic agents**
Definition	Inhibits the growth of undesirable microorganisms (yeast, fungi, bacteria)	Inhibit atmospheric oxidation. Mainly used for the products that contains unsaturated fatty acids, oils, and lipids	Prevent natural ripening process and oxidative deterioration of food by inhibiting the bacteria, fungi, parasite
Mechanism	Creates unfavourable environment for microorganisms by reducing moisture content and increasing acidity.	Oxidation of unsaturated fats produces free radicals; aldehyde and ketones are produced which result in the rancid taste of foods.	Blocks enzymatic process in the food that continue to metabolize after harvest. Metal chelating agents can remove the metal cofactor that many enzymes need
Applications	Sorbic acid and potassium sorbet for the preservation of cheese, bakery products, vegetable-based products dried fruits, beverages, and smoked fish, margarine, salad creams, and mayonnaises. Nitrite, Lactic acid for the preservation of meat	Butylated hydroxyl anisole, for the preservation of butter, lard, meats, beer, baked goods, snacks, potato chips, nut products, dry mix for beverages. Butylated hydroxyl toluene (BHT) in fats and oils processing.	Citric acid for the preservation of foods, beverages, dairy products, and pharmaceutical. EDTA (ethylenediamine tetra acetic acid) in food processing

	Anti-microbial agents	Anti-oxidant agents	Anti-enzymatic agents
	Sulphur dioxide, sodium sulfite for the preservation of dried fruits, certain fruits juices, potatoes, and wines.	Sulfites for the preservation of beer, wines, dried foods.	Polyphosphates for the preservation of fresh peeled fruits and vegetables

Adapted from (Amit *et al.*, 2017)

Additive

Chemical Additive

Antioxidants

Oxidation causes raw apples and potatoes to go brown. Therefore, antioxidants are as added as food additives, which act as oxygen scavengers to prolong the shelf-life of the food. Ascorbic acid (vitamin C) is used as an antioxidant in beers, cut fruits, dried potatoes and jams to prevent discoloration of these food due to oxidation.

Sweeteners

Sweeteners are added to enhance sweet taste with and without any extra calories. Artificial sweeteners are synthetic sugar substitutes, and may be derived from naturally occurring substances, including herbs and sugar itself. Artificial sweeteners, enhance the sweet taste of the food without adding many calories.

Colour Additives

A colour additive are intended for colour enhancement of foods to add to their aesthetic appeal. They are present in many foods like, snack, margarine, cheese, jams and jellies, desserts, drinks, etc.

Flavour Enhancer

Flavour enhancers, are flavourless chemical additives, added to savoury foods. Natural flavour enhancers can increase stability of food. Inosinic acid is the most commonly used flavour enhancers.

Anti-Caking Agents

Anti-caking agent prevents clumping food constituents together after packaging either by absorbing moisture or acting as sealant to repel water and oil. Sodium ferrocyanide, Decahydrate, Propylene glycol, Magnesium silicate are the most commonly used anti-caking agents.

Emulsifiers

Emulsions act as stabilizers and are used as salad dressings, in baked goods and chocolates. Ammonium salt of PolyGlycerol Ester (PGE), Calcium stearoyl dilaciate (CSL) are used extensively as emulsifiers in various food items. Some examples of additives and their purpose in food processing are listed in Table 21 and 22.

Table 21: List of Some common and important additives used in food processing.

Antioxidants		**Sweeteners**		**Colour Additives**		**Flavour Enhancer**		**Anti-caking Agents**		**Emulsifiers**	
Chemical	Code	Chemical	Code	Chemical	Code	Chemical	Code	Chemical	Code	Chemical	Code
Butylated hydroxyanisole	E320	Sucralose	E420	Curcumin	E110	Disodium-5-ribonucleotides	E635	Bentonite	E558	PG Ester	PGME
Ascorbyl palmitate	E304	Alitame	E956	Brilliant blue FCF	E133	Monosodium Glutamate	E621	Calcium aluminum silicate	E556	Sorbitan Ester	SOE
Calcium ascorbate	E302	Aspartame	E951	Tartrazine	E102	Monopotassium Glutamate	E623	Calcium silicate	E552	Phosphatidic acid	E442
Ascorbic acid	Vitamin C	Saccharin	E954	Sunset yellow	FCF	Calcium Di glutamate	E626			Sorbitan monosterate	E491
Hypochlorous acid	HClO			Orange yellow	S	Guanylic acid	E627			Lecithins	E332

Table 22: Some type of food additives

Type of additives	Purpose	Example
Emulsifiers, stabilizers and thickeners	Impart a consistent texture to products; prevent separation of food	Algin, carrageenan
Anticaking agents	Enable products such as table salt to flow freely	Calcium silicate
Nutrients	Enrichment (replacement of nutrients lost during processing) and fortification adding to the nutritional value of foods)	Folic acid, beta carotene, vitamin D, iron, iodine, etc.
Preservatives	Retard spoiling, prevent fats and oils from becoming rancid, prevent fresh food from turning brown	Nitrates, parabens, BHA, BHT, etc.
Leavening agents	Cause bread and baked goods to rise during baking	Sodium bicarbonate
Flavouring agents	Enhance flavour of foods	Monosodium glutamate
Sweeteners	Add sweetness with or without extra calories	Sucrose
Colouring agents	Impart colour to foods	Caramel
Fat replacers	Impart texture and creamy 'mouth feel' to food	Cellulose gel

Adapted from (Amit *et al*., 2017)

Side-effects of Food Preservation Methods

Demerits of Using Additive and Preservatives

Although, chemical food additives are largely considered safe, nevertheless, many of them are detrimental to human health. Commonly used preservative, nitrates get converted to nitrites and nitrosamines upon ingestion, which can be harmful to a foetus *(https://www.womensnutritionalhealthcenter.com/food-preservatives/)*. International Agency for Research of Cancer (IARC) considers nitrates and nitrites as probable carcinogens (Nogardy, 2013). Nitrites react with haemoglobin to form met-haemoglobin, which can lead to unconsciousness and death, particularly, in infants. Nitrites and benzoates are unsafe for pregnant women. Benzoate and nitrite consumptions decrease serum bilirubin and increase serum urea., which affect lower the mean weight and length of the foetus (Mowafy *et al*., 2001). Sodium nitrite consumption reduces haemoglobin and haematocrit values in pregnant women. Artificial food colourings, like, tartrazine, allura red, ponceau, and benzoate preservatives, have detrimental effects on brain, leading to abnormal hyperactive behaviour in infants (Kent, 2015). Individuals with asthma are also intolerant to preservatives. Sulfites, such as, sodium bisulfite, sodium meta-bisulfite, and potassium bisulfite, present in wine, beer, and dried fruits can trigger asthmatic responses and lead to migraines in sensitive individuals. Table 23 discusses the excerpts of negative effects of harmful food preservatives (Kannall, 2017; Marcola, 2010; Oishi, 2002; Panday & Upadhyay, 2012; Soomro *et al*., 2002) .

Table 23. Possible negative effects of food preservatives

Preservatives	Where found	Possible negative effects
Sodium benzoate (E211)	Carbonated drinks, pickles, sauces, certain medicines (even some 'natural and homeopathic' medications for kids)	Aggravates asthma and suspected to be a neurotoxin and carcinogen, may cause fetal abnormalities. Worsens hyperactivity
Sulfur dioxide (E20)	Carbonated drinks, dried fruit juices, cordials, potato products	May include gastric irritation, nausea, diarrhoea, asthma attacks and skin rashes. Destroys vitamin B1. Causes fetal abnormalities and DNA damage in animals.
Sodium meta-bisulfite	Preservative and antioxidant	May provoke life- threatening asthma
Potassium nitrate (E249)	Cured meats and canned meat products	May lower oxygen carrying capacity of blood; may combine with other substances to form nitrosamines that are carcinogens
P-hydroxy benzoic acid esters (parabens)	Preserved foods and pharmaceuticals	These compounds exert a weak estrogenic activity. Butyl parabens adversely affects the secretion of testosterone and the function of the male reproductive system
Lactic acid bacteria	Fermented foods	Listeria monocytogenes may growth in raw milk, meat, and vegetables during fermentation process. This pathogen is responsible for causing foodborne illness
Mono sodium glutamate (MSG)	All frozen foods, canned tuna and vegetables	Eating too much MSG can cause general weakness, flushing, heart palpitations or numbness
Aspartame	Used as a low-calorie sweetener in gum, drinks, pudding, and yogurt	It causes allergy and migraine headache
Sodium nitrite and sodium nitrate	Processed meats and fish to retain red colour and avoid botulism	Consuming high amount of bacon, hot-dog, sausage containing nitrite or nitrates may cause type-1 diabetes. The risk is too prominent for pregnant women and children. These salts may also cause irritation to digestive system including mouth, oesophagus, and stomach.
Sodium sulphite (E221)	Used in wine making and others processed foods	Increase the risk of asthma and in extreme case may cause cardiac arrest
Potassium bromate	White flour, bread, and rolls	This salt is considered as carcinogenic and its presence in bread may cause harmful effects to human

Adapted from (Amit *et al.*, 2017) Under Creative Commons Attribution 4.0 International License (http://creativecommons.org/licenses/by/4.0/)

Demerits of Using Non-Thermal Food Processing Techniques

In the following section, the demerits associated with the selected non-thermal food processing technologies covered in this study have been discussed.

Pulsed Electric Field (PEF)

PEF requires huge start-up costs which can be considered as a disadvantage as some of the required equipment can cost between 250,000–2,000,000 USD (Kumar *et al.*, 2015; M. Pal, 2017), with subsequent upgradation incurring more expenses (Hernández-Hernández, H. Moreno-Vilet & Villanueva-Rodríguez, 2019; Shahbaz *et al.*, 2018). PEF is not full proof as some bacteria (vegetative and spore) have developed resistance against this method, with the looming risk to public health (Arroyo *et al.*, 2012; Bahrami *et al.*, 2020). Besides, PEF is ineffective in treating solid foodstuffs. Additionally, scaling up is a challenge with this technique and is still at an investigative stage as evident from handful of currently available reports regarding its industrial aspects (Arshad *et al.*, 2020; Hernández-Hernández *et al.*, 2019; Priyadarshini *et al.*, 2019).

High-Pressure Processing (HPP)

HPP is unsuitable for dehydrated and porous foodstuffs, because use of water is indispensable in this preservation technique. Besides, HPP treated foodstuffs require refrigeration for storage because unlike, high pressure and high temperature treatment which efficiently eliminate microbes, only high pressure is less effective in this regard Strict requirement of plastic as packing material for food meant for HPP treatment is yet another demerit associated with this technique because several other eco-friendly packing materials are automatically excluded from consideration. HPP requires packaging materials that can undergo at least a 15% compression, and plastic materials meet this requirement quite well (Huang *et al.*, 2017; La Peña *et al.*, 2019). HPP is reportedly unsuitable for dairy products, as this technique can reduce casein micelle magnitude, raise free fatty acid levels and alter the natural qualities of whey (Liepa *et al.*, 2016). Lastly, huge initial costs add to the list of disadvantages associated with HPP. Equipment costs as high as 7000 USD needs to be taken into account if considering setting up HPP facility (Pinto *et al.*, 2016), although the operational costs are comparatively low (Verma *et al.*, 2020). Moreover, appropriate operational skillsets are required for handling the complicated operation of the technique.

Ultraviolet (UV) Radiation

UV irradiation experiences an entirely different type of challenge, and it has more to do with the acceptability of it safety among the consumers rather than any aspect of the technique *per se*. (Hernández-Hernández, H. Moreno-Vilet & Villanueva-Rodríguez, 2019; M. K. Khan *et al.*, 2018). Consumers are apparently sceptical about UV radiation leading to radioactive materials in foods, which may subject them to serious health issues (Bahrami *et al.*, 2020; Xuan *et al.*, 2017). While this apprehension is unfounded, nevertheless, handling UV-C does pose health risks to the handler. It can cause severe skin burns and eye injuries (photokeratitis) *(https://www.fda.gov/medical-*

devices/coronavirus-covid-19-and-medical-devices/uv-lights-andlamps-ultraviolet-c-radiation-disinfection-and-coronavirus). However, on hindsight, this UV-C limitation implicates its ability to alter chemistry of the food stuffs exposed to it. Besides, UV radiation induced isomerization and oxidation of lycopene especially with increased radiation concentrations and contact times, cannot be disregarded (Bhattacharjee *et al.*, 2019). Turbidity of liquid food items limits the effect of UV-c irradiation (Pinto *et al.*, 2016). Due to its low penetrability, food stuffs with indefinite shape and structure are less affected by UV radiation. However, this drawback can be removed by combining UV irradiation with other non-thermal processes (Bahrami *et al.*, 2020). Like other non-thermal preservation techniques, requirement of bulky initial investment is also a limiting factor for UV radiation process (Hernández-Hernández *et al.*, 2019).

Ionizing Irradiation (IOR)

IOR is capital intensive and requires establishment dedicated facilities. IOR doses need to be optimized for each food item based on the geometry/smoothness and density of the latter (Indiarto *et al.*, 2020). Higher than required doses of IOR can deteriorate sensory properties of foods such as colour and aroma. Moreover, preservation through IOR expediates lipid oxidation (Indiarto *et al.*, 2020). IOR has the ethical disadvantage of posing occupational hazard to the manpower involved in its operation (Syaza *et al.*, 2018).

Demerits of Using Cold Treatment

Chilling

Like IOR, Chilling comes with very high infrastructure costs. Besides, it reduces inherent crispiness of food items and dehydrates unwrapped food surfaces (Amit *et al.*, 2017).

Double-Dead Meat/Botcha

The meat obtained from animals who have died due to diseases is called double-dead meat. It is a Fillipino appellation, also known as hot meat or *botcha* (Lopez, 2007) (https://www. thepigsite.com/news/2009/03/double-dead-pork- seized-in-balintawak -1). Since, the appropriate method of dealing with diseased animal carcasses is disposal by burial or burning (Lazaro, 2007), therefore, they are illegal. Mainly, the double-dead meats are from dead pigs illegally butchered for sale. Double dead meats fail to meet the quality requirements for edibility. Nevertheless, the struggling economies of the third world countries do not allow the luxury of disregarding these double-dead meats as food, they come at a substantially cheap price (Jaypee, 2008).

Legal Issue of Double-Dead Meat/Botcha

Due to incredible sub-standard for an edible item, double-dead meat market is a bonafide illegal entity. Phillipino government has taken strict and stricter measures from time to time to curb and altogether ban this market. Under the Phillipino law, Republic Act 9296, anybody engaged with double-dead meat business is punishable with a fine of Php 1,000.00 and Php 10,000.00 plus six months to five years imprisonment *(https://web.archive.org/web/20090303204525/http://www.gmanews.tv/story/127155/(Update)-Authorities-charge-man-caught-transporting-double-dead-meat-in-Pasay).* The National Meat Inspection Service (NMIS) of Phillipines is entrusted to monitor the entire production chain of meat industry from slaughterhouse to market (Jennifer, 2008) and have the authority to confiscate any meat that fails to pass their standards to bury them to prevent from re-entering the market (Aurelio, 2008). Certain local governments in Phillipines are in fact going a step ahead to revocate business permits ("QC to Raise Penalties for Double- Dead Meat," 2007). But, due to enfeebled economy, demand for this kind of sub-standard meats are ever-increasing, thereby fuelling the growth of double-dead meat market.

Traditional Method for Preserving Meat

With the growth of supermarkets, that require transporting meats over long distances before they are sold, preservation has taken a centre stage to ensure that the meat is up to the mark in terms of appearance, nutrition and edibility (Nychas GJE, Skandamis PN, Tassou CC, 2008). However, traditional methods of meat preservation still hold good in meeting the modern requirements. These techniques are elaborated in subsequent paragraphs.

Chilling/Refrigeration

Chilling/refrigeration at 2 to 5°C is the most popular choice of preservation technique for for short-term storage of meat and helps in retaining nutritional quality and edibility of the food (Cassens, 1994; Zhou *et al.*, 2010). Carcasses are passed through chilled coolers (15°C) to remove their body heat, before storing in holding coolers (5°C). Proper spacing between carcasses is maintained to allow proper air circulation. Chilling is either done by either immersing the meat in chilled water or by misting with misting with chilled water (Carroll & Alvarado, 2008). The subsequent processing steps of holding, cutting, transportation, retail, display and even storing in the customer household before the ultimate use are done under refrigeration. Loss of moisture is ensured with a 90% relative humidity in the surrounding. Shelf-life of the refrigerated meat depends on the species of origin, initial microbial load, packaging temperature and humidity conditions during storage. For example, pork and poultry has comparatively high microbial load to begin with. Nevertheless, caution should be exercised during preservative handling of the meat to avoid any further increase in the microbial load (Mahendra Pal & Devrani, 2018).

Freezing

Freezing keeps the original characteristics of fresh meat. Therefore, freezing is the ideal preservation technique for meat. Depending on the species, 50-75% of weight of meat is water which turns in to ice crystals during freezing, almost 98 % at -20°C and completely at -65°C. (Heinz G, 2007). Freezing significantly reduces microbial load and cellular metabolism and enzymatic reactions at -12°C and -18°C, respectively, while retaining most of the nutritive values. Water that drips during thawing process does not lose much nutrient, thereby, making freezing an ultimate choice for preserving meats. However, in order to avoid freeze burns, fresh meats should be wrapped up in suitable packaging films before freezing. Freeze burns occur due to increasing surface dehydration that leads accumulation of meat pigments. Fast freezing is preferable because small ice crystals are uniformly formed within the meat unlike, slow freezing wherein, larger ice crystals result, causing physical damage to the muscle fibres (Perez-Chabela & Mateo-Oyague, 2004; Rosmini *et al*., 2004). Although, freezing as a preservation technique for meat is unmatchable, nevertheless, it has certain lacunae. For example, ice crystals within the meat can in the long run lead to spoilage. Likewise, despite 60 % of microbial load gets destroyed during freezing, the surviving microbes gradually increase over time, ultimately, leading to meat spoilage. But the spoilage is significantly delayed (M. S. Rahman, 1999; Zhou *et al*., 2010).

Curing

Heavy salting of meat and fish is an old technique of preservation. By using excess amount of certain permissible, edible substances/salts, the external environment of the food is made hypertonic. This helps in extracting out excess moisture from within the meat and fish, thereby reducing the water activity in the micro-environment. Additionally, this reduces microbial load through change in osmotic pressure and thus, reduces the rate of spoilage. Among the popular curing agents are salt and sugar, while some other more recent options being, Sodium nitrate and Sodium nitrite. It can be dry curing, pickle curing, injection curing and direct curing. Some bacteria can be inhibited by concentrations as low as 2% (Mahendra Pal & Devrani, 2018). A concentration of 20% of sodium chloride is high enough to inhibit many food spoilage yeasts including *Debaryomyces hansenii, Yarrowia lipolytica, Kloeckera apiculata, Kluyveromyces marxianus, Pichia anomala, Pichia membranaefaciens, Saccharomyces cerevisiae, Yarrowia lipolytica, Zygosaccharomyces bailii,* and *Zygosaccharomyces rouxii* (Praphailong & Fleet, 1997). Sodium and potassium nitrites are the preferred forms of nitrites in the preservation industries. Nitrites are effective in colour stabilization and retarding rancidity, lipid oxidation and controlling anaerobic bacteria, besides, providing cured flavour to the meat. The current safety limit of nitrite in food is 156 ppm in US, and 200 ppm in Canada for meat products (Mahendra Pal & Devrani, 2018).

Smoking

Smoke generated from slow combustion of saw dust or wood contains many organic compounds, like, aldehydes, ketones, organic acids and phenols, most of which are known for their anti-microbial property. Therefore, smoking has been and is still used to preserve. In addition, smoking imparts a typical smoky flavour to the meat which is a mostly liked by the customers. Besides, being effective against the microbial load of the meat, smoking leads to surface dehydration and reduction of surface pH. The antioxidant property of smoke makes it a great choice against lipid oxidation in meat. Smoking is done at 300°C.

Currently, many liquid smoke preparations are commercially available in the developed countries. Liquid smoke is generally prepared from hard wood wherein polycyclic-hydrocarbons are removed by filtration. Application of liquid smoke on the product surface before cooking imparts it a smoky flavour, which is very much liked by the consumer (Pal, 2014).

Canning

Thermal sterilization of food held in a sealed container is called canning. It efficiently preserves appearance, flavour and texture of the meat products offering a prolonged shelf life of at least two years at ambient temperature. Different steps involved in canning precooking, filling, exhausting, seaming, thermal processing, cooling, and storage (Pal, 2014).

Dehydration

The extent to which moisture within a food item is available to microorganisms is called the water activity of the food. Dehydration reduces the moisture content of the food considerably, thereby reducing the water activity and concomitant spoilage due to microbial growth. Dehydration is an ancient food preservation technique. Meat chunks used to be sun dried after hunts for future consumption. But such dehydrated meats could not be rehydrated. The modern version of dehydration involves mechanical drying by passage of hot air under controlled humidity. But rehydration is still an issue. Rehydration issue can be overcome through freeze drying. In addition, freeze drying efficiently retains nutritive qualities increasing the acceptability of meat products. Freeze drying removes water from a food by sublimation. Water within the food is first frozen and then vapourised directly without passing through liquid phase. Freeze drying involves three steps of pre-freezing, primary drying and secondary drying (Pal, 2014). In pre-freezing, meat is first frozen at -40°C, followed by drying under vacuum for 9-12 hours at low temperature at 1-1.5 mm pressure of Hg. At this stage, ice crystals sublime to water vapours effectively removing 90-95% of total moisture content of meat, without any rise in temperature. The remaining 4-8% water gets removed during secondary drying. The dried products are packed under vacuum and have good stability through the shelf life. Freeze drying is used for preparing dehydrated meat soup mixtures (Mahendra Pal & Devrani, 2018).

Health Risks Associated with Sub-standard Meat Consumption

(https://opentextbc.ca/meatcutting/chapter/diseases-associated-with-meat/)

Meats are extremely rich in nutrients for both humans and microbes. So, if the microbes start feasting on these nutrients before humans, the end result is a hot bed of disease-causing agents. And if such meat is consumed, human health can be affected irreversibly. Some of the deadly diseases associated with meat consumptions are mentioned below:

- **BSE (bovine spongiform encephalitis)** from beef cattle
- **Scrapie** from lamb and mutton
- **Trichinosis** from pork

Following paragraphs are dedicated towards descriptive accounting of these above-mentioned diseases. Some of the microbes responsible for these diseases come from the animals whereas, some others get contributed by the processing steps. Regardless of the source of these microbes, the severity of their health impact is doubtless.

Bovine Spongiform Encephalopathy (BSE) is a fatal brain-degenerative disease (encephalopathy) of cattle. In addition to brain, spinal cord and gut are also affected. Due to plaque formation, the brain becomes spongy. With a debilitated brain, the animals behave erratically, rightfully earning the disease the common name 'mad cow disease'. Meat industry feed their cattle on the meat and bone meals which make the healthy animals susceptible to the disease, and when humans consume meat from these infected animals, they develop the human variety, the Creutzfeldt-Jakob disease (CJD). Due to a long incubation period of two-and-a-half to eight years, it is difficult to detect BSE in cattles before they reach four to five years of age. Worst part is that all breeds are equally susceptible, and the infection is easily transmissible across species barrier countries like, USA, UK and Canada has experienced outbreaks of CJD. CJD is untreatable and since symptomatically overlap with other treatable neurodegenerative disorders, jeopardize prognosis of these other diseases (Haldar *et al.*, 2013).

Scrapie, like BSE, when transmitted to humans, cause CJD. It is a fatal disease affecting the central nervous system of sheep and goats. Scrapie is a transmissible spongiform encephalopathy (TSE). Unlike BSE, scarpie is not caused by the animal's feed, and its exact cause is still elusive, making its origin sporadic Scrapie results in the accumulation of misfolded prion protein in the brain of the animals and humans.

7.3.3 Trichinosis, is caused by the parasitic nematode, *Trichinella* (round worm). *Trichinella* cysts (roundworm larvae) containing undercooked or raw pork and raw dry cured pork products transmit this infection into humans. Trichinosis is not contagious unless infected human muscle is eaten. However, almost all carnivores (meat eaters) or omnivores (meat and plant eaters), can both become infected with *Trichinella* and, if eaten, can transmit the disease to other carnivores and omnivores. In rare instances, larvae in cattle feed can infect cattle.

Various Detection Methods for Meat Adulteration

Between 20 - 35 % of dietary protein comes from meat, thus, being a rich source for all essential amino acids (lysine, threonine, methionine, phenylalanine, tryptophan, leucine, isoleucine and valine). Besides, it is also provides fats and a considerable amounts of different micronutrients (Aida *et al.*, 2005). Meat adulteration has now become a global issue for the industries with significant adverse effects on its market economy. Adulteration of ethically, ethnically and religiously approved edible meat when adulterated with cheaper, undesirable and/or objectionable meats, raises serious concerns impacting the meat industry market significantly. Therefore, it is of utmost importance that any kind of adulteration in the meat products be detected for legal, health, economic, ethical, ethnical and religious compliances (Spink & Moyer, 2011). Consumption of unchartered meat can have serious health issues in the form of allergic reactions (Ong *et al.*, 2007). Given the increasing consumer awareness regarding the food sources, it is important that meat industries ensure utmost conformation to the issued legalities (Murphy *et al.*, 2014). Proper quality compliance of the meat products is solely dependent on the efficiency of the food testing laboratories in detecting adulterations in the meat and its related products (Luo *et al.*, 2008). In a nutshell, detection of meat adulteration is essential for establishing quality trading control management, health safety, safeguarding religious sentiments, maintaining product hygiene through detection of veterolegal cases and conservation of implemented laws, like, Prevention of cow slaughter act 1954 in India which is in effect since march, 2015 (Chappalwar *et al.*, 2020).

The most common adulterations are cat meat for chicken or rabbit meat, goat meat (chevon) for sheep meat (mutton), mutton for venison (deer meat), dog and cat meat for chevon, sheep meat (mutton) for beef, horse meat for beef (mostly in Britain and Ireland) and beef for kangaroo meat in Australia. Meat origin, that includes geographical habitat, gender, breed, feed intake, age of slaughtering, growth environment (wild versus farmed), farming conditions (conventional versus organic) and meat cuts of the source animal are certain parameters which creates difficulties in establishing the authenticity of the meat. Other parameters adding to this difficulty are nutrient content substitution through species switching, employed meat processing techniques and addition of non-inherent ingredients of meat, like additives and water. However, the most obnoxious out of all these parameters is the meat adulteration through species substitution which runs very high chances of violating the sensitive religious restrictions on food restrictions (Chappalwar *et al.*, 2020). Therefore, major focus has been on developing techniques that can identify the animal sources of the meat from the processed product.

In this regard, as spectrum of simple to sophisticated analytical techniques, based on simple morphology of meat to gene-based identification, are used to identify of meat species to protect consumers from the fraudulent marketing practices. These techniques can be broadly divided into physical, chemical, immunological and molecular categories and have been described in details in the following paragraphs (Chappalwar *et al.*, 2020).

Physical methods

These methods of identifying the animal species of the meat depends on external features and general appearance of the carcass, such as, colour, consistency, odour and marbling. These characteristics can only provide a primary idea of the meat species. For example, poultry meat is white with firm consistency and absence of marbling, whereas chevon is light red with very firm consistency and typical goaty odour. Likewise, pork is distinguishable by its greyish white to light red colour and very soft consistency with urine like odour. Physical methods of identifying the species of the meat are undoubtedly economic and time-saving but can be used only if the adulteration is not done with closely related species. Some of the physical techniques used to identify the species of meat are described below:

a. Anatomical methods: These methods employ the species-specific anatomical features, such as, bone, rib numbers, sterna ribs and bone percentage, which give a fairly good idea for the identification of the species of the animal from which the meat sample has been produced. Personnel skilled in using anatomy to differentiate between closely related species is required to employ this method. Commonly, this technique is used to distinguish beef from buffalo meat.

b. Histological method: It mainly use the characteristic species-specific features of the muscle fibres to distinguish between different types of meats. This technique involves extensive microscopic histological analysis muscle tissues from meat samples. For example, carabeef can be differentiated from beef from its angular muscle striations more muscle fibre/ mm2 and smaller muscle fibre diameter.

Chemical Methods

The biochemical profiles those of the lipids and the carbohydrate, particularly, fatty acids and glycogen, respectively, vary significantly from the species sources of the meat. For example, horse meat has higher glycogen content than chevon, pork, mutton, carabeef and beef. Likewise, linoleic acid content of horse meat is between 1-2 %, which is quite higher when compared to the 0.1 % present in chevon, mutton, beef and buffalo meat. Out of all these meats, mutton has the highest intra-muscular fat content of 13.3 %. Besides the carbohydrate and fats, the contents of several natural coloration pigment in the meats also vary from source to source and can be altogether absent in some sources altogether. For example, the distinct yellow coloration of the beef is due to its high carotene content, whereas carabeef, that lacks carotene is white in colour. Even the physical properties of these biochemical components of the meat vary depending on its source species. Iodine content and refractive index of horse fat is higher than those of other species.

All the above-mentioned methods of assessing adulteration in meat are applicable only for raw meat, because, processing alters these parameters and therefore, cannot be reliably used for the purpose of source authentication of the meat sample. In order to circumvent this shortcoming, immunological methods are used, which are suitable for source authentication of even heat-treated meat samples.

Immunological methods: Immunological techniques rely on the antibody-mediated specific detection of proteins, characteristic to particular meat sources. Some of the commonly used immunological tests used for source authentication of meats are described below.

- **Ring precipitation test:** As the name itself suggests, this read-out of this test is a qualitative one in the form of a ring formed due to the antigen/protein-antibody interactions. Antibodies against specific against particular protein specific for the adulterating source is allowed to interact with the target protein, that eventually appears as a ring, confirming adulteration. However, this test is prone to offering false positive results and the rings formed are short-lived and get diffused, therefore, failing to give a permanent visual proof of the test.

- **Double Immunodiffusion Test:** This test is an improvised version of the above-mentioned ring precipitation test, sharing the same scientific rationale with the latter. The improvisation includes the use of complements to enhance the intensities of the band/ring arising from the antigen/protein-antibody interactions. This offers the scope for quantification of the interactions, besides, providing prolonged visible proof of the test. Additionally, this test is applicable even for meat cooked at 80°C for 10 min.
- **Enzyme Linked Immunosorbent Assay (ELISA):** ELISA, as a meat adulteration detection technique is a rapid, high throughput one that can give the results in 2-3 hour. Like the above two techniques, ELISA is also dependent on antigen-antibody interactions, but is more sensitive being able to distinguish between two closely related source species of meat samples and genders within the same species. Besides, it can be used for meat samples pressure cooked at 133°C for 20 min.
- **Isoelectric Focusing (IEF):** This technique profiles the protein content of a meat sample based on the differences in their electrophoretic migratory behaviours in response to a pH gradient. It helps to generate signature profiles for different source species of fresh meat samples as well as those cooked at 100°C. However, IEF does not have the sensitivity to differentiate between related species (Skarpeid *et al.*, 1998). Coomassie brilliant blue dye-stained IEF-PAGE can distinguish between whole muscles of different species. IEC of specific proteins, like, phosphoglucomutase enzyme can identify low levels of buffalo, pig or horse meat in beef. Similarly, IEF of adenylate kinase enzyme can identify presence of kangaroo or horse meat in beef which high sensitivity, whereas, phosphor gluconate dehydrogenase (PGD) can differentiate mutton from chevon (King, 1984).

Based on the basic immunological principle of antigen-antibody interaction, several identification kits are commercially available which are listed in below (Table 24).

Table 24: Overnight Rapid Identification Test

Test	Species Identification	Principle
PROFIT	Poultry	Blank+ Precast Agar+ Overnight PPT
ORBIT	Beef	Blank+ Precast Agar+ Overnight-PPT
Dot Blot Techniques	Beef, Pork, Poultry, sheep, Horse and deer meat	Antigen+ nitrocellulose or cynogen nitrocellulose containing antibodies bromide activated
MULTI-SIFT	Beef, Pork, Poultry, sheep, Horse and deer meat	Blank+ Precast Agar+ Overnight-PPT

Recreated and adapted from (Chappalwar 2020)

Although some of the above-mentioned immunological techniques can be used for heat treated meat samples, yet, since, the outcome of these tests are dependent on the protein-antibody interactions and heat tends to alter protein structure, therefore, these techniques have their limitations while handling meat samples that have undergone extensive thermal processing. For such samples, a more robust biomolecule, DNA, that can withstand heat and chemical treatments better, is used as a target identifier. The techniques developed for using DNA from meat samples to detect adulteration come under the category of molecular methods, and are discussed below.

Molecular Methods

Besides, having a stabler disposition towards rigorous heat and chemical processing's, the DNA content of an organism is its most specific natural identity, that is actually helpful in differentiating between very closely related species with high sensitivity. In this regard, not only nuclear, but the molecular DNA is equally useful in sorting out the source of meat samples (Girish *et al.*, 2005). Mitochondrial DNA genes, like, *cyt b*, *D loop*, *12S rRNA* and *16S rRNA* are very reliable species characterizing markers. Besides an unmatched relaibility, DNA based methods are rapid and cost-effective (Luo *et al.*, 2008).

- **Polymerase Chain Reaction (PCR):** Polymerase chain reaction is probably the most reliable DNA detection method that offers the most undoubted read-out for species adulteration in meats. Through mere amplification of specific genes from DNA samples from adulterated meat, followed by their sequence analysis, the contaminating species can be identified (Yin *et al.*, 2009). With unparallel speed and accuracy, PCR is a technique reliable for closely related species, for detecting the gender of the source of the meat. for fresh or processed meat and with samples with very small amount of adulteration (up to 1%) (V. P. Singh *et al.*, 2014).

- **Real-time PCR:** This is an advanced version of the PCR technique mentioned above with significant improvisation made in the rapidity of the test. With this technique the read-outs are obtained in real time during the running of the tests, making it way faster than the original PCR technique. In addition, very sample starting material is sufficient to give a reliable read out making it a superior option in terms of accuracy (Köppel *et al.*, 2009). Most recently, real-time PCR has been used to assess adulteration in beef, pork, lamb, horse, chicken, turkey and duck.

Non-Destructive Detection Methods

- **Near-Infrared Spectroscopy Technique**

The first non-visible spectral area discovered in the absorption spectrum is known as near-infrared spectroscopy, or NIR. Distinct substances exhibit different spectral curves in the near-infrared region, with wavelengths ranging from 780 to 2,526 nm between visible and mid-infrared light. Therefore, the analysis (identification and quantification) of the chemicals and their components can be done by using the correlation between the original spectral data of the samples across the entire wavelength range and the corresponding physical and chemical index values (function relationship). Thus, it can be inferred that the degree of putrefaction in meat storage and the physical-chemical characteristics and parameters (such as moisture, protein, fat, water retention, gravy loss, etc.) during processing can be determined using the fundamentals of NIR spectroscopy and the NIR detection system. Moisture, protein, and fat levels are strongly correlated with meat deterioration, spoiling, and decreasing freshness. These changes in organic components like fat and protein in fresh beef may be objectively seen using NIR spectroscopy.In chilled pork, a visible and NIR (Vis-NIR) transmittance spectroscopy to identify fat, protein, and water. The assessment model performed well, with correlation coefficients for fat, protein, and water of 0.95, 0.92, 0.71, and 0.46, respectively (Shi *et al.*, 2021).

• Raman Spectra Technique

Raman spectroscopy is a method of spectrum analysis created on the basis of the Raman scattering phenomenon. It is produced by the modification in polarizability brought on by the vibration of the sample molecules and can reveal information about the vibration or rotation of molecules. When analysing the molecular structure, each functional group molecule in the meat emits a distinct Raman spectral signal that is complementary to the infrared spectrum. Therefore, the link between the molecular structure and different radical groups in meat can be qualitatively studied, and meat quality may then be recognised and assessed. The approach of chemometrics can be used to obtain representative information from the Raman spectrum of meat. Raman spectroscopy has been used more frequently recently to assess meat quality. In order to predict the shear force (SF) of fresh lamb, a portable Raman probe was employed. The SF prediction model was shown to have a high degree of accuracy. Tenderness and Raman data were correlated using the partial least squares regression (PLSR) approach. Following SF measurements, it was discovered that rough and tender samples could be distinguished with, respectively, 70%-88% and 59%-80% accuracy (Shi *et al.*, 2021).

• Ultrasonic Imaging Technique

In practical applications, ultrasonic may be separated into two types: power ultrasonic waves and detecting ultrasonic waves. Power ultrasonic produces low-frequency, high-energy ultrasonic waves that are frequently employed in the processing of food, including food sterilisation, thawing, drying, filtering, and homogenization. Food quality is frequently assessed and detected using high-frequency, low-energy ultrasonic generated by detecting ultrasonic waves. The study of changes in acoustic characteristic parameters for predicting meat composition, muscle thickness, fat thickness, etc. forms the basis of ultrasonic approaches for meat quality detection. Without altering the internal characteristics of the meat, quick and non-destructive detection and grading evaluation for meat quality are achieved (Shi *et al.*, 2021).

Although, there is a whole spectrum of adulteration assessment techniques exits, yet the choice is dictated by the trade-off between the sample conditions and the technique that can give the most reliable read-out with the condition of the given meat test sample.

Analysis of Market Economy of Preserved Foods: Global Perspective

Food processing industry influences global economy in a prominent way. The processed food market is constantly evolving at a matching pace to that of lifestyle changes and concomitant alterations in the food consumerism sector. Technological developments and increasing demand of processed foods have fuelled the growth of this industry in both developed and developing countries (Debnath & Khan, 2017; Islam *et al.*, 2017; Regmi & Gehlhar, 2005). There is a likelihood that in near future, food processing industry will grow at a faster pace in the developing countries, like, India and China, rather than the developed countries, like USA. There is noticeable shift in the food consumerism pattern in these two types of countries: inhabitants of developing countries are moving towards more and more processed foods while majority of the developed countries are trying to go back to feeding on fresh produces. Processed fruit and vegetable market is expected to expand at 3.3 % annual rate making the resultant revenue grow at 3.0% *(https://www.ibisworld.com/united-kingdom/market-research-reports/fruit-vegetable-processing-industry/).* Majority of the global fresh produces of fruits and vegetables come from the developing countries *(https://www.ibisworld.com/united-kingdom/market-research-reports/fruit-vegetable-processing-industry/).* According to the Food and Agriculture Organization of the United Nations, nearly half of the world's vegetables and one-third of the world's fruits are produced in China *(https://www.ibisworld.com/united-kingdom/market-research-reports/fruit-vegetable-processing-industry/).*

Table 25: Contribution of different regions in global processed fruit and vegetable production

Region	**Percentage**
Europe	42.3
North Asia	20.1
North America	11.5
South America	9.0
South East Asia	7.2
India and Central Asia	4.7
Africa and Middle East	3.2
Oceania	2.0

Adapted from (Amit *et al.*, 2017)

Conclusion

The knowledge of food preservation was indeed one of the revolutionary changes in human history. With the advent of this knowledge, men could devote more time towards inventive activities instead of producing/hunting/growing food. The historical knowledge of food preservation, was, however, very primitive with rooms for significant improvement. Nevertheless, the traditional techniques of drying, chilling, freezing, and fermentation have performed marvellously over centuries in preserving foods at a domestic scale. With the increasing modernization of farming and cattle rearing, the amount of food being produced through these sources started increasing steeply, prompting the need for more sophisticated industrial level food preservation strategies. This required understanding the mechanism behind food spoilage, and this requirement was central to designing of robust food preservation strategies. Gradually, food preservation became a subject of higher education. This was a significant positive move towards expansion of modern food processing industry. This chapter aims at holistic familiarization of professionals and researchers intending to work on food processing and food safety with the important aspects of food processing.

References

Abusaloua, A., Mohamed, G., Ali, A., & Zahmol, W. (2019). Food Additives and Preservatives As Slow Poisons. Journal of Applied Science, 2(3), 42–48. https://doi.org/10.47891/sabujas.v2i3.42-48

Agrahar-Murugkar, D., & Jha, K. (2010). Effect of drying on nutritional and functional quality and electrophoretic pattern of soyflour from sprouted soybean (Glycine max). Journal of Food Science and Technology, 47(5), 482–487. https://doi.org/10.1007/s13197-010-0082-5

Ahmed, J., & Rahman, M. (2012). Handbook of food process design (1st editio). Wiley-Blackwell.

Aida, A. A., Man, Y. B. C., Wong, C. M. V. L., Raha, A. R., & Son, R. (2005). Analysis of raw meats and fats of pigs using polymerase chain reaction for Halal authentication. Meat Science, 69(1), 47–52. https://doi.org/10.1016/j.meatsci.2004.06.020

Amit, S. K., Uddin, M., Rahman, R., Islam, S. M. R., & Khan, M. S. (2017). A review on mechanisms and commercial aspects of food preservation and processing. Agriculture & Food Security, 6(51), 1–22. https://doi.org/10.1186/s40066-017-0130-8

Arcand, Y., & Boye, J. (2012). Green Technologies in Food Production and Processing. (1st editio). Springer.

Arora, A. K., Mahajan, S., Gupta, P., & Kapoor, S. S. (2010). Is chemical safety to food hazardous? - Dangers of food preservatives. J Indian Acad Forensic Med, 31(4), 419–422.

Arroyo, C., Cebrián, G., Condon, S., Pagán, R., & Auré, G. C. (2012). Development of resistance in Cronobacter sakazakii ATCC 29544 to thermal and nonthermal processes after exposure to stressing environmental conditions. J. Appl. Microbiol., 112(3), 561–570. https://doi.org/10.1111/j.1365-2672.2011.05218.x

Arshad, R.N.; Abdul-Malek, Z.; Munir, A.; Buntat, Z.; Ahmad, M.H.; Jusoh, Y.M.; Bekhit, A.E.-D.; Roobab, U.; Manzoor, M. F. ., & Aadil, R. M. (2020). Electrical systems for pulsed electric field applications in the food industry : An engineering perspective. Trends in Food Science & Technology, 104, 1–13. https://doi.org/10.1016/j.tifs.2020.07.008

Arvanitoyannis, I. (2010). Irradiation of food commodities: techniques, applications, detection, legislation, safety and consumer opinion. (1st editio). Elsevier. https://doi.org/10.1016/C2009-0-01843-1

Aurelio, J. M. (2008). "Double-dead' meat seized in QC". Philippine Daily Inquirer.

Bahrami, A., Baboli, Z. M., Schimmel, K., Jafari, S. M., & Williams, L. (2020). Efficiency of novel processing technologies for the control of Listeria monocytogenes in food products. Trends Food Sci. Technol., 96, 61–78. https://doi.org/10.1016/j.tifs.2019.12.009

Baker, C. G. J. (1997). Industrial Drying of Foods (1st edition). Springer.

Baker, C. G. J., Ranken, M. D., & Kill, R. C. (1997). Food Industries Manual (24th edition). Springer.

Balasubramanian, S., & Viswanathan, R. (2010). Influence of moisture content on physical properties of minor millets. Journal of Food Science and Technology, 47(3), 279–284. https://doi.org/10.1007/s13197-010-0043-z

Barbosa-Cánovas, G. V., Altunaker, B., & Mejía-Lorío, D. J. (2005). Freezing of fruits and vegetables.

Barnwal, P., Singh, K. K., Mridula, D., Kumar, R., & Rehal, J. (2010). Effect of moisture content and residence time on dehulling of flaxseed. Journal of Food Science and Technology, 47(6), 662–667. https://doi.org/10.1007/s13197-010-0113-2

Battcock, M., & Azam-Ali, S. (1998). Fermented frutis and vegetables. A global perspective. FAO Agricultural Services Bulletin No. 134,Food and Agriculture Organization of the United Nations Rome. http://www.fao.org/3/x0560e/x0560e00.htm

Bengtsson, G. (2003). Keeping quality of fresh foods. Workshop on Food Safety and Quality.

Berk, Z. (2013). Food process engineering and technology. In Food Science and Technology (2nd editio). Academic Press. https://pubs.acs.org/doi/10.1021/acsami.9b03822

Bhat, R., & Alias, A. K. (2012). Progress in food preservation. Wiley.

Bhattacharjee, C., Saxena, V., & Dutta, S. (2019). Novel thermal and non-thermal processing of watermelon juice. Trends Food Sci. Technol., 93, 234–293. https://doi.org/10.1016/j.tifs.2019.09.015

Blum, D. (2012). Food that lasts forever. TIME Magazine.

Brennan, J. G. (2006). Food processing handbook. WILEY-VCH Verlag GmbH & Co. KGaA. https://doi.org/10.1002/3527607579.ch14

Brown, A. C. (2007). Understanding Food: Principles and Preparation. Wadsworth Publishing. https://books.google.com/books?id=Mt7KAgAAQBAJ&pgis=1

Carroll, C. D., & Alvarado, C. Z. (2008). Comparison of air and immersion chilling on meat quality and shelf life of marinated broiler breast fillets. Poultry Science, 87(2), 368–372. https://doi.org/10.3382/ps.2007-00213

Cassens, R. G. (1994). Meat Preservation, Preventing Losses and Assuring Safety (1st editio, Vol. 0). Food and Nutrition Press.

Cavazos-Garduño, A., Serrano-Niño, J. C., Solís-Pacheco, J. R., Gutierrez-Padilla, J. A., González-Reynoso, O., García, H. S., & Aguilar-Uscanga, B. R. (2016). Effect of Pasteurization, Freeze-drying and Spray Drying on the Fat Globule and Lipid Profile of Human Milk. Journal of Food and Nutrition Research, 4(5), 296–302. https://doi.org/10.12691/jfnr-4-5-5

Chappalwar, A. M., Pathak, V., Goswami, M., Sharma, B., Singh, P., & Mishra, R. (2020). Recent novel techniques applied for detection of meat adulteration and fraudulent practices. Indian J. Vet. Public Hlth., 7(1), 01–06.

Clark, S., Costello, M., Drake, M., & Bodyfelt, F. (2008). The sensory evaluation of dairy products. In The Sensory Evaluation of Dairy Products (2nd editio). Springer. https://doi.org/10.1007/978-0-387-77408-4

Dagoon, J. D. (1993). Applied nutrition and food technology, revised edition. Rex Printing Company Inc.

Dar, H. Y., Chaturvedi, S., Srivastava, K., Azam, Z., Anupam, R., Mondal, R. K., Tomar, G. B., Singh, A. K., Mishra, P. K., & Srivastva, R. K. (2017). Immunomodulatory Effects of Food Additives. International Journal of Immunotherapy and Cancer Research, 3(2), 019–031. https://doi.org/10.17352/2455-8591.000015

Debnath, M., & Khan, M. S. (2017). Health concerns of pesticides, In pesticide residue in foods: sources, management, and control (pp. 103–118). Springer International Publishing.

Desrosier, N. W., & Singh, R. P. (2018). Food preservation. Encyclopaedia Britannica, Inc., Chicago. https://www.britannica.com/topic/food-preservation

Dunne, C. P. (2007). High pressure processing of foods (1st editio). Blackwell.

Enfors S-O. (2008). Food microbiology. KTH-Biotechnology.

Fabunmi, O. A., Osunde, Z. D., Alabadan, B. A., & Jigam, A. A. (2015). Influence of moisture content and temperature interaction on mechanical properties of DESMA (Novella pentadesma) SEED. J Adv Food Sci Technol, 2(2), 81–85.

Fellows, P. J. (2009). Food processing technology:Principles and Practice (3rd editio). Woodhead Publishing. https://doi.org/10.1533/9781845696344

French, S. (2004). What does "sustainability" mean to the food industry? Food Technol., 58(9), 160.

Friis, R. H. (2012). Essentials of environmental health (2nd editio). Jones & Bartlett.

Garg, N., Garg, K. L., & Mukerji, K. G. (2010). Laboratory Manual of Food Microbiology. International Publishing House Pvt. Ltd. https://books. google.com.my/books?id= 8h4Ze5s6sFUC&pg = PA63&dq=The+purpose+of+ making+a+smear +is+to+fix +the+bacteria+onto+the+slide+ and+to+ prevent+the+sample+from+being+ lost+ during+a+ staining+ procedure&hl=en&sa=X&ved = 0ahUKEwjurfD Jgc7UAh VGv Y8KHXImCs8Q6AEIJzAA

George, M. (2008). Food Biodeterioration and Preservation (Gary S. Tucker (Ed.)).Blackwell. https://doi.org/10.1002/9780470697849

Girish, P. S., Anjaneyulu, A. S. R., Viswas, K. N., Shivakumar, B. M., Anand, M., Patel, M., & Sharma, B. (2005). Meat species identification by polymerase chain reaction-restriction fragment length polymorphism (PCR-RFLP) of mitochondrial 12S rRNA gene. Meat Science, 70(1), 107–112. https://doi.org/10.1016/j.meatsci.2004.12.004

Gould, G. W. (1989). Introduction. In: Mechanisms of Action of Food Preservation Procedures. In Elsevier Applied Science.

Haldar, S., Beveridge, J., Wong, J., Singh, A., Galimberti, D., Borroni, B., Zhu, X., Blevins, J., Greenlee, J., Perry, G., Mukhopadhyay, C. K., Schmotzer, C., & Singh, N. (2013). A low-molecular-weight ferroxidase is increased in the CSF of sCJD cases : CSF ferroxidase and transferrin as diagnostic biomarkers for sCJD. Antioxid Redox Signal, 19(14), 1662–1675. https://doi.org/10.1089/ars.2012.5032

Heinz G, H. P. (2007). Meat Processing Technology. For Small-to Medium Scale Producers. Food and Agriculture Organization of the United Nations Regional Office for Asia and the Pacific.

Heldman, D. R., & Moraru, C. I. (2010). Food encyclopedia of agricultural, food, and biological engineering. CRC Press. https://doi.org/10.1201/9780429257599

Hernández-Hernández, H. Moreno-Vilet, L., & Villanueva-Rodríguez, S. (2019). Current status of emerging food processing technologies in Latin America: Novel non-thermal processing. Innov. Food Sci. Emerg. Technol., 58, 102233.

Hoff, J. E., & Castro, M. D. (1969). Chemical Composition of Potato Cell Wall. Journal of Agricultural and Food Chemistry, 17(6), 1328–1331. https://doi.org/10.1021/jf60166a058

Huang, H. W., Wu, S. J., Lu, J. K., Shyu, Y. T., & Wang, C. Y. (2017). Current status and future trends of high-pressure processing in food industry. Food Control, 72(12), 1–8.

Igarashi, Y., Eroshkin, A., Gramatikova, S., Gramatikoff, K., Zhang, Y., Smith, J. W., Osterman, A. L., & Godzik, A. (2007). CutDB: A proteolytic event database. Nucleic Acids Research, 35(SUPPL. 1). https://doi.org/10.1093/nar/gkl813

In't Veld, J. H. . (1996). Microbial and biochemical spoilage of foods: An overview. International Journal of Food Microbiology, 33(1), 1–18. https://doi.org/10.1016/0168-1605(96)01139-7

Indiarto, R., Abdillah, M., & Qonit, H. (2020). A Review Of Irradiation Technologies On Food And Agricultural Products. Int. J. Sci. Technol. Res., 9(01), 4411–4414.

Indiarto, R., Pratama, A. W., Sari, T. I., & Theodora, H. C. (2020). Food Irradiation Technology : A Review of The Uses and Their Capabilities. Int. J. Eng. Trends Technol, 68(12), 91–98. https://doi.org/10.14445/22315381/IJETT-V68I12P216

Indira, V., & Sudheer, K. P. (2007). Post Harvest technology of horticultural crops (Horticultu). New India Publishing Agency.

Inetianbor, J., Yakubu, M., & Ezeonu, C. (2015). Effects of Food additives and Preservatives on Man. Asian J. of Sci. and Tech., 6(2), 11118–11135. https://doi.org/10.47891/sabujas.v2i3.42-48

Islam, M. N., Bint-E-Naser, S. F., & Khan, M. S. (2017). Pesticide food laws and regulations (M. S. Khan & M. S. Rahman (Eds.); Pesticide). Springer International Publishing. https://doi.org/10.1007/978-3-319-52683-6

James, S. (2008). Food Biodeterioration and Preservation. Blackwell.

Jangam, S. V., CL, Law, C. L., & Mjumder, A. S. (2010). Drying of foods, vegetables and fruits (1st ed.).

Jay, J. M., Loessner, M. J., & Golden, D. A. (2000). Modern Food Microbiology (6th editio). Gaithersburg: Aspen Publishers.

Jaypee. (2008). QC officials step up drive vs "hot meat." Local Government of Quezon City.

Jennifer, N. A. (2008). "Govt to monitor markets, slaughterhouses to prevent sale of double-dead meat." Buisness Mirror.

Kader, A. A., Zagory, D., & Kerbel, E. L. (1989). Modified atmosphere packaging of fruits and vegetables. Critical Reviews in Food Science and Nutrition, 28(1), 1–30. https://doi.org/10.1080/10408398909527490

Kanatt, S. R., Chander, R., & Sharma, A. (2006). Effect of radiation processing of lamb meat on its lipids. Food Chemistry, 97(1), 80–86. https://doi.org/10.1016/j.foodchem.2005.03.024

Kannall, E. (2017). The Effects of Food Preservatives on the Human Body. Chron: The Hearst Newspaper, LLC;, 1–5.

Karel, M., & Lund, D. (2003). Physical Principles of Food Preservation. In Physical Principles of Food Preservation. https://doi.org/10.1201/9780203911792

Karmas, R., Buera, M. P., & Karel, M. (1992). Effect of glass transition on rates of nonenzymic browning in food systems. 40(5), 873–879. https://doi.org/10.1021/jf00017a035

Katz, F. (2001). Active cultures add function to yoghurt and other foods. Food Technol., 55, 46–49.

Kent, L. T. (2015). Food additive side effects. In Live Strong.

Khan, M. K., Ahmad, K., Hassan, S., Imran, M., Ahmad, N., & Xu, C. (2018). Effect of novel technologies on polyphenols during food processing. Innov. Food Sci. Emerg. Technol., 45, 361–381. https://doi.org/10.1016/j.ifset.2017.12.006

King, N. L. (1984). Species identification of cooked meats by enzyme-staining of isoelectricfocusing gels. Meat Science, 11(1), 59–72. https://doi.org/10.1016/0309-1740(84)90017-2

Kirk-Othmer. (2007). Food and Feed Technology (1st editio, Vol. 0). WileyInterscience.

Knechtges, P. L. (2012). Food Safety: Theory and Practice (1st editio). Jones and Bartlett. https://books.google.com/books?id=FwU1pejLEJsC&pgis=1

Köppel, R., Zimmerli, F., & Breitenmoser, A. (2009). Heptaplex real-time PCR for the identification and quantification of DNA from beef, pork, chicken, turkey, horse meat, sheep (mutton) and goat. European Food Research and Technology, 230(1), 125–133. https://doi.org/10.1007/s00217-009-1154-5

Koutchma, T., Popović, V., Ros-Polski, V., & Popielarz, A. (2016). Effects of Ultraviolet Light and High-Pressure Processing on Quality and Health-Related Constituents of Fresh Juice Products. Comprehensive Reviews in Food Science and Food Safety, 15(5), 844–867. https://doi.org/10.1111/1541-4337.12214

Kumar Kar, B. (2014). Multi-Stakeholder Partnership in Nutrition: An Experience from Bangladesh. Indian J Comm Health, 26(1), 15–21. http://citeseerx.ist.psu.edu/viewdoc/download?doi=10.1.1.671.8026&rep=rep1&type=pdf

Kumar, Y., Patel, K. K., & Kumar, V. (2015). Pulsed Electric Field Processing in Food Technology. Int. J. Eng. Stud. Tech. Approach, 01, 6–7.

Kutz, M. (2008). Handbook of Farm, Dairy and Food Machinery (1st editio). William Andrew.

La Peña, M. M.-D., Welti-Chanes, J., & Martín-Belloso, O. (2019). Novel technologies to improve food safety and quality. Curr. Opin. Food Sci., 30, 1–7. https://doi.org/10.1016/j.cofs.2018.10.009

Laudan, D. (2009). Food and nutrition: lifespan, human to pesticides. Marshall Cavendish.

Lazaro, R. E. R. (2007). Bulacan holds lechon festival to show safety of pork meat.

Leniger, H. A., & Beverloo, W. A. (1975). Food Process Engineering. Springer.

Levine, H. (1988). Principles of "cryostabilization" technology from structure/property relationships of carbohydrate/water systems—a review. Cryo Letters, 9(1), 21–63.

Levine, Harry, & Slade, L. (1986). A polymer physico-chemical approach to the study of commercial starch hydrolysis products (SHPs). Carbohydrate Polymers, 6(3), 213–244. https://doi.org/10.1016/0144-8617(86)90021-4

Lewin, A. (2012). Food Fermentation : Preserving Whole Fresh Food with Live Cultures in Your Home Kitchen (4th editio). Quarry Books.

Liepa, M., Zagorska, J., & Galoburda, R. (2016). High-Pressure processing as novel technology in dairy industry: A Review. Res. Rural Dev., 1, 76–83.

Light, N., & Walker, A. (1990). Cook-chill catering: technology and management. Elsevier Science Publishing co. Inc.

Lopez, A. (2007). "400 kilos of 'double-dead' meat seized from Divisoria." Inquirer.Net. Archived from the Original on 2008-01-26.

Lund, B. D., & Karel, M. (2005). Physical Principles of Food Preservation (2nd editio). Taylor & Francis; 2005.

LUO, J. qin, WANG, J. qi, BU, D. pan, LI, D., WANG, L., WEI, H. yang, & ZHOU, L. yun. (2008). Development and Application of a PCR Approach for Detection of Bovis, Sheep, Pig, and Chicken Derived Materials in Feedstuff. Agricultural Sciences in China, 7(10), 1260–1266. https://doi.org/10.1016/S1671-2927(08)60173-X

Magnussen, O. M., Haugland, A., Torstveit Hemmingsen, A. K., Johansen, S., & Nordtvedt, T. S. (2008). Advances in superchilling of food - Process characteristics and product quality. Trends in Food Science and Technology, 19(8), 418–424. https://doi.org/10.1016/j.tifs.2008.04.005

Marcola, J. (2010). Top 10 food additives to avoid. Food Matters. Food Matters International Pty Ltd;, 8–10.

Mathavi, V., Sujatha, G., Ramya, S. B., & Dev, B. K. (2013). New trends in food processing. Int J Adv Eng Technol., 5(2), 176–187.

Meyer, A. S., Suhr, K. I., Nielsen, P., & Holm, F. (2002). Natural food preservatives. In: Ohlsson T, Bengtsson N (Eds.) Minimal processing technologies in the food industry, chap 6. (pp. 124–174). Woodhead Publishing. https://doi.org/10.1016/s0023-6438(03)00019-7

Miller, G. D., Jarvis, J. K., & Mcbean, L. D. (2007). Handbook of Dairy Foods and Nutrition Book (3rd editio). CRC Press.

Mohammed, M. E. A., Eissa, A. H. A., & Aleid, S. M. (2016). Application of Pulsed Electric Field for Microorganisms Inactivation in Date palm Fruits. Journal of Food and Nutrition Research, 4(10), 646–652. https://doi.org/10.12691/jfnr-4-10-3

Moniruzzaman, M., Alam, K., Biswas, S. K., Pramanik, K., Islam, M. M., & Uddin, G. M. S. (2016). Irradiation to Ensure Safety and Quality of Fruit Salads Consumed in Bangladesh. Journal of Food and Nutrition Research, 4(1), 40–45. https://doi.org/10.12691/jfnr-4-1-7

Mowafy, A. R., Darwish, A. M., El-Kholy, S. A., & Abdel-Mohsen, S. H. (2001). Effect of food preservatives on mother rats and survival of their offspring. The Journal of the Egyptian Public Health Association, 76(3–4), 281–295.

Murphy, K. J., Parker, B., Dyer, K. A., Davis, C. R., Coates, A. M., Buckley, J. D., & Howe, P. R. C. (2014). A comparison of regular consumption of fresh lean pork, beef and chicken on body composition: A randomized cross-over trial. Nutrients, 6(2), 682–696. https://doi.org/10.3390/nu6020682

Nielsen, H. B., Sonne, A. M., Grunert, K. G., Banati, D., Pollák-Tóth, A., Lakner, Z., Olsen, N. V., Žontar, T. P., & Peterman, M. (2009). Consumer perception of the use of high-pressure processing and pulsed electric field technologies in food production. Appetite, 15(1), 115–126.

Nogardy, B. (2013). The hard facts of food additives. ABC Health and Wellbeing.

Nummer, B. A. (2002). Historical Origins of Food PreservatioHistorical Origins of Food Preservationn. National Center for Home Food Preservation, May, 9–11. http://nchfp.uga.edu/publications/nchfp/factsheets/food_pres_hist.html

Nychas GJE, Skandamis PN, Tassou CC, K. K. (2008). Meat spoilage during distribution. Meat Science, 78, 77–89. https://doi.org/10.1016/j.meatsci.2007.06.020

Ohlsson, T., & Bengtsson, N. (2002). Minimal processing technologies in food industry. Woodhead Publication.

Oishi, S. (2002). Effects of propyl paraben on the male reproductive system. Food Chem Toxicol., 40(12), 1807–1813. https://doi.org/10.1016/s0278-6915(02)00204-1

Ong, S. B., Zuraini, M. I., Jurin, W. G., Cheah, Y. K., Tunung, R., Chai, L. C., Haryani, Y., Ghazali, F. M., & Son, R. (2007). Meat molecular detection: Sensitivity of polymerase chain reaction-restriction fragment length polymorphism in species differentiation of meat from animal origin. International Food Research Journal, 14(1), 51–59.

Pal, M. (2014). Preservation of various foods.

Pal, M. (2017). Pulsed Electric Field Processing : An Emerging Technology for Food Preservation. Journal of Experimental Food, 3, 2–3. https://doi.org/10.4172/24

Pal, Mahendra, & Devrani, M. (2018). Application of Various Techniques for Meat Preservation. Journal of Experimental Food Chemistry, 04(01), 1 6. https://doi.org/10.4172/2472-0542.1000134

Panda, H. (2003). Herbal Foods and Its Medicinal Values (N. Institute & of I. Research (Eds.)).

Panday, R. M., & Upadhyay, S. K. (2012). Food Additive. In El-Samragy Y (Ed.) Food Additive (Vol. 3, Issue 2, pp. 1–30). InTech. https://doi.org/10.5772/34455.In

Perez-Chabela, M. L., & Mateo-Oyague, J. (2004). Frozen meat: Quality and shelf life. In I. Hui YH, P Cornillon & K. N. W. Legaretta, MH Lim, KD Murrell (Eds.), Handbook of Frozen Foods. Marcel Dekker Inc. https://doi.org/10.1201/9780203022009

Pinto, L., Baruzzi, F., & Ippolitto, A. (2016). Recent advances to control spoilage microorganisms in washing water of fruits and vegetables: The use of electrolyzed water. In S. Ippolito, A., Sanzani, S.M., Wisniewski, M., Droby (Ed.), III International Symposium on Postharvest Pathology: Using Science to Increase Food Availability. International Society for Horticultural Science (ISHS): Leuven, Belgium. https://doi.org/10.17660/ActaHortic.2016.1144.72

Potter, N. N., & Hotchkiss, J. H. (1999). Food Science (5th edition). Springer.

Praphailong, W., & Fleet, G. H. (1997). The effect of pH, sodium chloride, sucrose, sorbate and benzoate on the growth of food spoilage yeasts. Food Microbiology, 14(5), 459–468. https://doi.org/10.1006/fmic.1997.0106

Priyadarshini, A., Rajauria, G., O'Donnell, C. P., & Tiwari, B. K. (2019). Emerging food processing technologies and factors impacting their industrial adoption. Crit. Rev. Food Sci. Nutr., 59(19), 3082–3101. https://doi.org/10.1080/10408398.2018.1483890

Pruthi, J. S. (1999). Quick Freezing Preservation of Foods: Foods of Plant Origin. (1st edition, Vol. 02). Allied Publishers Limited.

QC to raise penalties for double- dead meat. (2007). The Manila Times.

Rahman, M. (2007). Handbook of food preservation. In Food science and technology. (2nd ed.). CRC Press.

Rahman, M. S. (1995). Food properties handbook. (M. S. Rahman (Ed.); 2nd edition). CRC Press.

Rahman, M. S. (1999). Purpose of food preservation and processing. In: Handbook of Food Preservation. (Marcel Dek). CRC Press; 2007.

Rahman, M. S., & Velez-Ruiz, J. F. (1999). Food Preservation by Freezing. CRC Press. https://doi.org/10.1201/9780429091483-44

Rahman, R. (2014). Food Preservation. In National Encyclopedia of Bangladesh.

Ramaswamy, S., & Tung, M. A. (1984). A Review on Predicting Freezing Times of Foods. Journal of Food Process Engineering, 7(3), 169–203. https://doi.org/10.1111/j.1745-4530.1984.tb00302.x

Rayaguru, K., & Routray, W. (2010). Effect of drying conditions on drying kinetics and quality of aromatic Pandanus amaryllifolius leaves. Journal of Food Science and Technology, 47(6), 668–673. https://doi.org/10.1007/s13197-010-0114-1

Regmi, A., & Gehlhar, M. (2005). New directions in global food markets. Agriculture Information Bulletin Number 794.

Reid, D. S. (1990). Optimizing the quality of frozen foods. Food Technol, 44(7), 78–82.

Rodriguez-Gonzalez, O., Buckow, R., Koutchma, T., & Balasubramaniam, V. M. (2015). Energy Requirements for Alternative Food Processing Technologies-Principles, Assumptions, and Evaluation of Efficiency. Comprehensive Reviews in Food Science and Food Safety, 14(5), 536–554. https://doi.org/10.1111/1541-4337.12142

Rodriguez, F., & Mesler, R. (1984). Some drops don't splash. J Colloid Interface Sci., 106(2), 347–352.

Rogers, L. D., & Overall, C. M. C. (2013). Proteolytic post-translational modification of proteins: Proteomic tools and methodology. Molecular and Cellular Proteomics, 12(12), 3532–3542. https://doi.org/10.1074/mcp.M113.031310

Rosmini, M. R., Perez-Alvarez, J. A., & Fernandez, L. J. (2004). Operational Processes for Frozen Red Meat. In P. Hui YH & K. M. and W. K. N. Cornillon, IG Legaretta, MH Lim (Eds.), Handbook of Frozen Foods. Marcel Dekker Inc. https://doi.org/10.1201/9780203022009.ch10

Sagar, V. R., & Suresh Kumar, P. (2010). Recent advances in drying and dehydration of fruits and vegetables: A review. Journal of Food Science and Technology, 47(1), 15–26. https://doi.org/10.1007/s13197-010-0010-8

Salvato, J. A., Nemerow, N. L., & Agardy, F. J. (2003). Environmental Engineering, (5th editio). Wiley. https://doi.org/10.2134/jeq2004.1162

Saravacos, G., & Kostaropoulous, A. E. (2002). Handbook of food processing equipment: food engineering series. Kluwer Academic/Plenum Publishers.

Sebranek, J. G. (1996). Poultry and poultry products. In Freezing Effects on Food Quality. Marcel Dekker.

Shahbaz, H. M., Kim, J. U., Hyoung, S., & Park, J. (2018). Advances in Nonthermal Processing Technologies for Enhanced Microbiological Safety and Quality of Fresh Fruit and Juice Products. In A. M. Grumezescu, A.M., Holban (Ed.), In Food Processing for Increased Quality and Consumption, Handbook of Food Bioengineering; (pp. 179–217). Academic Press.

Sharma, S. (2015). Food Preservatives and their Harmfull Effect. International Journal of Scientific and Research Publications, 5(4), 5–6. www.ijsrp.org

Shenga, E., Singh, R. P., & Yadav, A. (2010). Effect of pasteurization of shell egg on its quality characteristics under ambient storage. J Food Sci Technol, 47(4), 420–425. https://doi.org/10.1007/s13197-010-0069-2

Shi, Y., Wang, X., Borhan, M. S., Young, J., Newman, D., Berg, E., & Sun, X. (2021). A review on meat quality evaluation methods based on non-destructive computer vision and artificial intelligence technologies. Food Science of Animal Resources, 41(4), 563–588. https://doi.org/10.5851/KOSFA.2021.E25

Shivasankar, B. (2002). Food Processing and Preservation. Prentice Hall of India Pvt Limited. https://doi.org/10.1201/9781482270341-14

Silva, M., & Lidon, F. (2016). Food Preservatives- overview on applications and side effects. Emirates Journal of Food and Agriculture, 28(6), 366–377.

Singh, R. P. (1994). Scientific principles of shelf life evaluation. In Shelf Life Evaluation of Foods. Springer. https://doi.org/10.1007/978-1-4615-2095-5_1

Singh, V. P., Pathak, V., Nayak, N. K., & Verma, K. (2014). Recent Developments in Meat Species Speciation-a review. Livestock Sci., 5, 49–64.

Skarpeid, H. J., Kvaal, K., & Hildrum, K. I. (1998). Identification of animal species in ground meat mixtures by multivariate analysis of isoelectric focusing protein profiles. Electrophoresis, 19(18), 3103–3109. https://doi.org/10.1002/elps.1150191811

Smith, J. (2004). Technology of Reduced Additive Foods. In International Journal of Food Science and Technology (2nd editio). Wiley-Blackwell. https://doi.org/10.1111/j.1365-2621.2004.00938.x

Smith, J. S., & Pillai, S. (2004). Irradiation and food safety. Food Technology, 58(11), 48–55.

Sommers, B. A. (2010). Irradiation: food. encyclopedia of agricultural, food, and biological engineering. 122–135. https://doi.org/10.1016/b978-0-12-815781-7.00015-9

Soomro, A. H., Masud, T., & Kiran, A. (2002). Role of Lactic Acid Bacteria (LAB) in Food Preservation and Human Health – A Review. Pak J Nutr, 1(1), 20–24. https://doi.org/10.3923/pjn.2002.20.24

Sperber, W. H., & Doyle, M. P. (2009). Compendium of the Microbiological Spoilage of Foods and Beverages. In M. P. Sperber, William H., Doyle (Ed.), Food microbiology and food safety. New York: Springer. https://doi.org/10.1007/978-1-4419-0826-1

Spink, J., & Moyer, D. C. (2011). Defining the Public Health Threat of Food Fraud. Journal of Food Science, 76(9). https://doi.org/10.1111/j.1750-3841.2011.02417.x

Steele, R. (2004). Understanding and Measuring the Shelf-Life of Food. In 1st Edition (Ed.), Understanding and Measuring the Shelf-Life of Food. Woodhead Publishing Limited and CRC Press LLC. https://doi.org/10.1201/9781439823354

Sun, D.-W. (2014). Emerging Technologies for Food Processing (2nd editio). Academic Press. https://doi.org/10.1016/B978-0-12-676757-5.X5000-2

Syamaladevi, R. M., Tang, J., Villa-Rojas, R., Sablani, S., Carter, B., & Campbell, G. (2016). Influence of Water Activity on Thermal Resistance of Microorganisms in Low-Moisture Foods: A Review. Comprehensive Reviews in Food Science and Food Safety, 15(2), 353–370. https://doi.org/10.1111/1541-4337.12190

Syaza, S., Umar, R., Hazmin, S., Kamarudin, M., Hassan, A., & Juahir, H. (2018). Non-ionizing radiation as threat in daily life. J. Fundam. Appl. Sci., 9, 308. https://doi.org/10.4314/jfas.v9i2s.21

Tamime, A. Y. (2009). Dairy fats and related products (1st edition). WileyBlackwell.

Tianli, Y., Jiangbo, Z., & Yahong, Y. (2014). Spoilage by Alicyclobacillus Bacteria in Juice and Beverage Products: Chemical, Physical, and Combined Control Methods. Compr Rev Food Sci Food Saf., 13(5), 771–797. https://doi.org/10.1111/1541-4337.12093

Tucker, G.S. (2007). Food Biodeterioration and Preservation (1st edition). Wiley-Blackwell.

Van Boekel, M. A. J. S. (2008). Kinetic Modeling of Food Quality: A Critical Review. Comprehensive Reviews in Food Science and Food Safety, 7(1), 144–158. https://doi.org/10.1111/j.1541-4337.2007.00036.x

Verma, D. K., Thakur, M., Kumar, J., Srivastav, P. P., Al-Hilphy AR, S., Patel, A., & Suleria HA, R. (2020). High Pressure Processing (HPP): Fundamental Concepts, Emerging Scope, and Food Application. In Emerging Thermal and Nonthermal Technologies in Food Processing (1st editio, pp. 225–257). Apple Academic Press: Palm Bay.

Walter, R., & Taylor, S. (1991). The chemistry and technology of pectin. In J Food Science and Technology (1st editio, Vol. 5, Issue 4). Academic Press. https://doi.org/10.1016/0889-1575(92)90067-t

White, G. W., & Cakebread, H. (1966). The glassy state in certain sugar-containing food products. International Journal of Food Science & Technology, 1966(1), 75–82. https://doi.org/https://doi.org/10.1111/j.1365-2621.1966.tb01031.x

Wilkinson, J., & Roch, R. (2008). Agri-Processing and Developing Countries.

World Food Market Overview Marketing Essay. (2013). http://www.ukessays.com/essays/marketing/world-food-market-overview-marketingessay.php?cref=1

Xuan, X. T., Ding, T., Li, J., Ahn, J. H., Zhao, Y., Chen, S. G., Ye, X. Q., & Liu, D. H. (2017). Estimation of growth parameters of Listeria monocytogenes after sublethal heat and slightly acidic electrolyzed water (SAEW) treatment. Food Control, 71, 17–25. https://doi.org/10.1016/j.foodcont.2016.06.018

Yeung, C. K., & Huang, S. C. (2016). Effects of High-pressure Processing Technique on the Quality and Shelf Life of Chinese Style Sausages. Journal of Food and Nutrition Research, 4(7), 442–447. https://doi.org/10.12691/jfnr-4-7-5

Yin, R. H., Bai, W. L., Wang, J. M., Wu, C. D., Dou, Q. L., Yin, R. L., He, J. B., & Luo, G. B. (2009). Development of an assay for rapid identification of meat from yak and cattle using polymerase chain reaction technique. Meat Science, 83(1), 38–44. https://doi.org/10.1016/j.meatsci.2009.03.008

Zhou, G. H., Xu, X. L., & Liu, Y. (2010). Preservation technologies for fresh meat - A review. Meat Science, 86(1), 119–128. https://doi.org/10.1016/j.meatsci.2010.04.033

4

Food Additives as Food Adulterant A Perspective or Food Additives – Importance, Scope, and Trends in Adulteration of Major Food Groups

Neha Thakur[1], Vandana Chaudhary[2], Ruby Siwach[3]

[1]Department of Livestock Product Technology, Lala Lajpat Rai University of Veterinary and Animal Sciences, Hisar, Haryana, Inida
[2,3]College of Dairy Science and Technology, Lala Lajpat Rai University of Veterinary and Animal Sciences, Hisar, Haryana, Inida

Introduction

Since ancient times, addition of chemicals to food and food products has been done in order to to accomplish specific functions. Spices, smoke, vinegar, and alcohol were utilised to preserve foods more than 10,000 years ago. Usage of additives in food increased dramatically in the early 1900s, in tandem with advances in food chemistry and preservation. New, tasty, convenient, and healthy meals continue to be in high demand. As a result, by the early 1960s, over 2500 different types of additives were identified and utilized by developing countries in their diets. Basic foods are additive-free, but as we increase the level of processing for conversion into an array of products, variety of additives are incorporated to increase the shelf life; to retain the nutritional value as well as to achieve higher standard of sensorial characteristics.

Definition of food additives as per the Food Safety and Standards Act of India 2006 is :

"food additive" means any substance not normally consumed as a food by itself or used as a typical ingredient of the food, whether or not it has nutritive value, the international addition of which to food for technological (including organoleptic) purpose in the manufacture, processing, preparation, treatment, packing, packaging, transport or holding such food results, or may be reasonably expected to result (directly or indirectly) in it or its by-products becoming a component of or otherwise affecting the characteristics of such food but does not include "contaminants" or substances added to food for maintaining or improving nutritional qualities.

Although many people are in favour of using these chemicals, , it is not without debate. We'll examine some of the most significant benefits and drawbacks of employing additives, classification of additives and their roles in food, in this chapter. Additionally, the fraudulent practices in food trade and adulteration pertaining to food additives will discussed with a brief information about major food recall incidents in the recent years.

Classes of Food Additives

Based on their frequency/occurence, origin, and use in various categories of food products, food additives are divided into many groups..

1. On basis of occurrence

1.1. Intentional food additives: These are the ones which are deliberately added to a given food product to serve a particular beneficial function e.g. preservatives, flavor enhancers, stabilizers, emulsifiers, etc.

1.2. Unintentional Food Additives (contaminants): These are substances that unintentionally and accidentally enter food products; instead of serving a useful purpose, they may represent a health risk or hazard or even cause the food to decay. Examples include toxins, pesticides, heavy metals, and others.On basis of source the food additives are classified as:

2.2.1 Natural food additives: These are the ones which are derived from organic resources such as animals, plants, micro-organisms etc. such as Lycopene from tomatoes and the dye betain from beets can both be employed as colouring agents in culinary goods.2.2.2 Synthetic food additives: These are the ones that are artificial food additives and are chemically synthesized. For example, azodicarbonamide is not naturally present in foods and it is artificially synthesized to be used as a flour improver.

2.2.3 Nature Identical food additives: These are the ones produced artificially, although chemically speaking, they are equivalent (similar) to those derived from natural sources. For example, benzoic acid is naturally found in foods but is also made synthetically and a common preservative.

2.3 On basis of functions

Numbering System (INS) for Food Additives (CAC/GL 36-1989), which offers an accepted global numbering system for identifying food additives, are the foundations for the functional classes for food additives. The goal of the food additive functional classes is to provide customers with information through product labelling. Under Codex, there are 27 functional classes in all, of food additives as listed below in Table 1:-

Table 1: Classes and role of food additives

Sr. No.	Functional Classes	Definition and role	Additive name
1.	Acidity regulator	Used to control and adjust the acidity or alkalinity and buffering activity of a food.	Acids and sodium salts of sorbic acid, acetic acid, benzoic acid, and propionic acid.
2.	Anticaking agent	Used to reduce the tendency of components/particles of food to adhere or stick to one another thereby serving as antisticking, drying and dusting agent to ease packaging, transport, flowability and consumption.	Calcium carbonate, Magnesium carbonate, Sodium aluminosilicate, Sodium ferrocyanide, Potassium ferrocyanide, Calcium ferrocyanide, Sodium silicate, Magnesium trisilicate, Magnesium stearate, Sodium silicate, Calcium silicate
3.	Antifoaming agent	Used to stop or lessen foaming while preparing or serving food with the goal of reducing effusion or effervescence.	Polydimethylsiloxanes and other silicones, insoluble oils, certain alcohols, stearates and glycols.
4.	Antioxidant	Are added to restrict oxidative deterioration in foods.	Acorbic acid, Tocopherol, Butylated hydroxyl anisole, Butylated hydroxytoluene, Propyl gallate, sodium citrate, calcium disodium EDTA, phosphates, TBHQ (tertiary butyl hydroxyquinone)
5.	Bleaching agent	Used to decolorize food (non-flour use). Bleaching agents do not include pigments.	Chlorine gas, Chlorine dioxide, Nitrogen dioxide, Calcium and benzoyl peroxides, Azodicarbonamide, L-cysteine, potassium bromate, benzoyl peroxide, bromine, acetone peroxide, calcium bromate, hydrogen peroxide, sodium stearyl fumarate
6.	Bulking agent	Utilized as filler to thicken out food without dramatically increasing the amount of energy it contains.	Guar gum, psyllium husk and starch
7.	Carbonating agent	Used to provide carbonation in a food	Carbon dioxide
8.	Carrier	Utilized to disperse, dilute, or physically alter a food additive or nutrient without changing how it works (or having any technological impact of its own, such as an encapsulating agent) in order to make it easier to handle, apply, or use.	Triacetin, propylene glycol, glycerol, sorbitol, mannitol, agar, guar gum, xanthan gum, pectin, cellulose, polysorbates

Sr. No.	Functional Classes	Definition and role	Additive name
9.	Color	Utilised to improve or add colour to culinary products.	Carotene, beet powder, annatto, cochineal, oleoresins, titanium dioxide, beta-apo-8, carotenal, beta carotene, canthaxanthin, caramel, dried algeal meal, paprika, grape skin extract, iron oxide, turmeric, saffron, synthetic food colors (FD and C)
10.	Color retention agent	Used to stabilize, retain or intensify the color of a food.	Sodium bisulphite, Ascorbic acid
11.	Emulsifier	Used to create or keep a stable, homogeneous emulsion in a food of otherwise immiscible phases.	Propylene glycol, propylene glycol monoesters, sorbitan monostearate, cholic acid, glycocholic acid, deoxycholic acid, lecithin, sorbitan, polysorbate 60, mono- and di-glycerides, fatty acid esters, dioctyl sodium sulfosuccinate
12.	Emulsifying salt	Used to rearrange proteins in processed food manufacturing to prevent fat separation.	Sodium lactate, calcium lactate, potassium citrate, sodium citrate, calcium citrate, potassium phosphates, diphosphates, triphosphates, polyphosphates
13.	Firming agent	Utilized to develop or maintain firm, crisp fruit or vegetable tissues, or to work with gelling chemicals to strengthen or create gels.	Calcium chloride, calcium sulphate, aluminum sulfate, ammonium aluminum sulphate, potassium aluminum sulphate and sodium aluminum sulfate
14.	Flavor enhancer	Used to enhance the actual taste and/or odor of a edible product.	Monosodium glutamate, sodium lactate, disodium guanylate, disodium inosinate, hydrolyzed vegetable protein, nucleotides
15.	Flour treatment agent	Used in flour or dough to boost the baking quality and/ or color.	Ammonium chloride, ascorbic acid
16.	Foaming agent	Used as an aerating and whipping agent to create or maintain a consistent dispersion of a gaseous phase in a liquid or solid food thus acting.	Whey proteins, hydrolyzed yeast or vegetable proteins
17.	Gelling agent	Used in formation of a gel to give good texture to food product.	alginate, pectin, carrageenan, gellan, gelatin, agar, modified starch, methyl cellulose

Sr. No.	Functional Classes	Definition and role	Additive name
18.	Glazing agent	Used to provide a shielding effect or to give food's exterior a shiny appearance	Stearic acid, Beeswax, Candelilla wax, Carnauba wax, Shellac, Microcrystalline wax, Crystalline wax, Lanolin, Oxidized polyethylene wax, Esters of colophonium, Paraffin,
19.	Humectant	Used to counteract the effects of a dry environment by acting as a moisture retention or wetting agent to keep food from drying out.	Fructose, corn syrup, honey, sorbitol, propylene glycol, glycerine, glycerine monostearate
20.	Packaging gas	It is introduced into a container before, during or after filling with food to protect the food from physical, chemical or microbial spoilage.	Nitrogen gas, Argon
21.	Preservative	Used to prolong a food's shelf life by preventing microbiological degradation and spoiling.	Benzoates, Sorbates – including potassium sorbate, calcium sorbate and sodium sorbate, Propionates, Nitrites, Sulfites, including sodium sulfite, sodium bisulfite, sodium metabisulfite, potassium bisulfite and potassium metabisulfite. Vitamin E (tocopherol) Vitamin C (ascorbic acid) Butylated hydroxyanisole,
22.	Propellent	A food is expelled from a container using this gas.	Nitrous oxide, carbon dioxide, propane, n-butane and isobutane
23	Raising agent	Used to liberate gas and thereby to increase the volume of a dough or batter.	Yeast, Baking powder, baking soda, baking ammonia
24.	Sequestrant	Used to control the availability of a ions for various biochemical reactions in foods.	EDTA, citric acid, polyphosphates, calcium disodium EDTA
25.	Stabilizer	Used to keep two or more components evenly dispersed.	Gum arabic, gum acacia, modified starches, calcium carbonate, sodium caseinate, magnesium stearate, agar, carboxymethyl cellulose, alginates, pectins, carrageenan, locust bean gum, ammonium alginate, arabinogalactan, carob bean gum, cellulose, gelatin, gum ghatti, karaya, larch gum, mannitol, propylene glycol, tragacanth, furcellaran

Sr. No.	Functional Classes	Definition and role	Additive name
26.	Sweetener	Employed to set forth sweet taste to the food item (other than a mono- or disaccharide sugar)	Non-nutritive sweeteners: Saccharin, cyclamate, aspartame, acesulfame K, ammoniated glycerrhizin and nutritive sweeteners: Saccharin, cyclamate, aspartame, acesulfame K, ammoniated glycerrhizin
27.	Thickener	Harnessed to surge the viscosity of a food product.	Xanthan gum, carboxymethylcellulose, methyl cellulose, hydroxypropyly methyl cellulose, gum Arabic, galactomannans

Source: Food and Agriculture Organization of the Unites Nations (FAO), (2019). http://www.fao.org/gsfaonline/reference/techfuncs.html

3. Safety Aspects of Food Additives

The natural food additives are being used for centuries in foods however the major concerns about safety of using food additives are associated with the synthetic food additives. The necessity of food additive safety has increased dramatically as a result of ever rising consumption of processed foods and increased consumers' awareness about health risks associated with use of food additives.

The two global organizations, namely, World Health Organization (WHO) and the Food and Agriculture Organization (FAO) are associated with quality evaluation and risk assessment of food additives. An impartial, international expert scientific committee called the Joint FAO/WHO Expert Committee on Food Additives conducts risk assessments of food additives (JECFA). Only food additives that have passed a JECFA safety evaluation and were judged to not significantly endanger consumer health may be utilized. This holds true regardless of whether the food additives are artificial or natural. National authorities can then approve the use of food additives at defined amounts for certain foods based on the JECFA assessment or a national assessment (Pressman *et al.*, 2017).

The following are some factors related to the safety of food additives

Evaluation of Safety of Food Additives

The Codex GSFA specifies the guidelines for using food additives (General Standards for Food Additives).These include: the application of the General Principles for the Use of Food Additives, Good Manufacturing Practices, and the maximum level of use, the food category and any special conditions. JECFA evaluations are based on scientific reviews of all relevant biochemical, toxicological, and other data on a specific additive, taking into account human observations, research studies, and required animal testing. The acute, short-term, and long-term studies required by JECFA assess how the food additive is absorbed, distributed, and excreted as well as any potential negative effects of the chemical or its byproducts at specific exposure levels. 3.2 Acceptable Daily Intake (ADI)

It refers to the maximum daily intake of a food and an additive expressed as a function of body weight that does not pose a significant risk to health. Using products that comply with this requirement fall within the guidelines of good manufacturing practise (GMP). The international food safety authority known as Joint Food Agriculture Organization / World Health Organization Expert Committee on Food Additives (JECFA) is the responsible for collecting and evaluating scientific data related to additives in food and allocates acceptable daily intake (ADI)) to the food additives.

Maximum Use Level

The Codex Alimentarius Commission, a joint FAO and WHO intergovernmental agency that sets food standards, uses the safety evaluations conducted by JECFA to determine levels for the maximum use of additives in foods and beverages. Consumers worldwide can have confidence that the food they eat complies with agreed-upon standards for safety and quality, regardless of where it was produced, thanks to the Codex standards, which serve as the reference for national standards for consumer protection and for international trade in food.The maximum usage level of an additive is the highest concentration of the addition that has been demonstrated to be functionally effective in a food or food category and has been deemed safe according to Codex guidelines. It is typically given as mg/kg of food, and under GMP, the maximum use level shall not always equate to the ideal, recommended, or normal level of use. Depending on the intended technical effect, the particular food in which the additive would be used, the type of raw material, food processing, post-manufacture storage, transport, and handling by distributors, retailers, and consumers, the optimum, recommended, or typical use level will vary for each application of an additive.

Permitted Food Categories

Keeping in view the intended consumers such as infants, children, adults or aged persons, a given food supplement/ additive may be permitted in a particular food group and may not be permitted in another food group as per various national and international standards. No additives are permitted in dairy products like milk and buttermilk (plain). Many additives are not permitted in infant food formulae.

Food Grade Quality of Food Additives

The additives in food shall be of food grade quality and shall conform to the applicable Specifications of Identity and Purity as recommended under Codex regulations including their manufacture, storage and transport accordance with Good Manufacturing Practice (GMP) to ensure safety. For example only food safe colors are permitted in different food products.

Carry-over of Food Additives

Additives are added intentionally in a given food product to serve certain functions, however, it may be present in a food as a result of carry-over from a raw material or ingredient used food manufacturing and processing, this is known as carry-over of food additives and as per provisions of Codex regulations is permitted only when:

- The use of additives in the raw materials or other substances is allowed and permitted. The maximum use level of the additive is not exceeded in the raw materials or other constituents.
- The food into which the addition is introduced does not contain the additive in an amount greater than that which shall be introduced through the use of raw materials, ingredients, or manufacturing processes under appropriate technological conditions.
- Foods like infant formula, follow-up formula, formula for special medical purposes for infants, and complementary foods for infants and young children are examples of foods where carrying over a food additive from a raw material or ingredient is not permitted, unless a food additive provision in the specified category is mentioned in Codex regulations.

GRAS Status and Safe Limits

GRAS stands for "generally recognized as safe" and is a United States Food and Drug Administration (FDA) designation that a food additive added to food is safe when consumed. GRAS substances are those whose use is generally recognized by experts as safe, under the conditions of its intended use based on their extensive history of use in food or based on published scientific evidence. Salt, garlic and its derivatives, beta-carotene, sugar, spices, vitamins, agar agar etc. are classified as GRAS substances. Although most food additives are considered to be without any potential adverse effects, there have been problems concerning the safety of some of these chemicals.

Analytical Methods for Detection and Quantification of Food Additives

Utlization of food additives offers many advantages to food industry and consumers by sustaining or enhancing various technological, physical, chemical and sensory attributes of different food products. However, in recent times the synthetic food additives owing to low cost and easy production at large scale have gradually replaced the natural food additives. This has raised concerns over potential risks posed to human health due to indiscriminate use of synthetic additives, use in excess of permitted levels and use of non-permitted additives. Therefore, it is highly important to conduct qualitative and quantitative analysis of the food additives to ensure compliance with regulatory standards which in turn ensures safety of the consumers. The analytical methods for qualitative and quantitative estimation of food additives include traditional chemical methods to several analytical techniques such as spectroscopy, gas chromatography and liquid chromatography to recent modern immunoassays and other biotechnological methods (Wu *et al.*, 2021). Therefore, the methods for detection of additives in a food can be categorized into two types:

1. Qualitative methods (to detect presence or absence of food additive in a food)
2. Quantitative methods (to determine the exact amount of a food additive in a food)

Table 2: Methods for analysis of food additives

Class	Additive	Name of method	Type	Principle
Preservative	Benzoic acid	Ferric chloride test	Qualitative	Diethyl ether extract of food product gives Salmon color precipitate of ferric benzoate with ferric chloride solution which indicates the presence of benzoic acid.
		Modified Mohler's test	Qualitative	The aqueous solution of the diethyl ether extract of food product when heated in presence of sodium hydroxide, sulphuric acid, potassium nitrate and ammonium sulphide, formation of Red brown ring indicates presence of benzoic acid. Upon heating the color finally changes to greenish yellow.
		Titrimetric method	Quantitative	Benzoic acid is separated from a known quantity of the sample by saturating with sodium chloride and then acidifying with dilute hydrochloric acid and extracting with chloroform. The chloroform layer is made mineral acid free and the solvent is eliminated by evaporation. The residue is dissolved in neutral alcohol and the amount of benzoic acid is determined by titration against standard alkali.
		Spectrophotometric method	Quantitative	Benzoic acid is extracted from prepared sample using diethyl ether and the absorbance of the ether layer is measured at 272 nm, 267.5 nm and 276.5 nm in the UV region. From the corrected absorbance and the calibration graph obtained using standard benzoic acid solution, the amount of benzoic acid is determined.
		HPLC Method	Quantitative	Benzoic acid is extracted and separated by liquid chromatography (LC) on C18 column, detected by ultra violet absorbance at 230 nm, and quantitated by standard calibration plot.
Preservative	Sorbic acid	UV Spectrophotometric Method	Quantitative	Sorbic acid is extracted from the sample using the solvent mixture of diethyl ether and petroleum ether (1:1) and absorbance of the extract is measured at 250 nm. Sorbic acid in another aliquot is destroyed with permanganate and absence of the peak at 250 nm is taken as confirmation of the presence of sorbic acid in the sample.

Class	Additive	Name of method	Type	Principle
		Steam distillation method	Quantitative	This analysis involves stem distillation of the free acid, oxidation to malonaldehyde using dichromate and reaction with thiobarbituric acid to form a red complex. This is determined spectrophotometrically at 532 nm.
Preservative	p- hydroxy benzoates (parabens)	Thin layer chromatography (TLC)	Qualitative	The sample is acidified and extracted with diethyl ether. The concentrated ethereal extract is subjected to TLC. U.V. or Deniges reagent is used for visualization.
		Deniges reagent	Quantitative	The 4-hydroxy benzoic acid esters present in the sample are hydrolyzed using alkali and is extracted with diethyl ether after acidification of the sample. After re-extraction with sodium hydroxide from ether, color is developed with Deniges reagent and the absorption is read at 518 nm.
Non-nutritive sweeteners	Saccharin	1. Phenolsulphuric acid test 2. Resorcinol sulphuric acid test	Qualitative	These tests are depend upon the observation of color in the sample when treated under conditions as per these methods. In phenolsuplhuric acid test if saccharin is present a magenta or reddish-purple color develops. In resorcinol sulphuric acid test if saccharin is present, a green fluorescence is developed.
		Phenol-sulphuric acid colorimetric method	Quantitative	Saccharin is extracted from the acidified sample with chloroform and the solvent evaporated. The residue obtained is treated with phenol sulphuric acid and heated at 175°C for 2 hours. After making alkaline with sodium hydroxide, the absorbance is read at 558 nm.
Non-nutritive sweeteners	Aspartame, cyclamate, saccharin, Acesulphame -K	HPLC method	Quantitative	Extraction of sample with water or eluent, if necessary, clarification on solid phase extraction column or with Carrez reagent, chromatography at an HPLC reversed phase column and spectrophotometrical determination at a wavelength of 220 nm
Natural color	Caramel	Fiehe's reaction	Qualitative	Extract the sample solution with 50 mL ether and evaporate it in a porcelain dish. To the residue add 3 drops of 1% solution of resorcinol in hydrochloric acid. Addition of caramel is indicated by appearance of rose colour.

Class	Additive	Name of method	Type	Principle
Natural color	Annatto		Qualitative	Shake the melted fat or oil with 2% sodium hydroxide solution and pour the aqueous extract on moistened filter paper. The filter paper will show a straw color which will remain with a gentle wash with water. Dry the paper and add a drop of 40% stannous chloride solution and dry carefully. If the color turns purple, the presence of annatto is confirmed.
Synthetic food colors	Carmoisine, Ponceau 4 R, Erythrosine, Fast green FCF, Indigo carmine, Brilliant blue FCF, Tartrazine, Sunseyt yellow FCF	Column chromatography	Quantitative	Each color has its own unique absorption maxima and after extracting the color from sample it is estimated either using standard curve or the extinction coefficient.
Antioxidants	Butylated hydroxy toluene (BHT), Butylated hydroxy anisole (BHA), Tert-butyl hydroquinone (TBHQ), Propyl gallate (PG)	Spectrophotometric method	Quantitative	Oil or melted fat is dissolved in petroleum ether and extracted with ammonium acetate solution and water. The absorbance of the colored solution is read at 540 nm for PG and 610 nm for BHA. The amount of antioxidant present in the sample is calculated from the calibration graph.
		HPLC method	Quantitative	Antioxidants are extracted into acetonitrile and the extract is concentrated and diluted with 2-propanol. Antioxidants are separated by liquid chromatograph and measured by UV detection at 280 nm.

Class	Additive	Name of method	Type	Principle
Emulsifiers	Stearyl tartrate, glycerol esters, propylene glycol esters, sorbitan esters of fatty acids	TLC	Qualitative	Emulsifier together with fat is extracted by blending with chloroform and methanol. On adding a calculated amount of water, chloroform containing fat and emulsifier separates. The chloroform layer is evaporated to dryness and emulsifier present extracted with methanol. The methanol extract is subjected to TLC.
Flavors	Menthol	Gas chromatography	Quantitative	Menthol along with other flavoring matter is steam distilled into chloroform. The dry chloroform extract is directly subjected to gas chromatography. The amount of menthol present in the sample is determined using peak height/area of sample and standard.
Flavor enhancer	Mono sodium glutamate (MSG)	Ion exchange chromatography	Quantitative	Glutamic acid is extracted from foods using water, separated from other amino acids by using an ion-exchange resin chromatography and titrated potentimetrically using 0.1N sodium hydroxide.

Sources

1. The Association of Official Analytical Chemists. (2000). *Official Methods of Analysis.* 17th Edition, Gaithersburg, MD, USA.
2. Food Safety and Standards Authority of India. (2016). *Manual for analysis of food additives.*
3. Food and Agriculture Organization of the Unites Nations. (1986). *Manual of food quality control.*
4. Food and Agriculture Organization of the Unites Nations (FAO), (2019). *Food additive functional classes.* http://www.fao.org/gsfaonline/reference/techfuncs.html
5. Kirk, R., & Sawyer, R. (1989). *Pearson's composition and analysis of foods, 9th Edition.* Longman Scientific, London

Extraction Methods of Food Additives for Detection Of Contaminant

In today's world food additives are very often used in all foods, regardless of whether they have been previously standardized through law making bodies or not. Food additives can maintain food safety, freshness, taste, texture, or appearance but sometimes their excessive use may be harmful to human health. Sometimes when used in excess, they may be proven hazardous to health and therefore banned in some countries.

Sample preparation is a crucial step in a food analytical procedure since it consumes the majority of the overall analysis time, contributes significantly to the total cost of analysis, and has a significant impact on the analysis' outcomes. There are a number of options of sample extraction techniques and the technique chosen is determined by the type of food matrix and the analyte of interest. In general the steps involved in extraction are depicted in figure 1. Choice of appropriate solvent and the technique are crucial factors to be considered in order to get the desired yield of the contaminants. There are various methods which are used to detect the presence of food additives in foods and many advanced technologies are there to describe the exact concentration of that additive. Every technique comes with its own sets of merits and demerits which are summarized in Table 3 with their principle involved.

Table 3: Summarization of various techniques utilized for extraction of food additives for detection of contaminants

Technique used for sample preparation	Principle Involved	Potential benefits	Pitfalls	References
Soxhlet Method	Used to extract analytes that are thermally stable. By employing boiling and condensation, the extraction solvent is continuously cycled through the food matrix, resulting in the collection of sample in the hot solvent.	1. Highly efficient 2. Complete extraction of the analyte 3. Extraction can be reoeated many times to ensure complete extraction.	1. Techniques is not selective. 2. Requires further concentration of samples. 3. Time consuming. 4. Conventional soxhlet method require substantial amount of solvent as compared to automatic systems.	Hirondart *et al.*, 2020; Heleno *et al.*, 2016; Simko, 2002
Maceration	Involves the transformation of intact food matrix into a fine textured material.	1. Extraction of thermo-labile constituents is possible	1. Slow method 2. Raw material is not fully exhausted 3. Large volume of solvent required	Abubakar and Haque, 2020; Azmir *et al.*, 2013
Hydro-distillation	The food matrix is submerged in boiling water. The volatile analytes are transported away with water vapours and then pooled following condensation and decantation.	1. Easy separation 2. Lower risk of decomposition	1. Combustion of sample 2. Degradation of volatile compounds	Wu *et al.*, 2015; Petigny *et al.*, 2014
Solid phase extraction	This technique is utilized for preparing and purifying samples selectively for chromatographic analysis. The process involves isolation of analyte(s) by extraction followed by partitioning or adsorption (or both) on amsolid stationary phase	1. Lesser amount of organic solvent used. 2. Higher selectivity and reproducibility 3. Faster sample preparation time 4. Ease of operation.	1. May lead to lower recovery rate of analytes owing to the interactions between the food components	

Technique used for sample preparation	Principle Involved	Potential benefits	Pitfalls	References
Pressurized Fluid Extraction/ High Pressure Solvent Extraction	Employs a combination of high temperature and high pressure with solvents. High pressure is usually applied so that the solvent remains in liquid state even at temperatures above boiling point. High temperature aid in breakdown of interactions between the analyte and the matrix, thereby increasing diffusion rate and solubility of analytes and hence extraction efficiency	1. Environment friendly technique 2. Clean technology 3. Takes lesser time 4. Cheaper	1. Expensive setup 2. Decreased specificity Dilution of analyte may take place	Hirondart *et al.*, 2020; Vazquez-Roig and Pico, 2015
Liquid-liquid extraction	Based on the relative solubility of an analyte in between the sample matrix and immiscible solvent. Extraction is achieved by the differences in polarity of the two phases.	1. High capacity of extraction 2. High selectivity of separation 3. Low cost technique	1. Multiplication of extraction steps to obtain optimum output 2. Emulsion difficulties which hinders full recovery 3. Labour intensive 4. Low reproducibility 5. Extracts are often diluted with water 6. Time consuming	Delgado-Blanca *et al.*, 2020; Shahrestani *et al.*, 2018
Ultrasound assisted extraction	Based on the working principle of acoustic or ultrasonic cavitation, aiding in non-fabrication of food matrix	Minimum extraction time	1. Limited extraction, efficiency 2. Time and solvent consumption	Laurora *et al.*, 2021; Singla and Sit., 2021; Barba *et al.*, 2015

Technique used for sample preparation	Principle Involved	Potential benefits	Pitfalls	References
Microwave assisted extraction	Microwaves are utilized to assist in heating of the solvent, enhancing the solvent dispersibility into the sample, also rampeneing up the separation of the compounds in the solvent.	1. Simple instrumentation 2. Reduced solvent consumption 3. Minimum extraction duration 4. Rapid technique	1. Instrument cost is high 2. Absorbing material requires filtration after extraction 3. Only applicable to thermally stable analytes 4. Choice of solvents is limited	Moret *et al*., 2019; Barba *et al*., 2015
Supercritical fluid extraction	Separation of extrants from its food matrix by supercritical fluids as solvents for extraction. Extraction is carried on the basis of difference in solubility.	1. Lower operating temperature 2. Improved yield 3. Lower production cost 4. Can be performed on small sample	1. Elevated pressures required 2. Relative high cost of investment 3. Complicated phase behaviour 4. Matrix dependent 5. Require optimization of a number of parameters 6. A phase of clean-up is required	Manjare, and Dhingra, 2019; Oroian and Escriche, 2015; Hajeb *et al*., 2015
High pressure assisted extraction	Based on Le Chatlier principle. Implication of high pressure to trigger a number of phenomenon like phase transition, altering reaction kinetics and alterations in internal structure of the food matrix, promoting decrease in volume thereby enhancing the extraction yield.	1. User friendly 2. Increased efficiency 3. Safe and rapid extraction	1. High cost of investment 2. For analytical use only	Khan *et al*., 2018; Garcia-Mendoza *et al*., 2015

Food Additives in Excess and Mislabelling: A Gamble Between Taste and Good Health

The food processing business was shaken in the previous ten years by a biggest meat fraud scandal that is the European horse meat scandal. It was discovered that meat products sold to various big brands as beef contained horse meat. This was simply one of the most advertised, acknowledged illustration of food fraud by mislabelling in the food industry. Food fraud is a broad term that refers to a variety of intentional deceptions. Food fraud is the premeditated mislabelling of food products for financial gain with the intent of deceiving the consumer regarding what is actually inside the package.Mislabelling, whether intentional or inadvertent, can have major health ramifications, not simply financially. A food product can be mislabeled in a variety of ways like false marketing; not labelling the constituents purposely or accidentally on the food packet; non listing of potential allergens contained in the food product or false claim regarding an ingredient which is not contained in it. Discrepancies between food content and label can cause adverse reactions in people with hypersensitivity to particular food components. There are approximately around 160 food items that might trigger allergic reactions. Some of them are: eggs, peanuts, sesame seeds, milk, fish, soyabean, wheat and so on.

False marketing occurs when the manufacturer claims something which it can't back up. Olive oils that claim to be "extra virgin" and imported from Italy are often blended with other, less expensive oils and come from the country wherever its manufacturing cost is lowest. In accordance to a report published by Forbes in 2016 that 80% of the Italian extra virgin olive oil on the market is fraudulent (www.highrivertimes.com).

Labels like "craft," "artisan," "handmade," and "natural" have set in as a new trend and have become more prevalent in the food business. When it comes to having legally specified criteria that corporations must adhere to while describing their products, the lines are printed blurred on the package. Such foods are more expensive because they are thought to be of premium quality and require considerably more labour-intensive and sophisticated techniques to produce. For instance Cargill Inc., the company that owns Truvia sweetener, has settled a class action lawsuit alleging that the product was incorrectly branded as "natural." Cargill was sued in two separate class action lawsuits for its Truvia marketing, advertising, and labelling. Cargill has been ordered to issue cash refunds to qualified consumers as part of the settlement (www.consumerlawchicago.com). Another instance for food mislabelling was reported for The Ghirardelli Chocolate Company. The company has reached an agreement to settle a class action lawsuit alleging that it misrepresented the ingredients and advantages of its white chocolate baking chips. These bags formerly stated that they contained "Premium Baking Chips — Classic White," while the product actually did not contain white chocolate. There were also questions about whether Ghirardelli could properly refer to its chips as "All Natural" (www.consumerlawchicago.com)

Due to perceived risk or side effects, food additives such as sulfites and MSG must be clearly noted on a food label. These additives are frequently listed as "flavourings" or "spices" rather than by their precise names by some producers. Some of them are trade secrets that they don't want their competition to know about. For example, mono sodium glutamate (MSG), may cause headaches and nausea in some people, may be included in a product but disguised on the label as "natural flavor" or "hydrolyzed yeast" (https://greenliving.lovetoknow.com). Indian food authority FSSAI, banned Nestle India product Maggi in 2015, after the tests revealed that it contains high levels of lead and for misleading labelling stating "No added MSG". Nestle had to go for a major food recall after this ban. Later on after the upliftment of the ban the company reintroduced Maggi into the retailer chain with removal of the label from the packet (https://economictimes.indiatimes.com).

Seafood is one of the most widely traded food commodities in the world, and it is also one of the food groups most vulnerable to fraud. 283 samples were collected from 180 fast food restaurants from 23 European nations. It was exhibited from the results obtained by DNA barcoding that 26% of the total samples were mislabelled, and 31% of the shops sold mislabelled seafood. Moreover Spain, Iceland, Finland, and Germany had the highest rate of mislabelling, with nearly half of the retailers evaluated selling mislabelled seafood (Pardo *et al.*, 2018). In a recent study in 2018, it was revealed that more than fifty percent of roasted "Xue Yu" fish fillets in China were incorrectly labelled (www.foodnavigator-asia.com).

Consumers can help protect themselves from food fraud by taking certain measures or steps. Whole, unprocessed food products are less likely to be adulterated in comparison to processed one. For example buying whole fruits instead of juices. When it comes to processed foods, buying from trustworthy sources and brands with a vested interest in maintaining their reputation is a good idea. Whenever food fraud incidents are revealed, they lead to significant financial losses and causes considerable damage to a company's reputation. As a result, organisations with a big and trustworthy brand and those having a genuine interest in maintaining their good name will take proactive measures to protect their products.

Health Implications

Practices such as mislabelling, contamination and adulteration lead to serious health issues in the consumers. The chemical nature of the food additives and adulterants is responsible for onsetting a chain of immune responses in the body which are termed as food allergy. Food allergies among infants are more pronounced in the Western countries (Vale *et al.*, 2021) and food allergy is set to become a serious paediatric health challenge in modern times (Prescott *et al.*, 2013). Additionally, food induced anaphylactic reactions are also coming up in Asia, though a lot of variations exists. For example, wheat allergy is a common finding in Japan and Korea, whereas it is less reported from other Asiatic regions (Lee *et al.*, 2013).Allergens in food tend to have a lifelong effect upon the human body and the sensitive individual needs to discontinue the intake of the given food on a permanent basis (Nishe *et al.*, 2021). Some dietary additives may cause intolerance or allergic reactions in some people, which may have short-term or long-term repercussions. For some people with allergic disorders, a small amount of sulphur dioxide may trigger bronchoconstriction and an asthmatic reaction. (Huber, 2021). "Chinese Restaurant Syndrome" upon consumption of

Chinese food products containing Mono Sodium Glutamate (MSG) - a flavor enhancer ischaracterized by burning or tingling sensation on face, neck and head, tightness, stiffness or pressure in chest and facial muscles (Wong, 2021).

Excess amounts ofacrylamide have been detected in various foods like potato chips, french fries, bread, cookies, coffee, panela, fried banana and traditional fritters. Acrylamide is a potent carcinogen with neurotoxic, genotoxic and mutagenic effects. It undergoes bio-accumulation in organs like brain, heart, liver, kidney, breast milk and lungs(Cortes *et al.*, 2021). Bisphenol A used in food and beverages packaging is also a toxic compound that affects reproductive, respiratory, nervous and immune systems, along with being a potent carcinogen (Barzegar *et al.*, 2021). Nitrosamines in canned meat products are another class of cancer-causing agents that are becoming a public health concern (Omar *et al.*, 2021). Polycyclic aromatic hydrocarbons (PAHs) are another class of important carcinogenic chemical contaminants in smoked meats (Asamoah *et al.*, 2021). After enzymatic action, biogenic amines are created in meat and meat products, dairy and dairy products, vegetables, and fruits.Their presence in food products can cause various symptoms like nausea, hot flashes, palpitations, erythema, headaches, hypertension or hypotension in affected individuals (Behera *et al.*, 2021).

Reduced tolerance to foods can be cited as amajor reason for increase in food allergens other than fraudulent practices. In order to beat the problem of food allergies or other health implications, an early introduction of babies to foods can be practiced. This would help in achieving a better oral and intestinal tolerance of the food allergens to prevent immunoglobulin E (IgE)-mediated (type I hypersensitivity) or non-IgE-mediated (type IV hypersensitivity) (Yang *et al.*, 2021).

Case Studies and Food Recalls: Incidents and Advisories

Undeclared allergen, microbial contamination, presence of foreign matter, biotoxin, chemical/contaminant, mislabelling are the major driving forces for recalls or terminations. Important food recalls/terminations over the recent years have been tabulated in Table 4 with the details of the incidents.

Table 4: Major food recall/ terminations/advisories across various continents in recent years

Year	Country	Incident	Details	Reference
Incidents related to Milk and Milk Products				
2007	United States	Peter Pan butter recall	• Butter from Peter Pan company was linked to cases of Salmonellosis leading to a massive recall of the product	Bakhtavoryan *et al.*, 2014
2008	China	Melamine Milk Scandal	• Fresh milk spiked with Melamine • Tainted milk used in manufacture of infant formula powders • Kidney stones and renal damage in infants • 6 babies died from kidney stones, 54,000 hospitalized. Estimated victims 294,000 victims	Pei *et al.*, 2011
2013	New Zealand	Food recall on soft cheeses	• Jindi cheeses from New Zealand involving a variety of brands sold in different packaging were linked to cases of Listeriosis in people leading to illness and putting pregnant women at risk.	FSANZ, 2021
Incidents related to Meat, Eggs and seafoods				
2012	United States	Raw Scraped Ground Tuna product Salmonellosis outbreak	• A total of 425 persons infected with the outbreak strains of *Salmonella* Bareilly (410 persons) or *Salmonella* Nchanga (15 persons) were reported from 28 states and the District of Columbia. • 55 ill persons were hospitalized, and no deaths were reported • Raw Nakaochi Scrape tuna product from Moon Marine USA Corporation was recalled	Hoffmann *et al.*, 2016
2013	Europe	Horse-meat scandal	• Horse meat was found in meat products in beef burgers • Replacement levels were as high as 100% in some cases • Risk of entry of anaesthetic, phenylbutazone in food chain Human at risk of Equine infectious anaemia	Smith and McElwee, 2021
2015	China	Zombie-meat scandal	• Smuggling of frozen meat over 40 years old • Meat was soaked in hydrogen peroxide to preserve it Rotten meat included chicken's claws, chicken feet	Giusto, 2017

Year	Country	Incident	Details	Reference
2017		Apricot kernels (raw)	• The sale of raw apricot kernels is prohibited under the Food Standards Code. They contain cyanogenic glycosides that can convert to a type of cyanide when eaten.	FSANZ, 2021
2019	Australia	Salmonella Enteritidis (SE) linked to eggs	• A rare strain of Salmonella Enteritidis associated with human illness traced to egg consumption • Recall notices were served to various firms	FSANZ, 2021
2020	Malaysia	Fake-halal meat scandal	• Cargos containing non-halal meat were being certified as halal • Incorrect documentation and certification had been ongoing for a period of more than 40 years	Ruzulan *et al.*, 2021
Incidents related to Fruits/Vegetables/Cereals				
2010	Australia	Reformulated Bonsoy soy milk with kombu seaweed	• Soy milk product containing kombu seaweed was recalled following nine cases of people with thyroid problems. • The product was subsequently reformulated to make it safe to consume and was returned to sale	FSANZ, 2021
2011	United States	Contaminated cantaloupes linked to Jensen Farms of Holly, Colorado	• There was a widespread outbreak of *Listeria monocytogenes* food poisoning across 28 US states • 33 deaths, 1 miscarriage and total 147 illnesses due to Listeriosis were recorded	McCollum *et al.*, 2011
2017	Australia	Apricot kernels (raw)	• The sale of raw apricot kernels is prohibited under the Food Standards Code. They contain cyanogenic glycosides that can convert to a type of cyanide when eaten.	FSANZ, 2021
2015	United States	Caramel apples outbreak	• Multistate outbreak of *Listeria monocytogenes* infections was linked to whole apples used in commercially produced, pre-packaged caramel apples • 35 people were taken ill in 12 US states out of which 7 died	Angelo *et al.*, 2017

Year	Country	Incident	Details	Reference
2016	United States and Canada	Leafy green salads outbreak	• An outbreak of listeriosis associated with packaged leafy green salads. • A total of 09 cases were reported • Led to a recall of packaged leafy green salads made at the processing facility • Measures to prevent contamination of fresh produce with *Listeria monocytogenes* contamination	Self *et al.*, 2019
2016	Australia	Rockmelon linked to salmonella	• Food Standards Australia New Zealand (FSANZ) recalledwhole rockmelons from Red Dirt, a rockmelon grower in the Northern Territory • It was linked to cases of Salmonella Hvittingfoss infection, a relatively uncommon strain	FSANZ, 2021
2020	United States	Red Onions Salmonellosis outbreak	• A total of 1,127 people infected with the outbreak strain of *Salmonella* Newport were reported from 48 states. There were 167 hospitalizations and no deaths reported • Epidemiologic and traceback evidence showed that red onions from Thomson International Inc. were the likely source of this outbreak.	CDC, 2020
2021	United States	Dried wood ear mushrooms Salmonellosis outbreak	• A total of 55 people infected with the outbreak strain of Salmonella Stanley were reported from 12 states. • 6 hospitalizations were reported. No deaths were reported. • Epidemiologic, laboratory, and traceback information showed that wood ear mushrooms distributed by Wismettac Asian Foods, Inc., were the likely source of this outbreak. • On September 23, 2020, Wismettac Asian Foods recalled external icon dried fungus due to possible Salmonella contamination.	CDC, 2021
2021	United States	Dried apricots	• Undeclared suplhites • Product recalled from Lipari Foods	FDA, 2021
2021	United States	Salt & Vinegar Flavored Potato Chips	• Undeclared milk • Product sold by Shearer's Foods LLC under the trade name Raley'swas terminated	FDA, 2021

Year	Country	Incident	Details	Reference
Spices/ extracts/ supplements				
2005	United States	Sudan I food dye scare	• Chilli powder containing illegal food dye Sudan I found in Worcester Sauce • Sudan I dye is a potent carcinogen • More than 470 food products had to be recalled as Worcester Sauce was ingredient of many • The scare cost around £100m	Everstine *et al.*, 2013
2021	United States	Dietary supplement products by Living Free	• Undeclared soy lecithin • Product recalled from Bea Lydecker's Naturals, Inc.	FDA, 2021
2021	United States	Lipodrene w/25mg Ephedra Extract Dietary Supplement	• Presence of 1,4-dimethylamylamine (DMAA) • Product recalled from Hi-Tech Pharmaceuticals. Inc	FDA, 2021
2021	United States	Prebiotic Fiber Supplement	• Potential for plastic pieces or shavings • Product sold by GSK Consumer Healthcare under the trade name Benefiber was terminated	FDA, 2021

Conclusion

Food additives are an important class of agents that enhance the overall performance and acceptability of the food product that they are added into. But recently, the practice of adulteration in foods is on the rise outcome of mislabelling, counterfeiting, claim violations, tampering, fraudulent substitution, concealment etc (Visciano and Schirone, 2021). Multiple sources of food trade and different price in different markets adds to the woes of consumers. Stringent regulation of food trade and distribution with international compliance; along with scrutiny over supply chains and a better system of traceability is required to minimize the risks associated with food safety. Routine inspections and audits, use of new technologies like Nanotech warning systems, Near-infrared spectroscopy and use of data science in predictive analytics is the key to combat adulteration (Charlebois *et al.*, 2021). Packaging practices too need to be reformed with stringent control over labelling and use of active-intelligent packaging. And finally, imparting training and education in relevant areas, better safety compliance and developing the organizational culture with consumer safety in mind. All these measures will pave the way for a better food safety system where additives play the role of upgrading the food, like they were meant to be in the first place and not cause harm to the end-consumer.

References

Abubakar, A. R., & Haque, M. (2020). Preparation of Medicinal Plants: Basic Extraction and Fractionation Procedures for Experimental Purposes. Journal of Pharmacy & Bioallied Sciences, 12(1), 1–10. https://doi.org/10.4103/jpbs.JPBS_175_19

Angelo, K. M., Conrad, A. R., Saupe, A., Dragoo, H., West, N., Sorenson, A., ... & Jackson, B. R. (2017). Multistate outbreak of Listeria monocytogenes infections linked to whole apples used in commercially produced, prepackaged caramel apples: United States, 2014–2015. Epidemiology & Infection, 145(5), 848-856.

Asamoah, E. K., Nunoo, F. K. E., Addo, S., Nyarko, J. O., &Hyldig, G. (2021). Polycyclic aromatic hydrocarbons (PAHs) in fish smoked using traditional and improved kilns: Levels and human health risk implications through dietary exposure in Ghana. Food Control, 121, 107576.

Azmir, J., Zaidul, I. S. M., Rahman, M. M., Sharif, K. M., Mohamed, A., Sahena, F., & Omar, A. K. M. (2013). Techniques for extraction of bioactive compounds from plant materials: A review. Journal of Food Engineering, 117(4), 426–436.

Bakhtavoryan, R., Capps, O., &Salin, V. (2014). The impact of food safety incidents across brands: the case of the Peter Pan Peanut Butter Recall. Journal of Agricultural and Applied Economics, 46(4), 559-573.

Barba, F.J., Grimi, N., &Vorobiev, E. (2015). New Approaches for the Use of Non-conventional Cell Disruption Technologies to Extract Potential Food Additives and Nutraceuticals from Microalgae. Food Engineering Reviews. 7, 45–62.

Barzegar, F., Kamankesh, M., & Mohammadi, A. (2021). Recent Development in Formation, Toxic Effects, Human Health and Analytical Techniques of Food Contaminants. Food Reviews International, 1-27.

Behera, S. S., Panda, S. K., & Ray, R. C. (2021). Biogenic amines in fermented vegetables: food safety issues. Microbial Biotechnology in Food and Health, 165-195.

Charlebois, S., Juhasz, M., Music, J., & Vézeau, J. (2021). A review of Canadian and international food safety systems: Issues and recommendations for the future. Comprehensive Reviews in Food Science and Food Safety.

Cortés, W. R. B., Mejía, S. M. V., & Mahecha, H. S. (2021). Consumption study and margin of exposure of acrylamide in food consumed by the Bogotá population in Colombia. Journal of Food Composition and Analysis, 100, 103934.

Delgado-Blanca, I., Llorent-Martínez, E.J., Ruiz-Medina, A., & Pilar, O. (2020). Automated on-line liquid-liquid extraction in a multisyringe flow injection analysis manifold for migration studies in food-contact materials: analysis of 4,4′-dihydroxybiphenyl, Food Additives & Contaminants: Part A. 37(1), 174-182. DOI: 10.1080/19440049.2019.1678769

Everstine, K., Spink, J., & Kennedy, S. (2013). Economically motivated adulteration (EMA) of food: common characteristics of EMA incidents. Journal of food protection, 76(4), 723-735.

FDA. 2021. https://www.fda.gov/safety/recalls-market-withdrawals-safety-alerts

Food and Agriculture Organization of the Unites Nations (FAO), (2019). Food additive functional classes. http://www.fao.org/gsfaonline/reference/techfuncs.html

Food and Agriculture Organization of the Unites Nations. (1986). Manual of food quality control.

Food Safety and Standards Authority of India. (2016). Manual for analysis of food additives.

FSANZ. Food Standards Australia New Zealand. https://www.foodstandards.gov.au/industry/foodrecalls/recalls/Pages/default.aspx

Garcia-Mendoza, M.P., Paula, J.T., Paviani, L.C., Cabral, F.A., & Martinez-Correa, H.A. (2015). Extracts from mango peel by-product obtained by supercritical CO_2 and pressurized solvent processes. LWT-Food Science and Technology, 62(1), 131–137.

Giusto, A. (2017). Language and Food Safety: the "Zombie Meat" scandal. ВестникСанкт-Петербургскогоуниверситета. Востоковедение и африканистика, 9(1), 70-95.

Hajeb, P., Selamat, J., fsah-Hejri, L., Mahyudin, N. A., Shakibazadeh, S., &Sarker, M.Z.I. (2015). Effect of Supercritical Fluid Extraction on the Reduction of Toxic Elements in Fish Oil Compared with Other Extraction Methods. Jornal of Food Protection. 78(1): 172–179.

Heleno, S.A., Diz, P., Prieto, M.A., Barros, L., Rodrigues, A., Barreiro, M.F., & Ferreira, I. C. (2016). Optimization of ultrasound-assisted extraction to obtain mycosterols from Agaricusbisporus L. by response surface methodology and comparison with conventional Soxhlet extraction. Food Chemistry, 197, 1054–1063.

Hirondart, M., Rombaut, N., Fabiano-Tixier, A. S., Bily, A., &Chemat, F. (2020). Comparison between Pressurized Liquid Extraction and Conventional Soxhlet Extraction for Rosemary Antioxidants, Yield, Composition, and Environmental Footprint. Foods (Basel, Switzerland), 9(5), 584. https://doi.org/10.3390/foods9050584

Hoffmann, M., Luo, Y., Monday, S. R., Gonzalez-Escalona, N., Ottesen, A. R., Muruvanda, T., ... & Brown, E. W. (2016). Tracing origins of the Salmonella Bareilly strain causing a food-borne outbreak in the United States. The Journal of Infectious Diseases, 213(4), 502-508.

https://economictimes.indiatimes.com/industry/cons-products/fmcg/nestle-india-to-contest-fine-in-maggi-noodles-case/articleshow/71073426.cms?from=mdr

https://greenliving.lovetoknow.com/dangers-food-additives-preservatives

https://www.consumerlawchicago.com/class-action-litigation/three-notable-food-mislabeling-class-action-lawsuits/

https://www.foodnavigator-asia.com/Article/2018/02/13/Fish-food-fraud-rife-in-China-with-more-than-half-of-fillets-mislabelled-Study https://www.highrivertimes.com/life/getting-to-the-truth-about-olive-oil

Huber, K. (2021). Environmental Influences on Pediatric Asthma: The Role of Social Determinants of Health in Costa Rica.

Indian Council of Medical Research. (1990). Manual on Methods of Analysis for Adulterants and Contaminants in Foods.

Khan, S. A., Aslam, R., &Makroo, H. A. (2018). High pressure extraction and its application in the extraction of bio-active compounds: A review. Journal of Food Process Engineering. e12896.

Kirk, R., & Sawyer, R. (1989). Pearson's composition and analysis of foods, 9th Edition. Longman Scientific, London.

Laurora, A., Lund, M. N., Tiwari, B. K., & Poojary, M. M. (2021). Application of ultrasound to obtain food additives and nutraceuticals In Design and Optimization of Innovative Food Processing Techniques Assisted by Ultrasound: Developing Healthier and Sustainable Food Products. pp 111-141

Lee, A. J., Thalayasingam, M., & Lee, B. W. (2013). Food allergy in Asia: how does it compare?. Asia Pacific Allergy, 3(1), 3-14.

Manjare, S.D., & Dhingra, K. (2019). Supercritical fluids in separation and purification: A review. Materials Science for Energy Technologies. 2(3), 463-484.

McCollum, J. T., Cronquist, A. B., Silk, B. J., Jackson, K. A., O'connor, K. A., Cosgrove, S., ... & Mahon, B. E. (2013). Multistate outbreak of listeriosis associated with cantaloupe. New England Journal of Medicine, 369(10), 944-953.

Moret, S., Conchione, C., Srbinovska, A., &Lucci, P. (2019). Microwave-Based Technique for Fast and Reliable Extraction of Organic Contaminants from Food, with a Special Focus on Hydrocarbon Contaminants. Foods. 8, 503.

Nishie, M., Masaki, K., Okuzumi, S., Mochimaru, T., Kabata, H., Miyata, J., & Fukunaga, K. (2021). Successful treatment of a patient with adult food allergy and severe asthma using omalizumab. Asia Pacific Allergy, 11(3).

Nyarko, E., Kniel, K. E., Millner, P. D., Luo, Y., Handy, E. T., Reynnells, R., & Sharma, M. (2016). Survival and growth of Listeria monocytogenes on whole cantaloupes is dependent on site of contamination and storage temperature. International journal of food microbiology, 234, 65-70.

Omer, A. K., Mohammed, R. R., Ameen, P. S. M., Abas, Z. A., &Ekici, K. (2021). Presence of Biogenic Amines in Food and Their Public Health Implications: A Review. Journal of Food Protection, 84(9), 1539-1548.

Oroian, M., & Escriche, I. (2015). Antioxidants: Characterization, natural sources, extraction and analysis. Food Research International, 74, 10–36.

Pardo, M. Á., Jiménez, E., Viðarsson, J. R., Ólafsson, K., Ólafsdóttir, G., Daníelsdóttir, A. K., & Pérez-Villareal, B. (2018). DNA barcoding revealing mislabeling of seafood in European mass caterings. Food Control, 92, 7–16.

Pei, X., Tandon, A., Alldrick, A., Giorgi, L., Huang, W., & Yang, R. (2011). The China melamine milk scandal and its implications for food safety regulation. Food policy, 36(3), 412-420.

Petigny, L., Périno, S., Minuti, M., Visinoni, F., Wajsman, J. and Chemat, F. (2014). Simultaneous microwave extraction and separation of volatile and non-volatile organic compounds of boldo leaves. From lab to industrial scale. International Journal of Molecular Sciences, 15(5), 7183–7198.

Prescott, S. L., Pawankar, R., Allen, K. J., Campbell, D. E., Sinn, J. K., Fiocchi, A., ... & Lee, B. W. (2013). A global survey of changing patterns of food allergy burden in children. World Allergy Organization Journal, 6(1), 1-12.

Pressman, P., Clemens, R., Hayes, W., & Reddy, C. (2017). Food additive safety: A review of toxicologic and regulatory issues. Toxicology Research and Application, 1, 1–22.

Ruzulan, Z., Jamaludin, M. A., & Ishak, A. H. (2021, May). Meat and meat-based products: Challenges and opportunities in halal food security. In IOP Conference Series: Earth and Environmental Science (Vol. 756, No. 1, p. 012017). IOP Publishing.

Self, J. L., Conrad, A., Stroika, S., Jackson, A., Whitlock, L., Jackson, K. A., ... & Basler, C. (2019). Multistate outbreak of listeriosis associated with packaged leafy green salads, United States and Canada, 2015–2016. Emerging infectious diseases, 25(8), 1461.

Shahrestani, M., Tehrani, M.S., Shoeibi, S., Azar, P.A., & Husain, S.W. (2018). Comparison between Different Extraction Methods for Determination of Primary Aromatic Amines in Food Simulant. Journal of Analytical Methods in Chemistry. Article ID 1651629. https://doi.org/10.1155/2018/1651629

Simko, P. (2002). Determination of polycyclic aromatic hydrocarbons in smoked meat products and smoke flavouring food additives. Journal of Chromatography B. 770 (1–2), 3-18

Singla, M., & Sit, N. (2021). Application of ultrasound in combination with other technologies in food processing: A review. Ultrasonics sonochemistry, 73, 105506. https://doi.org/10.1016/j.ultsonch.2021.105506

Smith, R., & McElwee, G. (2021). The "horse-meat" scandal: illegal activity in the food supply chain. Supply Chain Management: An International Journal.

The Association of Official Analytical Chemists. (2000). Official Methods of Analysis. 17th Edition, Gaithersburg, MD, USA.

Vale, S. L., Lobb, M., Netting, M. J., Murray, K., Clifford, R., Campbell, D. E., & Salter, S. M. (2021). A systematic review of infant feeding food allergy prevention guidelines–can we AGREE?. World Allergy Organization Journal, 14(6), 100550.

Vazquez-Roig, P., & Pico, Y. (2015). Pressurized liquid extraction of organic contaminants in environmental and food samples. Trends in Analytical Chemistry.71, 55-64.

Visciano, P., &Schirone, M. (2021). Food frauds: Global incidents and misleading situations. Trends in Food Science & Technology.

Wong, J. (2021). The truth about MSG. New Scientist, 249(3323), 22.

Wu, L., Zhang, C., Long, Y., Chen, Q., Zhang, W., & Liu, G. (2021). Food additives: From functions to analytical methods. Critical Reviews in Food Science and Nutrition. DOI: 10.1080/10408398.2021.1929823

Wu, C., Wang, F., Liu, J., Zou, Y., & Chen, X. (2015). A comparison of volatile fractions obtained from Lonicera macranthoides via different extraction processes: Ultrasound, microwave, Soxhlet extraction, hydrodistillation, and cold maceration. Integrative Medicine Research, 4(3), 171–177.

Yang, X., Liang, R., Xing, Q., & Ma, X. (2021). Fighting Food Allergy by Inducing Oral Tolerance: Facts and Fiction. International Archives of Allergy and Immunology, 1-11.

5

A Perspective on Food Contact Packaging Materials as a Source of Food Adulterants, Toxicants and Contaminants

***Narender Raju Panjagari*[1], *Sangita Ganguly*[2]**

[1]*Food Packaging Laboratory, Dairy Technology Division*
[2]*Functional Ingredients & Applications Laboratory, Dairy Technology Division*
ICAR-National Dairy Research Institute, Karnal, Haryana, India

1. Introduction

The materials widely used for food contact applications include paper and coated-paper products, cellulose products, cellophane, metals, ceramics, glass, rubber, plastics, and miscellaneous materials such as wood, fabric, etc. Most of these materials have been in vogue for several years. Among them plastics are widely used in the food industry as they offer many advantages due to their high cost-to-benefit ratio. However, food safety issues in terms of migration of additives is a cause of concern. Plastics contain a number of additives either deliberately added during their manufacture or inevitable due to technical requirement. Polymers due to their very high molecular weight, mostly being inert, and of limited solubility, are unlikely to be migrated into food to a significant level. Even if fragments were accidentally swallowed, they would not react with body fluids present in the digestive system. The number and quantity of additives incorporated in a resin vary with resin type and application. The presence of additives raises the concern of additive migration from the food contact surfaces. Most additives diffuse within the food contact materials especially plastics and often tend to migrate to the surface. When a packaged product is in direct contact with a compounded plastic, there may be migration of the additive to the product. However, this depends on several favorable factors such as polarity of the food, contact period, etc. Excessive intake of such additives may cause toxicity in human beings.

Compounds Contributing to Contamination

Packaging contaminants from plastic food contact materials primarily arise from two sources namely polymerization residues and processing aids. Polymerization residues including monomers, oligomers (MW < 200), catalysts, solvents, emulsifiers, wetting agents, raw material impurities, plant contaminants, inhibitors, decomposition and side reaction products. Fillers, lubricants, plasticizers, pigments, mould-releasing agents, heat stabilizers, UV light stabilizers, antioxidants, antistatic agents, etc. are all important processing aids in plastics manufacture. Volatile organic compounds (VOCs) of different monomers such as vinyl chloride, ethylene, etc. at higher concentration usually dissipate or decrease with time. However, their very low levels may endure for long time with VOCs such as acrylonitrile and vinyl benzene (styrene) being the most difficult ones to eliminate. In most cases, the VOCs belonging to polyolefin group remain. Scientific efforts and technological developments led to the reduction in vinyl chloride monomers to obtain the purest form. Due to the technological and application requirements, a number of compounds are incorporated in the master batch (purest form of polymer in which the concentrated additives are incorporated) of which some are likely to be present in greater amounts than the polymer fragments, which needs strict compliance. In general, the above mentioned processing aids are normally listed in the list of approved chemicals of the regulatory bodies for food contact materials (FCMs).

Chemical Compounds Incorporated into Plastics

Pure resins are rarely processed into final products without the addition of selected compounds, called *additives* that are incorporated during the process of extrusion and moulding of a plastic resin or applied externally on the formed material. Resins without additives are sometimes called as *barefoot resins. Compounding* is the process of uniformly mixing the additives into the resin, to make a homogenous material. It starts with dry mixing and usually ends with an extruder producing pellets. A *blend*, by definition, is formed from two or more polymer resins, which have been mixed together. Concentrate of an additive dispersed in a polymer is termed as *master batch.* Blending of the master batch with the barefoot resin to achieve the desired additive concentration is called *letting down*. The purposes for the use of additives in a resin are: (a) to improve the processing conditions, (b) to improve or increase the resin's stability to oxidation, (c) to obtain better impact resistance, (d) to increase or decrease hardness, (e) to control surface tension, (f) to facilitate extrusion and melting, (g) to control blocking, (h) to reduce cost, and (i) to increase flame resistance. In general, a plethora of additives are commonly incorporated into plastics used for packaging applications. However, they may be classified based on their functional properties into antioxidants, heat stabilizers, UV stabilizers, antiblocking agents, plasticizers, slip agents, anti-slip agents, lubricants, mould release agents, anti-fogging agents, colourants, anti-static agents, nucleating agents, fillers and reinforcements. Some of the important additives are briefly described below:

Antioxidants

Polymeric materials chemically deteriorate during fabrication, processing and storage due to a series of complex chemical oxidation reactions with atmospheric oxygen. The major result is (a) chain scission, which can yield a significant decrease in molecular weight and/or (b) crosslinking, which increases molecular weight and affects flow properties. To prevent the damage to the polymer caused by the oxidative degradation, chemical additives called antioxidants are incorporated into the polymer. Two main types of antioxidants are available commercially are called primary and secondary antioxidants. Primary antioxidants work by removing free radical species while secondary antioxidants work by interfering with the formation of free radicals. In packaging industry, polyethylene (PE), polypropylene (PP) and high impact polystyrene (HIPS) account for the major plastics market for antioxidants. Irganox™ (phenolic antioxidant), Irgafos™ (phosphate antioxidant), butylated hydroxyl toluene (BHT), α-tocopherol, etc. are commonly used as antioxidants in plastics processing.

Heat Stabilizers

They are used to prevent degradation of polymers when exposed to heat especially during processing. For most polymers, oxidation is primary mode of reaction and hence, antioxidants function effectively as heat stabilizers. For certain polymers such as polyvinyl chloride (PVC), other reactions are critical such as dehydrochlorination yielding hydrochloric acid and resulting in formation of double bond in the main polymer chain. This reaction imparts a visible colour to PVC, first yellow, then amber, reddish, brown and finally black. Organometallic compounds, salts derived from lead, cadmium, barium, zinc and tin, epoxides and phosphites are most common heat stabilizers used.

UV light Stabilizers

UV radiation from outdoors light and γ-radiation used for sterilization can cause photo-oxidation in PS, PP, PVC and other polymers. High energetic UV photons can be captured by a polymeric chain resulting in the breaking of covalent bonds and production of free radicals. Changes in colour, loss of flexibility and gloss, and scission of polymer and the production of fragments (low MW components) are some of the effects that can be caused by photo-oxidation. There are three types of UV light stabilizers, (i) those which absorb the harmful UV radiation and emit harmless radiation of larger wavelength and lower energy (UV absorbers), e.g. hydrobenzophenone, (ii) those which bring the polymeric chain from an excited state to a stable one by absorbing the energy (UV quenchers), e.g. organosalts of nickel, and (iii) free radical scavengers which accept free radicals and prevent them from reacting with other polymer molecules, e.g. hindered amine light stabilizers (HALS).

Plasticizers

These are the substances that are incorporated into a rigid plastic to increase its flexibility, workability and extensibility. By reducing the glass transition temperature

(T_g) and increasing chain lubricity, they also improve processing and extrusion characteristics, reduce the minimum required processing temperature, reduce hardness and improve low temperature flexibility. A plasticizer may not be required in all polymers but in certain cases it may inevitable such as PVC for the desired properties. PVC is a semi-crystalline plastic in the absence of a plasticizer, which results in brittleness and difficult to process. At lower concentrations, the plasticizer helps to reduce processing temperatures and minimize the thermal degradation of the polymer. At higher concentrations, plasticizer reduces hardness and increases flexibility of the finished material besides improving processing conditions. The polar nature of PVC gives it a high affinity for plasticizers and therefore, PVC accounts for more than 80% of the total use of plasticizers. Most plasticizers are liquids at room temperature and have high boiling points and low vapour pressure, hence they are not easily lost from the plastic during processing. The most common plasticizers are the phthalates and among all, diethylhexyl phthalate is most widely used. For food packaging films, only legally approved plasticizers can be used. Most often, an adipate plasticizer, normally di-(2-ethylhexyl) adipate combined with epoxidized soybean oil is used.

Colorants

Colorants give an opaque appearance that can contribute to light protection of a packaged product and also influence consumers purchasing considerations. Hundreds of different colorants are used in plastics packaging industry. Three major categories of colorants used in plastics are dyes, organic pigments and inorganic pigments. Inorganic pigments are widely used in plastics industry as they are less expensive, more opaque and more stable at high temperature and have less tendency to migrate. However, most of them are extremely toxic since they are oxides of heavy metals such as chromium, lead, cadmium or nickel.

Toxicological Aspects of Constituents of Plastic Packaging Materials

In 1969, the International Agency for Research on Cancer of the WHO initiated a programme to evaluate the carcinogenicity of chemicals and their risk to humans. Relevant biological data and results of epidemiological studies on humans were reviewed from in a timely manner to ascertain the safety of several monomers and additives used in plastics manufacturing processes. Toxicological or safety data of certain compounds present in food contact materials are presented here:

Vinyl Chloride

In different studies, animal models (rats, mice, and hamsters) have been exposed, either by inhalation or by oral administration, to different concentrations of vinyl chloride monomer (VCM) and reported production of a number of tumors at different sites including angiosarcomas of the liver due to the inhalation of VCM (50 to 10,000 ppm) in air for 4 hours a day, 5 days in a week for one year. Also, low doses of oral administration of VCM (3.33 mg/kg body weight) were found to be carcinogenic. Further, VCM has also been shown to be mutagenic. The principal toxic effects of VCM in humans included histopathological changes in the liver and spleen as well as lesions of the bones in the terminal joints of the fingers and toes. Reports indicated that VCM has long latent period for tumor development (about 20 years) for the symptoms to appear and long-term exposure gave rise to a rare form of liver cancer (angiosarcoma) as well as tumors of the brain, lung, and hematolymphopoietic systems in humans.

Vinylidene Chloride

Reports have indicated that the monomer levels of vinylidene chloride in air of plants manufacturing PVDC were found to be 90-100 μg/m^3 with workers getting exposed to very low concentrations (20 mg/m^3). Little toxicological information is available regarding the safety of vinylidene chloride. However, it has been reported that the main pathway of excretion is through the lungs followed by kidneys for other metabolites. The LD_{50} values for rats and mice are around 1500 and 200 mg/kg body weight, respectively. It was reported that it affects the activity of several rat liver enzymes and decreases the glutathione stores and some tumors after prolonged exposure in rats or rabbits.

Styrene

Styrene or vinyl benzene is a monomer of polystyrene. Psychological and neurological disturbances or disorders are among the most frequently observed ill effects of styrene exposure. The major urinary metabolites of styrene, mandelic acid and phenylglyoxylic acid, are regarded as indices of exposure to styrene. Both styrene and its oxide are metabolized in the liver and others such as kidney, intestine, lung and skin. The oral LD_{50} value of styrene for rats is 5 g/kg body weight. It is metabolized to styrene oxide, a potent mutagen. Styrene vapors are absorbed through the lungs (60-90%) and is soluble in blood and has been found accumulated in fat tissue. Styrene and styrene-metabolized products have been shown to produce chromosomal aberrations under certain conditions.

Acrylonitrile

Acrylonitrile is considered more toxic than the vinyl chloride or vinylidene chloride monomers with evidence of carcinogenicity in animals and potentially humans. It has an LD_{50} value in the range of 80-90 mg and 27 mg /kg body weight in rats and mice, respectively. After metabolic activation, it converts in to a mutagen by the liver enzymes. In animals, acrylonitrile is metabolized to cyanide. However, it is later converted in to thiocyanate and excreted through urine.

Ethylene

It has been found that ethylene inhaled at a dose of 11.5 g/m^3 (10,000 ppm) for 4 hours is acutely hepatotoxic in rats. High concentrations of ethylene can cause asphyxia by lowering oxygen concentration in humans.

Propylene

The massive production of both propylene and polypropylene and the general use of the polymer over the past several decades cause exposure of both workers engaged in their production and general population to these compounds. No epidemiological studies or case reports relating to the carcinogenicity of these compounds are available.

Polyethylene Terephthalate (PET) Oligomer

PET due its high melting point (about 265 °C, especially the crystalline PET) is used for the production of food contact materials suitable for thermal processing and microwave treatments. The main volatile substance found in PET is acetaldehyde, which is of concern due to its effect on odour quality of the food or beverages. However, it has been shown that PET contains small amounts of oligomers ranging from dimers to pentamers which could have been released during thermal processing or microwave processing. Depending on the type of PET (amorphous or crystalline), these oligomers have been detected at levels that range between 0.06 to 1.0%.

Caprolactam

Caprolactam is one of the constitutional units of a type of nylon. Studies have indicated that nylon-6 oligomers and caprolactam residues (monomer of nylon-6) migrate into boiling water. No acute toxic effects of caprolactam were reported up on oral administration. However, studies have indicated that it can cause undesirable bitter taste in foods and may exert a minor but sustained effect on thermo-regulation. Further, it was proved that occurrence of irritation symptoms and burning of eyes, nose, throat and skin, as well as headaches, malaise, confusion and nervous irritation could be due to inhalation of high levels of ε-caprolactam. According to a study, in workers who were exposed to ε-caprolactam for long time, immunological and dermatological changes, and gastrointestinal, neurological, and cardiovascular effects were reported.

Food Safety Concerns of Additives used in Plastics

Plasticizers such as di-2-ethylhexyl phthalate, dioctyl phthalate and di-2-ethylhexyl phthalate (DEHA) belonging to the group of esters of phthalic acid and adipic acid are the most common ones. They are used in plastisol (cap-sealing resins) and gaskets of bottled food for sealing, PVC films and other plastic packaging materials. In general, they do not persist in the outdoor environment due to biodegradation, photodegradation and anaerobic degradation. However, depending on their respective volatility, some may be present in higher concentrations in air compared to others. PVC films are commonly used as "stretch films" or "cling" films for wrapping fresh fruits, vegetables, poultry meat, red meat and cheeses at retail level. As mentioned earlier, PVC is almost always processed with compounds such as adipates and pthalates. Goulas *et al.* (2000) studied the migration of DEHA from food grade PVC film used to wrap *Kefalotyri*, *Edam* and *Feta* cheeses. It was reported that at the end of 20 days (240 h) of contact time at 4 °C, the migration of DEHA was approximately 18.9 mg/dm^2, 12.2 mg/dm^2 for Edam and 7.3 mg/dm^2, respectively for *Kefalotyri*, *Edam*, and *Feta* indicating that the loss of DEHA from the PVC film. The migration kinetics of diphenylbutadiene (DPBD) from low density polyethylene (LDPE) film (444 μm) into whole milk powder (WMP), skim milk powder (SMP), toast with fat, wheat flour, honey and rice was studied by Silva *et al.* (2009) who reported that migration of DPBD was highest in

WMP followed by toast with fat, flour and rice and was negligible in honey and SMP. It was concluded that fat content greatly affected the migration of DPBD in dry foods. Thermal stabilizers often added to plastic food contact materials and their degree of purity affects toxicity. For example, residual ethylene oxide from epoxidised vegetable oil is a highly toxic material.

Safety Concerns of Paper-based Packaging Materials

Polychlorinated dibenzo-*p*-dioxins and polychlorinated dibenzofurans, are referred to as *Dioxins*, are characterized by high toxicity with the most toxic isomer being 2,3,7,8-tetrachlorodibenzo-*p*-dioxin. Dioxins are used in the manufacturing paper-based packaging materials. As a photo initiator for inks and lacquers treated with UV radiation and as a wetting agent for pigments or as a flow rate enhancer, benzophenone is widely used. UV light is widely used for treating printing inks since it helps in fast online cutting and folding. Studies have found that benzophenone and 4-methoxybenzophenone can act as carcinogens, as they have estrogenic effects and high toxic effects on genes. *Nitrosamines* and some of its precursors, found in certain papers and wax coated containers, are known to be genotoxic carcinogenic substances. N-nitrosomorpholine and morpholine can lead to formation of nitrosamines inside the human body when ingested by interacting with ingested or salivary nitrite.

Safety Concerns of Metal Packaging Materials

Food packed in tin cans with soldered side seams are a source of number of heavy metals such as lead, chromium, cadmium, etc. Technological developments such as introduction of two-piece cans (cans without side seams) and cans with electro-welded side seams have been very beneficial in reducing the intake of lead from this source. The highly toxic effects of lead can significantly affect the central nervous system. It is often reported that infants are much more susceptible to the toxic effects of lead with a high percentage of lead being accumulated in bones and brain in comparison to adults. Subacute ingestion of lead can lead to convulsions, encephalopathy, and mental retardation in children.

As a result of the use of tinplate for food and beverage packaging, it is obvious that some tin will dissolve into the food content, particularly when plain uncoated internal surfaces are used. Lacquering of the inside of cans greatly reduce the contamination of foods by tin. Tin (*Sn*) exist in two different chemical oxidation states, namely divalent tin (Sn^{2+}, stannous tin) and tetravalent (Sn^{4+}, stannic tin). In addition to the *Sn* metal, dissolution of metallic tin from the inside of the can body into food content will result in presence of tin in the divalent form. In general, the levels of tin in foods, which are responsible for acute symptoms such as gastro-enteritis *i.e.* nausea, abdominal cramps, vomiting, headaches, diarrohea and fever are between 250 and 2000 mg/L. According to the Codex, the tolerable weekly intake for tin is 14 mg/kg body weight while the recommended maximum permissible levels of tin in food are typically 250 mg/kg for solid foods and 150 mg/kg for beverages. The toxicological response of tin is affected by its chemical nature with the divalent tin being most important in in a canned food product. Inorganic tin is said to be poorly absorbed from the gastrointestinal tract in humans and animals, possibly due to low solubility and hence, inorganic tin has been termed as "essentially non-toxic". In human subjects exposed to tin at a concentration of 1370 ppm through ingestion, nausea and/or diarrhoea were reported.

The leaching of aluminum from cooking vessels used to prepare high-acid foods has been known for a long time. A study showing an association between Alzheimer's disease and aluminum levels in drinking water raised concerns about all sources of aluminum. When substantial leaching into acid foods occurs, it may highly contribute to the daily intake of aluminum apart from drinking water. In general, aluminum intake should not regularly exceed the FAO/WHO provisional tolerable weekly intake of 7 mg/kg body weight.

Can lacquers or enamels such as epoxy resins contain two or more epoxide groups per molecule and can be formed by condensing epichlorohydric and bisphenol A (BPA). Depending on the different degrees of condensation, bisphenyl A diglycidyl ethers (BADGE) are formed. BPA is an endocrine disruptor that has been connected with estrogenic action. Their toxic effects are basically dependent on fractional concentration of unreacted epoxy groups.

Safety Aspects of Glass and Ceramics

Glass has been used for several years and has raised very few toxicological concerns. The only major concern is the leaching of lead from high-quality crystal glass, which may contain up to 30% lead oxide. The lead content of wines and spirits stored in lead crystal decanters increases over time and may reach to a level of 1-2 mg/mL after 3-4 months storage or exceptionally up to 21 mg/mL after several years of storage. Like glass, ceramic materials only caused problems through leaching of heavy metals from enamel and pottery glazes, particularly when comes in contact with acidic beverages such as fruit juices and cider. Lead toxicity in humans may occur due to exposure to lead glazed ceramic ware for a few months. Although cadmium from ceramic glazes has not been identified as a definite cause of toxicity, damage to the kidney from ingestion over a period of years could occur.

Migration Aspects of the Migrants

Food packaging must protect food from contamination until it is consumed and help to keep the food fresh. It should also fulfil other requirements, such as to convey information about the food stuff and present it in an appealing manner. Plastics are able to fulfil the functional requirements and offer unparalleled advantages compared to other materials. It is equally important that the food contact material is inert and does not contaminate the food stuff through migration or transfer of substances used in its manufacture. Also, food may interact with packaging materials and this may change the initial properties and safety of the product. The interaction may be classified into three phenomena: migration, permeation and sorption. Migration of potentially toxic substances from packaging material to contact phase is also a major concern in the selection and use of materials for food packaging because of the possible effect upon human health. There are two types of migration *viz.* global and specific migration.

Global migration refers to total transfer i.e. the quantity of all substances migrating from the package into the packaged food while *specific migration* is related to the transfer of one or more identifiable substances that is a constituent of the packaging material. The main factors that control the migration are:

i. ***Diffusion process of migrants***: Migration of chemical substances is a diffusion process subject to both kinetic and thermodynamic control, and can be described by diffusion mathematics derived from Fick's Law. The mathematics describe the diffusion process as a function of time, temperature, thickness of the material, amount of chemical in the material, partition coefficient and distribution coefficient. The kinetic dimension of migration dictates how fast the process of migration occurs. The thermodynamic dimension dictates how extensive the transfer of substances will be when migration is finished i.e. when the system is at equilibrium.

$$\frac{dC_P}{dt} = D_P \times \frac{d^2C_P}{dx^2}$$

$$\frac{dC_S}{dt} = D_S \times \frac{d^2C_S}{dx^2}$$

where, D_p and D_S are diffusion coefficients (diffusivity) of migrant in the polymer ($m^2\ s^{-1}$) and food phases ($m^2\ s^{-1}$), respectively; C_p and C_S are concentrations of migrants in the polymer and food phase, respectively; x is the space coordinates measured normal to the polymer-food interface (m) and t is the time (s).

ii. ***Composition of packaging material***: The packaging material is the source of chemical migration. The extent of any migration depends on the concentration of the chemical in the packaging material. If a substance is present in the packaging material, other things remaining equal, migration levels will be higher if the concentration of the substance in the packaging material is increased and *vice versa*.

iii. ***The nature and extent of contact***: This depends on the physical properties of the food and, the size and shape of the pack. Further, if a chemical that may migrate is located in one layer of the packaging material but is separated from the food by an intervening layer, then this barrier layer may retard or prevent migration from occurring.

iv. ***Nature of the food***: If the packaging is not compatible with a given type of food then there can be a strong interaction leading to an accelerated release of chemical substances. Examples are the interaction of fats and oils with certain plastics that leads to swelling of the plastic and leaching of substances from that plastic.

v. ***Temperature of contact***: The migration of chemicals is accelerated by heat. Migration will occur faster if the temperature is raised.

vi. ***The duration of contact***: Materials suitable for short duration contact may not be suitable for longer service times. The kinetics of migration is first-order in that the extent of migration increases according to the square-root of the time of contact: $M \propto t^{1/2}$.

vii. ***Mobility of the chemicals in the packaging material***: The mobility of a chemical in the packaging material depends on the size and shape of the molecule, any interaction it experiences with the material, and the intrinsic resistance to mass transfer that the material presents. It is assumed that the chemical is compatible with the material. If the chemical is not compatible then it could "bloom" to the surface and give enhanced migration. Permeable materials such as plastics, rubbers and elastomers offer some limited resistance to migration, but this occurs not only from the surface but also from the interior of the material. The resistance to mass transfer depends on structure, density, crystallinity, etc. of the material.

Table 1: Recommended stimulants for determination of migration into food stuffs

Sr. No.	Stimulant	Description
1	A	Distilled water or water of equivalent quality
2	B	3 per cent acetic acid (*w/v*) in aqueous solution (using stimulant 'A')
3	C^1	10 per cent ethanol (*v/v*) in aqueous solution for food stuffs having alcohol less than 10 per cent (*v/v*) (using the stimulant 'A')
4	C^2	50 per cent ethanol (*v/v*) in aqueous solution for food stuffs having alcohol more than 50 per cent (*v/v*) (using the stimulant 'A')
5	D	*n*-heptane (freshly distilled)
6	E*	Rectified olive oil or mixture of synthetic triglycerides or sun flower oil

* Stimulant 'E' suggested by EEC for fatty foods is not considered at present as the methodology of estimation is not yet developed.

Determination of global migration of constituents from plastics

The Bureau of Indian Standards has laid down standard procedures for estimating the overall/global migration of constituents of plastics materials and articles intended to come in contact with foodstuffs (IS 9845: 2016R). The standard is applicable for determination of overall migration of constituents of single or multi-layered heat sealable films, single homogenous non-sealable films, finished containers and closures for sealing as lids in the finished form, preformed or converted form. The determination is to be carried out by using stimulants (Table-1) for different foods categorized in Table 2. The maximum extraction value of the material shall not exceed 10 mg/dm^2 or 60 mg/Kg.

Table 2: Stimulating solvents for different food products

Sr. No.	Type	Description	Examples	Stimulants
1	I	Aqueous, non-acidic foods (pH > 5) without fat	Honey, mineral water, **skimmed milk**, ***rasogolla***, *murabba*, yeast paste, etc.	A
2	II	Aqueous, non-acidic foods (pH 5) without fat	Fruit juices, squashes, jams, jellies, carbonated beverages, sauces, RTS beverages, soups, lemonade, vinegar, etc.	B
3	III	Alcoholic beverages: i. Alcohol concentration less than 10 per cent ii. Alcohol concentration above 10 per cent	 Beer and some pharmaceutical syrups Wine, brandy, whiskey, arrack and other alcoholic drinks	 C^{1} C^{2}
4	IV	Oils, fats and processed dry foods with surface fat or volatile oil	Vegetable oils, ghee, vanaspati, cocoa butter, lard, spice powder, snacks and savoury, chocolate, caramels, malted foods, egg powder, fried and roasted nuts, tea, coffee powder, etc.	D
5	V	Non-acidic foods (pH>5) or high fat and having high moisture content	Butter, bread, pastry, cakes, shrikhand, milk-based sweets, ice-cream, moist and fatty confectionery products	A and D
6	VI	Acidic foods (pH<5) or high fat and having high moisture content	Pickles, ketchup, cheese, curd, fresh and processed meat and fish products, sauces having fat, frozen foods, mayonnaise, etc.	B and D
7	VII	Dry processed foods without fat	Cereals and pulses, dehydrated vegetable and fruits, dried yeast, corn flakes, salt, sugar, milled products, barley powder, oats, vermicelli, spaghetti, etc.	No end test

References

Goulas AE, Anifantaki KI, Kolioulis DG and Kontominas MG (2000). Migration of di-(2-ethylhexyl) adipate plasticizer from food grade polyvinyl chloride film into hard and soft cheeses. Journal of Dairy Science, 83: 1712-1718.

IARC (1979). Monograph on the Evaluation of the Carcinogenic Risk of Chemicals to Humans: Some monomers, Plastics and Synthetic Elastomers and Acrolein. International Agency for Research on Cancer, Lyon. Vol. 19.

IARC (1986). Monograph on the Evaluation of the Carcinogenic Risk of Chemicals to Humans: Some Chemicals used in Plastics and Elastomers International Agency for Research on Cancer, Lyon. Vol. 39.

IS 9845 (1998). Determination of the overall migration of constituents of plastics materials and articles intended to come in contact with food stuffs-method of analysis. Bureau of India Standards, New Delhi.

Silva AS, Freire JMC, Sendon R, Franz R and Losada PP (2009). Migration and diffusion of diphenylbutadiene from packages into foods. Journal of Agriculture and Food Chemistry, 57: 10225-10230.

6

Foodomics for Detection of Adulteration: Principles and Emerging Trends

Janifer Raj Xavier and Gopal Kumar Sharma

Defence Food Research Laboratory, Defence Research and Development Organization, Mysore, Karnataka, India

Introduction

Food adulteration compromises quality of foods and causes human health hazards is a menace mankind had been facing since ages. Spice adulteration scandals in the year 2004, Horse gate scandal in Europe during 2013 are famous examples of food fraud. Various forms of food adulteration and cheating such as substitution of cheaper food components in place of original product, addition of non-permitted substances, and false claims by misbranding or mislabeling are prevalent from time immemorial (Spink *et al* 2019). Detection of adulteration of foods has an imperative role to guarantee safe, healthy and quality food. Food fraud accounts to 40 billion US dollars per annum and gains obtained by fraudsters are considered no less than profits of drug trafficking (PwC, 2016; Esteki *et al* 2019). Traditional analytical methods in food science and technology has taken a new model, by using contemporary approach by integration of various latest technologies for food quality testing, safety, traceability, interaction of food components as well as their implications on human health and wellness. In this context, foodomics has emerged as a relevant platform with integration of advanced technologies such as genomics, proteomics, transcriptomics, metabolomics to achieve the goal of food safety by rapid and precise detection of food adulteration. Novel technologies based on deoxyribonucleic acid DNA (gene), messenger RNA (transcriptome), protein, metabolites and their integration and application in food science and technology is termed as foodomics. Traditional methods to ensure food safety are reliable and precise however, has limitations in terms of consumption of long time to obtain precise results while cutting edge 'omic' technologies are quick and achieve objectives which are impossible by other analytical methods.

Foodomics as an advanced comprehensive strategy is gaining importance in tackling various features of food quality, traceability, safety, bioactivity in the context of food adulteration control and mitigation in the modern world (Figure 1). World as a global village is facing tremendous challenges in the form of food security owing to growing population, increased need for food resources, emerging cases of adulteration for economic benefits and their impact on sustainable food production systems as well as compromised health of mankind. Recent breakthrough technologies and their applications in food systems such as identification of origin of foods, biomarkers for nutrients and toxins, bioactivity pattern recognition and detection of contamination by genetically modified produce may pave way to advancements in food science and technology (Alvarez-Rivera *et al* 2019;Corsaro *et al* 2021).

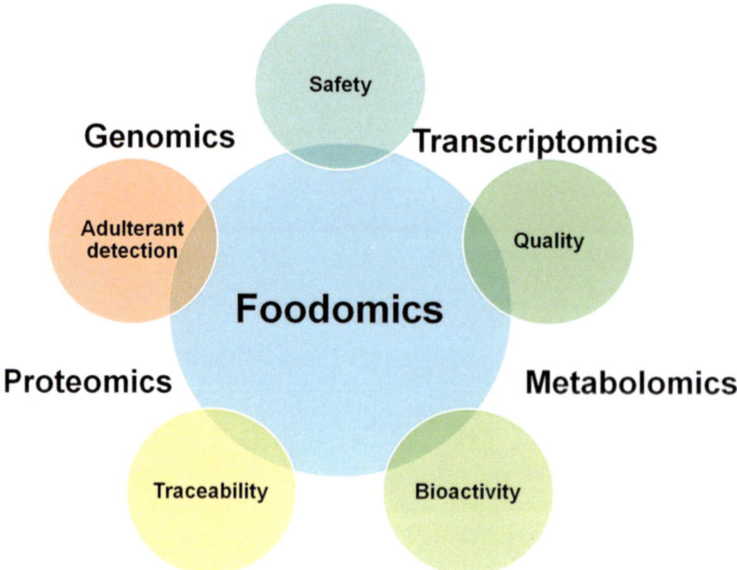

Fig. 1: Foodomics for food applications

In the present scenario, requirement for safe food is growing everyday and detection of adulteration either intentional or non-intentional ways are considered essential. Physical parameters of food such as colour, appearance, flavor, etc., and chemical parameters specific for foods are taken into consideration to certify the food as safe. Parameters for quality checks used from time immemorial are used as targets of various 'omics' approaches to determine level of adulteration (Cifuentes, 2021). Need for extended knowledge on interaction and complexity of food components of a specific food on health, has opened up venues for development of modern technologies and their incorporation in food related research. Amalgamation of modern analytical techniques and latest extensive data mining tools with field of foodomics has resulted in research fields such as food microbiome, food allergome, food digestome etc. which would primarily be utilized for studying food-health interaction. Data generated by various approaches of foodomics need to be unified and analysed in a meaningful manner to derive fruitful conclusions and understand globally obtained results (Leon *et al* 2018).

Foodomics driven research has generated vast data with help of bioinformatics which could be exchanged between research groups as well as food processing industries in order to design, develop, optimize and validate methods suitable for safe and desirable development of food products and processes. Collaborative ventures by public and private institutions in the field of foodomics and establishment of centers of excellence for omic technologies may result in generation of enormous scientific data useful for knowledge exchange between academia and industries. Zhang *et al* (2020) created FADB-China, the worldwide first database based on detection of food adulteration using molecular tools. Cases of food adulteration of about 961 were collected and analyzed for contaminants over a period of two decades (1998 to 2019) as per records from literature and government sources. Illegal food additives and their molecular fingerprints are stored in this database which could be used to predict around 1919 chemicals commonly used for adulteration. Moreover, these predictors derived based on *in silico* studies are potential markers to prevent food adulteration.

Availability of household approaches for detection of food adulteration is mostly qualitative in nature while complex procedures hosting omics technologies are precise and quantitative in nature (Table 1). This chapter summarizes the various foodomics approaches available for detection of food adulteration in different commodities as well as their advantages and disadvantages. Recent advances reported in research studies based on genes, proteins and metabolites have been reviewed in detail, with emphasis on those to determine food authenticity, traceability, quality and safety.

Table 1: List of selected foods and modern foodomic technologies available for detection of adulteration

S.No.	Food category	Parameter tested	Detection method	Reference
A. Plant based foods				
	Italian Durum Wheat Pasta	Wheat cultivated outside Italy	Fourier-transform and near-infrared (FT-NIR) spectroscopy	De Girolamo *et al* (2020)
	Rice	Authentication	Ambient mass spectrometry and vibrational Raman spectroscopy	Uawisetwathana and Karoonuthaisiri (2019)
	Thailand Hom Mali rice	Geographical origin	Fourier-transform infrared spectroscopy with attenuated total reflection (FTIR-ATR) and near-infrared (NIR) spectroscopy	Srinuttrakul *et al* (2021)
	Edible oil	Authentication	Advanced analytical techniques	Sudhaka *et al* (2021)
	Edible oil	Quality, authenticity, and adulteration	Fourier-transform infrared (FT-IR) spectroscopy	Mousa *et al* (2021)
	Coconut oil	Palm kernel oil and palmolein	Reverse phase-Liquid Chromatography (RP-HPLC) based triacylglycerol profile	Komaram *et al* (2021)
	Coconut milk	Water	Fourier-transform infrared (FT-IR) spectroscopy	Sitorus *et al* (2021)
	Dye contaminated vegetables	Copper sulfate, malachite green, and Sudan red	^{1}H-nuclear magnetic resonance (NMR) relaxometry	Shomaji *et al* (2021)
	Cassava	Flavone, phenylpropanoids and alkaloids	Transcriptome and metabolome profiles	Xiao *et al* (2021)
	Apple juice concentrate	Glucose syrup, fructose syrup, and date concentrate	High-performance liquid chromatography coupled with a refractive index detector (HPLC-RID)	Yeganeh-Zare *et al* (2021)
	Lime juice	Detection of fraud	Fourier-transform infrared (FT-IR) spectroscopy	Mohammadian *et al* (2021)
	Pistachio	Green pea and peanut	Portable Fourier transform mid-infrared (FT-MIR) and Fourier transform near-infrared (FT-NIR)	Aykas and Menevseoglu (2021)
	Chocolate	Cocoa butter substitutes	Differential scanning calorimetry (DSC) and fatty acid composition analysis	Yazdi *et al* (2021)

S.No.	Food category	Parameter tested	Detection method	Reference
	Coffee	Common adulterants: chicory, barley, and flours	Untargeted high-performance liquid chromatography with ultraviolet and fluorescence detection (HPLC-UV-FLD) fingerprinting	Núñez *et al* (2021)
	Turmeric (*Curcuma longa*)	Other turmeric species	Loop-mediated isothermal amplification (LAMP) ITS2-26S rDNA	Sheu *et al* (2021)
	Turmeric powder	Sudan Red, starch and metanil yellow fraud	Infrared (IR) spectroscopy	Khodabakhshian *et al* (2021)
	Ginger powder	Grade 3 chickpea powder	Convolutional neural networks of powder image samples	Jahanbakhshi *et al* (2021)
	Vanilla pods, tinctures, paste (inner part), oleoresin and powders	Synthetic ethyl vanillin	Planar chromatography	Morlock *et al* (2021)
	Saffron	Chemical Element Compositions	Statistical analysis	Templ and Templ (2021)
	Bee products	Cheaper adulterants	Analytical techniques based on DNA and spectroscopy	Wang *et al* (2021)
	Honey (Manuka and Jarrah)	Sugar syrups (rice, corn, golden, treacle, glucose and maple syrups	High-Performance Thin-Layer Chromatography (HPTLC)	Islam *et al* (2021)
	Stingless bee honey	Vinegar and floral honey	Untargeted ^{1}H NMR metabolomics	Yong *et al* (2022)
	Australian poly floral table honey and New Zealand *Leptospermum scoparium* honey	Geographical origin	Next Generation Sequencing (NGS) - small RNAs	Smith *et al* (2021)
	Honey	Corn syrup	Ultra-rapid real-time PCR	Truong *et al* (2021)
	Essential oils	Vegetable oils	^{13}C NMR Spectroscopy	Truzzi *et al* (2021)
	Turkish delights (lokum)	Corn syrup adulteration instead of sucrose	Fast Field Cycling Nuclear Magnetic Resonance Relaxometry	Pocan *et al* (2021)

S.No.	Food category	Parameter tested	Detection method	Reference
	Japanese Chardonnay wines	Geographical origin	Oxygen stable isotope and multi-element analyses	Akamatsu *et al* (2022)
B.Dairy based foods				
	Milk	Riboflavin for checking spoilage and urea adulteration	Fluorescence spectroscopy	Pandey and Joshi (2021)
	Milk	Melamine	Excitation-emission matrix (EEM) fluorescence spectroscopy	Barreto *et al* (2021)
	Powdered milk	Melamine	Isotope Dilution Liquid Chromatography-Mass Spectrometry (ID LC-MS)	Strashnov *et al* (2021)
	Infant milk formula	Mineral fortification	Inductively coupled plasma mass spectrometry	Cilla *et al* (2021)
	Human, cow, donkey, mare, goat, sheep milk	Monophosphorylated metabolite	^{31}P NMR metabolite profile of milk serum	Bruschetta, *et al* (2021)
	Cheese	Cheese ripening	Next Generation Sequencing (NGS)	Afshari *et al* (2020)
	PDO feta cheese (prepared from mixture of sheep/ goat milk)	White cheeses (prepared from cow milk	Untargeted protein based Matrix-assisted Laser Desorption/Ionization -Time-Of-Flight Mass Spectrometry (MALDI-TOF-MS)	Kritikou *et al* (2021)
	Rice, milk, and different ready-to-eat meals	Cereulide dipeptide toxin *Bacillus cereus*	Matrix Assisted Laser Desorption/Ionization-Time of Flight (MALDI-TOF) mass spectrometry	Ducrest *et al* (2019)
C. Meat based foods				
	Sheep (*Ovis aries*) meat	Mutton-in-beef adulteration	Alkaline lysis (AL) and loop-mediated isothermal amplification (LAMP) technique	Mounika *et al* (2021)
	Beef	Chicken, duck, and pork meat	Electronic tongue (e-tongue)	Lu *et al* (2021)
	Beef	Offal	Laser induced breakdown spectroscopy	Velioglu *et al* (2018)
	Meat	On-site detection of food authenticity	CRISPR-/Cas12a-Mediated Liposome-Amplified Strategy	Liu *et al* (2021)
	Meat products	Cold cut meat detection	Proteomics	Huang *et al* (2020)

S.No.	Food category	Parameter tested	Detection method	Reference
	Halal foods: Beef luncheon meat and spiced beef	Discrimination of pork meat from beef, mutton, and chicken	Amplification-free and mix-to-read CRISPR-Cas12-based nucleic acid method	Wu *et al* (2021)
	Cow, goat, donkey and chicken	Protein biomarkers for meat quality	^{1}H-NMR Nuclear Magnetic Resonance spectroscopy	Akhtar *et al* (2021)
	Edible bird's nest	Cheaper ingredients nutrient agar, collagen, gelatine, karaya gum and melamine	^{1}H-NMR Nuclear Magnetic Resonance spectroscopy	Yong *et al* (2022)
	Seafood, meat, edible mushroom and insect	Authentication	MALDI-TOF MS profiling	Pavlovic *et al* (2021)
	Fresh sea foods	Frozen-Thawing	Near-infrared (NIR) spectroscopy	Hassoun *et al* (2020)
	Wild tropical tuna	Highly migratory large pelagic fish	^{1}H-NMR Nuclear Magnetic Resonance spectroscopy	Bodin *et al* (2022)
	Mi-iuy croaker *Miichthys miiuy* Republic of Korea	*Sciaenops ocellatus*	Portable ultra-fast PCR	Hong *et al* (2021)
D. Fermented foods				
	Microbial fermentation products	Antimicrobial resistance (AMR) genes	Shotgun metagenomics	Buytaers *et al* (2021)
	Food-related microbial isolates and mixed microbial communities in foods	Effect on gut microbes	High-Throughput Sequencing	Walsh *et al* (2017)
	Fermented sausage and salami products	Identification of beef, chicken, and pork	Protein-based Laser Induced Breakdown Spectroscopy (LIBS)	Sezer *et al* (2021)

Detection of adulteration: A foodomics approach

Foodomics is defined as "the collection of all compounds present in any investigated food sample and/or in any biological system interacting with the investigated food at a given time". In early 2009's Cifuentes defined foodomics as "a discipline that studies the food and nutrition domains through the application and integration of advanced omics technologies to improve consumer's well-being, health, and confidence". Foodome is defined as "the collection of all compounds present in any investigated food sample and/or in any biological system interacting with the investigated food at a given time" (Cifuentes, 2017). Foodomics as an integrated approach of various omics technologies relevant to food science are genomics, transcriptomics, proteomics, and metabolomics which are capable of exploring various expressions levels of DNA, mRNA, protein data and metabolites, respectively (Figure 2). Molecular level studies on food has enabled aspects of nutrition, their composition, interaction with other factors and influence on health and changed perception of food to a global level (Zheng and Chen, 2014). Álvarez-Rivera *et al* (2021) described in detail concepts of foodomics, state-of-the-art as well as application of omics technologies in food science. Moreover, unification of omic technologies ensured a broader coverage of foodome in detail.

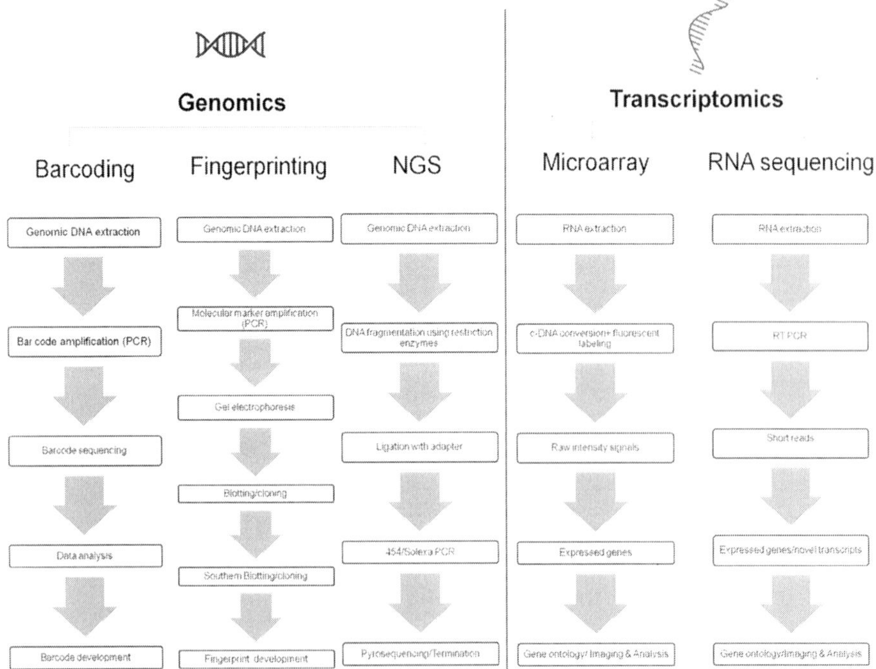

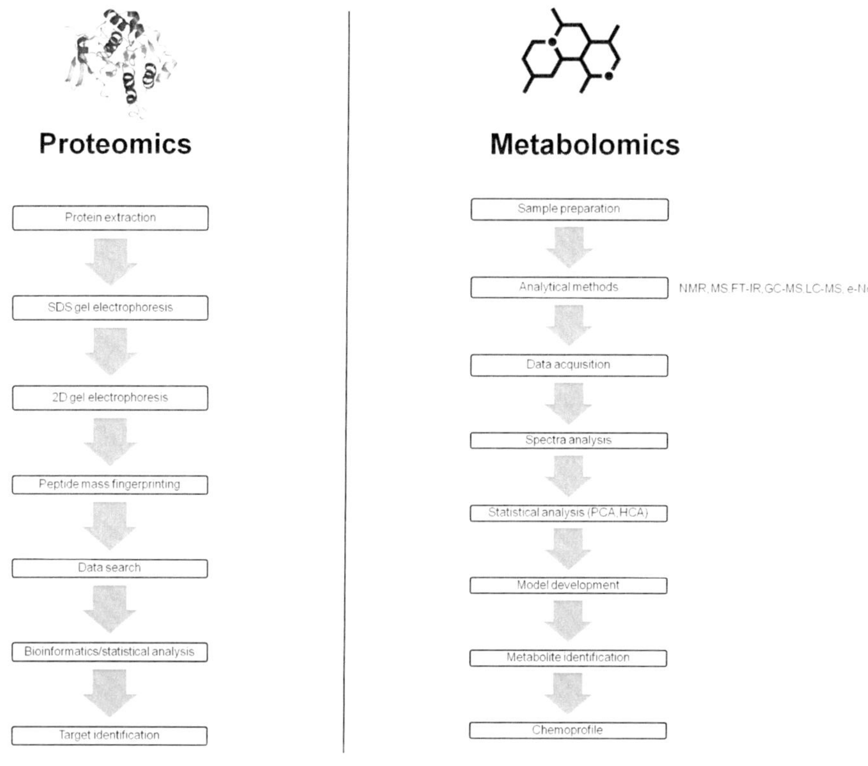

Fig. 2: Workflow of a) genomics, b) transcriptomics c) proteomics and d) metabolomics methods for food applications

Moreda-Piñeiro and Moreda-Piñeiro (2021) emphasized the importance of sample preparation and recent advances in the field for achieving a comprehensive foodomics based studies. Food being a complex substance rich in carbohydrate, fats and proteins, interferences during extraction and analysis are commonly encountered. Pre-concentration of target metabolites either concurrently or orderly, enrichments using sorbents, microextraction techniques based on solid, liquid, membranes or clean-up strategies were discussed. Microwave, ultrasound, high pressure, supercritical fluid and QuEChERS (quick, easy, cheap, effective, rugged and safe) combination methods for effective extraction and sample preparation for further analysis were elaborated. Green chemistry and their applications in field of foodomics for a cleaner scientific analysis were reported. Use of green alternative deep eutectic solvents for analysis as well as approaches through biorefinery for novel methods in food product development and analysis were also elaborated. Tools based on proteomic and metabolomics as eco friendly alternatives of commonly used raw materials for a greener approach to foodomics were also discussed.

Casas-Agustench and Andres-Lacueva (2020) reviewed the importance of systems biology and its integration with omic technologies to study diet-related disease progression and regulation of metabolic pathways, signaling and personalized nutrition. Physiology of living systems and complexity of processes could be studied using latest techniques such as transcriptomics, genomics, proteomics, metabolomics and chemometric data analysis. Lifestyle disease progression and influence of nutrition on regulation of human health and wellness may be explored in detail using techniques such as nutriproteomics, nutrimetabolomics, nutrigenomics, nutrigenetics and nutrimicrobiomics.

Recent technologies of foodomics are found superior in identification of bio-derived toxins in comparison to conventional methods. These toxins cause food borne illness and are secreted by food borne pathogens and could be identified by sensitive, accurate and robust assays by methods of genomics, proteomics and metabolomics. Furthermore, these high-throughput techniques when used in integration are capable of conducting real time surveillance of causative agents of food borne pathogens and detect or predict outbreaks and poisonings caused by biotoxins due to consumption of contaminated foods (Rešetar and Pavelić, 2018).

i. Genomics

Studies based on variability of DNA sequences and their assessments are termed as genomics. Techniques involving molecular markers such as Simple sequence repeat (SSR), Inter Simple Sequence Repeat (ISSR), Single nucleotide polymorphism (SNP), Amplified fragment length polymorphism (AFLP), TaqMan based real time PCR, high throughput sequencing techniques viz. Next generation sequencing (NGS), Whole-genome sequencing (WGS), other methods such as digital droplet PCR (ddPCR), High Resolution Melting (HRM), Padlock Probe Ligation and Microarray Detection (PPLMD), Capillary Electrophoresis (CE) are used to study the differences in DNA. Microarrays and NGS are described as important analytical techniques for gene based omic technologies. Targeted and untargeted analysis is possible by DNA based methods as genomic material is highly resistant for changes, large amount of details are generated with high specificity and sensitivity. Disadvantages of DNA based methods include complex procedures and requirement of trained personnel to perform experiments, consumption of long time and destructive procedures. DNA fingerprinting in form of barcodes is a proven method to determine the authenticity of foods all over the globe. Barcoding information obtained from molecular marker analysis act as tags on foods comes handy when detailed analytical procedures of samples are not feasible. Integration of nanotechnology and DNA method introduced nanotracer, which is capable of determining food authenticity by naked-eye. Molecular traceability using this method was found useful in identifying substitutes used for food adulteration as well as presence of their diluted quantity (Valentini *et al* 2017).

Hong *et al* (2017) reported use of DNA based methods in combination with physico-chemical and sensory parameters to identify adulteration in six major classes of food prone to practices of food fraud. Detection of adulterants in foods such as cereal foods, spices, oils, beverages, honey, animal food, sea food and organic dietary supplements based on suitable chromatography, spectroscopy and DNA methods along with chemometrics was elaborated. Böhme *et al* (2019) reviewed authentication of foods and their components based on tools of genomics such as polymerase chain reaction (PCR), high throughput sequencing etc. Advantages of DNA based technologies to detect claimed ingredients in food include their ability to multiplex, enhanced sensitivity, cost effectiveness and ability to handle multiple parameters at one time. Fanelli *et al* (2021) elaborated on recent molecular tools widely used for authentication and traceability of food products and agricultural produce in their fresh and processed forms. Molecular marker based genomic tools are well established for fingerprinting of living material. Recent advances in DNA based methods such as metabarcoding and Loop-mediated isothermal amplification of DNA (LAMP) method involving detection through naked eyes are being used in agri-food sector. Even though, latest technologies have certain limitations they could contribute in reassuring consumers of safe food as well as producers to promote their products based on specific ingredients and their easy identification and authentication. Galimberti *et al* (2019) described tools such as DNA barcoding and meta barcoding for confirming authenticity of food products. These techniques act as diagonistic markers specific to ingredients of a particular food to identify them correctly. Aspects of healthy and safe nutrition for personalized nutrition would require tools based on DNA for reliable health benefit on their consumption.

Traditional culture dependant methods have limitations in studying the microbial profile of uncultivable microorganisms present in fermented foods. Complex bioactive production in fermented olives may be attributed to consortia of microorganisms used for development of final product. Fermented table olives form an important part of Mediterranean diet and are considered rich in bioactive compounds. Omic technologies have taken investigations on micro organisms, which play a major role in food fermentations to higher levels of sophistication in deciphering the functionality in food systems such as table olives (Vaccalluzzo *et al* 2020). Walsh *et al* (2017) reviewed the potential of omic technologies to study food microbiology. Studies on microbial strains used in preparation of fermented foods as well as mixed cultures present in fermented foods using latest approaches of transcriptomics, metatranscriptomics, genomics and metagenomics were elaborated. High throughput sequencing technologies and their use in understanding the mechanism of action of probiotic cultures, their safety for use in foods and composition of human microbiota on intake of probiotic rich foods were also explained. Buytaers *et al* (2021) described metagenomic tools used for differentiation of genetically modified and non-genetically modified microorganisms for development and production of fermented foods. Short and long read sequencing was carried out using shot gun approach and a proof-of-concept study was reported to identify genetically modified microorganism in tested product. Resistance genes for antibiotics, vector fragments as well as unnatural associations of genetically modified regions were used to locate genetically modified microbes in microbial fermented food products. Meta-omics technologies and their application to study cheese, a fermented dairy product with characteristic flavor and taste was elaborated under the term cheesomics. Microorganisms used for fermenting milk into cheese determine flavor, taste and colour of the end product. Gene sequencing of ribosomal RNA (rRNA), analysis and metagenomics tools and their role in decoding markers specific to cheese ripening for prediction of quality of obtained cheese were elaborated. Metaomics in form of metaproteomics, metatranscriptomics and metabolomics derived biomarkers specific for cheese quality and safety were described (Afshari *et al* 2020).

Amane and Ananthanarayan (2019) reported barcoding of food products based on black gram (*Vigna mungo* L.) for their authentication and determination of adulteration. Wheat and white pea flour are the common adulterants in black gram flour and barcodes namely ITS, rbcL and trnH-psbA were used for barcoding. Market samples of black gram products such as *papad*, *medu vada* mix, flour used for preparation of papad and flour were analysed using the primers which could amplify barcodes ITS, rbcL and trnH-psbA with amplicon sizes of 680, 600 and 380 base pairs (bp), respectively on PCR amplification. This method of multiplexing showed 100 percent results in identifying adulterant species in black gram based food products. Honey is a special food known for its health benefits from time immemorial. Small RNA profiling of honey by next generation sequencing for authentication has been described. Honey is being adulterated with sugar syrup and mislabeled by providing wrong information of its geographical origin. Polyfloral honey sourced from Australian and New Zealand were sequenced for their RNA and small RNA markers such as miRNA, non-coding RNA and some tRNA fragments were identified specific for type of honey. Unique small RNA signatures identified opened up novel targets for authentication of honey based on their geographical origin (Smith *et al* 2021).

El Sheikha (2019) reported use of DNAFoil, a portable DNA based on-site test kit that function without hi-tech laboratory and PCR equipments for rapid detection of food adulteration within a time span of 30 minutes. Detection of food adulteration using such novel approaches of DNA testing aims in creating impacts on health and economy as well as ensure safe foods to customer. Xiao *et al* (2020) reported a chicken-PCR-sensor for rapid identification of adulteration in chicken. Primers targeting mitochondrial DNA specific to chicken were designed and amplification reaction using PCR method was standardized. Rapid identification of chicken meat within 2-3 minutes up to 0.01 percent sensitivity was achieved in commercial meat samples.

ii. Protein Based Omics

Proteomics has evolved as a powerful technology to understand proteins, the complex biological macromolecules in detail. Food science and related applications of proteomics to understand composition, quality and safety of raw material has been elaborated. Proteomic studies based on mass spectrophotometric techniques such as mass spectrometry (MS), electrospray-ion trap (ESI-IT) MS, matrix-assisted laser desorption/ionization time of flight (MALDI-TOF) etc., characterized proteins based on distinct molecular signatures. Complex proteins present in food, interactions and their detailed nutritional function has been deciphered based on bioinformatics analysis of the MS generated data (Carrera *et al* 2021). Valletta *et al* (2021) elaborated recent trends in mass spectrometry and its tools for food fraud identification based on protein fingerprinting. Instrumentation used in MS such as Liquid chromatography-electrospray ionization tandem mass spectrometry (LC-ESI MS/MS), Matrix-assisted laser desorption ionization mass spectrometry (MALDI-TOF MS) and their use in discovery of biomarkers (proteins and peptides) has been elaborated. Geisslitz and

Scherf (2021) reviewed proteomic approaches to understand complexity of plant and animal based food stuffs and their quality control. Protein rich foods such as eggs, meat, cereals, milk and dairy have been used as raw material for various food products. Proteomics in combination with other foodomics approaches revealed structure, function and bioactivity of proteins present in foods as well as interaction of protein with other substances during digestion, absorption and modulation of immune system.

Identification of protein based biomarkers for purpose of food authenticity is used in immunochemical and MS methods, based on ability of the biomarker peptide to overlap linear epitopes of the polypeptide as well as a conserved segment for the tested food product (Minkiewicz *et al* 2015). Pavlovic *et al* (2021) reviewed non-targeted approaches of food authentication using Matrix-assisted laser desorption ionization time-of-flight mass spectrometry (MALDI-TOF MS) profiling of proteins in foods such as edible mushroom, meat , seafood and insects. Species identification for authenticity of foods using latest MS techniques as well as sample preparation and, methods available for comparison with available database has been elaborated. Information on protein component of foods is essential to study their complex interactions with food especially in the omics era. Functions of protein as well as the possibility of using them as markers in detection of adulteration using data derived from mass spectrophotometry was reviewed in detail. Workflow from sample preparation, analysis, quantification and recent studies involving use of MS-based proteomics were also elaborated (Andjelković and Josić,2018).

Public and private databases play an important role in providing information necessary for studying the proteins in detail based on their structure and functionality. Pathway Studio, Cytoscape, ingenuity pathway analysis and pathguide (Nikitin *et al* 2003) are some of the databases available for studying proteins based on their structure, function and pathways involved. Applicability of proteomics in food science and technology is a recent trend and Pathway Studio has protein networks for maize and rice while Reactome has pathways of baker's yeast (*Saccharomyces cerevisiae*), wild boar (*Sus scrofa*), cow (*Bos taurus*) and fowl (*Galloanserae* sp.) in their databases (Croft *et al* 2011).

Maestri *et al* (2016) elaborated the significance and abundance of biologically active peptides in foodstuffs with focus on peptides with immuno-modulatory property. Proteomic data of crop plants indicated that about 6000 protein molecules may harbor bioactive peptides. Gene ontology based approaches would help in narrowing down immuno-modulatory peptides suitable for the therapeutic applications. Availability of crop based proteomic data, analysis, and interpretation as well as finding biomarkers are made possible through latest omics technology such as proteomics. Martini *et al* (2021) defined food peptidomics as study of all peptides present in food which allows tracking the bioactive peptides released during digestion of foods in human body. Detailed information of peptides through mass spectrometric approaches as well as proteomics in combination would pave way to study components of food derived after hydrolysis of polypeptides available in food. Furthermore, peptidomics in relevance to valorization of protein rich food wastes has also been highlighted.

Whey is the popular body building supplement due to the presence of high quality whey protein. Bovine whey supplements are prone to get adulterated with cheaper sources such as potato, soybean, chicken, maize and wheat proteins. Shotgun proteomics methods detected adulteration in ten out of seventeen samples of whey based products using soy, rice and wheat. Furthermore, proteome and its identification specific to food products could serve as tools for product quality control (Garrido *et al* 2016). Health and safety of fish consumption has been jeopardized in recent past due to false substitution of valuable species and mislabeling. Biomarkers derived by proteomics would result in quick, in-line authentication of fish and avoid frauds related to fish trade. Mazze and Siciliano (2016) reported authentication of fish species used for preparation of fish based products such as baby food, fillets and sticks using proteomics. Powders produced using marine derived peptides are being subjected to adulteration with low-cost additives. Amino acid profiles were derived for peptide powders developed using raw materials such as sea cucumber, oyster, fish skin and Antarctic krill. Protein content as well as presence of amino acids such as lysine, glycine and taurine were used to determine authenticity of marine peptide powders. Developed method was 100 percent accurate in predicting specific marine peptides from mixture of samples (Wang *et al* 2021).

Arena *et al* (2016) described a peptide profiling based on MALDI-TOF-MS protein to detect freshness of buffalo milk specific to the geographical region Italy. It is used for production of a soft, stretched curd cheese named Mozzarella di Bufala Campana registered under Protected Designation of Origin of EU legislation limits. Markers for peptides viz. β-casein-derived phosphopeptides, γ2-, γ3- and γ4-fragments of β-casein, α-lactalbumin, β-lactoglobulin and phosphopeptides of GLYCAM1 were found to determine freshness of milk significantly. Moreover, studies pertaining to identification of biomarkers for determining the storage parameters of raw materials play a major role in ensuring authenticity of final produce. Kandasamy *et al* (2021) reported authentication of mozzarella cheese sourced from Korean farmsteads using matrix-assisted laser desorption ionization-time-of-flight mass spectrometry (MALDI-TOF MS) and multivariate analysis of obtained data. Seven protein markers were identified while two of the identified markers i.e. *m/z* 7407.8 and 11,416.6 were specific to farmstead sourced cheese. Protein profiling, creation of database and multivariate analysis could pave way for mitigating food frauds pertaining to cheese with respect to their origin.

Colgrave *et al* (2016) reported a method for determination of barley (*Hordeum vulgare*) contamination in processed foods. Barley cultivars (12 nos.) were studied to derive nine peptide markers using LC-MS/MS methodology. Barley contamination was studied using samples commonly adulterated with barley namely breakfast cereals, rye, oats, corn, rice, chia, amaranth, buckwheat and wheat up to 0.08 percent sensitivity. Methods involving chromatography may lead to replacement of expensive serological methods such as ELISA for identification of specific biomarkers of barley. Amane and Ananthanarayan (2019) also reported protein fingerprinting using 13 peptides of six Indian varieties of black gram using 2D-GE. Commercially available software PD Quest™ was employed to analyze the obtained protein fingerprints.

Protein biomarkers were identified for refined flour and white pea with molecular weight of 15.0 and 21.5 KDa, respectively to locate adulteration of black gram flour products with cheap substitutes even at level of 5.0 percent. Furthermore, this method also determined six unique proteins for varieties viz. TAU-1 and PDKV Black Gold.

Fasoli and Righett (2015) reported use of combinatorial peptide library for detection of specific proteins for authentication of fruit juices and alcoholic and non- alcoholic beverages. Commercially available beverages and their proteomic signatures consisting of abundant, non-abundant and trace proteins were compared to ensure genuineness as well as check for food allergens. Fruit beverages based on goji berries, mango, lemon, alchoholic beverages such as beer,champagne (France),amaro Braulio, Barnzi, Cynar liqueur as well as orgeat syrup, ginger ale, almond milk and cola were tested using proteomic and MS approaches. Zheng *et al* (2021) described proteomic signatures for honey sourced from geographical regions such as Changhua, Taichung and Tainan and botanical origin from *Litchi chinensis* (litchi) and *Dimocarpus longan* (longan) using proteomic analysis based on sequential window acquisition of all theoretical fragment ion spectra mass spectrometry (SWATH-MS). Chemometric analysis also identified adulteration of honey with cheaper sugar syrups up to 20 percent using the described platform.

iii. Chemometrics/Metabolomics

Profiling foods based on related metabolites such as polyphenols, pigments, peptides is termed as metabolite profiling while metabolic fingerprinting refers to approaches including comparison of phenoptypes grown or obtained from different locations, different diet patterns and relevance to variations caused by environmental factors. Metabolomics studies thus aim in profiling the specific compounds which are responsible for bioactivity as well as the changes in metabolite profile due to presence or absence of bioactives (Bayram *et al* 2018). Trimigno *et al* (2015) reviewed use of NMR based techniques as a tool for food quality assurance in terms of authenticity, chemical composition during food processing operations, storage and marketing of the same. Corsaro *et al* (2021) reviewed use of nuclear magnetic resonance (NMR) and their applications in the field of foodomics. NMR methods rely upon the chemical shifts caused by various functional groups present in the analyte and are useful in determining the authenticity of samples. Complete blueprint of metabolites/chemometric data present in the food sample obtained using NMR technique would confirm the chemical profile of the tested sample. Therefore, accurate information about the tested food sample ensures its composition, authenticity, bioactivity, safety and functionality. Sobolev *et al* (2017) reviewed NMR as a tool to monitor authenticity of food based on origin, genetics as well as adulteration by identification of biomarkers or profiling of entire metabolome. Characterization of metabolome with help of NMR and MS is gaining popularity due to the efficiency of these techniques in precise fingerprinting of food metabolites. Studies on interaction of food and microbiome, identification of dietary markers for nutrition and adulteration as well as nutritional interventions for personalized nutrition were also elaborated (Mancano *et al* 2018).

Alcoholic beverages are obtained by yeast fermentation and their characteristic properties are based on origin, additives, process conditions, starter culture and other natural parameters. Metabolomics based on NMR methods are found to be authentic in detection of adulteration due to reliable fingerprints obtained during analysis (Anne *et al* 2021). Bodin *et al* (2022) reported metabolomic approach to trace tuna, the most traded fish species. H_1 NMR derived fingerprint distinguished tunas based on condition of storage, size, species as well as geographical region. Metabolites targeted were dipeptides such as anserine, carnosine, organic acids creatine, lactate to discriminate white muscle while red muscle tuna markers were found to be free amino acids, choline, derivatives of choline, phosphatidylethanolamine along with anserine, carnosine and organic acids.

Lacalle-Bergeron *et al* (2021) reviewed literature published during five years period (2017-2020) on advances in untargeted metabolomic methods based on GC, LC and HRMS for detection of biomarkers in food. Chromatographic techniques coupled with mass spectrophotometry provided high resolution of results and proved to be highly sensitive and selective in nature. Application of these recent techniques on food research especially to identify adulteration based on derived biomarkers were also elaborated. Chemoprofiling or chromatographic fingerprinting of chemicals present in foods has been established as a reliable and innovative tool to identify and authenticate food in the field of metabolomics. Applications of suitable chromatographic methods for profiling of chemical components, their profiling by means of data acquisition followed by fingerprinting the food based on analysis of obtained data were discussed in detail. Gas chromatography and liquid chromatography has been explained in detail for purpose of food identification and authentication (Cuadros-Rodríguez *et al* 2021).

Foodomics studies reported based on ambient mass spectrometry (AMS) techniques involving desorption electrospray ionization (DESI), direct analysis in real time (DART), paper spray mass spectrometry (PS-MS) and easy ambient sonic spray ionization mass spectrometry (EASI-MS) were reviewed (Klampfl, 2018). Gilbert-López *et al* (2017) elaborated the use of targeted analysis of food metabolites using liquid chromatography (LC) and liquid chromatography-mass spectrometry (LC-MS) based techniques in context of foodomics. Use of amalgamation of chromatography and spectrophotometry techniques and their applications for exploring various parameters of food quality, safety and health were reviewed in detail. Two-dimensional liquid chromatography (2D-LC) and its applications in foodomics perspective have been reviewed in detail. 2D-LC is considered to a robust technique to be study complex matrices of food with high resolution in identifying potential components of food. Recent developments on usage of 2D-LC in foodomics, challenges and limitations have also been described (Herrero *et al* 2019).

Azcarate *et al* (2017) described use of metabolomics and its applications in detection of adulteration in infant formula prepared using milk. Nutritional content with respect to 24 minerals was studied using inductively coupled plasma mass spectrometry. Presence of metals namely Cu, Mn, S and Zn at higher concentrations in neat infant formula and fortified infant formula were detected using chemometric tools for data classification. Wang *et al* (2021) reviewed recent advances in detection techniques and methods for bee products in China. Growing demand for bee products for health benefits and even corporate gifting in form of royal jelly, honey bee pollen and propolis as well as globalization of markets lead to fraud in honey bee products. Different methods of sensory, genetic material, spectrophotometry such as isotope ratio mass spectrometry and MS has been reviewed for rapid detection of adulteration in context of China. Li *et al* (2022) reported differentiation of tartary and common buckwheat seed using metabolic profiling using Ultra High Performance Liquid Chromatography coupled to triple-quadrupole Mass Spectrometry (UHPLC-QqQ-MS). Tartary buckwheat indicated presence of two fold increased flavonoid content (61 nos.) and non-flavanoids (94 nos.) in comparison to common buckwheat. Andjelković *et al* (2017) reviewed foodomics approach to detect food allergens based on MS based approach. DNA and protein based methods and their role in allergen detection has been elaborated in various studies and their limitations related to reliability, accuracy and sensitivity led to development of chromatographic methods for food allergen detection. Patel *et al* (2021) elaborated recent advances and potential of capillary electrophoresis mass spectrometry for analysis of food reported in the previous decade (2010 to 2020). Various combinations/interfaces such as CE-ESI-MS, CE-ICP-MS and offline CE-MALDI MS and their application in foods of solid and liquid phase were described.

Kalogiouri *et al* (2020) reviewed in detail the studies related to authentication of extra virgin olive oil using mass spectrometric methods of high resolution coupled with chemometric techniques. Organoleptic properties, type of cultivation, origin based on biology and geography as well as addition of common adulterants were the major parameters studied using advanced techniques for chemo profiling and identification of biomarkers for olive oil of type extra virgin. Data analysis using supervised (Linear Discriminant Analysis (LDA), Support Vector Machine (SVM),Partial Least Square Discriminant Analysis (PLS-DA), Orthogonal Projection to Latent Structure-Discriminant Analysis (OPLS-DA), Counter Propagation Artificial Neural Networks(CP-ANNs), Self-Organizing Maps (SOMs) and Random Forest (RF)) and non-supervised (Principal Component Analysis (PCA) and Hierarchical Clustering Analysis (HCA)) tools and their interpretation proved to yield reliable results for identification of foods.

Beć *et al* (2021) reviewed in detail on use of hyperspectral imaging techniques which are promising quick, non-destructive method with minimal sample preparation requirements for food traceability. Data generated by hyper spectral imaging is vast in a short time period and are superior to traditional methods. Basics of instrumentation, work flow, methodology, image analysis and the chemometrics obtained for food samples are explained in detail. Recent examples on use of hyper spectral imaging have been quoted along with challenges faced during use in foodomics applications. Badaró *et al* (2021) insisted on significance of linear and non-linear multivariate data analytical tools to analyze and derive conclusions using data generated by various omic tools in the field of food science and technology. Vibrational spectroscopy based tools used for generation of data for food quality assurance, processing and fusion of generated data for prediction of consistent results has been described. Modeling based on constraints and kinetics was used to model data generated from metabolomics studies. Enormous data generated through multi-omic based research and their interpretations in meaningful manner are essential to understand the molecular mechanisms underlying various biological systems which are unique and complex. Foodomics related data derived by metabolomics were modeled and reviewed of their future applications (de Mas *et al* 2021). Rodriguez-Saona *et al* (2020) elaborated high throughput rapid screening of foods using miniaturization techniques. Recent availability of spectral libraries and food fingerprints as a result of the evolved metabolomics field has opened up options for portable and rugged miniature devices as green eco-friendly options for food quality assurance by on-site detection of contaminants and adulterants.

Methods of Near Infra Red spectroscopy (NIR) coupled with chemometrics are gaining importance in context of food adulteration testing and detection. Raw material testing based on targeted approach have to be replaced with latest technologies which compare fingerprints of tested food with spectroscopically obtained data. Testing of raw materials which form ingredients in variety of food products termed as process analytical technology at industries would lead to rapid detection of adulterants at initial stages of food production (Sørensen *et al* 2016). Liu *et al* (2021) reviewed literature quoting techniques for authentication of wheat, the staple food consumed globally. Identification of geographical origin, organic wheat and adulteration of wheat flour with low cost alternatives using various chemo metric tools such as NIR, FT-IR, HIS along with discriminatory analysis using data analysis tools were elaborated. Weng *et al* (2020) reported visible/near-infrared (Vis/NIR) reflectance spectroscopy based rapid identification of beef adulteration with pork. Adulteration of different forms of beef such as frozen, thawed and minced variants using turkey meat was detected using FT-NIR spectroscopy with multivariate analysis. Inter species adulteration is a growing menace and development of omic technologies aid in determination of meat authenticity using tools of chemo metrics (Alamprese *et al* 2016). Zheng *et al* (2019) reported identification of alleged adulteration of mince lamb meat using duck meat and its detection by visible near-infrared hyperspectral imaging. The developed method was rapid and accurate in detecting the adulteration. Fengou *et al* (2021) studied detection of beef and pork minced meat adulteration using bovine offal and chicken, respectively in frozen and thawed forms. Sensors based on spectroscopy viz. visible light, fluorescence and multi spectral imaging methods were tested on adulterated meat and results derived from multi spectral imaging method was found to be accurate up to 100 percent in detecting adulteration. Visible spectra was accurate in range of 57 to 97 percent while fluorescence spectra obtained provided the least accuracy. Hence, usage of spectral methods for adulteration of food products appears to be sensor specific and has to be chosen wisely for the intended purpose. Teixeira *et al* (2021) reported vibrational spectroscopy Near Infrared (FT-NIR) tool based identification of adulteration in goat milk based dairy products such as cheese and yogurt due to their health benefits. Cow's milk adulteration up to 10 percent level could be detected based on the described method which is portable and useful in on-site detection of adulterants.

Ndlovu *et al* (2021) reported a non-destructive rapid method based on visible to near-infrared (Vis-NIR) spectroscopy for detection of gluten in raw banana flour. Prediction and quantification of gluten concentration was reliable using the developed method for authentication of the raw banana flour. Kim *et al* (2021) described a reliable method based on second-Derivative FT-IR Spectroscopy to discriminate origin of red pepper powder samples available at Korean markets.

Sultanbawa *et al* (2021) reviewed in detail about the possibilities of combining machine learning with vibrational spectroscopy. Fruits are rich in bioactive metabolites and several compounds were identified using techniques such as Raman, near infra red and middle infra red spectroscopy. Electronic noses and color based sensors and their use in combination with spectra obtained from various tools are gaining popularity in quantification of different plant and fruit bioactives. Rizo *et al* (2020) elaborated applications of omics in studying fermented foods to understand the functionality of produced metabolites by microbial consortia involved in product formulation. Commercially sound fermented foods such as beer, wine, cheese and bread has been subjected to omic technologies to study impact of fermentation on quality of products. Traditional fermented foods and omics techniques to study phenomena of fermentation have been reviewed. Calixto *et al* (2021) reviewed in detail chemo metric fingerprinting of medicinal plants to determine authenticity of raw material used in herbal drug preparations based on origin, variety, plant part used etc.

Volatilomics is emerging as a promising technology using tools of GC-MS and e-nose. Signature of volatiles obtained from foods is being used to determine quality as well as confirm authenticity of specific foods. Extraction, analysis, data generation, multivatiate analysis for statistical grouping are the important steps in work flow of volatilomics. Specific volatiles determined in foods are used as biomarkers which could determine origin related to biology and geography (Lytou *et al* 2019). Lipidomics is an upcoming field of metabolomics and forming an essential component of foodomics in recent years. Lipid molecules present in biological systems eg. Food has been studied in detail by use of mass spectrometry based studies on lipids. Lipidomics provide comprehensive insights on function and nutritional aspects of lipids. Composition, structure and metabolism of food lipids and role in human health could be studied in detail when techniques of MS and powerful chemometrics are used in combination. Food frauds due to adulteration of oils could be prevented by confirmation of authenticity of lipids using lipidomics (Wu *et al* 2021). Wang *et al* (2021) reported chemometrics based detection of adulteration of flaxseed oil with cheaper alternatives such as soybean, sesame, sunflower, ground nut, corn and rapeseed up to 20 percent. Fatty acid fingerprint specific to flaxseed oil was obtained and used for comparison of samples to identify adulteration. Heat map based approach indicated linolenic acid as potential marker to identify adulteration of flax seed oil. PCA of derived data was useful in detecting other vegetable oils except corn and rapeseed oils.

Opportunities and Prospects

Challenges faced in field of food safety could be overcome by the use of foodomics approach which is capable of rapidly detecting important contaminants, allergens, pesticides, food borne pathogens and toxins using specific biomarkers derived from techniques of foodomics. Food safety assurance by detection of novel pathogens, unculturable micro organisms and novel toxins has become possible by the latest methods of 'omics'. Food attributes namely composition, flavour, colour and place of origin are preferred over other parameters to decide the level of acceptance of food quality. Analysis at real time and on-site play a major role, in detection of food adulteration in preventing fraud, ensure quality as well as traceability of food (Li *et al* 2021). Detection of GMOs, authentication at species and breed level, geographical origin, ingredient source, method of processing in fresh and processed food products is capable by utilising tools of proteomics and metabolomics by characterisation of food at molecular level. Bioactivity of food ingredients and their interaction with human health has gained importance in recent years. Personalized nutrition concept based on effect of individual food components during normal or diseased conditions, for enhancing health of individuals are being prescribed using modern analytical techniques and their scientific interpretations using molecular modelling (Maruvada *et al* 2020). Furthermore, genetic and protein engineering using latest editing methods may also open road to modify food components at source (microbes, plant, animals) and aid in production of novel and safe substances for food use (Pan and Barrangou, 2020).

Development of robust sample preparation and extraction methods to study complex food matrices to evade interferences caused by food components is essential to make use of modern omic based techniques for variety of foods and their components. Moreover, expensive equipments, biochemicals, highly skilled manpower requirement remain a larger challenge for using high throughput omic technologies based on gene, protein and metabolites in food research (Bianchi *et al* 2018). Enormous data generated by the above mentioned techniques for different kinds of ingredients and foods need to be shared in open source databases which may be possible only by extensive collaboration between research institutes working all over the globe. Software used for studying food interactions based on molecular modelling to derive useful and meaningful interactions need to be made open ware for wider usage and applications in food quality authentication.

References

Abbas, O., Zadravec, M., Baeten, V., Mikuš, T., Lešić, T., Vulić, A., ... & Pleadin, J. (2018). Analytical methods used for the authentication of food of animal origin. Food Chemistry, 246, 6-17. https://doi.org/10.1016/j.foodchem.2017.11.007.

Afshari, R., Pillidge, C. J., Dias, D. A., Osborn, A. M., & Gill, H. (2020). Cheesomics: the future pathway to understanding cheese flavour and quality. Critical reviews in food science and nutrition, 60(1), 33–47. https://doi.org/10.1080/10408398.2018.1512471

Afshari, R., Pillidge, C. J., Dias, D. A., Osborn, A. M., & Gill, H. (2020). Cheesomics: the future pathway to understanding cheese flavour and quality. Critical Reviews in Food Science and Nutrition, 60(1), 33-47.

Akamatsu, F., Shimizu, H., Igi, Y., Kamada, A., Koyama, K., Yamada, O., & Goto-Yamamoto, N. (2022). Prediction method for determining the carbon stable isotopic composition of berry sugars in the original must of Chardonnay wines. Food Chemistry, 369, 130854.

Akhtar, M. T., Samar, M., Shami, A. A., Mumtaz, M. W., Mukhtar, H., Tahir, A., ... & Kaka, U. (2021). 1H-NMR-Based Metabolomics: An Integrated Approach for the Detection of the Adulteration in Chicken, Chevon, Beef and Donkey Meat. Molecules, 26(15), 4643.

Alamprese, C., Amigo, J. M., Casiraghi, E., & Engelsen, S. B. (2016). Identification and quantification of turkey meat adulteration in fresh, frozen-thawed and cooked minced beef by FT-NIR spectroscopy and chemometrics. Meat Science, 121, 175–181. https://doi.org/10.1016/j.meatsci.2016.06.018

Álvarez-Rivera, G., Toro-Uribe, S., Ibañez, E., Cifuentes, A., & Bueno, M. (2019). Omics| Foodomics: Overview. Editor(s): Paul Worsfold, Colin Poole, Alan Townshend, Manuel Miró, 2019,

Álvarez-Rivera, G., Valdés, A., León, C., & Cifuentes, A. (2021). Foodomics–Fundamentals, State of the Art and Future Trends. pp. 1-53

Andjelković, U., & Josić, D. (2018). Mass spectrometry based proteomics as foodomics tool in research and assurance of food quality and safety. Trends in Food Science & Technology, 77, 100-119.

Andjelković, U., Gavrović-Jankulović, M., Martinović, T., & Josić, D. (2017). Omics methods as a tool for investigation of food allergies. TRAC Trends in Analytical Chemistry, 96, 107-115.

Arena, S., Salzano, A. M., & Scaloni, A. (2016). Identification of protein markers for the occurrence of defrosted material in milk through a MALDI-TOF-MS profiling approach. Journal of proteomics, 147, 56-65.

Aykas, D. P., & Menevseoglu, A. (2021). A rapid method to detect green pea and peanut adulteration in pistachio by using portable FT-MIR and FT-NIR spectroscopy combined with chemometrics. Food Control, 121, 107670.

Badaró, A. T., Pasquini, C., & Barbin, D. F. (2021). Food quality and NIR spectroscopy in the omics era. Editor(s): Alejandro Cifuentes, Comprehensive Foodomics, Elsevier, 2021, Pages 517-535, ISBN 9780128163962, https://doi.org/10.1016/B978-0-08-100596-5.22802-3.

Balkir, P., Kemahlioglu, K., & Yucel, U. (2020). Foodomics: A new approach in food quality and safety. Trends in Food Science & Technology.

Barreto, M. C., Braga, R. G., Lemos, S. G., & Fragoso, W. D. (2021). Determination of melamine in milk by fluorescence spectroscopy and second-order calibration. Food chemistry, 364, 130407. https://doi.org/10.1016/j.foodchem.2021.130407

Bayram, B., González-Sarrías, A., Istas, G., Garcia-Aloy, M., Morand, C., Tuohy, K., ... & Mena, P. (2018). Breakthroughs in the health effects of plant food bioactives: a perspective on microbiomics, nutri (epi) genomics, and metabolomics. Journal of Agricultural and Food Chemistry, 66(41), 10686-10692.

Beć, K. B., Grabska, J., & Huck, C. W. (2021). Issues in hyperspectral traceability of foods. Editor(s): Alejandro Cifuentes, Comprehensive Foodomics, Elsevier, 2021, Pages 517-535, ISBN 9780128163962, https://doi.org/10.1016/B978-0-08-100596-5.22802-3.

Beć, K. B., Grabska, J., & Huck, C. W. (2021). X-Ray, LIBS, NMR, and MS Applications in Food, Feed, and Agriculture. Portable Spectroscopy and Spectrometry 1: Applications, 2, 299-323.

Bianchi, F., Giannetto, M., & Careri, M. (2018). Analytical systems and metrological traceability of measurement data in food control assessment. TrAC Trends in Analytical Chemistry, 107, 142-150.

Bodin, N., Amiel, A., Fouché, E., Sardenne, F., Chassot, E., Debrauwer, L., & Canlet, C. (2022). NMR-based metabolic profiling and discrimination of wild tropical tunas by species, size category, geographic origin, and on-board storage condition. Food Chemistry, 371, 131094.

Böhme, K., Calo-Mata, P., Barros-Velázquez, J., & Ortea, I. (2019). Review of Recent DNA-Based Methods for Main Food-Authentication Topics. Journal of agricultural and food chemistry, 67(14), 3854–3864. https://doi.org/10.1021/acs.jafc.8b07016

Böhme, K., Calo-Mata, P., Barros-Velázquez, J., & Ortea, I. (2019). Recent applications of omics-based technologies to main topics in food authentication. TrAC Trends in Analytical Chemistry, 110, 221-232.

Bruschetta, G., Notti, A., Lando, G., & Ferlazzo, A. (2021). A promising ^{31}P NMR-multivariate analysis approach for the identification of milk phosphorylated metabolites and for rapid authentication of milk samples. Biochemistry and biophysics reports, 27, 101087. https://doi.org/10.1016/j.bbrep.2021.101087

Carrera, M., Mateos, J., & Gallardo, J. M. (2021). Data treatment in food proteomics. Cifuentes, A. (2020). Comprehensive Foodomics. Elsevier.

Casas-Agustench, P., & Andres-Lacueva, C. (2020). A Broader View on Omics and Systems Biology. In Principles of Nutrigenetics and Nutrigenomics (pp. 89-97). Academic Press.

Caterina, R. D., Martinez, J. A., & Kohlmeier, M. (2020). Principles of nutrigenetics and nutrigenomics.

Cifuentes, A., Foodomics: Advanced Mass Spectrometry in Modern Food Science and Nutrition in Foodomics, ed. A. Cifuentes, Wiley, 2013, pp. 1–13. ISBN: 978-1-118-16945-2.

Cifuentes, A. (2009). Advanced separation methods in food analysis. Journal of Chromatography, A, 1216(43), 7109-7358.

Cifuentes, A. (2017). Foodomics, foodome and modern food analysis. Trends Anal. Chem., 96, 1.

Cilla, A., Alegría, A., Barberá, R., & García-Llatas, G. (2021). Minerals and Trace Elements. In Handbook of Dairy Foods Analysis (pp. 557-587). CRC Press.

Colgrave, M. L., Byrne, K., Blundell, M., & Howitt, C. A. (2016). Identification of barley-specific peptide markers that persist in processed foods and are capable of detecting barley contamination by LC-MS/MS. Journal of Proteomics, 147, 169-176.

Corsaro, C., Fazio, E., & Mallamace, D. (2021). Direct Analysis in Foodomics: NMR approaches. Editor(s): Alejandro Cifuentes, Comprehensive Foodomics, Elsevier, 2021, Pages 517-535, ISBN 9780128163962, https://doi.org/10.1016/B978-0-08-100596-5.22802-3.

Croft, D., O'kelly, G., Wu, G., Haw, R., Gillespie, M., Matthews, L., & Stein, L. (2010). Reactome: a database of reactions, pathways and biological processes. Nucleic acids research, 39(suppl_1), D691-D697.Wang, X., Han, Y., Li, Y., Wang, S., & Wang, J. (2021). Detection of Qinghai-Tibet Plateau Flaxseed Oil Adulteration Based on Fatty Acid Profiles and Chemometrics. Food Control, 108332.

Cuadros-Rodríguez, L., Ruiz-Samblás, C., Valverde-Som, L., Pérez-Castaño, E., & González-Casado, A. (2016). Chromatographic fingerprinting: An innovative approach for food'identitation'and food authentication–A tutorial. Analytica Chimica Acta, 909, 9-23.

De Girolamo, A., Arroyo, M. C., Cervellieri, S., Cortese, M., Pascale, M., Logrieco, A. F., & Lippolis, V. (2020). Detection of durum wheat pasta adulteration with common wheat by infrared spectroscopy and chemometrics: A case study. LWT, 127, 109368.

de Mas, I. B. M., de Leeuw, M., Nouran, P. G., & Nielsen, L. K. (2021). Metabolic Modeling and Omics Data Integration: A Systems Biology Approach to Food Science. In Comprehensive Foodomics (pp. 396-413). Editor(s): Alejandro Cifuentes, Comprehensive Foodomics, Elsevier, 2021, Pages 517-535, ISBN 9780128163962, https://doi.org/10.1016/B978-0-08-100596-5.22802-3.

Ducrest, P. J., Pfammatter, S., Stephan, D., Vogel, G., Thibault, P., & Schnyder, B. (2019). Rapid detection of Bacillus ionophore cereulide in food products. Scientific reports, 9(1), 5814. https://doi.org/10.1038/s41598-019-42167-0

Fanelli, V., Mascio, I., Miazzi, M. M., Savoia, M. A., De Giovanni, C., & Montemurro, C. (2021). Molecular Approaches to Agri-Food Traceability and Authentication: An Updated Review. Foods (Basel, Switzerland), 10(7), 1644. https://doi.org/10.3390/foods10071644

Fasoli, E., & Righetti, P. G. (2015). Proteomics of fruits and beverages. Current Opinion in Food Science, 4, 76-85.

Fengou, L. C., Lianou, A., Tsakanikas, P., Mohareb, F., & Nychas, G. E. (2021). Detection of Meat Adulteration Using Spectroscopy-Based Sensors. Foods (Basel, Switzerland), 10(4), 861. https://doi.org/10.3390/foods10040861

Fyfe, S., Chaliha, M., Netzel, M. E., & Sultanbawa, Y. (2020). Ethnic foodomics: metabolomics studies of ethnic foods.

Gaião Calixto, M., Alves Ramos, H., Veríssimo, L. S., Dantas Alves, V., D Medeiros, A. C., Alencar Fernandes, F. H., & Veras, G. (2021). Trends and Application of Chemometric Pattern Recognition Techniques in Medicinal Plants Analysis. Critical reviews in analytical chemistry, 1–13. Advance online publication.

Galanakis, C. M. (Ed.). (2016). Innovation strategies in the food industry: Tools for implementation. Academic Press.

Galimberti, A., Casiraghi, M., Bruni, I., Guzzetti, L., Cortis, P., Berterame, N. M., & Labra, M. (2019). From DNA barcoding to personalized nutrition: the evolution of food traceability. Current Opinion in Food Science, 28, 41-48.

Garrido, B. C., Souza, G. H., Lourenço, D. C., & Fasciotti, M. (2016). Proteomics in quality control: Whey protein-based supplements. Journal of proteomics, 147, 48-55.

Geisslitz, S., & Scherf, K. A. (2021). Proteomics in Food Quality. Cifuentes, A. (2020). Comprehensive Foodomics. Elsevier.

Gilbert-López, B., Mendiola, J. A., & Ibáñez, E. (2017). Green foodomics. Towards a cleaner scientific discipline. TrAC Trends in Analytical Chemistry, 96, 31-41.

Hassoun, A., Shumilina, E., Di Donato, F., Foschi, M., Simal-Gandara, J., & Biancolillo, A. (2020). Emerging Techniques for Differentiation of Fresh and Frozen–Thawed Seafoods: Highlighting the Potential of Spectroscopic Techniques. Molecules, 25(19), 4472.

Hong, E., Lee, S. Y., Jeong, J. Y., Park, J. M., Kim, B. H., Kwon, K., & Chun, H. S. (2017). Modern analytical methods for the detection of food fraud and adulteration by food category. Journal of the Science of Food and Agriculture, 97(12), 3877–3896. https://doi.org/10.1002/jsfa.8364

Hong, Y., Kim, J. J., Yu, Y. C., Kim, H. S., Moon, G., & Park, E. M. (2021). Ultra-fast PCR method for the distinguishing between Miichthys miiuy and Sciaenops ocellatus. Food science and biotechnology, 30(9), 1225–1231. https://doi.org/10.1007/s10068-021-00954-4

Huang, C., Hou, C., Ijaz, M., Yan, T., Li, X., Li, Y., & Zhang, D. (2020). Proteomics discovery of protein biomarkers linked to meat quality traits in post-mortem muscles: Current trends and future prospects: A review. Trends in Food Science & Technology.105, 416-432.

Islam, M. K., Vinsen, K., Sostaric, T., Lim, L. Y., & Locher, C. (2021). Detection of syrup adulterants in manuka and jarrah honey using HPTLC-multivariate data analysis. PeerJ, 9, e12186. https://doi.org/10.7717/peerj.12186

Jahanbakhshi, A., Abbaspour-Gilandeh, Y., Heidarbeigi, K., & Momeny, M. (2021). Detection of fraud in ginger powder using an automatic sorting system based on image processing technique and deep learning. Computers in biology and medicine, 136, 104764. https://doi.org/10.1016/j.compbiomed.2021.104764

Kalogiouri, N. P., Aalizadeh, R., Dasenaki, M. E., & Thomaidis, N. S. (2020). Application of High Resolution Mass Spectrometric methods coupled with chemometric techniques in olive oil authenticity studies-A review. Analytica Chimica Acta. 1134 150-173

Kandasamy, S., Yoo, J., Yun, J., Kang, H. B., Seol, K. H., & Ham, J. S. (2021). Rapid Discrimination and Authentication of Korean Farmstead Mozzarella Cheese through MALDI-TOF and Multivariate Statistical Analysis. Metabolites, 11(6), 333. https://doi.org/10.3390/metabo11060333

Khodabakhshian, R., Bayati, M. R., & Emadi, B. (2021). An evaluation of IR spectroscopy for authentication of adulterated turmeric powder using pattern recognition. Food chemistry, 364, 130406. https://doi.org/10.1016/j.foodchem.2021.130406

Kim, M., Hong, J., Lee, D., Kim, S., Chun, H. S., Cho, Y. H., Kim, B. H., & Ahn, S. (2021). Discriminant Analysis of the Geographical Origin of Asian Red Pepper Powders Using Second-Derivative FT-IR Spectroscopy. Foods (Basel, Switzerland), 10(5), 1034. https://doi.org/10.3390/foods10051034

Klampfl, C. W. (2018). Ambient mass spectrometry in foodomics studies. Current Opinion in Food Science, 22, 137-144.

Komaram, A. C., Anjaneyulu, E., Goswami, K., Nayak, R. R., & Kanjilal, S. (2021). Detection and quantification of palmolein and palm kernel oil added as adulterant in coconut oil based on triacylglycerol profile. Journal of food science and technology, 58(11), 4420–4428. https://doi.org/10.1007/s13197-020-04927-z

Kritikou, A. S., Aalizadeh, R., Damalas, D. E., Barla, I. V., Baessmann, C., & Thomaidis, N. S. (2021). MALDI-TOF-MS integrated workflow for food authenticity investigations: An untargeted protein-based approach for rapid detection of PDO feta cheese adulteration. Food chemistry, 370, 131057. Advance online publication. https://doi.org/10.1016/j.foodchem.2021.131057

Lacalle-Bergeron, L., Izquierdo-Sandoval, D., Sancho, J. V., López, F. J., Hernández, F., & Portolés, T. (2020). Chromatography hyphenated to high resolution mass spectrometry in untargeted metabolomics for investigation of food (BIO) markers. TrAC Trends in Analytical Chemistry, 116161.

León, C., Cifuentes, A., & Valdés, A. (2018). Foodomics applications. In Comprehensive analytical chemistry (Vol. 82, pp. 643-685). Elsevier.

Li, S., Tian, Y., Jiang, P., Lin, Y., Liu, X., & Yang, H. (2021). Recent advances in the application of metabolomics for food safety control and food quality analyses. Critical Reviews in Food Science and Nutrition, 61(9), 1448-1469.

Liu, H. Y., Wadood, S. A., Xia, Y., Liu, Y., Guo, H., Guo, B. L., & Gan, R. Y. (2021). Wheat authentication-An overview on different techniques and chemometric methods. Critical reviews in food science and nutrition, 1–24. https://doi.org/10.1080/10408398.2021.1942783

Liu, J., Chen, J., Wu, D., Huang, M., Chen, J., Pan, R., Wu, Y., & Li, G. (2021). CRISPR-/Cas12a-Mediated Liposome-Amplified Strategy for the Surface-Enhanced Raman Scattering and Naked-Eye Detection of Nucleic Acid and Application to Food Authenticity Screening. Analytical chemistry, 93(29), 10167–10174. https://doi.org/10.1021/acs.analchem.1c01163

Lu, B., Han, F., Aheto, J. H., Rashed, M., & Pan, Z. (2021). Artificial bionic taste sensors coupled with chemometrics for rapid detection of beef adulteration. Food science & nutrition, 9(9), 5220–5228. https://doi.org/10.1002/fsn3.2494

Lytou, A. E., Panagou, E. Z., & Nychas, G. J. E. (2019). Volatilomics for food quality and authentication. Current Opinion in Food Science, 28, 88-95.

Maestri, E., Marmiroli, M., & Marmiroli, N. (2016). Bioactive peptides in plant-derived foodstuffs. Journal of Proteomics, 147, 140-155.

Mancano, G., Mora-Ortiz, M., & Claus, S. P. (2018). Recent developments in nutrimetabolomics: from food characterisation to disease prevention. Current Opinion in Food Science, 22, 145-152.

Martini, S., Solieri, L., & Tagliazucchi, D. (2020). Peptidomics: new trends in food science. Current Opinion in Food Science. 39 51-59

Martinović, T., Andjelković, U., Gajdošik, M. Š., Rešetar, D., & Josić, D. (2016). Foodborne pathogens and their toxins. Journal of Proteomics, 147, 226-235.

Maruvada, P., Lampe, J. W., Wishart, D. S., Barupal, D., Chester, D. N., Dodd, Djoumbou-Feunang, Y., Dorrestein, P. C., Dragsted, L.O., Draper., J. Duffy, L.C., Dwyer, J. T., Emenaker, N. J., Fiehn, O., Gerszten, R. E., Hu, F. B.,. Karp, R. W,. Klurfeld, D. M, Laughlin, M. R., Little, A. R., Lynch, C. J., Moore, S. C., Nicastro, H. L., Brien, D. M. O, Ordova´s, J. M., Osganian, S.K., Playdon, M., Prentice, R., Raftery, D., Reisdorph, N., Roche, H. M., Ross, S. A., Sang, S., Scalbert, A., Srinivas, P. R., & Zeisel, S. H. (2020). Perspective: dietary biomarkers of intake and exposure—exploration with omics approaches. Advances in Nutrition, 11(2), 200-215.

Mazzeo, M. F., & Siciliano, R. A. (2016). Proteomics for the authentication of fish species. Journal of Proteomics, 147, 119-124.

Melanie Pavlovic, Ingrid Huber, Ulrich Busch, 1.53 - MALDI-TOF MS Profiling Based Identification of Food Components,

Minkiewicz, P., Darewicz, M., Iwaniak, A., Sokołowska, J., Starowicz, P., Bucholska, J., & Hrynkiewicz, M. (2015). Common amino acid subsequences in a universal proteome—Relevance for food science. International Journal of Molecular Sciences, 16(9), 20748-20773.

Mohammadian, A., Barzegar, M., & Mani-Varnosfaderani, A. (2021). Detection of fraud in lime juice using pattern recognition techniques and FT-IR spectroscopy. Food science & nutrition, 9(6), 3026–3038. https://doi.org/10.1002/fsn3.2260

Montero, L., & Herrero, M. (2019). Two-dimensional liquid chromatography approaches in Foodomics–A review. Analytica chimica acta, 1083, 1-18.

Morlock, G. E., Busso, M., Tomeba, S., & Sighicelli, A. (2021). Effect-directed profiling of 32 vanilla products, characterization of multi-potent compounds and quantification of vanillin and ethylvanillin. Journal of Chromatography. A, 1652, 462377. https://doi.org/10.1016/j.chroma.2021.462377

Mounika, T., Girish, P. S., Shashi Kumar, M., Kumari, A., Singh, S., & Karabasanavar, N. S. (2021). Identification of sheep (Ovis aries) meat by alkaline lysis-loop mediated isothermal amplification technique targeting mitochondrial D-loop region. Journal of Food Science and Technology, 58(10), 3825–3834. https://doi.org/10.1007/s13197-020-04843-2

Mousa, M., Wang, Y., Antora, S. A., Al-Qurashi, A. D., Ibrahim, O., He, H. J., Liu, S., & Kamruzzaman, M. (2021). An overview of recent advances and applications of FT-IR spectroscopy for quality, authenticity, and adulteration detection in edible oils. Critical reviews in food science and nutrition, 1–19. Advance online publication. https://doi.org/10.1080/10408398.2021.1922872

Nikitin, A., Egorov, S., Daraselia, N., & Mazo, I. (2003). Pathway studio—the analysis and navigation of molecular networks. Bioinformatics, 19(16), 2155-2157.

Núñez, N., Saurina, J., & Núñez, O. (2021). Authenticity Assessment and Fraud Quantitation of Coffee Adulterated with Chicory, Barley, and Flours by Untargeted HPLC-UV-FLD Fingerprinting and Chemometrics. Foods (Basel, Switzerland), 10(4), 840. https://doi.org/10.3390/foods10040840

Ortea, I., O'Connor, G., & Maquet, A. (2020). Review on Proteomics for Food Authentication. In Proteomics for Food Authentication (pp. 3-36). CRC Press.

Pan, M., & Barrangou, R. (2020). Combining omics technologies with CRISPR-based genome editing to study food microbes. Current opinion in biotechnology, 61, 198-208.

Pandey, G., & Joshi, A. (2021). Riboflavin as an internal marker for spoilage and adulteration detection in milk. Food chemistry, 357, 129742. Advance online publication. https://doi.org/10.1016/j.foodchem.2021.129742

Pavlovic, M., Huber, I., & Busch, U. (2021). MALDI-TOF MS Profiling Based Identification of Food Components.

Pocan, P., Knapkiewicz, M., Rachocki, A., & Oztop, M. H. (2021). Detection of Authenticity and Quality of the Turkish Delights (Lokum) by Means of Conventional and Fast Field Cycling Nuclear Magnetic Resonance Relaxometry. Journal of agricultural and food chemistry, 10.1021/acs.jafc.1c00943. Advance online publication. https://doi.org/10.1021/acs.jafc.1c00943

Rešetar, D., & Pavelić, S. K. (2018). Application of Foodomics for Food Toxins. Pages 53-62, ISBN 9780081019849, https://doi.org/10.1016/B978-0-12-409547-2.14005-3.

Rizo, J., Guillén, D., Farrés, A., Díaz-Ruiz, G., Sánchez, S., Wacher, C., & Rodríguez-Sanoja, R. (2020). Omics in traditional vegetable fermented foods and beverages. Critical reviews in food science and nutrition, 60(5), 791–809. https://doi.org/10.1080/10408398.2018.1551189

Rodriguez-Saona, L., Aykas, D. P., Borba, K. R., & Urtubia, A. (2020). Miniaturization of optical sensors and their potential for high-throughput screening of foods. Current Opinion in Food Science, 31, 136-150.

Sébédio, J. L., & Malpuech-Brugère, C. (2016). Implementation of Foodomics in the food industry. In Innovation Strategies in the Food Industry (pp. 251-269). Academic Press.

Sezer, B., Bjelak, A., Murat Velioglu, H., & Hakkı Boyaci, I. (2021). Identification of meat species in processed meat products by using protein based laser induced breakdown spectroscopy assay. Food chemistry, 372, 131245. Advance online publication. https://doi.org/10.1016/j.foodchem.2021.131245

Sheu, S. C., Wu, Y. C., Lien, Y. Y., & Lee, M. S. (2021). Specific, sensitive and rapid Curcuma longa turmeric powder authentication in commercial food using loop-mediated isothermal nucleic acid amplification. Saudi Journal of Biological Sciences, 28(10), 5931–5936. https://doi.org/10.1016/j.sjbs.2021.06.057

Shomaji, S., Masna, N., Ariando, D., Deb Paul, S., Horace-Herron, K., Forte, D., Mandal, S., & Bhunia, S. (2021). Detecting Dye-Contaminated Vegetables Using Low-Field NMR Relaxometry. Foods (Basel, Switzerland), 10(9), 2232. https://doi.org/10.3390/foods10092232

Sitorus, A., Muslih, M., Cebro, I. S., & Bulan, R. (2021). Dataset of adulteration with water in coconut milk using FTIR spectroscopy. Data in brief, 36, 107058. https://doi.org/10.1016/j.dib.2021.107058

Smith, C., Cokcetin, N., Truong, T., Harry, E., Hutvagner, G., & Bajan, S. (2021). Cataloguing the small RNA content of honey using next generation sequencing. Food Chemistry: Molecular Sciences, 2, 100014.

Sobolev, A. P., Circi, S., Capitani, D., Ingallina, C., & Mannina, L. (2017). Molecular fingerprinting of food authenticity. Current Opinion in Food Science, 16, 59-66.

Sørensen, K. M., Khakimov, B., & Engelsen, S. B. (2016). The use of rapid spectroscopic screening methods to detect adulteration of food raw materials and ingredients. Current Opinion in Food Science, 10, 45-51.

Srinuttrakul, W., Mihailova, A., Islam, M. D., Liebisch, B., Maxwell, F., Kelly, S. D., & Cannavan, A. (2021). Geographical Differentiation of Hom Mali Rice Cultivated in Different Regions of Thailand Using FTIR-ATR and NIR Spectroscopy. Foods (Basel, Switzerland), 10(8), 1951. https://doi.org/10.3390/foods10081951

Strashnov, I., Karunarathna, N. B., Fernando, B. R., Dissanayake, C., & Binduhewa, K. M. (2021). An isotope dilution liquid chromatography-mass spectrometry method for detection of melamine in milk powder. Food additives & contaminants. Part A, Chemistry, analysis, control, exposure & risk assessment, 1–12. Advance online publication. https://doi.org/10.1080/19440049.2021.1937709

Sudhakar, A., Chakraborty, S. K., Mahanti, N. K., & Varghese, C. (2021). Advanced techniques in edible oil authentication: A systematic review and critical analysis. Critical reviews in food science and nutrition, 1–29. Advance online publication. https://doi.org/10.1080/10408398.2021.1956424

Sultanbawa, Y., Smyth, H. E., Truong, K., Chapman, J., & Cozzolino, D. (2021). Insights on the role of chemometrics and vibrational spectroscopy in fruit metabolite analysis. Food Chemistry: Molecular Sciences, 100033.

Tabago, M. K. A. G., Calingacion, M. N., & Garcia, J. (2020). Recent Advances in NMR-Based Metabolomics of Alcoholic Beverages. Food Chemistry: Molecular Sciences, 100009.

Teixeira, J. L. D. P., Caramês, E. T. D. S., Baptista, D. P., Gigante, M. L., & Pallone, J. A. L. (2021). Rapid adulteration detection of yogurt and cheese made from goat milk by vibrational spectroscopy and chemometric tools.

Templ, M., & Templ, B. (2021). Statistical Analysis of Chemical Element Compositions in Food Science: Problems and Possibilities. Molecules (Basel, Switzerland), 26(19), 5752. https://doi.org/10.3390/molecules26195752

Trimigno, A., Marincola, F. C., Dellarosa, N., Picone, G., & Laghi, L. (2015). Definition of food quality by NMR-based foodomics. Current Opinion in Food Science, 4, 99-104.

Truong, A. T., Kim, S., & Yoon, B. (2021). Determination of honey adulterated with corn syrup by quantitative amplification of maize residual DNA using ultra-rapid real-time PCR. Journal of the science of food and agriculture, 10.1002/jsfa.11411. Advance online publication. https://doi.org/10.1002/jsfa.11411

Truzzi, E., Marchetti, L., Benvenuti, S., Ferroni, A., Rossi, M. C., & Bertelli, D. (2021). Novel Strategy for the Recognition of Adulterant Vegetable Oils in Essential Oils Commonly Used in Food Industries by Applying ^{13}C NMR Spectroscopy. Journal of agricultural and food chemistry, 69(29), 8276–8286. https://doi.org/10.1021/acs.jafc.1c02279

Uawisetwathana, U., & Karoonuthaisiri, N. (2019). Metabolomics for rice quality and traceability: feasibility and future aspects. Current Opinion in Food Science, 28, 58-66.

Vaccalluzzo, A., Pino, A., Russo, N., De Angelis, M., Caggia, C., & Randazzo, C. L. (2020). FoodOmics as a new frontier to reveal microbial community and metabolic processes occurring on table olives fermentation. Food microbiology, 92, 103606. https://doi.org/10.1016/j.fm.2020.103606

Valentini, P., Galimberti, A., Mezzasalma, V., De Mattia, F., Casiraghi, M., Labra, M., & Pompa, P. P. (2017). DNA Barcoding Meets Nanotechnology: Development of a Universal

Colorimetric Test for Food Authentication. Angewandte Chemie (International ed. in English), 56(28), 8094–8098. https://doi.org/10.1002/anie.201702120

Valletta, M., Ragucci, S., Landi, N., Di Maro, A., Pedone, P. V., Russo, R., & Chambery, A. (2021). Mass spectrometry-based protein and peptide profiling for food frauds, traceability and authenticity assessment. Food Chemistry, 130456.

Velioglu, H. M., Sezer, B., Bilge, G., Baytur, S. E., & Boyaci, I. H. (2018). Identification of offal adulteration in beef by laser induced breakdown spectroscopy (LIBS). Meat Science, 138, 28-33.

Vijay, D. P., & Sutherland, K. (2021). Capillary Electromigration Techniques Coupled to Mass Spectrometry: Applications to Food Analysis. TrAC Trends in Analytical Chemistry, 116240.

Walsh, A. M., Crispie, F., Claesson, M. J., & Cotter, P. D. (2017). Translating omics to food microbiology. Annual Review of Food Science and Technology, 8, 113-134.

Wang, Q., Wang, Y., Jiang, X., Ma, L., Li, Z., Chang, Y., Wang, Y., & Xue, C. (2021). Amino Acid Profiling with Chemometric Analysis as a Feasible Tool for the Discrimination of Marine-Derived Peptide Powders. Foods (Basel, Switzerland), 10(6), 1294. https://doi.org/10.3390/foods10061294

Wang, Z., Ren, P., Wu, Y., & He, Q. (2021). Recent advances in analytical techniques for the detection of adulteration and authenticity of bee products - A review. Food additives & contaminants. Part A, Chemistry, analysis, control, exposure & risk assessment, 38(4), 533–549. https://doi.org/10.1080/19440049.2020.1871081

Weng, S., Guo, B., Tang, P., Yin, X., Pan, F., Zhao, J., Huang, L., & Zhang, D. (2020). Rapid detection of adulteration of minced beef using Vis/NIR reflectance spectroscopy with multivariate methods. Spectrochimica acta. Part A, Molecular and biomolecular spectroscopy, 230, 118005. https://doi.org/10.1016/j.saa.2019.118005

Wu, B., Wei, F., Xu, S., Xie, Y., Lv, X., Chen, H., & Huang, F. (2020). Mass spectrometry-based lipidomics as a powerful platform in foodomics research. Trends in Food Science & Technology.

Wu, Y., Dong, Y., Shi, Y., Yang, H., Zhang, J., Khan, M. R., Deng, S., He, G., He, Q., Lv, Y., & Deng, R. (2021). CRISPR-Cas12-Based Rapid Authentication of Halal Food. Journal of agricultural and food chemistry, 69(35), 10321–10328. https://doi.org/10.1021/acs.jafc.1c03078

Xiao, L., Cao, S., Shang, X., Xie, X., Zeng, W., Lu, L., & Yan, H. (2021). Metabolomic and transcriptomic profiling reveals distinct nutritional properties of cassavas with different flesh colors. Food Chemistry: Molecular Sciences, 2, 100016.

Xiao, M., Chen, Y., Chu, H., & Yin, R. (2020). Development of a polymerase chain reaction-Nucleic acid sensor assay for the rapid detection of chicken adulteration. LWT, 131, 109679.

Yazdi, T. K., Mogaddam, M., Mousazadeh, M., Monajjemzadeh, F., Tamizi, E., & Nemati, M. (2021). Differential scanning calorimetry and fatty acid composition analysis of chocolates marketed in Iran as an alternative method for identification of cocoa butter adulteration. Journal of AOAC International, qsab086. Advance online publication. https://doi.org/10.1093/jaoacint/qsab086

Yeganeh-Zare, S., Farhadi, K., & Amiri, S. (2021). Rapid detection of apple juice concentrate adulteration with date concentrate, fructose and glucose syrup using HPLC-RID incorporated with chemometric tools. Food chemistry, 370, 131015. Advance online publication. https://doi.org/10.1016/j.foodchem.2021.131015

Yong, C. H., Muhammad, S. A., Aziz, F. A., Nasir, F. I., Mustafa, M. Z., Ibrahim, B., Kelly, S. D., Cannavan, A., & Seow, E. K. (2022). Detecting adulteration of stingless bee honey using untargeted ^{1}H NMR metabolomics with chemometrics. Food chemistry, 368, 130808. https://doi.org/10.1016/j.foodchem.2021.130808

Yong, C. H., Muhammad, S. A., Ng, J. S., Nasir, F. I., Adenan, M. N. H., Moosa, S., ... & Seow, E. K. (2022). Detection of adulteration activities in edible bird's nest using untargeted 1H-NMR metabolomics with chemometrics. Food Control, 132, 108542.

Zhang, J., Hu, Q., Yu, Q., Chen, Y., Zhao, Y., & Qie, M. (2021). Metabolomics Analysis in Food Authentication. Editor(s): Alejandro Cifuentes, Comprehensive Foodomics, Elsevier, 2021, Pages 517-535, ISBN 9780128163962, https://doi.org/10.1016/B978-0-08-100596-5.22802-3.

Zhang, M., Li, Y., Zhang, Y., Kang, C., Zhao, W., Ren, N., Guo, W., & Wang, S. (2021). Rapid LC-MS/MS method for the detection of seven animal species in meat products. Food chemistry, 371, 131075. Advance online publication. https://doi.org/10.1016/j.foodchem.2021.131075

Zheng, X., Li, Y., Wei, W., & Peng, Y. (2019). Detection of adulteration with duck meat in minced lamb meat by using visible near-infrared hyperspectral imaging. Meat science, 149, 55–62. https://doi.org/10.1016/j.meatsci.2018.11.005

7

Nanotechnology A Promising Venture to Ensure Food Quality and Safety

Mrithula Mahalakshmi Madhan Kumar[1] and Janifer Raj Xavier[2]

[1]Department of Food Process Engineering, University of New South Wales Sydney, Australia
[2]Defence Food Research Laboratory, Defence Research and Development Organization, Mysore, Karnataka, India

Introduction

The act of substituting food or drink with low grade quality ingredients thereby the final product will fail to the meet the regulation or standards and in most cases even unfit for human consumption (Augustin *et al.*, 2009). With demand for profit based manufacturing market, there will be increase in points of adulteration with cheaper alternatives or inferior grade ingredients in the production of food or drink (Banti, 2020)]. Adulteration is highly prevails in developing countries such as India, where about 70% of milk supply is adulterated on day to day basis (Bittar *et al.*, 2017). It is every nation and each individual's responsibility to provide safe and nutritious food to the consumers. In India, since the implementation of the Food Safety and Standards Act (FSS) – 2006, there has been increased awareness among producers and consumers regarding the judicial regulation of food adulteration through nationwide campaigning (Dhara, 2021). The increased awareness among consumers has risen the demand for high quality food which is free of hazardous contaminants or adulterants (Table 1). In 2011, Food Safety and Standards Authority of India (FSSAI) has regulated the manufacturing, processing, storage, distribution, sale and importation of food to ensure distribution and availability of safe food for human consumption (Goyal, 2012). The application of nanotechnology is endless, especially in food industry where it has revolutionised various areas such as preservation, packaging, nutrition delivery and nano sensing (Kumar *et al.*, 2016). Devices at the size of nano which consists of biological or chemical sensory points to measure highly sensitive and specific quantitative transmissions (Spink *et al.*, 2017). Nanosensors are highly advantageous due to its unique properties and lower toxicity. Some key benefits are as follows: light weight, low production cost, lesser energy consumption, environment friendly, controllable optical, mechanical, and electrical properties, specific target recognition and repeatable signal transduction. Nanotechnology engineers the application of nanomaterials in different fields for the benefit of living beings. Nanomaterials are utilized for measuring mechanical, chemical, and physical changes directly related to the indicators for sensing specific properties (Kumar *et al.*, 2017). Nanosensors are gaining attention in the field of detecting food adulterants and swiftly replacing other conventional methods to ensure food safety (Figure 1). Conventional methods are highly expensive, long analysis time, highly complicated technical instruments requiring skilled personnel and thereby creating bottle neck in the supply chain and reducing the efficiency of adulterant detection. Some common conventional methods are gas chromatography, mass spectrometry, high-performance liquid chromatography (HPLC) (Nambiar *et al.*, 2018). With growing demand in food supply chain has led to various innovative concepts for ensuring safe food production all throughout the year with enhanced quality control measures.

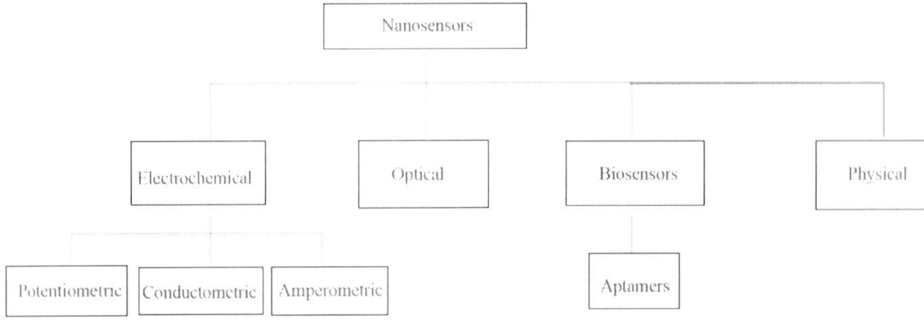

Fig. 1: Classification of Nano sensors based on functionality

Table 1: In house test methods for some of the most common adulterants found in food (Banti *et al.*, 2020)

Food Substance	Adulterant	In house tests
Rice	Urea	To 30 numbers of parched rise add 5ml distilled water. Mix well and rest the container. After 5 minutes filter the water and add 1/2 teaspoon of soybean powder n it. Followed by dipping a red litmus paper after 30s, appearance of blue colour indicates the presence of urea in parched rice.
Milk	Urea	Add 1/2 teaspoon of soybean powder to a teaspoon of milk and shake well for 5 minutes followed by dipping a red litmus paper in it. Observe colour change after 30s, if red colour changes to blue it indicates the presence of urea in milk.
Milk	Formalin	To 10 ml of milk, add 5 ml of conc sulphuric acid onto sides of the container without shaking. If violent or blue ring appears it shows the presence of formalin.

Food Substance	**Adulterant**	**In house tests**
Milk	Detergent	Shake 10ml of milk if adulterated with detergent, it forms dense lather if adulterated.
Rice grain	Colour/ Dye	Add 2 teaspoon of grain in water and mix well. Adulterated grain leaves colour immediately.
Chilli Powder	Colour/ Dye	Sprinkle chilli powder on surface of water in a glass tumbler. Artificial colourants will immediately descend forming colour streaks.

Table 2: Most common adulterant and relevant nanosensors used for its detection.

Adulterant	**Description, Source &Nano-sensors Used for detection**	**References**
Sudan I (1-phenylazo-2-napthol) – Synthetic azo-colorant dye	Adulterated in chilli powder, curry powder, sauces Sensors measure oxidation current corresponding to Sudan I quantity using electrochemical method. • Carbon electrode modified with cyclic voltametric (CV) magnetite NPs • Glassy carbon electrode modified with graphene, RGO/AU • NPs (DPV) and electrochemical RGO. • Platinum NPs decorated graphene-β-cyclodextrin modified electrode with LOD of 1.6 nM. La3+ doped Co3O4 nanocubes, ZnO-CuO nanoplate modified screen-printed electrode sensor. • Bismulthutungstate (Bi2WO6) nanosheets modified glassy carbon electrode with LOD of 0.002 μM.	Tajik *et al.*, 2019; Yang *et al.*, 2010.
Sunset Yellow & Tartrazine - Toxic synthetic dyes	Spherical carbon dots of 5 nm diameter are applied for Sunset Yellow & Tartrazine. • Hexadecyl trimethyl ammonium bromide functionalized GO/multiwalled CNTs modified glassy carbon electrode - simultaneously detects both Sunset Yellow & Tartrazine with LOD of 1×10-8 M.	Gómez et a., 2016; Kolozof *et al.*, 2020.
Sunset Yellow -Petroleum derived orange azo dye.	Electrochemical assay using voltammetry for quantifying Sunset Yellow in food. • Electrochemical sensor based on β-cyclodextrin/ ionic liquid/Au NPs functionalized magnetic GO. • Graphene/ poly (L-phenylalanine) • Spark generated molybdenum NPs modified graphite electrodes	

Adulterant	Description, Source &Nano-sensors Used for detection	References
Urea	Adulterated in milk to show higher nitrogen content. Regulated Urea value should be between 0.2 - 0.4 mg/mL • Enzymatic bio sensors - expensive. Therefore, a non-enzymatic electrochemical. Glass carbon electrode modified using Ag NPs decorated nitrogen doped SWCNTs and a layer of Nafion. • Graphene nanoplatelets conjugated with urease immobilized on a CNTs-covered electrode. Urease-immobilized graphene nanoplatelets and graphitized nanodiamonds nanocomposites. • ENFs modified with ZnO NPs and enzyme urease. Electrochemically analysis hydrolysis reaction of Urea by urease, to produce ammonium and carbonic ions.	Kumar *et al.*, 2018; Velychko et l., 2016.
Preservatives - Parabens	• Polymer nanofilm modified screen printed Au electrode. Electrochemical sensor with LOD of 0.706 μM. Parabens are chemically called as Parahydroxybenzoates or Parahydroxybenzoic acid esters	Mohammadi *et al.*, 2020
Preservatives - Sulfite (inorganic anion)	• Cetyltrimethylammonium bromide fluorescent organic NPs (CTAB-FONs). Sulfites are used in Tea industry as an antibacterial agent.	Mohammadi *et al.*, 2020
Sugars saccharide/ sugar analysis	• Semiconductor SWCNTs with a paradigmatic sensor configuration computationally electrochemical biocomposite of nanomaterials • Hydrogels from gum arabic and corn flour were reinforced by platinum NPs anchored with carboxyl functionalized CNTs. Followed by entrapping invertase and glucose oxidase in gum Arabic-corn flour hybrid matrix. Screen-printed carbon electrodes - enzyme-nanocomposite matrix.	Zeng *et al.*, 2015
Melamine (1,3,5-triazine-2,4,6-triamine, C3H6N5) .	• AU NPs - inherent properties. Colloidal solution of Au Nps has a red color, indicating proper dispersion of Au Nps. After exposure to melamine, these NPs aggregate and as a result solution color changes to blue. • Ag NPs - result in yellow to red colour change due to shift of plasmon band to lower wavelengths. rapid sensing - Dispersed Ag NPs gives strong absorption in UV- Vis spectra. unmodified Ag NPs and green synthesized Ag NPs. Melamineis used to compensate total protein value in milk or dairy products, egg, candies, yoghurt. Regulated level is 1.0ppm in infant formula; 2.5ppm in animal feeds.	Bittar *et al.*, 2017; Du *et al.*, 2015; Kumar *et al.*, 2014

Rapid Home Test for Detecting Adulterants

Food additives could have severe detrimental effects on human health and therefore controlling its usage is one the critical aspects of health monitoring. Conventionally adulteration in food could be detected using the following methods – thin-layer chromatography, adsorptive voltammetry, Differential Pulse Voltammetry (DPV), capillary electrophoresis, high-performance liquid chromatography, colorimetry, and conventional chromatography. Inefficiency and longer procedure with high time consumption are some of the disadvantages of conventional methods. Dyes are the most common food additives used as an alternate to compensate natural color loss during thermal process, preservation, or storage (Alves *et al.*, 2008). The use of synthetic dyes in food is highly controlled and restricted by various regulatory bodies such as the Food Safety and Standards Authority of India (FSSAI), Food and Drug Administration (FDA), Food Standards Australia New Zealand (FSANZ) & European Union (EU) (Mohammadi *et al.*, 2020).

Nano-sensors for Detecting Food Adulterants

In the year 1962, Clarke & Lyons developed the first biosensor consisting of an oxygen electrode modified with glucose oxidase overlapping with Teflon membrane. The initial design was larger in structure, where the electrode is similar in size of a pH probe. The structure consists of electrode connected to potentio-stat which further linked to pen recorder (Chelliah *et al.*, 2021). Later in the year 1972, a more compact unit called the Yellow Spring 24 M glucose analyzer was developed by refining the Clarke electrode. This revolutionary sensor in the size of a credit card was used for detecting blood glucose level using a strip containing screen printed electrode with an enzyme layer and mediator immobilized on the anode (Juneye *et al.*, 2009). Major disadvantage of the sensor being lower precision with SD >15%, thereby introducing risk among patients self-monitoring. However, these hardships were eliminated with the discovery of nanoparticles with high specificity to capture target combined with higher sensitivity capable of transducing signal with smaller sample volume (Kaushik *et al.*, 2009). Adulteration of milk could be detected using a smartphone-based device where a nanosized indicator paper (~10−9 m diameter) which on dipping into milk measures the level of acidity (Kumar *et al.*, 2016). The indicator paper made of nylon fibres are produced by electrospinning process and incorporated with three dyes which changes colour when dipped into milk according to the level of acidity (Ibrȧhim *et al.*, 2017). The colour change indicating the accurate acidic level will be displayed onto a smartphone. The pH ranges pure (6.9), acidic (<6.9) and basic (>6.9) will be captured on the smart phone using an image-based classifier (Paniel *et al.*, 2010). The levels of saccharin, potassium acesulfame, and cycla-mate could be detected using nanosensors made of bilayer lipid membranes. The amount of glucose in alcoholic beverages can be quantified using amperometric biosensor (Ping *et al.*, 2012). The sensor made up of carbon nanotube paste electrode modified with osmium functionalized polymer consisting of glucose dehydrogenase and diaphorase co-immobilized with NAD (+) can detect glucose up to 10 micromol L−1 in alcoholic beverages (Yin *et al.*, 2011).

Presence of synthetic food dyes such as sunset yellow or tartrazine as minimum as 0.025 μM and 0.01 μM respectively can be detected in orange juice using Wang *et al.* (2016) method. Electrochemical sensor made up of gold nanoparticles deposited on carbon electrodes were used for detection of sunset yellow; on the other hand, carbon dots were used for tartrazine detection in food (CapitÃ¡n, 1998). These carbon dots where of 5nm in diameter with high quantum yield and capable of releasing bright yellow photoluminescence indicating the presence of tartrazine (Qu rt al., 2018). Alternatively, tartrazine could also be detected with electrochemical sensor using graphene oxide composite modified glassy carbon electrodes. The graphene oxide composites were reduced using titanium oxide and thereby increasing the electrochemical response when applied in carbonated beverages to detect tartrazine with a LOD of 8.0 × 10−9 mol/L (Xu *et al.*, 2015). Presence of hydrogen peroxide in milk can be detected using amperometric sensor consisting of multi- walled carbon nanotube and chitosan film coated glassy carbon electrode with LOD 0.463 AM−1 cm−1. Alternatively, Nafion/ exfoliated graphene oxide- cobalt oxide nanocomposite based amperometric sensor could be used for monitoring of hydrogen peroxide with LOD 0.3 μmol L(−1) in milk samples (Boyac at al., 2010). Additionally in the year 2015, another-amperometric sensor was developed using carbon nano- tubes and nano iron oxide with LOD of 3.7 nM. Thermal co-precipitation was used for manufacturing nano iron oxide for detecting hydrogen peroxide in milk by performing cyclic voltammetry and amperometry (Hayat *et al.*, 2014). Another amperometric biosensor developed in the year 2012 for detection of benzoic acid in beverages reported LOD of 0.03 μM. The sensor consists of tyrosinase entrapped in titania gel modified with multilayer carbon nanotubes, and when brough in contact with alcoholic beverages the inhibition of benzoic acid on the activity of tyrosinase and polyphenol oxidase will direct indicate the level of benzoic acid in the drink (Kitikul *et al.*, 2016).

Presence of food preservative, the benzoic acid in cola, sprite, milk and yoghurt could be detected using a polyaniline-polyacrylonitrile composite amperometric biosensor. The real time sensor detects inhibition activity of benzoic acid of the polyphenol oxidase enzyme's biocatalytic activity to catechol. This sensor developed in the year 2007 has a LOD of 2×10(−7) M, where as an improvised version was again developed in the year 2008, using immobilization of tyrosinase enzyme by calcium carbonate nanomaterials had LOD of 1061.4 mA M−1 cm−2. Higher detection limit measured on modified glassy carbon electrode was based on benzoic acid inhibition on the tyrosinase/ Calcium Carbonate nano particles (Huang *et al.*, 2011). Similarly, the presence of toxic food dyes – metanil yellow and fast green could be detected using an electrochemical nanosensor made up of glassy carbon electrode modified with calixarene and gold nanoparticles (Hughes *et al.*, 2016). Modification of glass carbon electrode has contributed towards significantly greater signal selectivity and thereby leading to a detection limit of 9.8 and 19.7 nM for fast green and metanil yellow respectively. Alternatively plasmon resonance sensor was developed by seed mediated method to detect boric acid in water using the gold nanoparticle's plasmonic response which was linearly proportionate to boric acid and the device was designed to have a LOD of 0.01 mM (Pérez-Juste, 2020). Detection of formaldehyde in fresh food especially dried/ wet fish and meatballs can be carried out either using an amperometric biosensor or electrochemical sensor. Later anamperometric biosensor was developed in the year 2012 using 1.0% ferrocene mediator and an alcohol oxidase as biocatalysts deposition on carbon screen printed electrodes with a LOD of 0.007 mM. The sensor was also based on entrapment of nanogold in poly-2-hydroxyl ethyl methacrylate membrane to ease electron transfer in formaldehyde (Noor *et al.*, 2016). Later in the year 2016, Aini *et al.* developed an electrochemical based sensor using glassy carbon electrode deposited with gold nanoparticles and chitosan for analysing fish samples with LOD of 0.1 ppm (Aghaei *et al.*, 2020). Melamine compound in food could be detected using nanotechnology based structures comprising of carbon dots, nanotubes, nanocomposites, nanocrystals, nanowires, nanorods, quantum dots, and nanoclusters. Li *et al.* 2010 study reported melanine assay linearity from 39.6 to 238 nM using bis (8-quinolinolato) zinc (II) complex nanorod array (Figure 2). In the same year Chen *et al.* developed a surface-enhanced Raman scattering (SERS) using ZnO/ Au nanoneedle arrays to detect 10 nM to 100 μM concentration (Li *et al.*, 2010). Composites of silver nanoparticles and glutathione capped zinc selenium quantum dots were used to detect melamine with LOD of 872 pM (Madhavan *et al.*, 2019). Quantum dots were further utilized in real-time screening approach combined with silver nanocomposite and ultrasensitive colorimetric to achieve a LOD of 62.6 pM. Another study using L-custeine-capped Mn-doped zinc sulfide quantum dots showed room-temperature phosphorescence indicating melamine with detection limit of 47nM in dairy products [24]. Furthermore another study showed molecularly-imprinted CdTe quantum dots performing ratio-metric discrimination of melamine with detection limit of 38nM (Borase *et al.*, 2015). In another study using CdTe quantum dots with a molecularly-imprinted polymer using a ratio-metric fluorescent probe revealed a visual detection technique of melamine in milk with a LOD of 103 nM (Ji *et al.*, 2005).

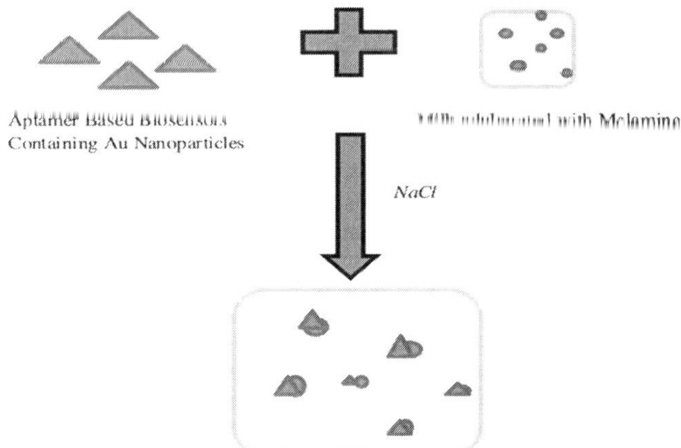

Fig. 2: Mechanism of Melamine detection in milk using aptamber based nano-sensor containing Au Nanoparticles (Huang *et al.*, 2011

Earlier horse-radish peroxidase (HRP) functionalized gold nanoclusters reported colorimetric detection of melamine based on the reversal of mercury (II) mediated inhibition (Kuang *et al.*, 2011). In another study with egg-white protected gold nanoclusters by Tang *et al.* in 2018 reported fluorescence mediated sensing of melamine by microwave technique achieved a limit of detection of 0.46 μM. Formalin the aqueous solution of formaldehyde, detection technique was developed by Hossian *et al.* 2019 using a numerical surface plasmon resonance biosensor made up of graphene- MOS2 with titanium oxide-silicon dioxide and achieved total reflection for monitoring surface plasmon resonance. Another highly preferred melamine detection tool due to lower toxicity, nano size and biocompatibility is Carbon dots (C-dots). Studies have showed the anti-quenching ability of Hg2+ combined with c-dots for real-time melamine determination is achievable to a LOD of 0.3μM. Adulteration of Pork in meatball could be detected using a nano biosensor (Stephen *et al.*, 2016). The sensor consists of 27-nucleotide AluI fragment of swine cytochrome b (cytb) gene and integrated to 3-mm dia citrate-tannate-coated gold nanoparticles (Singh *et al.*, 2015). The sensor could detect 1% pork in raw and cooked meatballs. On analyzing the hybridization kinetics,the biosensor was hybridized with a target DNA of several fold shorter than a PCR-template and thereby increasing the chances of detection and quantification of adulteration in highly processed meats (Subara *et al.*, 2018).

Classification and Functioning of Nano-sensors

Nanosensors consists of a transducer and a receptor, where receptors could be any inorganic or organic substance which interacts with analyte or derivatives releasing signals in the form of electrochemical, electrical, or optical which is then converted by transducers to determinable signals. Nanoparticles - between 1 and 100 nM in dia. Various shapes - spheres, needles, rods, and plates. The ratio of surface to volume and distance of energy level changes by varying the size of NPs (Rosi *et al.*, 2005). Which eventually causes various other NP properties to change. Main functions - catalysis of electrochemical reactions, increase in the electron transfer, immobilization of biological molecules and labelling of biological molecules (Qin *et al.*, 2010). Carbon compounds are considered as functional electron transfer intermediates as they possess low field emission, large potential windows, chemical inertness, and low-cost production. Excellent thermal stability, electrical conductivity, and high contact surfaces. Higher stability even with regular nanoscale structure – they are stable even when not functional (Table 3).

Table 3: Types of Nanoparticles, their properties, and applications

Type of Nanoparticles	Description	References
Metal NPs	Gold (Au), Silver (ag), Titanium Oxide (TiO_2), Zinc Oxide (ZnO) & Silica ($Si0_2$) are commonly used for manufacturing Metal MPs [70]. Applied for manufacturing sensors, adsorbent Materials, drug delivery systems, and antimicrobial materials. Optical property caused by Localized Surface Plasmon Resonance (LSPR) is the most common principle involved in utilizing Metal NPs. Metals with higher surface area will produce higher surface resonance according to the type of molecule or protein present in the environment due to alteration in dielectric properties.	Mohammadi *et al.*, 2020
Magnetic NPs	Magnetic MPs are made up of Iron, Nickel, Cobalt, Aluminium Oxide and Au NPs. Known for their Nano-dimensional properties; where three elements of iron, nickel and cobalt when combined with non-magnetic elements exhibit magnetic properties. Easy separation with external magnetic field, low toxicity and greater compatibility are key advantages. Fe_30_4 (superparamagnetic) is used in food application.	Li *et al.*, 2016
Quantum Dots QD	SQD are of 1 to 10 nM in size and interacts with biomolecules such as protein of 1- 20 nM in size. The phenomenon of larger gap between excitation and emission spectrum called as the stoke's shift is observed improving the tracking sensitivity and thereby providing biological stability. SQD are successfully applied as a fluorescent indicator for biological applications. QD are inorganic fluorophores where wavelength of emission increases with increasing size. Strong optical stability & absorption, narrow, symmetrical & bright spectrum are the key properties of QD Cadmium Sulfide (CdS), Cadmium Selenide (CdSe), Cadmium Telluride (CdTe), Zinc Selenide (ZnSe), Zinc Sulfide (ZnS), Lead (II) Sulfide (PbS) and Molybdenum Disulphide (MoS_2) are some of Quantum dot matrials [115]. Fluorescent dots (Semiconductor quantum dots - SQD, Carbon Quantum dots - CQD, Graphene Quantum dots - GQD) are used for detection of toxins and Antibiotics in natural/ organic substances.	Bagheri *et al.*, 2019;

Apart from the above mentioned, the Nanowires and Electron spun polymeric nanofibers are commonly used in detecting adulterants in food. Nanowires are extremely sensitive to detect molecular interactions at their surface due to their long length-to-diameter ratio, provide electronically interchangeable properties of high quality, signal free, electrically readable sensor semiconductor nanowires, easily adjusted to below 100 nm resulting in higher density, light weight (Table 4). One-dimensional nanowires have high performance in electron transfer and light activities due to exceptional change in their electrical properties by connecting a molecule to their surface. Modified glass carbon electrode with Au NWs to detect glutamate (Palanisamy *et al.*, 2016). Electron spun polymeric Nanofibers have high surface area relative to unit volume, high porosity, strong mech prop, high axial tensile with high flexibility, low weight, cost effectiveness are key advantages. There are various methods for production for producing electro spun polymeric nanofibers such as - wet spinning, dry spinning, melt spinning, sol-gel spinning and electrospinning (Qianwen *et al.*, 2019).

Table 4: Types of Carbon Nanomaterials, their properties, and applications

Type of Carbon Nanomaterials	Description	References
Carbon Nanotubes	CNTs are preferred for their smaller size, conductivity and semi conductivity, ballistic & thermal conduction, high Young modulus, high surface density and high tensile strength. These properties are highly dependent on CNTs length, presence of catalyst and chirality. Graphite Nanotubes is the most common Carbon Nanotube.	Majidi *et al.*, 2012
Single Walled CNTs	SWCNTs are regular hexagons shaped single layer of graphene cylinders. Advantages of SWCNTs include conductive or semiconductor ability, excellent mechanical, and electrical properties. While stabilizing of this property incurs a high production expense contributing as a major disadvantage. Applied in production of thermal & electrical, field effect devices.	Xiao *et al.*, 2008
Multi Walled CNTs	MWCNTs are graphene sheets (approx. 50) piped. Unlike SWCNT, MWCNT are easier and cheaper to produce thereby possess commercial advantage. Small proteins are trapped by physical adsorption into inner channel of opened nanotubes or also can be bounded to outer surface of nanotubes by hydrophobic and electrostatic interactions through covalent bonding or functionalization around nanotubes with polymer coatings.	Tak *et al.*, 2013
Graphene	Carbon atoms form a single, flat layer in 2-D hexagonal lattice structure made up of C-C bond with 0.142 nm in length. High electron mobility at RT (50,000 cm2/Vs), quantum hall effect, exceptional thermal conductivity (5,000 W/mK), high mechanical strength, larger surface area, low toxicity and high biocompatibility are key advantages of Graphene nanomaterials. Graphene oxide, reduced graphene oxide, graphene quantum dots. Due to quasi-conductivity, they are used in the construction of transistors, electroanalytical chemistry.	Yang *et al.*, 2014

Carbon Nanofibers	Cylindrical nanostructures of graphene sheets, either as small clumps with lattice surfaces or precise channels. Ribbon or herringbone, stacked cones, cups, plates, tubular. CNF have more functionalization than CNTs, extensive ratio of surface area of active groups to volume of nanomaterials. Improved thermal, electrical frequency are the key advantages. Used in synthesis of composites due to their properties and low cost	Qiawen *et al.*, 2019

Synthesis of Nanomaterials

Common methods for synthesis of nanoparticles are top-down and bottom-up approach where, in top- down approach the particles are produced in large scale patterns and then further reduced to lateral dimensions for forming nanostructures (Jain *et al.*, 2017). Whereas in bottom-up approach atoms or molecules are used to build nanostructures using synthetic chemicals and biochemical techniques (Figure 3). Commercial top-down approach is highly preferred for elaborating nanostructures (Kim *et al.*, 2007).

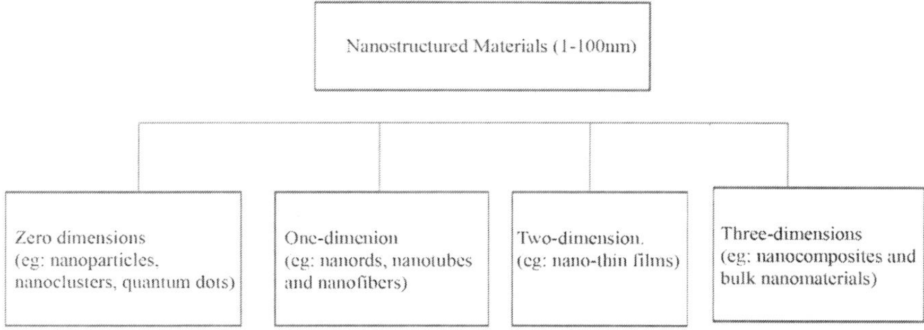

Fig. 3: Classification of Nanomaterials based on dimensions.

Top-Down Approach

Top-Down approach necessiates size reduction of large-scale structures using forces such as compression, impact, and shear. Various procedures such as dry milling, high-pressure homogenization, micro fluidization, membrane emulsification, Ultrasound (US) emulsification, supercritical fluids and electrified coaxial liquid jets are used in top-down method (Yu *et al.*, 2018). Simplest procedure is dry milling however ball and jet milling are most used process to produce ultrafine powders. The procedure involves high-pressure pump to induce high turbulence, shear, impact, decrease drop pressure and thereby converting liquids into a fine emulsion droplet. Micro fluidization is also very similar which utilizes very high pressure of 200-700 MPa creating cavitation, shear, and impact to produce emulsified fine particles of sub-micron range (Chaudhry *et al.*, 2011). Another method is Ultra-Sonic emulsification where two immiscible liquids in the presence of a surfactant and ultrasonic energy field inducing mass transfer between the two immiscible phases. Furthermore, in membrane emulsification technique where emulsions of narrow droplet-size distribution of the dispersed phase/ oil is formed through a membrane into a continuous phase/ water. US emulsification using supercritical fluids are highly advantageous due to absence of toxic residue due to high diffusivities and high mass transfer rate of the material. With electrospinning method, nanoscale ultrafine fibres can be produced from polymer solutions (Xue *et al.*, 209). However various research has been carried out to enhance the quality and improve functionality of nanofibers using coaxial electrospinning technique.

Bottom-Up Approach

Techniques such as phase separation, nanoprecipitation/ Self-Assembly and template method are few examples of bottom-up method. Self-assembly is the process by which nanoparticles gets organized spontaneously by hydrophobic or hydrophilic interactions or Coulomb and Van del Waals forces. For example, nanosized metal organic frameworks are produced by self-assembly of Fe (III) using its thermodynamic equilibrium. Self-assembly in the presence of a polymer, polymer solvent and non-solvent of the polymer is called as nanoprecipitation (Alexander *et al.*, 2000). Another bottom-up approach is phase separation technique where two kinds of biopolymer molecules possessing sufficient repulsive forces between them is used to separate into droplet phase and dispersed-phase. Electrostatic or steric repulsion is adjusted to achieve desirable droplet formation. Template method is used for active selective induction of nanoparticle deposition. Template method is a surface-modified substrates in 1,2 or 3 dimensions formed by arrangement of different particles into a structure morphologically complement to that of template (Ali *et al.*, 2016). Electrochemical sensors – These devices transduce biochemical energy from antigen-protein or enzyme-substrate reactions to electrical signals. Transducers can be potentiometric, amperometric or impedimetric where ion-selective principle or electrode-analyte interaction or receptor-analyte interaction are used receptively for biosensor analysis (Nikoleli, 2020). Immunochemical biosensors involve quantification of analytes using antigen – antibody reaction. An immunosensor is defined as the device that detects analytes or antigens of interest using immunoassay specific antibodies to detect stable immunocomplex (Zhuo *et al.*, 2005). DNA-based biosensors, immunosensors, affinity-based biosensors are few examples of immunochemical biosensors. Aptamer-based biosensors consists of single-stranded nucleic acids (aptamers) capable of specific binding to target molecules generating a secondary structure (Hosseini, 2015). Aptamers are ideal for biosensor fabrication due to their cost-effective production by chemical methods and easily modifiable structure improving signal efficiency (Weng *et al.*, 2016).

Toxicity of Nanomaterials and Nanosensors

Nanomaterials/ Nanoparticles are classified as natural and anthropogenic which was further divided into incidental and engineered nanoparticles. Incidental nanoparticles are formed because of natural human activity and therefore weak structure made up of different elements forming loose shape and size (Jin *et al.*, 2017). Whereas engineered nanoparticles mase up of polymeric/ organic/ inorganic bonds. Despite possessing too many novel properties, industrialization of nanomaterials has led to high concern regards to human health due to related potential health hazards (Nanotoxicology and Nanoecotoxicology Vol. 1, 2021). Thereby all engineered nanomaterials are assessed for their manufacturing/ processing parameters to identify and eliminate hazards if any before reaching the consumers. A minor change in processing parameters of nanoparticles will impose major changes in nutritional and chemical structure of the food substance giving rise to allergic reactions. Now, around 276 nanomaterials are available in the market, in which nano-encapsulated titanium dioxide and silver tops the list (Raghavan *et al.*, 2021). Where the Asian market contributes to nearly 50% of the world Nanofood market, globally contributing to half a trillion of US$ annually. Direct contact of food components with nanomaterials or migration of nanoparticles into food possess huge risk to human consumption. Once the nanoparticles get ingested into human body it passes through stomach, intestines and then later enters circulation. Later the ingested nanoparticles are distributed by the two important organs liver and spleen. The nanoparticles entering gastrointestinal tract excretes the particles immediately, contradicting to other ingestion routes such as inhalation and skin exposure. Magnesium oxide is found to affect the olfactory nerve and eventually the brain as well, by entering olfactory bundle through inhalation (Ahmed *et al.*, 2010). Zinc oxide nanoparticles showed genotoxic potential in epidermal human cells and similarly inhalation of high doses of nano-TiO2 caused chronic lung tumors in human (Alehosseini *et al.*, 2019). Studies have also proved that nanoparticles could not penetrate healthy or intact skin, however its impact on human body depends on exposure quantity, rate, and duration. Exposure of nanoparticles into bloodstream will affect the blood vessel lining causing blood clot formation and thereby leading to cardiovascular effects inducing platelet aggregation and vascular thrombosis. Once nanoparticles have affected the blood flow, they will begin to induce adverse effects on other organs of the body. In pregnant women, studies have shown that nanoparticles cross the blood brain barrier and thereby migrating into foetus through interaction with cells and organs. Chen *et al* (2006) studied the impact of nano copper on mice and reported that nano copper to be much more fatal compared to bulk copper and causing pathological damages in spleen, liver, and kidney. In various ocular studies, nanoparticles are found to produce behavioral changes in fish by producing biochemical changes from the interaction of apolipoproteins/ apoA-I with nanomaterials such as titanium dioxide and single walled carbon nanotubes (Bagal *et al.*, 2019). Various characteristics of nanomaterials such as surface chemistry, source of material used, state of agglomeration, preparation method and storage are all contributors of toxicity in nanomaterials (Crevillén *et al.*, 2007).

Future Prospects

In the last decade there has been tremendous advancement in the employment of nanotechnology in food processing, preservation, and enforcement of safety and quality. There is massive attention in the field of incorporating nanotubes in packaging to detect toxic proteins, microbe, or chemical adulterants. With the enhancement of nanotube application, the commercialization of active and intelligent packaging systems in our near future. However, the toxicity of nanoparticles on environment and human is unpredictable due to its novelty, continued studies in eliminating the toxicity will be highly helpful in broadening the areas of nanoparticles application. In the future with the application of nanotechnology we can enhance agricultural productivity by reducing chemical inputs to achieve eco-balance.

Acknowledgement

I would like to thank Dr. Gopal Kumar Sharma and Ms. Janifer Xavier from Defense Food Research Laboratory (DFRL), Mysore for their constant support and offering me with the opportunity to contribute towards the Manuscript titled "Nanotechnology – A promising venture to ensure food quality and safety".

References

Aghaei, Z., Ghorani, B., Emadzadeh, B., Kadkhodaee, R., & Tucker, N. (2020). Protein-based halochromic electrospun nanosensor for monitoring trout fish freshness. Food Control, 111, 107065. https://doi.org/10.1016/j.foodcont.2019.107065

Ahamed, M., Siddiqui, M. A., Akhtar, M. J., Ahmad, I., Pant, A. B., & Alhadlaq, H. A. (2010). Genotoxic potential of copper oxide nanoparticles in human lung epithelial cells. Biochemical and Biophysical Research Communications, 396(2), 578–583. https://doi.org/10.1016/j.bbrc.2010.04.156

Alehosseini, E., & Jafari, S. M. (2019). Micro/nano-encapsulated phase change materials (pcms) as emerging materials for the food industry. Trends in Food Science & Technology, 91, 116–128. https://doi.org/10.1016/j.tifs.2019.07.003

Alexandre, M., & Dubois, P. (2000). Polymer-layered silicate nanocomposites: Preparation, properties and uses of a new class of materials. Materials Science and Engineering: R: Reports, 28(1-2), 1–63. https://doi.org/10.1016/s0927-796x(00)00012-7

Ali, A., Zafar, H., Zia, M., ul Haq, I., Phull, A. R., Ali, J. S., & Hussain, A. (2016). Synthesis, characterization, applications, and challenges of iron oxide nanoparticles. Nanotechnology, Science and Applications, Volume 9, 49–67. https://doi.org/10.2147/nsa.s99986

Alves, S. P., Brum, D. M., de Andrade, É. C. B., & Netto, A. D. P. (2008). Determination of synthetic dyes in selected foodstuffs by high performance liquid chromatography with UV-DAD detection. Food chemistry, 107(1), 489-496.

Arora, K. (2018). Advances in nano based biosensors for food and agriculture. Nanotechnology, food security and water treatment, 1-52.

Augustin, M. A., & Sanguansri, P. (2009). Nanostructured materials in the food industry. Advances in food and nutrition research, 58, 183-213. Authentication in Meat and Meat Products. International Journal on Advanced Science, Engineering and Information Technology.

Bagal-Kestwal, D.R., & Chiang, B. (2019). Platinum nanoparticle-carbon nanotubes dispersed in gum Arabic-corn flour composite-enzymes for an electrochemical sucrose sensing in commercial juice. Ionics, 1-14.

Bagheri, M., Jafari, S. M., & Eikani, M. H. (2019). Development of ternary nano adsorbent composites of graphene oxide activated carbon, and zero-valent iron nanoparticles for Food Applications. Food Science & Nutrition, 7(9), 2827–2835. https://doi.org/10.1002/fsn3.1080

Banti, M. (2020). Food adulteration and some methods of detection, review. International Journal of Nutrition and Food Sciences, 9(3), 86-94.

Bittar, D. B., Catelani, T. A., Nigoghossian, K., Barud, H. D. S., Ribeiro, S. J. L., Pezza, L., & Pezza, H. R. (2017). Optimized synthesis of silver nanoparticles by factorial design with application for the determination of melamine in milk. Analytical Letters, 50(5), 829-841.

Borase, H. P., Patil, C. D., Salunkhe, R. B., Suryawanshi, R. K., Salunke, B. K., & Patil, S. V. (2015). Biofunctionalized silver nanoparticles as a novel colorimetric probe for melamine detection in raw milk. Biotechnology and Applied Biochemistry, 62(5), 652-662.

Boyaci, I. H., & Mutlu, M. (2012). Amperometric biosensors in food processing, safety, and quality control. Biosensors in Food Processing, Safety, and Quality Control.

Capitán-Vallvey, L. F., Fernández, M. D., De Orbe, I., & Avidad, R. (1998). Simultaneous determination of the colorants tartrazine, ponceau 4R and sunset yellow FCF in foodstuffs by solid phase spectrophotometry using partial least squares multivariate calibration. Talanta, 47(4), 861-868.

Chaudhry, Q., & Castle, L. (2011). Food applications of nanotechnologies: an overview of opportunities and challenges for developing countries. Trends in Food Science & Technology, 22(11), 595-603.

Chelliah, R., Wei, S., Daliri, E. B. M., Rubab, M., Elahi, F., Yeon, S. J., ... & Oh, D. H. (2021). Development of nanosensors based intelligent packaging systems: food quality and medicine. Nanomaterials, 11(6), 1515.

Crevillén, A. G., Ávila, M., Pumera, M., González, M. C., & Escarpa, A. (2007). Food analysis on microfluidic devices using ultrasensitive carbon nanotubes detectors. Analytical chemistry, 79(19), 7408-7415.

Dhara, D., Biswas, S., Das, S., & Biswas, O. (2021). Status of food safety and food security in india in the perspective of fssai. Indian J Anim Health, 60(2), 167-173.

Du, J., Wang, Y., & Zhang, W. (2015). Gold nanoparticles-based chemiluminescence resonance energy transfer for ultrasensitive detection of melamine. Spectrochimica Acta Part A: Molecular and Biomolecular Spectroscopy, 149, 698-702.

Gómez Arancibia, M., Arancibia Moya, V., Aliaga Miranda, M. E., Núñez, C., & Rojas Romo, C. R. (2016). Determination of Sudan I in drinks containing Sunset yellow by adsorptive stripping voltammetry.

Goyal, S., & Goyal, G. K. (2012). Nanotechnology in food packaging a critical review. Russian Journal of Agricultural and Socio-Economic Sciences, 10(10), 14-24.

Hayat, A., Catanante, G., & Marty, J. L. (2014). Current trends in nanomaterial-based amperometric biosensors. Sensors, 14(12), 23439-23461.

Hosseini, M., Khabbaz, H., Dadmehr, M., Ganjali, M. R., & Mohamadnejad, J. (2015). Aptamer-based colorimetric and chemiluminescence detection of aflatoxin B1 in foods samples. Acta Chimica Slovenica, 62(3), 721-728.

Huang H, Li L, Zhou G, Liu Z, Ma Q, Feng Y, Zeng G, Tinnefeld P, He Z. Visual detection of melamine in milk samples based on label-free and labeled gold nanoparticles. Talanta. 85: 1013-9.

Hughes, G., Westmacott, K., Honeychurch, K. C., Crew, A., Pemberton, R. M., & Hart, J. P. (2016). Recent advances in the fabrication and application of screen-printed electrochemical (bio) sensors based on carbon materials for biomedical, agri-food and environmental analyses. Biosensors, 6(4), 50.

Ibrahim, A. A., Ahmad, R., Umar, A., Al-Assiri, M. S., Al-Salami, A. E., Kumar, R., ... & Baskoutas, S. (2017). Two-dimensional ytterbium oxide nanodisks based biosensor for selective detection of urea. Biosensors and Bioelectronics, 98, 254-260.

Jain, A., Ranjan, S., Dasgupta, N., & Ramalingam, C. (2018). Nanomaterials in food and agriculture: an overview on their safety concerns and regulatory issues. Critical Reviews in Food Science and Nutrition, 58(2), 297-317.

Ji, X., Zheng, J., Xu, J., Rastogi, V. K., Cheng, T. C., DeFrank, J. J., & Leblanc, R. M. (2005). (CdSe) ZnS quantum dots and organophosphorus hydrolase bioconjugate as biosensors for detection of paraoxon. The Journal of Physical Chemistry B, 109(9), 3793-3799.

Jin, R., Li, L., Guo, L., Li, W., & Shen, Q. (2017). A graphene tip coupled with liquid chromatography tandem mass spectrometry for the determination of four synthetic adulterants in slimming supplements. Food Chemistry, 224, 329-334.

Jubete, E., Loaiza, O. A., Ochoteco, E., Pomposo, J. A., Grande, H., & Rodríguez, J. (2009). Nanotechnology: A Tool for Improved Performance on Electrochemical Screen-Printed (Bio) Sensors. Journal of Sensors.

Kaushik, A., Solanki, P. R., Ansari, A. A., Sumana, G., Ahmad, S., & Malhotra, B. D. (2009). Iron oxide-chitosan nanobiocomposite for urea sensor. Sensors and Actuators B: Chemical, 138(2), 572-580.

Kim, C., Jeong, Y. I., Ngoc, B. T. N., Yang, K. S., Kojima, M., Kim, Y. A., ... & Lee, J. W. (2007). Synthesis and characterization of porous carbon nanofibers with hollow cores through the thermal treatment of electrospun copolymeric nanofiber webs. Small, 3(1), 91-95.

Kitikul, J., Satienperakul, S., Preechaworapun, A., Pookmanee, P., & Tangkuaram, T. (2017). A simple flow amperometric electrochemical biosensor based on chitosan scaffolds and gold nanowires modified on a glassy carbon electrode for detection of glutamate in food products. Electroanalysis, 29(1), 264-271.

Kolozof, P. A., Florou, A. B., Spyrou, K., Hrbac, J., & Prodromidis, M. I. (2020). In-situ tailoring of the electrocatalytic properties of screen-printed graphite electrodes with sparked generated molybdenum nanoparticles for the simultaneous voltammetric determination of sunset yellow and tartrazine. Sensors and Actuators B: Chemical, 304, 127268.

Kuang, H., Chen, W., Yan, W., Xu, L., Zhu, Y., Liu, L., Chu, H., Peng, C., Wang, L., Kotov, N. A., & Xu, C. (2011). Crown ether assembly of gold nanoparticles: melamine sensor. Biosensors & Bioelectronics, 26(5), 2032–2037.

Kumar, N., & Kumbhat, S. (2016). Essentials in nanoscience and nanotechnology. John Wiley & Sons.

Kumar, N., Seth, R., & Kumar, H. (2014). Colorimetric detection of melamine in milk by citrate-stabilized gold nanoparticles. Analytical Biochemistry, 456, 43-49.

Kumar, T. H., & Sundramoorthy, A. K. (2018). Non-enzymatic electrochemical detection of urea on silver nanoparticles anchored nitrogen-doped single-walled carbon nanotube modified electrode. Journal of The Electrochemical Society, 165(8), B3006.

Kumar, V., Guleria, P., & Mehta, S. K. (2017). Nanosensors for food quality and safety assessment. Environmental Chemistry Letters, 15, 165-177.

Li, J., Wang, X., Duan, H., Wang, Y., Bu, Y., & Luo, C. (2016). Based on magnetic graphene oxide highly sensitive and selective imprinted sensor for determination of sunset yellow. Talanta, 147, 169-176.

Li, M., Yin, J. J., Wamer, W. G., & Lo, Y. M. (2014). Mechanistic characterization of titanium dioxide nanoparticle-induced toxicity using electron spin resonance. Journal of Food and Drug Analysis, 22(1), 76-85.

Li, X., Wang, T., Zhang, J., Zhu, D., Zhang, X., Ning, Y., ... & Yang, B. (2010). Controlled fabrication of fluorescent barcode nanorods. ACS Nano, 4(8), 4350-4360.

Madhavan, A. A., Qotainy, R., & Nair, R. (2019, March). Synthesis of functionalized silver nanoparticles and its application as chemical sensor. In 2019 Advances in Science and Engineering Technology International Conferences (ASET) (pp. 1-4). IEEE.

Majidi, M. R., Fadakar Bajeh Baj, R., & Naseri, A. (2012). Carbon Nanotube–Ionic Liquid (CNT–IL) nanocamposite modified sol-gel derived carbon-ceramic electrode for simultaneous determination of sunset yellow and tartrazine in food samples. Food Analytical Methods, 6(5), 1388–1397. https://doi.org/10.1007/s12161-012-9556-6

Mohammadi, Z., & Jafari, S. M. (2020). Detection of food spoilage and adulteration by novel nanomaterial-based sensors. Advances in Colloid and Interface Science, 286, 102297.

Mohammadi, Z., & Jafari, S. M. (2020). Nanoparticles/nanofibers for checking adulteration/ spoilage of food products. In Handbook of Food Nanotechnology (pp. 459-492). Academic Press.

Nambiar, A. P., Sanyal, M., & Shrivastav, P. S. (2018). Simultaneous densitometric determination of eight food colors and four sweeteners in candies, jellies, beverages and pharmaceuticals by normal-phase high performance thin-layer chromatography using a single elution protocol. Journal of Chromatography A, 1572, 152-161.

Nanotoxicology and Nanoecotoxicology Vol. 1. (2021). Environmental Chemistry for a Sustainable World.

Nikoleli, G. P. (2020). Advanced lipid based biosensors for food analysis. Advances in food and nutrition research, 91, 301-321.

Noor Aini, B., Siddiquee, S., & Ampon, K. (2016). Development of formaldehyde biosensor for determination of formalin in fish samples; malabar red snapper (Lutjanus malabaricus) and longtail tuna (Thunnus tonggol). Biosensors, 6(3), 32.

Palanisamy, S., Thangavelu, K., Chen, S. M., Thirumalraj, B., & Liu, X. H. (2016). Preparation and characterization of gold nanoparticles decorated on graphene oxide@ polydopamine composite: application for sensitive and low potential detection of catechol. Sensors and Actuators B: Chemical, 233, 298-306.

Paniel, N., Radoi, A., & Marty, J. L. (2010). Development of an electrochemical biosensor for the detection of aflatoxin M1 in milk. Sensors, 10(10), 9439-9448.

Pérez-Juste, J., Pastoriza-Santos, I., Liz-Marzán, L. M., & Mulvaney, P. (2020). Gold Nanorods: Synthesis, Characterization and Applications. In Colloidal Synthesis of Plasmonic Nanometals (pp. 15-88). Jenny Stanford Publishing.

Ping, H., Zhang, M., Li, H., Li, S., Chen, Q., Sun, C., & Zhang, T. (2012). Visual detection of melamine in raw milk by label-free silver nanoparticles. Food control, 23(1), 191-197.

Qianwen, M., Yaping, D., Li, L., Anqing, W., Dingding, D., & Yijun, Z. (2019). Electrospun MoS2 composite carbon nanofibers for determination of vanillin. Journal of Electroanalytical Chemistry, 833, 297-303.

Qin, P., Qiao, D., Xu, J., Song, Q., Yao, L., Lu, J., & Chen, W. (2019). Rapid visual sensing and quantitative identification of duck meat in adulterated beef with a lateral flow strip platform. Food chemistry, 294, 224-230.

Qin, Y., Ji, X., Jing, J., Liu, H., Wu, H., & Yang, W. (2010). Size control over spherical silver nanoparticles by ascorbic acid reduction. Colloids and Surfaces A: Physicochemical and Engineering Aspects, 372, 172-176.

Qu, J. H., Wei, Q., & Sun, D. W. (2018). Carbon dots: Principles and their applications in food quality and safety detection. Critical Reviews in Food Science and Nutrition, 58(14), 2466-2475.

Raghavan, V. S., O'Driscoll, B., Bloor, J. M., Li, B., Katare, P., Sethi, J., & Jenkins, D. (2021). Emerging graphene-based sensors for the detection of food adulterants and toxicants–A review. Food Chemistry, 355, 129547.

Rosi, N. L., & Mirkin, C. A. (2005). Nanostructures in biodiagnostics. Chemical Reviews, 105(4), 1547–1562.

Singh, P. K., Jairath, G., & Ahlawat, S. S. (2016). Nanotechnology: a future tool to improve quality and safety in meat industry. Journal of Food Science and Technology, 53, 1739-1749.

Spink, J., Ortega, D. L., Chen, C., & Wu, F. (2017). Food fraud prevention shifts the food risk focus to vulnerability. Trends in Food Science & Technology, 62, 215-220.

Inbaraj, B. S., & Chen, B. H. (2016). Nanomaterial-based sensors for detection of foodborne bacterial pathogens and toxins as well as pork adulteration in meat products. Journal of Food and Drug Analysis, 24(1), 15–28.

Subara, D., & Jaswir, I. (2018). Gold nanoparticles: Synthesis and application for halal authentication in meat and meat products. International Journal on Advanced Science, Engineering and Information Technology, 8(2018), 1633-1641.

Tajik, S., Aflatoonian, M. R., Shabanzade, R., Beitollahi, H., & Alizadeh, R. (2020). Amplified electrochemical sensor employing ZnO-CuO nanoplates for sensitive analysis of Sudan I. International Journal of Environmental Analytical Chemistry, 100(1), 109-120.

Tak, M., Gupta, V., & Tomar, M. (2013). Zinc oxide–multiwalled carbon nanotubes hybrid nanocomposite based urea biosensor. Journal of Materials Chemistry B, 1(46), 6392-6401.

Velychko, T. P., Soldatkin, O. O., Melnyk, V. G., Marchenko, S. V., Kirdeciler, S. K., Akata, B., ... & Dzyadevych, S. V. (2016). A novel conductometric urea biosensor with improved analytical characteristic based on recombinant urease adsorbed on nanoparticle of silicalite. Nanoscale Research Letters, 11(1), 1-6.

Xiao, F., Ruan, C., Liu, L., Yan, R., Zhao, F., & Zeng, B. (2008). Single-walled carbon nanotube-ionic liquid paste electrode for the sensitive voltammetric determination of folic acid. Sensors and Actuators B: Chemical, 134(2), 895-901.

Xiao, F., Ruan, C., Liu, L., Yan, R., Zhao, F., & Zeng, B. (2008). Single-walled carbon nanotube-ionic liquid paste electrode for the sensitive voltammetric determination of folic acid. Sensors and Actuators B: Chemical, 134(2), 895–901. https://doi.org/10.1016/j.snb.2008.06.037

Xu, H., Yang, X., Li, G., Zhao, C., & Liao, X. (2015). Green synthesis of fluorescent carbon dots for selective detection of tartrazine in food samples. Journal of Agricultural and Food Chemistry, 63(30), 6707-6714.

Xue, J., Wu, T., Dai, Y., & Xia, Y. (2019). Electrospinning and Electrospun Nanofibers: Methods, Materials, and Applications. Chemical Reviews, 119(8), 5298–5415.

Yang, D., Zhu, L., & Jiang, X. (2010). Electrochemical reaction mechanism and determination of Sudan I at a multi wall carbon nanotubes modified glassy carbon electrode. Journal of Electroanalytical Chemistry, 640, 17-22.

Yang, Y. J., & Li, W. (2014). CTAB functionalized graphene oxide/multiwalled carbon nanotube composite modified electrode for the simultaneous determination of ascorbic acid, dopamine, uric acid and nitrite. Biosensors and Bioelectronics, 56, 300-306.

Yin, H., Cui, L., Chen, Q., Shi, W., Ai, S., Zhu, L., & Lu, L. (2011). Amperometric determination of bisphenol A in milk using pamam–fe3o4 modified glassy carbon electrode. Food Chemistry, 125(3), 1097–1103.

Yu, H., Park, J. Y., Kwon, C. W., Hong, S. C., Park, K. M., & Chang, P. S. (2018). An overview of nanotechnology in food science: Preparative methods, practical applications, and safety. Journal of Chemistry, 2018.

Zeng, L., Li, Y., Wu, X., Zhang, J., Xie, J., & Sun, C. (2015). Simultaneous determination of 10 adulterants in antihypertensive functional foods using multi-walled carbon nanotubes-dispersive solid-phase extraction coupled with high performance liquid chromatography. Journal of Chromatographic Science, 53(9), 1611-1621.

Zhuo, Y., Yuan, R., Chai, Y., Tang, D., Zhang, Y., Wang, N., ... & Zhu, Q. (2005). A reagentless amperometric immunosensor based on gold nanoparticles/thionine/Nafion-membrane-modified gold electrode for determination of α-1-fetoprotein. Electrochemistry Communications, 7(4), 355-360.

8

Cereals, Millets, Flours and Their Products Detection of Common Adulterants, Contaminants and Safety

Twinkle Suri, Pradyuman Kumar

Department of Food Engineering and Technology, Sant Longowal Institute of Engineering & Technology, Longowal,Sangrur, Punjab, India

Introduction

About sixty percent of the world's food intake comes from the three major crops, i.e., wheat, rice and maize (Thielecke and Nugent, 2018). Depending upon the country of utilization and consumption, the products made from cereals such as barley, wheat, maize, triticale, spelt, buckwheat, rice etc. can vary in the range of pasta, bread, pastries, soups, chapatti, cooked rice etc. World cereal utilization in 2021/22 is forecasted at 2,811 million tonnes and the worldwide production of cereals is expected to reach 2,818 million tonnes by 2025 (FAO, 2020). With the increased utilization and demand as well as shortage and insecurity of grains in many countries (FAO, 2020), it becomes frequent to use practices that can increase production such as use of fertilizers, manures etc. and also the practice of adulteration to meet the demands of a particular grain and for monetary benefits. As a result, these grains become a major dietary contributor for contaminants and toxins in the human body. It has been estimated that each year, the post-harvest losses such as infestation by insects/ pests and fungi and their respective mycotoxins contribute majorly in the loss of more than 421 million tonnes of cereal grains globally during storage (Kaur *et al*., 2020). These contaminants can further include mycotoxins, pesticides, heavy metals, acrylamide, polycyclic aromatic hydrocarbons, and packaging material residues (Albero *et al*., 2020).

According to the CODEX ALIMENTARIUS, "the substances that have not been added intentionally to the food products, but become a part of the food chain at any processing step e it manufacturing, handling or storage and distribution, are termed as contaminants". They can also enter the food chain via environmental contamination through water, soil, air, bio-solids, dust, birds, rodents, insects, animals, microbes, humans etc. In order to make the food product safe for consumption, it becomes necessary to monitor their presence in the food material at almost every step of the processing chain. Each year, about 1 billion tonnes of foodstuff is lost due to mycotoxin contamination with about 255 of the crops affected by fungi, as per FAO statistics (Thielecke and Nugent, 2018). Several studies over the years have reported a high number of cases of mycotoxin outbreaks due to fungi infections in the grains, in countries such as Kenya, Nigeria, India, USA etc. Under organizations like CODEX ALIMENTARIUS, European Food Safety Agency (EFSA) and Food Safety and Standards Organization of India (FSSAI), 'maximum levels' for contamination from mycotoxins, metals and radionuclides have been established (CODEX, 2021) in different cereals and products. These can be further extended to different countries depending on the local/ common grains that they utilize for the production of various local products.

The first written account (possibly) of food fraud/ adulteration, published in the 19th century, was on the utilization of chalk and alum to make bread loaves white and the loaves were tipped in sawdust and plaster to make them heavier (Pastor *et al.*, 2019). Adulteration is defined as the method of intentional manipulation of the materials by deliberately substituting, diluting, counterfeiting or misinterpreting the food, ingredients or packaging material for financial profits as well as substitution for various advanced methods and technologies for earlier and easy production (CODEX, 2021). Adulteration can happen at any stage of the food supply chain starting from the primary production to processing, followed by storage and transportation. It now comes under the category of 'Food Fraud' by FAO, where the supplier is intentionally deceiving the consumers/ customers via the product or the process, having a humongous impact on the quality and safety of the food product (FAO, 2021). This can affect the confidence and trust of the costumer in the food businesses that can have longer run effects even if their systems tend to become safe in the future (Barnett *et al.*, 2016). In countries such as India, consumers are provided with on-line materials by FSSAI for the detection of common adulterants in the food products used regularly at home (FSSAI, 2011). It has been noted that some cereal adulterants comprise natural and organic adulterants with less harmful effects whereas other adulterants that are synthetic or inorganic can pose a serious health risk to consumers (Defernez and Wilson, 1995). CODEX ALIMENTARIUS food standards support national food safety authorities have undertaken various regulations and measures to address and deal with the same issue. Due to its direct impact on the human health, the national food safety frameworks and their criminal laws prohibit or criminalize the practice of adulteration in some parts of the world due to the increased incidences of disease outbreaks (FAO, 2021). The estimated cost of food fraud for global food industry is approximately 40 billion dollars a year, thus making it an extremely costly affair to respond to (Kamath, 2018). Prevention as well as proper labeling under various norms of guidance issued by the legalizing bodies in different countries and globally, have become a major factor of importance for the control.

For the detection of contaminants and adulterants, various methods have been developed so far (Pastor *et al.*, 2019). Although the traditional and standardized methods are still used often for separation and detection of these compounds, the growing demands of faster, cheaper and more authentic new methods has led to the modification old contemporary and analytical methods as well as the development of new techniques in the past few years with applications in the cereal industry (Pastor *et al.*, 2019).

Contaminants and Adulterants in Cereals and Products

Cereals are categorized as monocotyledonous plants (*Gramineae*) belonging to the grass family. The grains are considered important for the human diet due to the presence of micronutrients such as minerals and vitamins. Apart from the main (staple) cereals such as wheat, rice and maize, minor and pseudo-cereals such as buckwheat, oats, barley, spelt, rye, quinoa and amaranth are used in different forms for different products, that can prove as a simple and effective strategy for the replacement and enrichment of important dietary components such as fiber, resistant starch, polyphenols etc. (Pastor *et al.*, 2019) especially in the rural parts of many countries (Silletti *et al.*, 2019). Interestingly, EFSA (EFSA, 2009) Scientific Panel on Contaminants in the Food Chain (CONTAM) has regarded cereals and their products as the potential food group that can cause a serious exposure of cadmium to the human diet. The mycotoxin stability and concentration in cereals is further affected by the processing operations such as sorting, milling, cleaning and various thermal processes (Cheli *et al.*, 2010). It has been indicated that contamination concentration after milling may depend on the initial concentration in the unprocessed grain (Cheli *et al.*, 2010). Post-harvest grain losses can also arise due to the storage conditions leading to growth of molds and other pathogens which can have detrimental effects such as discoloration, loss of dry matter, protein and lipid degradation, production of volatile chemicals and mycotoxins and a reduction in germination, baking and malting qualities (Kaur *et al.*, 2020). Post-harvest mycotoxins are generally heat stable, and so, they are not easily removed and carry forward in the food chain (Kaur *et al.*, 2020). The main contaminants and adulterants found in cereals are given in Figure 1.

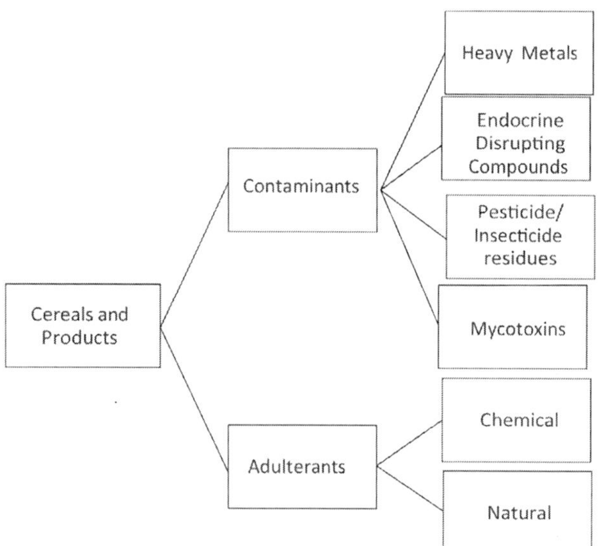

Fig. 1: Contaminants and adulterants in cereals and products

Heavy Metals

An increasing risk of food contamination comes from metal contaminants such as nickel (Ni), cadmium (Cd), mercury (Hg), lead (Pb), copper (Cu), arsenic (As) etc. Of these, trace metal elements can be categorized as essential and non-essential: Cu, Iron (Fe) and Zinc (Zn) are the essential elements only within the tolerable limits; As, Pb, Hg are considered as non-essential elements which can pose a health risk even at very low concentrations in the body (Tortoe *et al.*, 2018). The essential trace metals are vital for the growth and normal functioning of the body, but above the recommended levels, the accumulated amounts of Cu may result in nausea or vomiting, excessive amounts of Zn may damage the pancreas or interfere with protein metabolism (Tortoe *et al.*, 2018) or higher accumulation of Fe with Manganese (Mn) can increase the risk of Parkinson's disease (Skendi *et al.*, 2019). Having high atomic numbers, the contaminants will not get easily degraded (having half-life of about a year), and can easily accumulate in the body finding their paths at different steps through the food chain (Tortoe *et al.*, 2018; Khaniki *et al.*, 2005). They can contaminate the crops (especially Pb, Zn, Selenium, Cd and Ni) through soil or polluted air (which precipitates and falls as rain), and also via machinery used during processing operations (FAO, 1986; Tortoe *et al.*, 2018). The susceptibility of contamination of a specific food by a specific trace metal has found out to be dependent on the chemical and structural composition of the food/ crop (Hajeb *et al.*, 2014; Tortoe *et al.*, 2018). Organically grown crops also depend heavily on organic fertilizers (manures) and pose a risk of contamination with undesirable metal components (Thielecke and Nugent, 2018). Due to the retention of lead in the soil in considerable amounts, it is usually found in lower amount in cereals, but is still considered as a probable carcinogen and a physiological and neurological toxin to the human health transporting via sewage sludge, pesticides and atmosphere in the agricultural lands, depositing on the leaf surfaces, ears and other vegetative parts of the crop (Cheli *et al.*, 2010). Increased incidences of cadmium in the environment comes from battery emissions, coatings, pigments, phosphatic fertilizers and sewage sludge and is also dispersed by mining activities (Khaniki *et al.*, 2005). The metal can be taken up by the crop from the soil accumulating in the vegetative aerial parts of the crop and gets transported to the kernels within the concentrations mainly ranging from 0.1to 1- fold (Cheli *et al.*, 2010). These metals can accumulate in lungs, liver, bone, testes, brain and kidneys and can cause serious damage to the body organs and also the immune system and cardiovascular system (Cheli *et al.*, 2010; Thielecke and Nugent, 2018). Metals such as Cadmium have been categorized under category 1 (carcinogenic to humans) by the International Cancer Research Agency (IARC, 1993). Nickel contamination can occur in small amounts during cooking of food in nickel vessels. Arsenic can contaminate the crops via water uptake. Since the absorption of arsenic from the soil much more for rice than other grains, incidences of the crop being contaminated with high levels of arsenic have been found in areas where fields were irrigated with water having higher amount of the metal (Thielecke and Nugent, 2018; Tortoe *et al.*, 2018). Its toxicity is associated with the form it is present in: inorganic arsenic is considered more toxic than the organic form (Thielecke and Nugent, 2018). Arsenic in higher amounts in the body can affect the skin, lungs, brain, kidneys, liver, metabolic systems, immune system, cardiovascular system, endocrine system (Thielecke and Nugent, 2018). Lead, nickel, arsenic, and cadmium found in the cereal crops get carried up and easily contaminate the products such as flours. Further, Arsenic accumulates in the bran layer of rice and thus white rice contains 80% less arsenic in comparison to the whole grain (Milani, 2013). Mercury is used as seed dressings (FAO, 1986) and its organic form is considered as toxic (Thielecke and Nugent, 2018). Another food contaminant that can affect the crops is Nitrate, which comes from the nitrogen rich soil and also depends on other factors like soil and cultivar type, geographical conditions, climate, planting/ harvest dates, nitrate present in the irrigation water and groundwater and storage and postharvest processing (Skendi *et al.*, 2019). Conventional vs. organic cereals when studied to differentiate the amounts of toxic metal (Al, As, Pb), different cereals and products showed variable results in different countries due to variable conditions of weather, location, crop practices etc. (Vrček *et al.*, 2014; Thielecke and Nugent, 2018). Studies conducted on wheat in Belgium showed that the organically grown wheat and its products (flour, bread, breakfast cereals, dough and pastry) had less level of metal contaminants (Cd, Pb and Hg) in comparison to the conventional wheat and its products(Thielecke and Nugent, 2018). The levels of Ni and Zn have found out to be more in the organic cereals than in the conventional varieties. Crops that are organic in origin have been reported to contain 21% more Fe and 29% more Mg than the conventional ones (Skendi *et al.*, 2019). Organic products contain less nitrate amount than their conventional counterparts. Wet and dry milling processes of millets subject the milled products to metal contaminants such as iron, lead, nickel, chromium, magnesium, manganese, copper, molybdenum and silicon during processing operations or via using metal containers (Adu *et al.*, 2020). Cadmium and lead have been found in semolina (Cheli *et al.*, 2010).

Endocrine-Disrupting Compounds (EDC's)

These compounds interfere with the synthesis, secretion, transport, action or excretion of natural hormones in the body and result in developmental and reproductive abnormalities (Liao *et al.*, 2013a). EDCs such as bisphenol A (BPA), bisphenol F (BPF), methyl paraben, propyl paraben-linear chain nonylphenol in position 4 (NP), mixture of ring and chain isomers of NP and six pharmaceutical compounds (allopurinol, mefenamic acid, carbamazepine, paracetamol, diclofenac and ibuprofen) are found in in various cereal grains such as wheat, oat, barley and rice contaminated from various environmental sources such as air, irrigation water, fertilizers, biosolids etc. (Azzouz *et al.*, 2020; Albero *et al.*, 2020). Use of pesticides can lead to the crops contaminated with organophospholipids, carbamic, NP and ureic compounds (Gyllenhammar *et al.*, 2012). These compounds are linked with metabolic diseases such as diabetes and obesity and can cause breast, testicularand ovarian cancer (Azzouz *et al.*, 2020). Plastic materials and internal coatings of food and beverage cans are also the source of BPA, NP, 4-tert-butylphenol, andpentachlorophenol (banned by EC Directive 91/173/EEC) to the foodstuffs coming in contact with the material especially at high temperatures (Cao*et al.*, 2011). Both BPA and NP are mimic the action of the hormone oestrogen (Cao*et al.*, 2011; Gyllenhammar *et al.*, 2012). Contamination of whole wheat flour with EDCs such as parathion methyl, malathion, dichlorvos, chlorpyrifos and some more pesticides came in the reports from Brazil (Azzouz *et al.*, 2020). Low levels of BPA have been found in bread and cereal composite samples due to the availability of BPA in yeast and baking powder contaminated during production stage or either by can storage (Cao *et al.*, 2011). Replacement of BPA with bisphenol F (BPF) in the industry has led to its accumulation in the soil, also the use of recycled water and organic amendments made in the agricultural fields (Albero *et al.*, 2020). Parabens are alkyl esters of p-hydroxybenzoic acid which act as antimicrobial preservatives in foodstuffs in response to their stability, high water solubility, broad spectrum antimicrobial activity and low allergenicity (Liao *et al.*, 2013a). They are reported for accumulation in human breast tumors and sperm DNA damage in men (Liao *et al.*, 2013a). Cereals and products such as wheat flour, bread, rice, noodles, pie, pasta, pizza, corn products, cookies, cakes etc. have been reported to be contaminated with parabens through canned and plastic containers with incidences in China (Liao *et al.*, 2013b). The intrinsic complexity of matrices that contain EDCs makes it difficult to extract and isolate these chemicals from cereal grains and their products. Pressurized liquid extraction, ultrasound-assisted solid-liquid extraction, microextraction techniques, QuEChERS and other techniques have been applied so far for the purpose of extraction (Azzouz *et al.*, 2020). Very recently, for the estimation of phenols, parabens and triclosan together with pesticides, a new technique has been developed, in cereal-based foods (Azzouz *et al.*, 2020). Multi-residue technique is being used in practice to detect and quantify the antibiotics in cereals, with enrofloxacin detected in rice (Albero *et al.*, 2019, 2020). These are currently being examined and evaluated by the Scientific Committee for Food of the European Commission in the direction of generating regulations (Azzouz *et al.*, 2020). The specific migration limit in food or food stimulant for BPA was set at 0.6μg/g by the European Commission Directive in relation to plastic materials and articles that might come in contact with the foodstuffs (EC, 2004). USEPA (1993) has set up the maximum acceptable dose of BPA at 50μg/kg body weight/ day which is the same as the tolerable daily intake (TDI) set up by EFSA (2006). Danish researchers have proposed a TDI of 5μg/g for NP although limited by EU(Gyllenhammar *et al.*, 2012). For adults, the amount has been estimated to be 7.50μg/day in studies from Germany and around 30μg/day in studies from Taiwan (Gyllenhammar *et al.*, 2012). Joint FAO/WHO Expert Committee of Food Additives in 1974 has established the acceptable daily intake for the sum of 3 parabens: methyl-ethyl-propyl at 0-10 mg/kg body weight/day (JECFA, 1974). The use of 2 parabens: propyl and butyl, their isoforms and salts has been banned in children's products in Denmark since March 2011 (SCCS, 2011).

Pesticides/Insecticide Residues

Pesticides and insecticides used in crop production and agricultural practices are poisonous and have biochemical targets acting on both the crop and the humans. Various residues are found remaining in the crops after harvesting which can prove toxic and harmful for human health. Studies have shown that the critical effects of the exposure to pesticides occur in less than 24 hours; once ingested can induce toxicity in various organs of the body, especially liver, bile and intestines (with about 81.4% of pesticides absorbed in high yield (96%)) further affecting blood and lymph (Vasilachi *et al.*, 2020). Also, prolonged and persistent exposureto the pesticides can have carcinogenic orendocrine disrupting effects (Vasilachi *et al.*, 2020). Cereal products have been found to be contaminated with some post-harvest insecticides such as chlorpyriphos-methyl, dichlorvos and pirimiphos- methyl (Harcz *et al.*, 2007). Under normal circumstances, the interval between application ofpesticide and when the grain is consumed (in whatever form) as wellas the removal of different layers of grain prior to milling and eventful processingof the resultant flour is a contributing factor for decreasing the levels of pesticidespresent (Alldrick, 2017). Maximum residual levels have been provided for specific pesticides (such as mepiquat, quaternary ammonium pesticides etc.)/ food combinations given by the Codex Committee on Pesticide Residues (CCPR) and other federations such as EU Regulation 396/2005 (European Parliament andCouncil, 2005) and FSSAI (FSSAI, 2011) (Rausch *et al.*, 2021).

Mycotoxins

These are toxin, heat-resistant secondary metabolites generated by filamentous fungi (such as Aspergillus, Penicillium, Fusarium etc.) growing on various cereal crops (Soleimany *et al.*, 2012a; Vrček *et al.*, 2014). Contamination with these fungal species varies with pH, temperature, bioavailability of nutrients, light, insect and pest damage and water availability etc. (Milani, 2013; Thielecke and Nugent, 2018) with the mycotoxin contamination affected by the post- harvest processing, prolonged storage, adulteration and host-specific differences in susceptibility to pre harvest mycoflora and climatic factors (Kirincic *et al.*, 2015; Pitt *et al.*, 2013; Tahkapaa *et al.*, 2015; Hussain *et al.*, 2019; Wilson *et al.*, 2006; Thielecke and Nugent, 2018). The harmful effects besides toxicity include damaging the genetic information in the cell causing mutations (genotoxicity), ability to produce cancer (carcinogenicity), mutation of genetic information (mutagenesis) and abnormality causing fetal exposure during pregnancy (teratogenesis), which is further augmented by their capacity to be transferred along the alimentary chain from plant to animal/ human even after processing being heat stable (Vrček *et al.*, 2014). Their analysis is difficult due to their wide range of polarity strength and extraction via a mobile phase of variable degree of hydrophobicity is required (Soelimany *et al.*, 2012a, b). Further, special techniques are required for the detection of masked derivatives (e.g.α-ZOL and β-ZOL) (Vendl *et al.*, 2010). Before delving into each mycotoxin and its harmful effects, the optimum temperature, water activity (α) and the responsible micro-organisms for their development in the plant cropsare listed in the table below (Table 1) (non-exhaustive).

Table 1: Mycotoxins (major), factors for their production and contaminated products

Mycotoxin	**Temperature (°C)**	**Water Activity (aw)**	**Major Species**	**Found in (cancontaminate)**
Aflatoxins	33 (production), 35 (growth)	0.99 (production), 0.95 (growth)	A.flavus (B1, B2), G1, G2), A.oryzae	Maize, millet (pearl, fonio (Ezekiel *et al*., 2012, 2014)), sorghum, wheat, guinea corn, wheat flour, wheat bran, maize flour, rice (Odoemelam and Osu,2009)
Ochratoxin A (Walker, 2002)	25-30	0.96-0.98	Tropical and Subtropical Climate: *A.ochraceus, A.carbonarius, A.niger, A.alliaceus, A.albertensis, A.niger* Temperate climate: *P.verrucosum*	Barley, wheat, millet (pearl and fonio), wheat bran, wheat flour, maize, rice, guinea corn, oats, flour-based foods
Fumonisins	15-30	0.90-0.995	*F.Liseola, F,verticillioids, F.proliferatum F.moniliforme*	Maize and maize products (B1), wheat, sorghum, barley, oats
Zearalenone	25	0.96	*F.graminearum, F.culmorum, F.crookwellense, F.equiseti, F.semitectum A flavus A parasiticus*	Maize, corn-based products, wheat, wheat-based products (bread, flour, breakfast cereals, dough, pastry), fonio
Dooxynivalenol (DON)	26-30	0.995	*F.graminearum,F.culmorum, F.crookwellense F. chlamydosoprum*	Wheat, maize, barley, wheat-based products (bread,flour, breakfast cereals, dough, pastry), oats, rye (&flour)
Chirinin	20-30	0.75-0.85	*P.citrinum, Aspergillus species*	Rice, corn, wheat, barley

(Adapted from: Milani, 2013)

Fusarium (ZON/ DON)

Deoxynivalenol (DON) and zearalenone (ZON) are mycotoxins produced by the microbes of fusarium species (Milani, 2013; Vrček *et al.*, 2014). Their accumulation and concentration in the grain is generally dependent on factors such as substrate, and the growth and strain of fungi as well as extent of rainfall during the pre-harvest, harvest and post-harvest conditions (Milani, 2013; Pérez-Torrado *et al.*, 2010). The location for their infection in cereal grains is the cereal ears, which makes it difficult to remove them even after processing (Thielecke and Nugent, 2018; Vendl *et al.*, 2010). The retention level of the mycotoxins in products is found to be dependent on the variety of wheat and the penetration degree of the *Fusarium* molds (the growth being from surface to the interiorof the kernel) (Cheli, *et al.*, 2010). Studies conducted on wheat in Belgium showed that the organically grown wheat and its products (flour, bread, breakfastcereals, dough and pastry) had less level of DON and ZON in comparison to the conventional wheat and its products (Thielecke and Nugent, 2018).

DON/ nivalenol (NIV) is also produced by other filamentous fungi such as *Myrothecium*, *Stachybotrys*, and *Trichothecium* (Tanaka *et al.*, 2000). DON can withstand temperatures of 175- 350°C and comes under the category of type B trichothecene, acting as a protein synthesis inhibitor and can cause feed refusal, dizziness, diarrhea, nausea, headache, reduced body weight and vomiting as well as impairing the immune response and loss of productivity in humans and animalson consumption (Cheli *et al.*, 2010; Milani, 2013). DON has been found in concentrations two times or more in the bran than in the wheat kernels at the site of fungal growth (Cheli *et al.*, 2010) since bran has potential biochemical inhibitors to reduce the production of trichothecene (Cheli *et al.*, 2010). DON/ NIV inhibit macromolecular synthesis being cytotoxic in nature, with incidences of outbreaks of intoxication in human and animal species (Tanaka *et al.*, 2000). Also, NIVin combination with DON and also as such has found to induce inflammation and increase inflammatory response induced by lipopolysaccharide plus Interferon-γ in the non-tumorigenic intestinal epithelial cell line (Adesso *et al.*, 2017) and can induce apoptosis in HL-60 cells (Ueno *et al.*, 1995). DON can also act as a potential marker of occurrence of some other mycotoxins (Thielecke and Nugent, 2018). International Agency for Research on Cancer has categorized DON under Group-3 i.e. non classifiable to its carcinogen effect in humans (IARC, 1993).

ZON is found in nature as a co-occurrence of trichothecenes and is an estrogenic metabolite which causes hyperestrogenium in livestock (Tanaka *et al.*, 2000) and infects and proliferates in various agricultural commodities in the field and/ or during storage due to the presence of optimum growth temperatures (27°C) (Thielecke and Nugent, 2018; Pérez- Torrado *et al.*, 2010). It can withstand temperatures around 175°C, and its structure allows it to bind to mammalian oestrogen receptors in animals acting as a growth promoter for breast cancercells in humans especially with increasing age and other effects such as infertility,vulvae oedema and mammary hypertrophy in females (Massart *et al.*, 2008;

Ahamed *et al.*, 2001; Yu *et al.*, 2005; Pérez-Torrado *et al.*, 2010; Milani, 2013). IARC has categorized ZON in Group 3 and some other Fusarium toxins (Zinedine *et al.*, 2007; Pérez-Torrado *et al.*, 2010). Biosynthetic compounds of ZON have been found stored in corn, rice and wheat kernels (Milani, 2013).

DON and ZON are partially metabolized in the host plants and get convertedto various compounds such as DON-3-glucopyranoside (D3G), ZON-4-glucopyranoside (Z4G), α-ZOL- glucopyranoside (α-ZG) and β-ZG via glucosylation mechanism (Vendl *et al.*, 2010). Partial conversion of ZON leading to ZON-4-sulfate (Z4S) by various Fusarium species (*F. graminearum, F. culmorum, F. crookwellense, F. semitectum,* and *F. equiseti*) and to α- zearalenol (α-ZOL) and β-ZOL by different yeast strains during fungal metabolism affected bymicrobial proliferation (Vendl *et al.*, 2010; Milani, 2013). All the derivatives of DON and ZON (especially α-ZOL and β-ZOL) have similar toxicity as their precursors, and when ingested with food they can be enzymatically cleaved, fully or partially (e.g. D3G→ DON), during digestion due to functional groups attached to them such as glycosylic and sulfate residues (Vendl *et al.*, 2010).

Ochratoxin A (OTA)

OTA and its derivatives (B, C, α, β) are chlorinated isocoumarin compounds which are chemically stable, have a long half-life in blood and tissues and can enter the human body not only directly through cereals (e.g. millets) but also through pork and food products containing pig blood derived from animals fed on contaminated grains (Walker, 2002; Lasram *et al.*,2020). It can affect both humoral and cell-mediated immunity releasing its teratogenic and immunotoxic effects. The putative genotoxic effects of OTA can occur via secondary mechanisms such as DNA repair impairment, chronic cytotoxicity or induction of oxidative stress, enhancing its nephrotoxic and carcinogenic effects (Walker, 2002). It is also nephrotoxic, hepatotoxic, carcinogenic and immunosuppressive in nature (Milani, 2013).

Fumonosins

Fusarium species generate mycotoxins such as fumonisins, moniliformin, beauvericin, and T- 2/HT-2 toxins (Jurjevic *et al.*, 2007; Leslie *et al.*, 2005; Milani, 2013; Vismer *et al.*, 2019). Fumonisins (B1+B2) grow in hot and dry weather with high humidity environment (tropical and subtropical climate) (Milani, 2013; Thielecke and Nugent, 2018). F.moniliforme, which grows on maize, can produce mycotoxins such as beauvericin and Fumonisin B1, latter of which isa suspected human carcinogen, having cancer promoting activity with reported high incidence of esophageal cancer in South Africa (Wilson *et al.*, 2006; Milani,2013). T-2 toxin can induce apoptosis in HL-60 cells (Ueno *et al.*, 1995). Lowerlevels of toxins such as HT-1 and T-2 can also be found in oats grown organically w.r.t. the conventional way (Thielecke and Nugent, 2018).

A case study of pearl millet grown in USA when infected with the Fusarium species (F. semitectum and F. chlamydosporum) showed contamination with beauvericin and moniliforme (Wilson *et al.*, 2006). Maize, sorghum and porso and pearl millet can also be contaminated with fumonisins, moniliformin when subjected to various conditions and environmental factors responsible for growth of *Fusarium* species such as *F. proliferatum* and *F. pseudonygamai* (Vismer *et al.*, 2019).

Phytotoxins

These are secondary metabolites produced by the usual staple crops. The presence of these endogenous toxic materials is low in cereals, but tannins, phytate (phytic acid), pyrrolizidine and tropane alkaloids, under this category are potential toxins considered. They can also result as a reason of co-contamination with seeds of other plants co-harvested (Alldrick, 2017).

Phytic acid is found in the cereal kernels as its calcium or calcium-magnesium derivative. It is a major source of dietary phosphorous in grains and its location differs with grains: usually it is located in the maize germ, wheat aleurone layers and is uniformly distributed over millet.

Being resistant to normal digestive processes within the mammalian gut, as an unavailable source of dietary phosphorous and acting as a sequestering agent for the chemicals that are nutritionally significant (mainly Ca, Fe), it is considered as an anti-nutrient (Alldrick,2017).

Tannin, a botanical compound, having high molecular weight, ranges from 0.1% in brown rice to 0.4% in rye and wheat and up to 5% in some sorghum varieties, depending on different varieties. They are considered anti-nutrients since they have a very high number of phenolic hydroxyl groups and form cross links with proteins, inhabiting the digestive enzymes (esp. proteases and amylases) and other complex molecules. The amount of tannin can be reduced by growing 'reduced tannin' varieties as well as using methods such as decortication of seed coat containing tannin prior to milling and malting processes (Alldrick, 2017).

Pyrrolizidine usually comes from the co-harvested seeds and a range of flowering plants and species. 1,2-unsaturated pyrrolizidine alkaloids have been associated with hepatic veno-occlusive disease which in acute form exhibits cirrhosis of liver. Casesof outbreak of poisoning have been seen in Afghanistan and Slovenia, Australia, Germany, Netherlands (Alldrick, 2017).

Tropane alkaloids (scopolamine and hyoscyamine) are mainly the result of contamination from seeds of Datura stramonium (and some other species). It is available in high concentration different plant parts indicating multiple contamination pathways along the processing chain - from harvest to decortication (Begemann *et al.*, 2021). D. stramonium (Jimsonweed) has been detected with 26 different alkaloids such as TA atropine (a racemic mixture of L- and D-hyoscyamine) and scopolamine in varying proportions in different crops such as millets, corn and buckwheat and their products like popcorn, infant foods etc. (EFSA, 2013; Mulder *et al.*, 2015, 2016; Begemann *et al.*, 2021; Van Raamsdonik *et al.*, 2009; Caligiani *et al.*, 2011). Proso Millet has been found to be contamination with tropane alkaloids in different plantparts (Begemann *et al.*, 2021). Red millet has been found to be contaminated by poisonous tropane alkaloids such as scopolamine and atropine usually generated by D. stramonium, D. innoxia and D. ferox infestations (Van Raamsdonik *et al.*, 2009; Caligiani *et al.*, 2011). When digested in toxic amounts, these compounds are poisonous and affect the central nervous system with symptoms such as restlessness, disorientation, hallucinations and delirium (Müller and Wanke,1998; Berkov *et al.*, 2006; Krenzelok, 2010; Begemann *et al.*, 2021). When consumed in highly toxic doses, these can result in severe respiratory paralysis leadingto death (Begemann *et al.*, 2021). They are highly stable, not getting degraded in the processing conditions such as bread making (Begemann *et al.*, 2021). An outbreak involving more than 60 people was reported due to the consumption of buckwheat flour contaminated with these mycotoxins a few years ago in Solvenia(Perharic, 2005; Caligiani *et al.*, 2011). EFSA has jointly contained these toxins (scopolamine and hyoscyamine) for acute toxicity and has set an acute reference dose of 0.016μg/kg body weight/day (EFSA, 2015) and setting maximum limits: atropine (1.0μg/kg) and scopolamine (1.0μg/kg) for their presence in "processed cereal-based foods and baby foods for infants and young children, containing millet, sorghum, buckwheat or their derived products" (EC, 2016).

Aflatoxins

Aflatoxins B1, B2, G1, G2 are group of carcinogenic mycotoxins mainly generated in warm climate further affected by agents such as wounding during mechanical processes, types of infecting strains, location, late irrigation, delayed harvest, stress of hot/dry conditions, maturity status climatic changes, infestations by insects/rodents, catabolic activity, reduced coenzyme levels and metal ions etc. (Odoemelam and Osu, 2009; Milani, 2013; Sirma *et al.*, 2016; Battilani *et al.*, 2016). Aflatoxin B1, most native to cereal (especially maize) and cereal products, has shown multiple incidences of outbreaks (acute aflatoxicosis) around the globe, significantly in parts of South Africa and India, leading to childhood stunting, impaired growth and on chronic exposure, liver cancer (Wilson *et al.*, 2006; Sirma *et al.*, 2015, 2016; Thielecke and Nugent, 2018; Raghavender *et al.*, 2007). It has been associated as a risk factor for hepatocellular development and depression of immune responses and salivary immunoglobulin level (Wilson *et al.*, 2006; Odoemelam and Osu, 2009; Sirma *et al.*, 2015). Organically grown corn has higher aflatoxin levels compared to the one with conventional growth method (Thielecke and Nugent, 2018). Five different varieties of pearl millet were found contaminated with all aflatoxins (B1 > B2) more frequently noticed in seeds collected during the rainy seasonthan in winters, with G2 present andG1 absent in seeds collected during winter (Raghavender *et al.*, 2007). Further, these mycotoxins were constituted in stored and cooked varieties of pearl millet under different temperature conditions as well, posing as a potential threat to humans (Mishra and Daradhiyar, 1990). Incidence of A. flavus in infant based flours in the commercially sold products in the markets of Abidjan (Yapo-Crezoit *et al.*, 2019) has been reported. IARC has classified aflatoxin B1 under Group 1 human carcinogen mainly because of its ability to bind with DNA and proteins, forming adduct (Sirma *et al.*, 2016).

Citrinin

It is a secondary toxin benzopyran metabolite usually formed by the species of Aspergillus and Penicillium after harvesting of the grains under hot climate. Its toxins accumulate in the animal/ human kidneys and can lead to severe renal failure affecting liver and gastrointestinal tract (Milani, 2013).

Ergots

Ergots (sclerotium of *Claviceps purpurea*) contains lysergic acid (precursor for drug LSD) and high alkaloid content that acts as a mycotoxin for animals and humans with historical evidences of ergotism (human intoxication) dating back to 600BC in records from Assyria (Vermeulen *et al.*, 2013; Thielecke and Nugent, 2018). Ergot is a fungal disease primarily caused by the species of ascomycete fungus claviceps in cereal crops like wheat, sorghum, millet, barley with special incidence in rye (Thielecke and Nugent, 2018). Since transported to human feeding on animals, European Commission has limited the amount of ergots in all animal feedstuffs to 0.1% containing un-ground cereals (European Commission Directive 2002/32/EC). For regulations of ergots in cereal in foods, EC Regulation 689/92 has limited the concentration to 0.05% (EC, 1992). Also, recently EC has recommended its member states to analyze the samples of cereal and cereal products and feed/fodder/ grass for presence of six most prevalent ergot alkaloids (ergocristine, ergotamine, ergocryptine, ergometrine, ergosine, and ergocornine) (EC 2012) and for the simultaneous ergot body content determination, to establish the bond between the two (Vermeulen *et al.*, 2013). Analysis of Ergot alkaloids is possible with chromatography and immunochemical techniques and with multiplex dipstick assays (Scott, 2005, 2009; Krska and Crews, 2008; Contaminants in food and feed, 2008).

The detection of ergot bodies officially in cereals is done by weighing the fraction of the bodies further quantified by manual extraction of ergot sclerotia from the samples (European Commission Regulation 687/2008). Color retention and light microscopy have been used by International Association of Feedstuff Analysis for the detection of presence of ergot bodies (Vermeulen *et al.*, 2012, 2013). NIR hyperspectral imaging combined with chemometric tools has been used as a quick method for online determination and quantification of ergot bodies and differentiating them in higher sample volumes (Vermeulen *et al.*, 2013).

Process Toxicants

'Food process toxicants' can be produced by processing (e.g. thermal processes) during the preparation and manufacture of cereal products. The main process toxicants developed in the finished foods are polyaromatic hydrocarbons or acrylamide and chloropropanols (Alldrick, 2017).

Acrylamide is generated by processing if the cereal/ product to be processed contains carbohydrates (reducing sugars), amino acid asparagines (critical component) and heat. It is present as a processing by-product in foods like bread, breakfast cereals etc. (Ghazouani *et al.*, 2021). Asparagine content varies widely between the grain varieties and within one single grain and also growing (soil) field conditions (Thielecke and Nugent, 2018). Addition of antioxidants, glycine and using enzyme asparaginase can mitigate the formation (Horszwald *et al.*, 2010). Overall estimated consumption of acrylamide depending on population and age is about 0.3-2.0μg/kg body weight. IARC has classified acrylamide as a potential carcinogen causing gene mutation and DNA damage, recognized as a fact by WHO Consultation in 2002 and thus, the EU has updated the maximum limits for various food products whereas US FDA proposes acrylamide levels in food but does not regulate them (Thielecke and Nugent, 2018; Ghazouani etal., 2021).

Hydroxymethylfurfural (HMF), and furfural (FUR) are considered as potentially harmful components generated via millard reaction, caramelization and saccharide (sucrose and hexose for HMF, pentose for FUR) degradation during bread-making and some other heating processes such as roasting contributes to HMF and FUR formation especially with finer flour in case of rye (Horszwald *et al.*, 2010) andin breakfast cereals with roasted cereals (Ghazouani *et al.*, 2021). These heat generated compounds are associated with mutagenic, carcinogenic and cytotoxic effects (Ghazouani *et al.*, 2021). HMF in the human body can be metabolized into 2 products: 5- hydroxymethyl-2-furoic acid, which is excreted in urine and 5-sulfoxymethylfurfural, a potential toxin and carcinogen that can form a complex with DNA or protein (Ghazouani *et al.*, 2021).

Literature data has reported that food contaminated with HMF in high levels may also containacrylamide (Ghazouani *et al.*, 2021).

Polyaromatic Hydrocarbons contain two or more fused rings, some of which (esp. benzo[a]pyrene) are considered carcinogenic. It occurs mainly as a resultof the reactions taking place during the food processing operations. They can occur either via deposition through atmospheric pollution or used in appropriate fuel in grain drying or oven processes, leading to incomplete combustion and consequent contamination. Thus the control can be achieved by selecting the proper source of heat generation and controlling the temperature. Polyaromatic hydrocarbons when present in concentrations high in cereal products act as a toxic risk for contamination. EFSA (2008) estimated (oral carcinogenicity) that for8 different varieties of compounds under this category, the exposure was in the amount of 28.8-51.3ng/kg body weight/day with major sources being cereals and seafood (Alldrick, 2017).

Chloropropanols occur mainly as a reaction product between different ingredients in the food product. These compounds have been proved to be both in vivo gino-toxins and rodent carcinogens. Foods treated at high temperature with hydrochloric acid to speed up the process result in chloride reacting with glycerol backbone of lipids to produce 3-monochloropropne-1,2-diol (3-MCPD) (Alldrick, 2017). Lower concentrations of 3-MCPD have been found in bread, increased on toasting, but were not considered as a hazard to the consumer (Alldrick, 2017). Regulatory limits for hydrolyzed vegetable protein preparations have been set bya number of jurisdictions such as the European Community (EC, 2006).

Insect Infestations

Cereals are also inspected and analyzed for filth (uric acid), rodent hairs and excreta, infestation, damage (breakage, shriveling, fungal attack, insect damage), moisture (FAO, 1986). Insects such as weevils, lesser grain borers, different types of beetles and psocides are found in post-harvest cereals (Kaur *et al.*, 2020), mainly depending on the country of origin and storage. Insect infestation results in damage of the structure and nutritional profile of the grain and their respiration and body secretions create warm and moist spots within the grain that stimulate fungal spore germination and mold growth (Kaur *et al.*, 2020). Devices such as the one proposed in 2001, a high-speed integrated machine vision software using monochrome CCD camera with a personal computer, have been developed and could be used for using imaging as a high-speed detection device for pest, insects and other contaminants in cereals (Ridgway *et al.*, 1998). Rodent urine, a potential contaminant on grain, can be detected by UV light under which it shows fluorescence (FAO, 1986).

Relation between contaminants

Studies have been performed to establish a relation between the concentrationof heavy metals (Pb, Cd, As, Cu, Zn, Cr) and mycotoxin contaminants (total aflatoxin, aflatoxin B1 and Ochratoxin A) in the flours such as corn using Principle Component Analysis and Correlation Analysis (Algül and Kara, 2014). A very strong relationship was found between the presence of Cr and the aflatoxin production and between Cu and Zn concentration and Ochratoxin A production indicating that samples contaminated with these mycotoxins have also beenpolluted with the heavy metals (Algül and Kara, 2014). Earlier studies have also shown that chromium concentration can affect the aflatoxin yields and heavy metals present (such as Fe, Zn, Cu, B, Mo, Mn) have affected the production of ochratoxin in various agricultural produce (Algül and Kara, 2014).

Adulterants

Although there can be numerous adulterants used in different cereal and cereal products in different countries, some of the main adulterants used are listed below:

Chemical adulterants

Sudan I dye is a synthetic, red-colored chemical used as a coloring adulterant. Lardis commonly found in cake product formulations and is associated with health risks of hypercholesterolemia and coronary disease (Ambrose and Cho, 2014) Melamine is added as contaminant in wheat, soybean, corn and rice flours and asa protein booster in infant formulas due to its composition being similar to natural proteins (Ambrose and Cho, 2014).Although not established, few studies show thatmelamine might cause urolithiasis and bladder cancer in rodents as well as kidneyfailure after forming an insoluble salt with cyanuric acid (melamine cyanurate) (Ehling *et al.*, 2007). Wheat flour is adulterated with oxidizing agents for color improvement e.g. benzoyl peroxide, alloxan monohydrate (Lohumi *et al.*, 2019) and improvers such as L-cysteine in bakery products to reduce disulfide bonds increasing flow properties, dough elasticity and reduced mixing time (Lohumi *et al.*, 2019). Self-rising flour is made by addition of acid phosphate and sodium bicarbonate. Other adulterants such as ammonium persulphate, ammonium chloride, acetone peroxide, sulphur-dioxide, azodicarbonamide, potassium bromate, nitrosyl chloride, chlorine, chlorine-dioxide, potash alum, magnesium carbonate, sodium aluminium sulphate, calcium phosphate and di- and tri- calciumphosphate to assist the baking process and are subjected to legislation under various countries (FAO, 1986). Adulterants are mostly are unknown and thus it becomes difficult to use target screening techniques for the detection of the same(Lohumi *et al.*, 2019). Generally, the methods used in flours are HPLC, GCMS, HPLC- diode array detection, and ELISA (melamine detection) electrophoresis, chemiluminescence micro fluidic chip and surface plasma resonance (benzoyl peroxide detection), Ramanmicroscopy (L-cysteine detection) (Delwiche, 2016; Lohumi *et al.*, 2019).

Natural adulterants

There are a number of adulterants that get added to different cereal and their products mostly associated with the cheaper and easier availability. For different cereals and products, they are given as such:

. Wheat

Waxy wheat containing only trace amount of amylose is costly, produced via plant breeding and has potential use in baking, shelf-life extension of baked goods, ethanol production etc. Due to similar kernel resemblance, it is often adulterated with the common wheat (Delwiche, 2016). Italian and German markets now-a-days prefer the hulled wheat species (emmer and einkorn) for pasta and spelt wheat for bread, pasta, brewing and regional specialties in organic and natural food markets bioactive components and the ease of cultivation in less tolerant environmental conditions (von Büren *et al.*, 2001; Silletti *et al.*, 2019). Since their production is costlier than their regular counterparts, they are adulterated with flours of lower quality or different varieties. Another wheat variety, Einkorn (*Triticum monocum* L.) wheat, is produced in few regions of Europe and the US, in comparison to common wheat has higher levels of protein, unsaturated fatty acids, antioxidants (such as lutein, tocopherol, tocotrienol), fructans and trace elements (e.g. zinc and iron) and lower levels of β-amylase and lipoxygenase activities contributing excellent flour properties. But, due to its low yield and growth in limited areas, it is sold at a higher price. To generate unfair economic gains or to compensateits weakened gluten structure; einkorn flour tends to be adulterated with bread wheat flour (Ayvaz*et al.*, 2021).

Rice

In countries such as China, Japan etc. rice is considered as staple food of daily consumption, containing vitamins, minerals, fiber and many essential elements with variations occurring with geographical conditions and various varieties (Mendez *et al.*, 2019). Basmati rice (native variety to Indian and Pakistan), an expensive variety is adulterated with non-basmati rice for monetary benefits (Delwiche *et al.*, 2016). Another native premium variety of rice, Koshihikari, native to Japan, is also adulterated with lookalike varieties ((Delwiche *et al.*, 2016).). Talc, glycerol or oil containing trace of blue pigment is used as a contaminant and an adulterant when rubbed on rice for improvement in appearance, though it does not pose a health risk (FAO, 1986).

Quinoa

Quinoa is considered as a pseudo-cereal (FAO and CIRAD, 2015) and is widely cultivated in South America (esp. Andean region), Europe, Asia, North America, being resistant to weather climate and soil conditions (Ferreira *et al.*, 2015; Rodríguez *et al.*, 2019). In comparison to other cereals, Quinoa seeds have higher levels of proteins such as methionine and lysine and substantial amount of fiber and minerals (calcium, iron). It also contains higher amount of polyphenols, phenolic acid and flavanoids (chlorogenic acid, ferulic acid, vanillic acid, kaempferol, quercetin and rutin) providing antioxidant bioactivity and anti-inflammatory benefits (Filho *et al.*, 2017; Paśko *et al.*, 2008). Due to its immune-nutritional features such as improving iron absorption in the intestine and modulating the hepatic production ofinflammatory biomarkers, quinoa (white) flour has been considered as a substitute/ replacement for wheat flour in bread making (Stikic *et al.*, 2012; Laparra and Haros, 2018). But, the high cost of quinoa flour results in its adulteration with maize, soybean and wheat flours (Rodríguez *et al.*, 2019). Presence of flavonoids indicate the adulteration, especially in Quinoa (such as quercetin and kaempferol)(Xue *et al.*, 2021).

Whole grains

In recent years, the consumer attention has been turned over to the use of whole grains for a particularly 'healthy' and 'wholesome' diet. Whole grains basically consist of the intact, ground, cracked or flaked kernel (with bran, germ and endosperm) after the removal of the inedible parts (such as hull and husk)

(Thielecke and Nugent, 2018). Over the years, it has been seen that development of cardiovascular diseases, type 2 diabetes and some type of cancers can be reduced by consumption of whole grains having bran with greater carbohydrate quality, healthy fats, antioxidants, phyto-chemicals, minerals, B vitamins, proteins, superoxide anionic scavengers etc. (Liu, 2007; Thielecke and Nugent, 2018); thus leading to the scientific and governmental organizations to issue guidance and dietary recommendation for their consumption (Thielecke and Nugent, 2018). Further, there is some evidence that their consumption can potentially prevent some negative effects of contaminants and mycotoxin- induced damages with their binding abilities (Thielecke and Nugent, 2018). But the availability and different processing conditions leads to their adulteration.

'Gluten-free' Cereals

With the higher incidence of people with celiac disease (having gluten intolerance: glaidin and some prolamins in cereals such as rye, oats and barley) the consumption of gluten-free cereals and their products has increased (Størsrud *et al.*, 2003). During various stages of growth (cross- contamination, crop rotation), harvesting, processing or transportation, many gluten-free grains, seeds and flours can be contaminated with wheat, barley and/or rye via cross-contamination (Størsrud *et al.*, 2003; Thompson *et al.*, 2010). Since the cost of gluten-free products is high, they are often subjected to adulteration practices with a number of cases over the years in different countries (Størsrud *et al.*, 2003; Koerner *et al.*, 2013). In some countries, products of corn, sorghum, and oats are allowed certain percentage (e.g. 3-5% in Sweden) of other grains such as rye, barley and wheat and triticale (Størsrud *et al.*, 2003; Thompson *et al.*, 2010; US Department of Agriculture, 2009). The standards suggested by the Committee of Nutrition and Foods for Special Dietary Uses (CCNFSDU) under the FAO/WHO Food Standards Program in CODEX are 20 mg/kg of gluten (on dry matter basis) in natural gluten-free products and 200 mg/kg (on a dry matter basis) in products rendered "gluten free" which is the same for EU (Størsrud *et al.*, 2003; Delwiche, 2016).

Millets

Millets have higher amount of proteins, carbohydrates, fatty acids, fibers, vitamins and minerals and is now considered as a super food due to its various health benefits (Choi *et al.*, 2021). The majorly consumed variety is Pearl Millet, mainly due to its easy growth in acidic, relatively infertile and sandy, drought prone soils producing high-quality grains and fodder without irrigation (Wilson *et al.*, 2006). Millets are gluten-free, have more percentage of dietary fibre (in comparison to rice) and similar lipid percentage profile as that of maize. It also has a higher percentage of essential amino acids such as leucine, isoleucine and lysine in comparison to regularly consumed cereals such as wheat and rye. Millets are easy to process under the food processing operations such as milling, decorticating, germinating, fermenting, cooking and extruding to obtain a variety of food products such as flours, pasta, biscuits, porridge non-dairy probiotic beverages and other indigenous dishes & food products and can be used as a substitute grain for reducing the risk of diabetes and for measures such as weight management (Dias-Martins *et al.*, 2018). Further, this grain is a cost-effective substitute for preliminary product development of chemicals such as ethanol (Raghavender *et al.*, 2007). Fonio millet, which forms the part of staple diet in Nigeria and neighboring countries has high amino acid and mineral content as well as low glycemic index, making it suitable for diabetics consumption (Ezekiel *et al.*, 2014). These millet varieties are often adulterated with different cereals for monetary benefits. Cooked millet flour is similar in looks (color) and chemical composition to soybean flour. Since the price for millet is higher than the soybean, the millet flour is sometimes adulterated with soybean flour (Shao *et al.*, 2018).

Other products

Cereal flour itself is a processed product which is further used to make bread, cakes, cookies, pastries, pasta, noodles, snacks, baby foods and confectionary products (Brera *et al.*, 2013) that are easily prone to adulteration (Danezis *et al.*, 2016). Providing an alternative cheaper and easily produced flour variety, flours such as sorghum and millet are used in variable compositions now-a-days for the manufacture of bakery and confectionary products (Tortoe *et al.*, 2018). Flours with better quality and at higher price can sometimes be mixed with the cheaper ones to earn profits. Soy, wheat or maize flour can act as an adulterant in quinoa flour (Rodríguez *et al.*, 2019). Product such as semolina, obtained by grinding or milling the starchy endosperm of durum wheat (*Triticum durum*) after the bran and germ is removed (Cheli *et al.*, 2010) is often adulterated with the bread wheat, especially during pasta production (Casazza *et al.*, 2012; Pastor *et al.*, 2019). Durum wheat is about 20-25% costlier than the bread wheat. The cooking resistance is reduced with resultant lower quality pasta if adulteration is carried out (Girolamo *etal.*, 2020). According to the rules of the Italian law, pasta can be made only by durum wheat semolina and water (Cocchi *et al.*, 2006) with a maximum levelof 3% (mass fraction) of common wheat allowed as an unintentional cross contaminant. For export, this value can increase beyond 3% under proper labeling conditions (DPR n.187, 2001). Buckwheat derived products and byproducts have low allergen properties and are gluten free with a good protein and fiber content along with antioxidant properties (Caligiani *et al.*, 2011). They have a potential to be used in functional foods products and nutraceuticals as well as in baby foods and infant formulas (Caligiani *et al.*, 2011). Due to the difficult availability and processing conditions, they are often adulterated with cheap flours and products.

Separation, Identification and Detection Methods

The methods of separation, detection and identification of contaminants and adulterants can be classified as classical and modern methods. The classical methods include: liquid chromatography, gas chromatography, enzyme- linked immunosorbent assays (ELISA), and polymerase chain reaction (PCR). These techniques have limitations that they are offline requiring rigorous sample preparation (and destruction) that is time consuming and involves use of environmentally-unfriendly chemicals with low efficiency (Rammanee and Hongpattarakere,2011; Bahadoran *et al.*, 2016; Hussain *et al.*, 2019). The modern techniques which are non- destructive and rapid are: hyperspectral imaging, Raman imaging, fluorescence spectroscopy, nuclear magnetic resonance spectroscopy, laser light backscatter, soft ray imaging, magnetic resonance imaging, near infrared spectroscopy and FTIR etc. (Hussain *et al.*, 2019). Use of imaging techniques further provides excellent selectivity on agricultural products even at trace levels (Hussain *et al.*, 2019; Yaseen *et al.*, 2017). But these techniques also have limitations since they cannot deliver special distribution of the quality and safety characteristics (Barbin *et al.*, 2013), Raman Spectroscopy is affected by the strong interference of biotic fluorescence (Nekvapil *et al.*, 2018), while NIRS requires a set of 20-50 known concentrations of analytes for calibration before use (Cortés *et al.*, 2019). The techniques can be categorized as such (Figure 2):

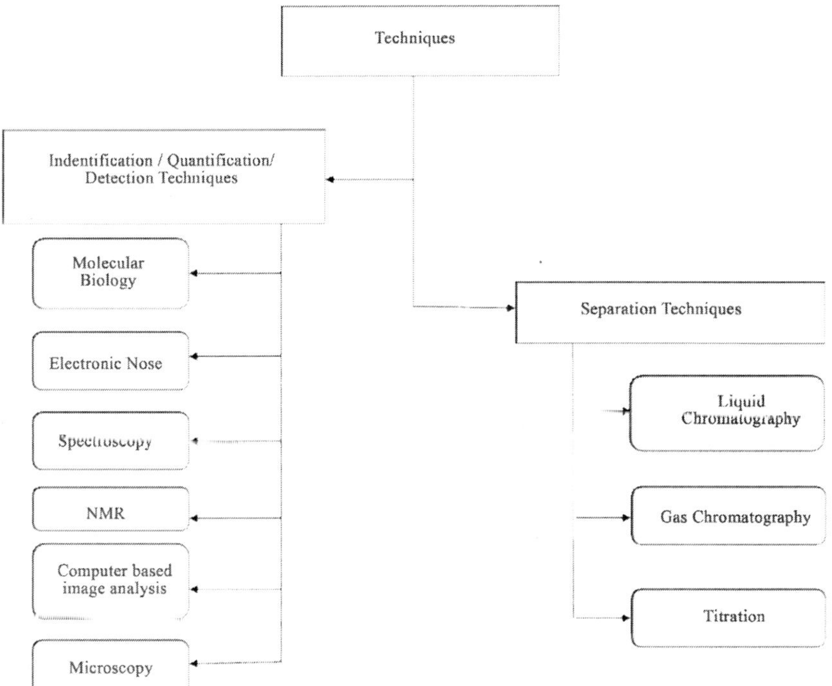

Fig. 2: Classification of techniques for separation and detection/ identification of contaminants and adulterants in cereals and products

Separation Techniques

Liquid Chromatography (LC) Methods

It is used to separate mixture of compounds for identification, quantification and purification of individual mixture components by applying high pressure to separate components for detection. Detectors used can be: photodiode array, UV/Vis, electrochemical, fluorescence, and mass spectrometry-based or their combination depending upon the sample (Vrček *et al.*, 2014; Kara *et al.*, 2015; Hussain *et al.*, 2019; Ambrose and Cho, 2014; Ghazouani *et al.*, 2021). Types of LC techniques are:

HPLC: Mainly used for plant-based sample analysis using reverse phased columns (highly hydrophobic)

Reverse phase HPLC: Uses diode array detectors, mass spectrometry or tandem mass spectrometry. It can be used for the detection of adulteration in pasta flour (De Noni *et al.*, 1994; Cocchi *et al.*, 2006)

High temperature LC: 2-D chromatography, used for complex sample

Ultra-high pressure LC: 2-D chromatography, used for complex samples

Ultra-performance liquid chromatography (UPLC): Used for the detection of ZON, NIV and DON in millet (Choi *et al.*, 2021), HT-2 and T-2 in differentcereals and detection of different wheat varieties (Soleimany *et al.*, 2012a; Russo *et al.*, 2014)

It is used to discriminate common wheat from durum wheat lipidome where characteristic curves are used to determine individual markers (digalactosyl diglycerides) for differentiating authentic durum wheat and its admixture down to3% levels. HPLC-ESI-TOF-MS has been used as profiling of phenolic content between old and new Italian wheat varieties; glaidin fractions used for distinction between wheat (has glutenin bound ω-glaidins) and spelt flours; alkylresorcinol content has been used to differentiate bread wheat and spelt (hexaploid), durum and emmer (tetraploid) and einkern (diploid) where distinction is based on concentration of saturated C17-, C19-, C21-, C23-, and C25- substituted alkylresorcinols (Knödler *et al.*, 2010; Hussain *et al.*, 2019; Ambrose and Cho, 2014). LC-MS/MS can be utilized for the detection of different cereals based on their peptide markers (sterol palmitates for durum vs common wheat); RP- HPLC combined with ESI-MS/MS or LC-MS/MS has been used to detect flour types, mycotoxins in cereals and products and to differentiate conventional to organic wheat based on their different carbohydrate content (Biselli and Hummert, 2005; Berthiller *et al.*, 2005; Cavaliere *et al.*, 2005;Delmulle *et al.*, 2006; Mulder *et al.*, 2015; Chala *et al.*, 2014; Ezekiel *et al.*, 2012; Kresse *et al.*,2019; Casazza *et al.*, 2012) using techniques such as pressurized liquid extraction (Pérez-Torrado *et al.*, 2010), stable isotope dilution assay (Al-Taher *et al.*, 2017) and 2D methods (Rausch *et al.*,2021). Bread types (whole wheat flour vs. refined wheat flour) can be detected using Fuzzy (FCMS) fingerprinting method; chemical composition profiling can be done using UHPLC-HRAM-MSn with ACPI in positive ionization mode (Pastor *et al.*, 2019). On combining with techniques such as triple quadrupole linear ion trap mass spectrometer (QTrap®), it can be used for the detection of a huge number of contaminants native to cereal groups for the differentiation between cereal kernels of various types (Vrček *et al.*, 2014; Skendi *et al.*, 2019; Kara *et al.*, 2015; Al-Taher *et al.*, 2017; Hussain *et al.*, 2019; Ghazouani *et al.*, 2021; Ziegler *et al.*, 2015; Koenig *et al.*, 2015; Geng *et al.*, 2016; Bönick *et al.*, 2017).

Gas Chromatography (GC) Methods

It is used to separate, quantify and identify volatile components in the sample. Separation is based on differences in vapor pressure the boiling point. Mobile phase is usually an inert gas (Argon, Helium, Nitrogen) and the stationary phase is usually in capillary packed columns where stationary phase is directly coated with a small diameter tube wall. Choice of detectors used mainly depends on selectivity and sensitivity, with most commonly used detectors are: thermal conductivity detectors, flame ionization detectors, thermionic detectors, flame photometric detectors, atomic emission detectors, GC-MS and electron capture detectors. It is used for detection of synthetic cereal adulterants, to quantify and classify metabolites in cereal plants such as maize, wheat, rye, rice, barley, oats etc. Further, application of chemometric analysis/ techniques is essential for sample extraction of adulterant molecular signature adding to expertise, labor and time (Ambrose and Cho, 2014) making it expensive (Hussain *et al.*, 2019; Ambrose and Cho, 2014). GC methods can be differentiated into two parts:

Gas solid chromatography: Stationary phase is a solid material: particulate silica, alumina or carbon used for separation of permanent gases and low boiling hydrocarbon compounds (Hussain *et al.*, 2019; Ambrose and Cho, 2014)

Gas- liquid chromatography: Stationary phase is applied as a thin film of liquid inside the column to help in partitioning the sample components between the stationary and mobile phase. (Hussain *et al.*, 2019; Ambrose andCho, 2014)

It is used for the detection of common wheat to durum wheat by component measurement of lipid fraction using LDA and QDA. High sensitive GC-tandem MS is used for one step detection of inorganic Arsenic in rice (Jung *et al.*, 2018). Chemical markers such as soluble sugars, fatty acids, nonsaponifiables, heterocyclic compounds, phytosterols, unsaturated C18 fatty acids of sterols etc. are used under GC-MS for distinction between cereals and pseudo-cereals, flour types and for determination of contaminants such as DON, NIV, trpoane alkaloids, ZON, EDCs in cereals and products (Albero *et al.*, 2020; Azzouz *et al.*, 2020; Hussain *etal.*, 2019; Pastor *et al.*,2016a, b, c, 2018; Pastor *et al.*, 2019; Ačanski *et al.*, 2015; Psodorov *et al.*, 2015; Caligiani *et al.*, 2011; Tanaka *et al.*, 2000). Capillary gas chromatography with ^{63}Ni electron capture detection has been used along with GC-MS for confirmation of mycotoxins in cereals (Tanaka *et al.*, 2000). Thin layer chromatography (TLC) is also been used for the detection of mycotoxins incereals (Rahmani *et al.*, 2009; Soleimany *et al.*, 2012).

Titration Technique

It is mainly used for differentiation of rye and wheat flour and for adulterant detection using where Karl Fischer titration kinetic parameters corresponding to surface and strongly retained water molecules (Hadaruga *et al.*, 2016).

Identification/ Quantification/ Detection Techniques

Molecular Biology Techniques

Differences between genomes of various species are used as a method of identification or distinction between different types. It includes techniques such as: PCR, RT-PCR, microsatellites or single nucleotide polymorphism (SNP) and ELISA. Durotest® Kit uses a monoclonal antibody specific to protein friabilin (found only in bread/ common wheat), for differentiation between wheat flours due to complex formation between the antibody and the protein (Cocchi *et al.*, 2006). But the main drawback of using protein based methods is the marker protein expression profile being influenced by the environmental growth conditions of the crop (Blumental *et al.*, 1993) and denaturation effects due to high temperature processes for product manufacturing forming large aggregates affecting the cooking quality of products such as pasta, thus making the accurate quantification of adulteration level difficult to determine (Aktan and Khan, 1992; Lamacchia *et al.*, 2007; Wagner *et al.*, 2011; Casazza *et al.*, 2012). Also, denaturation causes band broadening and a shift in elution time when glaidin proteins (markers) are subjected to HPLC, making accuratemarking of the contamination problematic (Bryan *et al.*, 1998).

DNA exhibits greater thermal resistance than protein, leaving fragments of sufficient length and integrity for PCR analysis, even after thermal treatments and a small amount of DNA is required for accurate detection for even low levels of adulteration making PCR analysis an advantageous and simple setup where one can handle numerous samples at a time (Casazza *et al.*, 2012). DNA techniques have found to have better sensitivity than the other techniques such as ELISA (Pastor *et al.*, 2019). Methods such as DNA fingerprinting have also been used for the detection of different types of flours in products such as pasta, breads and cookies (Casazza *et al.*, 2012; Silletti *et al.*, 2019). This method relies on the presence of intron-specific DNA length polymorphism in the plant β-tubulin gene highlighted through tubulin –based polymorphism profiling (capillary electrophoresis) and TBP light (Casazza *et al.*, 2012; Silletti *et al.*, 2019).

Enzyme Link Immunosorbent Assay (ELISA)

It is applied on analytical biochemical assay in which the biochemical reaction between antibody and enzyme substrate generates a signal which is visualized by the change in color of substrate, with intensity correlating with concentration of analyte. The assay mainly requires solid phase enzyme immunoassay to indicate the presence of a ligand (usually protein) in liquid samples with single or a pair of antibodies and has low sensitivity (thus used for detection of antigens present in large amounts in samples) (Hussain *et al.*, 2019; Ambrose and Cho, 2014). It is applied to detect aflatoxigenic molds in naturally contaminated corn, rice, and wheat. It can further be used with LC-MS/MS to enhance the performance of target detection (Ambrose and Cho, 2014). ELISA isa primary protocol for most commercially available kits to quantify trace levels of allergens in food products or to detect relative protein levels and amount of phosphorylation between numerous cell types (cell-based ELISA) (Hussain *et al.*, 2019; Ambrose and Cho, 2014). Different types of ELISA techniques are:

a) Direct ELISA is rapid but exhibits low sensitivity and is effective only when the samples contain large quantities of antigen
b) Indirect ELISA employs unlabeled primary antibody, which is combined with labeled secondary antibody making it more sensitive and flexible
c) Sandwich ELISA is mostly employed to detect the amount of antibody in high sensitivity and quantifies concentration of targeted molecules in given samples.
d) Competitive ELISA is effective in detecting compositional variance in a complex antigen mixture especially to screen small molecule antigens

It is also utilized for the detection of gluten in flour (with a limit of 20mg/kg), different types of wheat flours and mycotoxins such as OTA, ZON, and T-2 in numerous cereals and their products (Størsrud *et al.*, 2003; Stevenson *et al.*, 1994; Skendi *et al.*, 2019; Hussain *et al.*, 2019; Rosell *et al.*, 2014).

PCR/ Real Time PCR (RT-PCR)

A DNA-based technique, this is a highly specific and sensitive to molecular formation which is ideal for natural products with intact DNA (Pastor *et al.*, 2019). Quantification and classification of DNA at times is done by fluorescent dye (bisbenzimide H 33258) which binds tominor groove of dsDNA using established standards from DNA quantificationkit (Hussain *et al.*, 2019;Ambrose and Cho, 2014). But the extracted DNA amountis quite low and further contamination can occur by DNA polymerase inhibitors (Sonnante *et al.*, 2009). Further, cross-reactivity between these species when antibodies are raised to native purified proteins can occur as a disadvantage (Stevenson *et al.*, 1994). Microsatellites are sequences of repeated DNA analyzedby a single PCR reaction, providing short size amplicons (Pasqualone *et al.*, 2007). It is beneficial for the food samples which have undergone high temperatureand/ or mechanical treatments (e.g. pasta, bread) (Tautz *et al.*, 1989; Sonnante*et al.*, 2009) but has a disadvantage of binding all double-stranded DNA products (Sonnante *et al.*, 2009). With the help of fluorescent oligonucleotide probes, RT-PCR can be done to ensure increase in fluorescence results only from the accumulationof the desired product (Sonnante *et al.*, 2009). Another popular probe strategyis the assay based on dual-labeled probes that exploits the 5' to 3' exonuclease activity of Taq DNA polymerase to cleave a labeled hybridization probe during the extension phase of PCR (Holland *et al.*, 1991; Sonnante *et al.*, 2009).

The applications include (Alary *et al.*, 2002; Carloni *et al.*, 2016; Terzi *et al.*, 1999, 2002, 2003, 2005; Pegels *et al.*, 2015; Moon *et al.*, 2016; Pastor *et al.*, 2016; Pastor *et al.*, 2019; Tichy *et al.*, 2020; Pasqualone,2007, 2011; Sonnante *et al.*, 2009; Kaur *et al.*, 2015; Bryan *et al.*, 1998; Denery-Papini *et al.*, 2000**)**. Other applications are given as such (Table 2):

Table 2: Applications of PCR in Cereals and Products

Application	**Technique**	**Targeted gene**
Differentiation of cereals of poaceae species and flours	primers with oligonucleotide microarray (hybridization); DNA isolation by chaotropicsolid phase extraction and subsequent RT-PCR with primersand 5'-nuclease (Taqman) fluorescence probe;	chloroplast trnL intron sequence; puroindoline b; γ-hordein, gos9, and acetyl-CoA carboxylase sequence; Hor3 gene and12S seed storage protein,ribosomal ITS (barley and oats); low molecular weight glutenin region and ω-secalin (wheat, rye); MIPS gene (rye); gene encoding for the germ agglutinin isolectin A protein (Tri a 18 allergen); puroindoline-a (common wheat, barley), secaloindoline-a (rye), lipid transfer protein genes (durum wheat, rice, maize), thionin (oats); puroindoline-b; HMW-glutenin (wheat, spelt, rye); γ-glaidin gene GAG56D (Wheat, spelt)
	3 independent PCR assays; multiple-legion dependent amplification (MLDA); after hybridization probes are ligated and amplified by PCR; TaqMan RT-PCR sung 3 specificprimer/ probe systems 4th primer/probe system; High resolution melting (HRM) analysis + RT- PCR + fluorescent Evagreen dye; homologous target sequence; PCR-restriction fragment length polymorphism (PCR-RFLP)	
Soft wheat quantification	Lab-on-a-chip capillary electrophoresis (LOC-CE)	

Application	Technique	Targeted gene
Buckwheat adulteration		Primers corresponding to ITS region, 5.8S rRNA gene
Lupid flour adulteration	Quantitative PCR based on SYBR green (sensitivity of 0.1% or less in flour)	Lupin DNA
Agordino barley	SSR markers; multilocus PCR array	Genomic position across linkage groups; Polymorphicmarker alleles per locus
Detection of durum wheat from common(bread/ soft) wheat (pasta and flour)	PCR (RT-PCR), SSR markers; microsatellite sequence specific of the wheat D-genome; qualitative and quantitative Sybr green-based real-time experiments	amplification of RALyase; 'AERL' type (ω2 type glaidin) present in bread wheat

Electronic Nose (Hussain *et al.*, 2019; Ambrose and Cho, 2014)

It is mainly used for the detection of volatile adulterants in flour and other food products. An array of semi-selective chemical sensors and pattern recognition systems, based on mammalian olfactory sequencing, is used for generating fingerprints for interaction between the scent from the target sample and response from chemical sensors which are then compared to standard reference signals. A non- invasive technique, this is highly accurate technology used for rapid-detection, but has restricted applications due to its sensitivity to temperature, humidity and non-target gases and the requirement of highly trained and skilled operators for handling.

Spectroscopy Techniques

FTIR (Vibrational)

The spectral measurement with FTIR forms peaks, with the size of frequency peaks in the spectrum proportional to the amount of chemical constituents present in the sample in minor amount (Hussain *et al.*, 2019; Ambrose and Cho, 2014). Absorption bands indicate the functional groups present and their molecular confirmations based on their characteristic frequency (Sivam *et al.*, 2012, 2013). Fully functional in the near infrared range (4000-10,000 cm^{-1}) or mid-range (450-4000 cm^{-1}), this technique can be applied as a fingerprinting technique and is more reliable than conventional IR spectroscopy being suitable for both wet and dry samples. Application also included the detection of lard adulteration in cake formulations, differentiating two types of wheat bread with differing amount ofpectin (Ambrose and Cho, 2014). Attenuated total reflectance-FTIR (ATR-FTIR)used for discriminating flours of different maize varieties (Brazil); techniques such as PCA and ANN in FTIR region are used for genotype detection (Pastor *et al.*, 2019). FT-NIR can be used for the detection of adulterants such as ZON and for the classification of pasta wheat (De Girolamo *et al.*, 2019, 2020; Lee *et al.*, 2018; Shaikh *et al.*, 2017). FT-MIR is used for the detection of adulterants and contaminants e.g. ZON in different flours and other cereal products such as pasta,cakes, pies etc. (Rodríguez *et al.*, 2019; Sujka *et al.*, 2014, González-Muñoz *et al.*,2016; Durazzo *et al.*, 2018; De Girolamo *et al.*, 2019; Arslan *et al.*, 2020).

Near-IR Spectroscopy (Vibrational)

NIR radiation is a part of electromagnetic spectrum in the range 780-2500 nm involving C-H, N- H, S-H and O-H bonds. When the ray is incident on the target sample, it gets modified (absorption or transmission scattering) due to the interaction with sample influenced by physical and/ or chemical disturbances at specific wavelengths in the NIR spectrum. Canadian Grain Commission, Federal Grain Inspection Service (US) and AACC adopted this method for detection of adulterants in wheat. The working wavelength can be either short wave (780-1100 nm) or long wave (1100-2500 nm) which determines the type of detector used (Hussain *et al.*, 2019; Ambrose and Cho, 2014). PCA, PLS, PLS-DA, WILMA, SIMCA regression are used for chemometric formulations and multivariate statistical analysis to correlate the NIR spectrum (mainly to C-H, O-H, and N-H vibrations) with the quality attributes of interest (Ambrose and Cho, 2014). It is a non-destructive and rapid method which requires little to no sample processing. Advanced techniques like FTNIR, high speed NIRS have increased performance efficiency but improvement is required to find variables with lower cost, reduction in the time of analysis and to enhance the prediction efficiency (Ambrose and Cho, 2014). Applications include detection and distinction of different flours as well as wheat varieties (Cocchi *et al.*, 2005, 2006; Tao *et al.*, 2021; Ayvaz *et al.*, 2021; Delwiche *et al.*, 2011; Ziegler *et al.*, 2016; Saleh, 2012) and also for the detection of black rice varieties, organic vs. commercial rice (Delwich *et al.*, 2016;Mendez *et al.*, 2019).

Raman Spectroscopy

It is used for fingerprinting and investigating molecular distributions in the evaluated samples and can be used for the extraction of chemical information about the sample composition w.r.t. identification of chemical species and their localization (Lohumi *et al.*, 2019).Principle includes acquisition of 'hyper cube' which is done by collecting a large number of spectra at desired spatial positions of the sample, producing data cubes that contain the Raman signal intensity measured as a function of x and y (spatial) and spectral dimensions (Lohumi *et al.*, 2019). Its main drawback is that it is time intensive for large samples. Various types are: surface enhances RS (SERS), coherent anti-stroke RS (CARS) and resonance RS. Line scan (LTRS) extracts spectral data in a single shot from numerous samples stacked in a line; Non-target (NTRS) extracts data at several points. Selection of appropriate wavelength is required for a particular sample and multivariate stastical analysis is performed for signal pre-processing and data analysis (Hussain *et al.*, 2019; Ambrose and Cho, 2014). Raman spectroscopy with a chemical engineering component (hyperspectral) issued for the detection of adulterants at a large scale and for the authenticity analysis of flour using spectral angle mapping (Lohumi *et al.*, 2019).

UV-Vis spectroscopy (Hussain *et al.*, 2019; Ambrose and Cho, 2014)

It employs wavelength in visible range of the EM spectrum (400-800 nm) andis used for distinction based on color differences and other visual-sensorial characteristics. Thus, application involves detecting colored adulterants in cereals.

Fluorescence Spectroscopy (Electronic)

Main working principle is: when light is absorbed by the molecules, the electrons experience excitation and move onto the electronically excited state; the particles undergo a transition from upper excited state to the lower state without radiation causing inward vibrational conversion and since these molecules return to the groundstate, they emit light at longer wavelengths. It is a non- invasive method for determining the fluorescent properties in an object from both simple and complex biological matrices. An intense spectrum is obtained depending on the plant species and cultivars. The light sources used for excitation are lasers, LEDs, xenon arcs, mercury lamps etc. quantification of aflatoxin contamination can be done by one photon induced fluorescence in solid foods since two photon induced fluorescence creates weaker signals (Hussain *et al.*, 2019; Ambrose and Cho, 2014). PCA, PLS-DA and hierarchical cluster analysis (HCA) used to differentiate and classify flours using sync. intervals of 7 and 20nm. Excitation – emission matrices (EEM) ofcereal flours was measured in front face configuration, using four-component PARAFAC, over the UV-vis spectral range where amino acids, tocopherols, pyridoxine and 4-aminobenzoic acid showed intense spectra (both excitation and emission) (Pastor *et al.*, 2019). Front face synchronous fluorescence spectroscopy has been used for the detection of different flour types (Zandomeneghi, 1999; Karoui *et al.*, 2006) and for detection of adulterants in cereals and their products (Karoui and Blecker, 2011; Xue *et al.*, 2021; Lenhardt *et al.*, 2017; Zeković *et al.*, 2012).

Spectral Imaging/ Hyperspectral Imaging (HSI)

NIR wavelength is used for simultaneous acquisition of both spectral and spatial information about the main sample. Cereal and product composition can be understood by spatial information which allows mapping the chemical be extracted from a hypercube. Multivariate analysis is used for detection of adulteration and isconsidered superior to NIRS, providing information at the scanned point. They are used for hyperspectral data resolving with extraction of parameters like mean, median, minima, maxima, variance etc. from images at specific wavelengths (Hussain *et al.*, 2019; Ambrose and Cho, 2014). Linear discriminant analysis (LDA), quadratic discriminant analysis (QDA) and Mahalanobis analysis are mainly used for extracting useful analytical information about the different composition classes in a sample (Ambrose and Cho, 2014). The technique is used for detection of fungi in wheat grains and sprouted wheat kernels (Ambrose and Cho, 2014), classifying different kernels and flour types, the detection of contaminants such as fusarium species and aflatoxins in cereals and flours, adulterants in wheat flour and bread with cheap grains (sorghum, oats, corn) using short wave NIR-HIS technique; organic spelt flour adulteration detection with rye flour, organic wheat flour, spelt flour using HIS (900-1700 cm^{-1});geographical identification of rice (Delwich, 2016; Mendez *et al.*, 2019) and chemometric techniques such as PLS-DA, PLSR, PCR, SIMCA, MSCP are used for quantitative analysis of adulterant proportion (Vermeulen *et al.*, 2018; Ravikanth *et al.*, 2015; Barbedo *et al.*, 2015; Kandpal *et al.*, 2015; Shao *et al.*, 2019; Mishra *et al.*, 2015; Verdú *et al.*, 2016; Su *et al.*, 2016, 2017; Pastor *et al.*, 2019; Mendez *et al.*, 2019).

Analytical Techniques for Elemental Analysis (Pastor *et al.*, 2019)

Isotopic signatures are used for determining geographical origin of cereals and flours. Isotope- ratio ($^{13}C/^{12}C$ and $^{18}O/^{16}O$) mass spectrometry (IRMS) is used to analyze bread and flour samples (durum wheat vs. common wheat). Distinction between organic and conventional winter wheat, spring barley as well as detection of cadmium can be done by Inductively coupled plasma-optical emission spectrometry (ICP-OES) and inductively coupled plasma mass spectrometry (ICPMS) with chemometric analysis of multi-elemental fingerprints.

NMR

In a magnetic field, the orientation of spinning nuclei is modified when electromagnetic radiation is absorbed in the radio-frequency region of the spectrum. To relate the physical/ chemical properties and quality attributes of the product in consideration, sensors having highly sophistical mathematical models and computer software algorithms are required. NMR is used with PLS- DA using few latent models to detect several cereal grain (& product) adulterants. NMR contains less bias and is the usually employed technology for cereal-based metabolomics tests (Hussain *et al.*, 2019; Ambrose and Cho, 2014). Proton high resolution magic angle spinning nuclear magnetic resonance (1HHR-MASNMR) is used for distinction between soft and durum wheat varieties where the resonance is related to mobile molecules (mainly simple saccharides, lipids and polysaccharides) (Brescia *et al.*, 2003; Pastor *et al.*, 2019).

Computer-Based Image Analysis (CBI) (Ayvaz *et al.*, 2021)

Image signals are transformed to digital signals and which are processed using algorithms to mimic human discriminant criteria to overcome subjective differences and visual fatigue. It is mainly used for measuring and sorting purposes and is a fast, reproducible, and non-destructive method.

Microscopy Techniques (Pastor *et al.*, 2019)

Light microscopy and SEM is used to study the starch granule structure for distinguishing flour samples. Shape of starch granules depends on the plant's botanical origin so it helps in differentiating the samples of some plant speciesseparately, but not when flours are found in a mixture.

Other Techniques

Some of the latest techniques as well as the modification of old techniques have been used for the detection, separation and identification of contaminants and adulterants in cereals and their products. They are listed as such (Table 3):

Table 3: Other techniques for contaminations/ adulteration check

Technique	**Cereal/ Product Application**	**References (some)**
Graphite furnace atomic absorption spectrometry	Cadmium and lead detection in wheat grains, wheat after de-bran, semolina, wheat middlings and wheat-flour shorts	Cheli *et al*., 2010
Flame atomic absorption spectrometry	Detection of Lead, cadmium, nickel, arsenic, mercury, iron in Iranian flat breads, Composite flour, bakery products	Khaniki *et al*., 2005; Tortoe *et al*., 2018
Free zone capillary electrophoresis (FZCE)	Identification of Italian wheat cultivars; Detection of wheat flour adulteration	Bonetti *et al*., 2004
Handheld nicro-NIR spectrometer (Can be combined with chemometric methods such as PCA, SIMCA, PLS-DA, WILMA	Durum wheat flour to bread wheat flour difference; Waxy common wheat vs common wheat; Degree of adulteration in pasta, flours, breads replacement of wheat flour with cassava flour, einkorn flour, maize flours, emmer flour; Different types of wheat varieties	Cocchi *et al*., 2005, 2006; Tao *et al*., 2021; Ayvaz *et al*., 2021; Delwiche and Graybosch, 2014; Ziegler *et al*., 2016; Saleh, 2012
Near infrared hyper-spectral imaging (Can be combined with chemometric tools (PLSDA, SIMCA, PSR, PLS-DA, MSPC etc.)	Durum wheat to common wheat difference; differences in cereals; Detection of different kernels, flours (millet, corn, soybean, wheat, sorghum, oat, rye, spelt), ergot bodies, contaminants: foreign grains, dockage and animal droppings	Vermeulen *et al*., 2013, 2018; Ravikanth *et al*., 2015; Shao *et al*., 2019; Mishra *et al*., 2015; Verdú *et al*., 2016; Su *et al*., 2016, 2017
FTIR Spectroscopy (Can be used with PCA and artificial neural network (ANN)	Different varieties of pasta, cakes, pies; wheat bran; fried vermicelli; detection of DON; differentiating different types of soybeans (Chinese vs Korean)	De Girolamo *et al*., 2018, 2020; Lee *et al*., 2018; Shaikh *et al*., 2017
Fourier Transform – Mid infrared spectroscopy (FT-MIR) Can be used with PLS-DA, SIMCA)	Detection of adulterants and quality check of flours: Quinoa flour; soybean, maize, barley, oats, buckwheat, rye, triticale and wheat flours, cereal products (pasta, cakes, pies); wheat bran; detection of DON	Rodríguez *et al*., 2019; Cocchi *et al*., 2004; Sujka *et al*., 2014, González-Muñoz *et al*., 2016; Durazzo *et al*., 2018; De Girolamo *et al*., 2018; Arslan *et al*., 2019
Laser induced breakdown spectroscopy	Differentiation of flours: Wheat flour, oat flour, rye flour and their products (bread)	Akin *et al*., 2020

Technique	Cereal/ Product Application	References (some)
Front-face synchronous fluorescence spectroscopy (Can be used with PCA, PCA-LDA, PARAFAC)	Quinoa flour adulteration with maize and soybean flours	Karoui and Blecker, 2011; Xue *et al*., 2021; Lenhardt *et al*., 2017; Zeković *et al*., 2012
Fluorescence spectroscopy (Can be used with PCA, CA, PLS-DA, PARAFAC)	Differentiation in flours: Maize, rice, durum wheat and soft wheat flours; Pasta and semolina flours from kamut, common wheat, durum wheat; Wheat, corn, rye, buckwheat, rice, spelt, oat, barley flours	Zandomeneghi, 1999; Karoui *et al*., 2006;
Ultra-performance liquid chromatography	Detection of DON, NIV and ZEN in Millet (proso)	Choi *et al*., 2021
Triple quadrupole method used with ESI (in positive ion or both polarity modes) or APCI	Detection of mycotoxins, alfatoxins (AFB_1, AFB_2, AFG_1, AFG_2), ochratoxin A, zearalenone, DON and respective metabolites, Fumonisins (FB_2 and FB_2) and T-2/HT-2 toxins, Tropane alkaloids; metabolites; Pesticides, metabolites, plant growth regulators etc. in proso millet, wheat, cereal flours; cereal based foods; Sorghum, finger millet; Fonio millet; Infant cereals, corn	Choi *et al*., 2021; Soleimany *et al*., 2012; Levandi *et al*., 2014; Mulder *et al*., 2015; Chala *et al*., 2014; Ezekiel *et al*., 2012; Kresse *et al*., 2019; Al-Taher *et al*., 2017; Rausch *et al*., 2021
A-PAGE SDS-PAGE	Adulteration in pasta - Durum wheat flour to bread wheat flour difference (difference between different types of wheat flour)	Resmini, 1968; Cocchi *et al*., 2006; Barnwell *et al*., 1994; Bonetti *et al*., 2004; Igreias *et al*., 1999; Galleschi *et al*., 2003
Polymerase chain reaction (PCR) [real time or end time] and Taqman qPCR assays (DNA hybridization analysis)	Durum wheat flour to bread wheat and spelt wheat flour difference; Identification of barley, adlay, rice, wheat, oat, maize, rye etc. in wheat-based products; Determination of GM crop varieties (corn) and GMO; Quinoa flour; Gluten free to gluten based; Basmati rice vs. non-basmati rice	Alary *et al*., 2002; Carloni *et al*., 2016; Terzi *et al*., 1999, 2002, 2003. 2005; Pegels *et al*., 2015; Moon *et al*., 2016; Hübner *et al*., 2011; Kelly and Bhave, 2007; Sonnante *et al*., 2009; Pasqualone *et al*., 2007; Tichy *et al*., 2020

Technique	Cereal/ Product Application	References (some)
SSR markers (PCR) - DNA microsatellite-based test	Adulteration in pasta - Durum wheat flour to bread wheat flour difference; Basmati to non-basmati rice	Pasqualone, 2007, 2011; Sonnante *et al*., 2009; Archak *et al*., 2007; Vemireddy *et al*., 2007 Colyer *et al*., 2008; Kaur *et al*., 2015
DNA fingerprinting	Durum wheat flour to bread wheat flour difference in cereal products such as pasta, breads and cookies	Casazza *et al*., 2012; Silletti *et al*., 2019
Immunodetection: Enzyme-linked immunosorbent assay (ELISA) (Immunoblotting)	Durum vs. common wheat flour pasta; Gluten free to gluten -based foods; Organically produced cereal based foods; detection of OTA, ZEA, T-2	Størsrud *et al*., 2003; Skendi *et al*., 2019; Hussain *et al*., 2019; Rosell *et al*., 2014
Reverse phase-HPLC – for water soluble proteins	Durum wheat flour to bread wheat flour difference	De Noni *et al*., 1994; Sonetti *et al*., 2004; Cocchi *et al*., 2006
Durotest® Kit (Rhone Poulenc Diagnostic Ltd.), R-Biopharm AG, Germany	Detection of non-durum wheat adulteration - Durum wheat semolina	Cocchi *et al*., 2006
Spectral imaging	Adulteration in flours: Detection of common wheat flour, cassava flour and corn flour in Irish organic wheat flour; wheat kernels; corn cereals; Determination of presence of Fusarium, alfatoxin B-1; contamination of wheat flour with peanut traces	Su and Sun, 2017; Barbedo *et al*., 2015; Kandpal *et al*., 2015; Mishra *et al*., 2015; Verdú *et al*., 2016
Hyper-spectral imaging	Classification if flours: cooked millet flour, cooked soybean flour, wheat flour, wheat bread; Wheat types, oats, spelt, rice, corn etc.	Shao *et al*., 2018; Verdú *et al*., 2016; Mendez *et al*., 2019
Inductively coupled plasma mass spectrometry (ICPMS)	Detection of heavy metals; Differentiation of organic vs conventional varieties; different types of flours; Organic rice; wheat flour; corn flour; winter wheat, spring barley, cassava flour	Borges *et al*., 2015; Vrček *et al*., 2014; Laursen *et al*., 2011; Correia *et al*., 2017
ICP-OES	Detection of heavy metals and trace elements; Differentiation of organic vs conventional varieties; different types of flours; Organically produced cereal based foods; wheat flour; corn flour; winter wheat, spring barley, cassava flour, corn flour	Hongxing and Yu-Kui, 2011; Singh *et al*., 2010; Skendi *et al*., 2019; Vaishnavi *et al*., 2015; Laursen *et al*., 2011; Correia *et al*., 2017

Technique	Cereal/ Product Application	References (some)
EA-IRMS	Durum vs. common wheat products	Brescia *et al.*, 2003
Gas-chromatography-combustion-isotope ratio mass spectrometry (GC-C-IRMS)	Distinction between kernels based on species or geographical origin of food's organic compounds - Wheat; Koshihikari vs non-Koshihikari rice	Paolini *et al.*, 2015; Suzuki *et al.*, 2008; Delwiche, 2016
GC-MS	Detection of tropane alkaloids; Confirmation of trichothecene, mycotoxins (DON, NIV, ZON) - Buckwheat fruit, flour and related products (pasta, porridge, crackers, flakes); Cereals	Caligiani *et al.*, 2011; Tanaka *et al.*, 2000
Atmospheric pressure chemical ionization mass spectrometry (APCI-MS)	Detection of Moniliformin and Beauvericin in Pearl millet, corn	Wilson *et al.*, 2006
Atomic absorption spectrometry	Metal contaminants detection in milled maize products	Adu *et al.*, 2020
Instrumental Neutron Activation Analyzer technique	Metal contaminants detection in milled maize products	Adu *et al.*, 2020
Flame ionization detection	Detection of ZEA in corn and corn flour	Wu and Scott Smith, 2007
Time of flight (TOF) mass spectrometric	Detection of ZEA in corn and corn flour	Tanaka *et al.*, 2006
Raman hyperspectral imaging	Detection of adulterants: benzoyl peroxide, alloxan monohydrate and L-cysteine in wheat flour	Lohumi *et al.*, 2019
Computer based image analysis	Adulteration of flours: Einkorn flour, wheat flour	Ayvaz *et al.*, 2021
Analysis of alkylresorcinol composition	Durum wheat flour vs common wheat flour and pasta	KnÖdler *et al.*, 2010
Karl Fischer Titration	Rye flour vs. wheat flour	Hadaruga *et al.*, 2016

Maximum Limits as Set by Various Federations

Global as well as regional organizations (CODEX, FSSAI, EFSA, US FDA, Canadian Food Inspection Agency etc.) have set up the maximum limits (maximum permissible limits, MPL) and have outlined the measures to analyze, access and manage the risk for these contaminants and their consumption in cereals and products under the principles of 'Good Agricultural Practices' and 'Good Manufacturing Practices' and 'Codes of Practice' (Thielecke and Nugent, 2018). These maximum limits are set based on scientific advice established under the scientific committees and individual member states using these regulations are themselves responsible for food sampling for ensuring compliance (Thielecke and Nugent, 2018). For imported foodstuffs, country of origin is responsible for compliance with legislation of the country they're importing in and is checked systematically at the borders and in the market (Thielecke and Nugent, 2018). Although the different limits set in different countries/ continents may prove challenging for food importers and operating farmers, it is made sure that the compliance is met to provide safe and healthy food to the consumer pertaining to the levels of utmost standardization across the food chain. MPL for various contaminants and adulterants as set up are (Table 4):

Table 4: Maximum Permissible Limits of various contaminants and adulterants (GS, 2021; EC 2006; FSSAI, 2020)

Contaminant/ Adulterant	**CODEX**	**FSSAI**	**EU (1881/2006)**
Lead	0.2 mg/kg (cereal grains except buckwheat, canihua, quinoa), 0.01 mg/kg (infant formula, formula for special medical purposes for intended for infants and follow up formula)	0.2 mg/kg (Cereal grains except buckwheat, quinoa and canihua; infantfoods) 0.02 mg/kg (infant foods readyto use) 5 mg/kg drymatter basis (yeast products)	0.020 mg/kg wet weight (infant formulae and follow-on formulae), 0.20 mg/kg wet weight (cereals);
Copper		15 mg/kg and not less than 2.8 (Infant foods) 60 mg/kg dry basis (yeast products)	
Arsenic	0.35 mg/kg (husked rice), 0.2 mg/kg (polished rice) [Valuesfor inorganic Arsenic]	0.05 mg/kg (infantfoods)	
Tin		5 mg/kg (infantfoods)	50 mg/kg wet weight (canned baby foods, infant formulae, follow- on formulae and processed cereal-based foods in infants and young children, canned dietary foods for special medical purposes intended specifically for infants excluding dried powders and product)
Cadmium	0.1 mg/kg (cereal grains except buckwheat, canihua, quinoa, wheat and rice), 0.4 mg/kg (polished rice), 0.2 mg/kg (common wheat, durum wheat, spelt, emmer)	0.1 mg/kg (Cereal grains except buckwheat, quinoa and canihua; excluding wheat and rice and bran and germ) 0.4 mg/kg (polished rice) 0.2 mg/kg (wheat)	0.10 mg/kg wet weight (cereals excluding bran, germ, wheat, rice), 0.20 mg/kg wet weight (bran, germ, wheat, rice, soybean)

Contaminant/ Adulterant	CODEX	FSSAI	EU (1881/2006)
Benzo(a)pyrene			1.0 µg/kg wet weight (processed cereal-based foods and baby foods for infants and young children, infant formulae and follow-on formulae, dietary foods for special medical purposes specifically for infants)
Mercury		1 mg/kg	
Nitrate			200 mg NO_3/kg (processed cereal-based foods and baby foods for infants and young children)
Radionuclides	[For Infant foods] 1 Bq/kg (Pu-238, Pu-239, Pu- 240, Am-241) 100 Bq/kg (Sr-90, Ru-106, I-129, I-131, U-235) 1000 Bq/kg (S-35, Co-60, Sr-89, Ru-103, Cs-134, Cs-137, Ce-144, Ir-192, H-3, C-14, Tc-99)		
	[Foods other than infant foods] 10 Bq/kg (Pu-238, Pu-239, Pu-240, Am-241) 100 Bq/kg (Sr-90, Ru-106, I-129, I-131, U-235) 1000 Bq/kg (S-35, Co-60, Sr-89, Ru-103, Cs-134, Cs-137, Ce-144, Ir-192) 10000 Bq/kg (H-3, C-14, Tc-99)		
Total Aflatoxins (B1 + B2 + G1 +G2)		15µg/kg (cereal and cereal products)	4 µg/kg (all cereals and products derived from cereals + processed cereal products except maize and its usage); 10 µg/kg (maize to be subjected to sorting and other physical treatments before human consumption or use as an ingredient in food stuffs), 0.10 µg/ kg (processed cereal-based foods and baby foods for infants and young children, dietary foods for special medical purposes intended specifically for infants)

Contaminant/ Adulterant	CODEX	FSSAI	EU (1881/2006)
Aflatoxin B1		10µg/kg (cereal and cereal products)	2 µg/kg (all cereals and products derived from cereals + processed cereal products except maize and its usage); 5 µg/kg (maize to be subjected to sorting and other physical treatments before human consumption or use as an ingredient in food stuffs)
Ochratoxin A	5 µg/kg (raw common wheat, raw durum wheat, raw spelt, raw emmer, raw barley and raw rye)	20µg/kg (Wheat, rye, barley)	2 µg/kg (all cereals and products derived from cereals + processed cereal products except maize and its usage); 5 µg/kg (maize to be subjected to sorting and other physical treatments before human consumption or use as an ingredient in food stuffs)
Ochratoxin A	5 µg/kg (raw common wheat, raw durum wheat, raw spelt, raw emmer, raw barley and raw rye)	20µg/kg (Wheat, rye, barley)	5 µg/kg (unprocessed cereals), 3 µg/kg (all products derived from unprocessed cereals, including processed cereal products and cereals intended for direct human consumption except infant products and those for special dietary uses), 0.5 µg/kg (processed cereal-based foods and baby foods for infants and young children, dietary foods for special medical purposes intended specifically for infants)
Deoxynivalenol	200µg/kg (cereal-based foods for infants and young children), 1000µg/kg (flour, meal, semolina and flakes derived from wheat, maize or barley), 2000µg/kg (cereal grains: wheat, barley, maize destined for further processing)	1000µg/kg (Wheat)	1250 µg/kg (unprocessed cereals other than durum wheat, oats and maize), 1750µg/kg (unprocessed durum wheat and oats, unprocessed maize), 750 µg/kg (cereals intended for direct human consumption, cereal flours including maize flour, maize meal, maize grits, bran and germ as direct foods for consumption, dry pasta), 500µg/kg (bread including small bakery wafers, pastries, biscuits, cereal snacks and cereal breakfasts), 200 µg/kg (processed cereal based foods and baby foods for infants and young children)

Contaminant/ Adulterant	CODEX	FSSAI	EU (1881/2006)
Zearalenone			100 µg/kg (unprocessed cereals other than maize), 200 µg/kg (unprocessed maize, maize intended for direct human consumption, maize flour, maize meal, maize grits, maize germ), 75 µg/kg (cereals intended for direct human consumption, cereal flour, bran and germ as end market products), 50 µg/kg (bread including small bakery wafers, pastries, biscuits, cereal snacks, breakfast cereals), 20 µg/kg (processed cereal-based foods and baby foods for infants and young children)
Fumonisins (B1 + B2)	4000 µg/kg (raw maize grain), 2000µg/kg (maize flour and maize meal)		2000 µg/kg (unprocessed maize), 1000µg/kg (maize flour, maize meal, maize grits, maize germ), 400 µg/kg (maize-based foods for direct consumption), 200 µg/kg (processed maize based foods and baby foods for infants and young children)
Patulin			10 µg/kg (baby foods other than processed cereal-based foods for infants and young children)
Melamine	2.5 mg/kg (food and feed other than infant formulae), 1 mg/kg (powdered infant formula), 0.15mg/kg (liquid infant formula)	1 mg/kg (powder infant formula), 0.15mg/kg (liquid infant formula), 2.5 mg/kg (other foods)	
Acrylonitrile	0.02 mg/kg (all foods)		
Chloropropanols	0.4 mg/kg (liquid condiments containing acid hydrolyzed vegetable proteins)		

Reducing the Risk of Contamination

Mycotoxin contamination is mainly caused by fungal growth affected by the soil, humidity, temperature etc. during growing (Thielecke and Nugent, 2018) and interaction of numerous biotic and abiotic factors such as respiration rate, temperature, humidity, moisture content etc. during storage (Kaur *et al*., 2020). Farming practices such as delayed/ early harvest, draught and crop stresses such as insect/ bird/ rodent infestation and kernel damage as well as temperature and humidity modifications during storage and transportation should be avoided with proper crop rotation methods (CODEX) as well as proper/ controlled atmospheric conditions (Milani, 2013; Thielecke and Nugent, 2018; Kaur *et al*., 2020). Although the conventional approach to deal with the issue is prevention, if the crop is contaminated, there are ways to reduce the contamination as the product goes along the processing chain. To reduce the toxic metal contamination, molecular pathways responsible for accumulation should be understood in detail with inclusion of framing practices such liming the soil, organic farming as well as improved irrigation practices (Thielecke and Nugent, 2018). Mycotoxin contamination can be reduced by use of processing conditions such as sorting, cleaning, dehulling, de-branning etc. further enhanced by proper drying (< 14% moisture content) and storage (e.g. CO2 > 80% environment) conditions to arrest the development of any type of fungi (Bandyopadhyay *et al*., 2007; Thielecke and Nugent, 2018). The further practices of fermentation, cooking, ingredients and choice of grain used for particular food product affect the overall mycotoxin contamination and the nutritional profile enhancement in the final product and can be carefully manipulated (Thielecke and Nugent, 2018). Techniques such as cold plasma can be used to remove the biological post- harvest contaminants from the grains and cereals like wheat, rice, barley, corn and oats

Conclusion

The cereal grains make up a large part of the human diet, which makes thema major source of nutrients to the human body. In that case, contamination and adulteration practices not only tend to affect the crop or the product but also the human health on a vast scale. The consumer and the overall healthy diet consumption can play a significant role in mitigating the risk of contaminants in grains. Thus, it becomes important to ensure that the governmental policies and regulations are applied and followed by the various food agencies in different countries paying special attention to the contaminants and adulterants and their risk at various stages of food processing chain such as production, farming, distribution, storage etc.

References

Absorption Spectrometry and Instrumental Neutron Activation Analyser techniques. Food Control, 109: 106912.

Adesso, S., G. Autore, A. Quaroni, A. Popolo, L. Severino, and S. Marzocco. 2017. The food contaminants nivalenol and Deoxynivalenol induce inflammation in intestinal epithelial cells by regulating reactive oxygen species release. Nutrients, 9:1343.

Adu, J.K., D. Fafanyo, E. Orman, I. Ayensu, C.D.K. Amengor, and S. Kwofie. 2020. Assessing metal contaminants in milled maize products available on the Ghanian market with Atomic

Ahamed, S., J.S. Foster, A. Bukovsky, and J. Wimalasena. 2001. Signal transduction through the Ras/Erk pathway is essential for the mycoestrogen zearalenone induced cell-cycle progression in MCF-7 cells. Journal of Molecular Carcinogenicity, 30(2): 88-98.

Akin, P.A., B. Sezer, T. Sanal, H. Apaydin, H. Koksel, and İ.H. Boyaci. 2020. Multi-elemental analysis of flour types and breads by using laser induced breakdown spectroscopy. Food Chemistry, 92: 102920.

Aktan, B., and K. Khan. 1992. Effect of high-temperature drying of pasta on quality parameters and on solubility, gel electrophoresis, and reversed phase high performance liquid chromatography of protein components. Cereal Chemistry, 69: 288-295.

Al-Taher, F., J. Cappozzo, J. Zweigenbaum, H.J. Lee, L. Jackson, and D. Ryu. 2017. Detection and quantitation of mycotoxins in infant cereals in the U.S. market by LC-MS/MS using a stable isotope dilution assay. Food Control, 72: 27-35.

Alary, R., A. Serin, M.P. Duviau, P. Joudrier, and M.F. Gautier. 2002. Quantification of common wheat adulteration of durum wheat pasta using real-time quantitative polymerase chain reaction (PCR). Cereal Chemistry, 79 (4): 553-558.

Albero, B., J.L. Tadeo, and R.A. Pérez. 2020. Determination of emerging contaminants in cereals by gas chromatography-tandem mass spectrometry. Frontiers in Chemistry, 8: 571668.

Algül., I., and D. Kara. 2014. Determination and chemometric evaluation of total aflatoxin, aflatoxin B1, ochratoxin A and heavy metals content in corn flours from Turkey. Food Chemistry, 157: 70-76.

Alldrick, A. 2017. Chemical contaminants of cereals. In: Chemical Contaminants and Residues in Food. Campden BRI, Chipping Campden, United Kingdom.

Ambrose, A., and B.-K. Cho. 2014. A review of technologies for detection and measurement of adulterants in cereals and cereal products. Journal of Biosystems Engineering, 39(4): 357-365.

Arslan, F.N., G. Akin, Ş.N.K. Elmas, B. Üner, I. Yilmaz, H.-G. Janssen, and A. Kenar. 2020. FT-IR spectroscopy with chemometrics for rapid detection of wheat flour adulteration with barley flour. Journal of Consumer Protection and Food Safety, 15(3): 245-261.

Assaf, J.C., A. Atoui, A.E. Khoury, A. Chokr, and N. Louka. 2018. A comparative study of procedures for binding of aflatoxin M1 to Lactobacillus rhamnosus GG. Brazilian Journal of Microbiology, 49(1): 120–127.

Ayvaz, H., F. Korkmaz, H. Polat, Z. Ayvaz, and N.B. Tuncel. 2021. Detection of einkorn flour adulteration in flour and bread samples using computer-based image analysis and near-infrared spectroscopy. Food Control, 127: 108162.

Azzouz, A., L.P. Colón, L. Hejji, and E. Ballesteros. 2020. Determination of alkylphenols, phenylphenols, bisphenol A, parabens, organophosphorus pesticides and triclosan in different cereal-based foodstuffs by gas chromatography–mass spectrometry. Analytical and Bioanalytical Chemistry, 1-11.

Ačanski, M., D. Vujić, and D.B. Psodorov. 2015. Practical method for the confirmation of authentic flours of different types of cereals and pseudocereals. Food Chemistry, 172: 314–7.

Bahadoran, Z., P. Mirmiran, S. Jeddi, F. Azizi, A. Ghasemi, and F. Hadaegh. 2016. Nitrate and nitrite content of vegetables, fruits, grains, legumes, dairy products, meats and processed meats. Journal of Food Composition and Analysis, 51: 93–105.

Bandyopadhyay, R., M. Kumar, and J.F. Leslie. 2007. Relative severity of Aflatoxin contamination of cereal crops in West Africa. Food Additives and Contaminants, 24(10): 1109- 1114.

Barbedo, J.G., C.S. Tibola, and J.M. Fernandes. 2015. Detecting Fusarium head blight in wheat kernels using hyperspectral imaging. Biosystems Engineering, 131: 65-76.

Barbin, D.F., G. ElMasry, D.W. Sun, and P. Allen. 2013. Non-destructive determination of chemical composition in intact and minced pork using near-infrared hyperspectral imaging. Food Chemistry, 138(2-3): 1162-1171.

Barnett, M.J., W.R. Dripps, and K.K. Blomquist. 2016. Organivore or organorexic? Examining the relationship between alternative food network engagement, distorted eating, and special diets. Appetite, 105: 713-720.

Battilani, P., P. Toscano, H.J. Van der Fels-Klerx, A. Moretti, M. Camardo Leggieri, C. Brera, A. Rortais, T. Goumperis, and T. Robinson. 2016. Aflatoxin B1 contamination in maize in Europe increases due to climate change. Scientific Reports, 6(1): 1-7.

Begemann, J., S. Ostovar, and C. Schwake-Anduschus. 2021. Facing tropane alkaloid contamination in millet – Analytical and processing aspects. Quality Assurance and Safety of Crops & Foods, 13(2): 79-86.

Berkov, S., R. Zayed, and T. Doncheva. 2006. Alkaloid patterns in some varieties of Datura stramonium. Fitoterapia, 77: 179–182.

Berthiller, F., R. Schuhmacher, G. Buttinger, and R. Krska. 2005. Rapid simultaneous determination of major type A and B-trichothecenes as well as zearalenone in maize by high performance liquid chromatography–tandem mass spectrometry. Journal of Chromatography A, 1062: 209–216.

Bilge, G., B. Sezer, K.E. Eseller, H. Berberoglu, H. K€oksel, I.H. Boyacı,, 2016. Determination of Ca addition to the wheat flour by using laser-induced breakdown spectroscopy (LIBS). European Food Research and Technology, 242(10): 1685 1692.

Biselli, S., and C. Hummert. 2005 Development of a multicomponent method for Fusarium toxins using LC–MS/MS and its application during a survey for the content of T-2 toxin and deoxynivalenol in various feed and food samples. Food Additives and Contaminants, 22: 752–760.

Blumenthal, C.S., E.W.R. Barlow, and C.W. Wrigley. 1993. Growth environment and wheat quality: the effect of heat stress on dough properties and gluten proteins. Journal of Cereal Science, 18: 3-21.

Bonetti, A., I. Marotti, P. Catizone, G. Dinelli, A. Majetti, P. Tedeschi, and V. Brandolini. 2004. Compared use of HPLC and FZCE for cluster analysis of Triticum spp and for the identification of T. durum adulteration. Journal of Agricultural and Food Chemistry, 52: 4080-4089.

Borges, E.M., J.M.L.N. Gelinski, V.C. de Oliveira Souza, F. Barbosa Jr., and B.L. Batista. 2015. Monitoring the authenticity of organic rice via chemometric analysis of elemental data. Food Research International, 77: 299 309.

Brera, C., A. Peduto, F. Debegnach, E. Pannunzi, E. Prantera, E. Gregori, M. De Giacomo, and B. De Santis. 2013. Study of the influence of the milling process on the distribution of Deoxynivalenol content from the caryopsis to cooked pasta. Food Control, 32(1): 309–312.

Brescia, M.A., A. Sgaramella, S. Ghelli, and A. Sacco. 2003. 1H HR-MAS NMR and isotopic investigation of bread and flour samples produced in southern Italy. Journal of the Science of Food and Agriculture, 83(14): 1463–1468.

Bryan, G.J., A. Dixon, M.D. Gale, and G. Wiseman. 1998. A PCR-based method for the detection of hexaploid bread wheat adulteration of durum wheat and pasta. Journal of Cereal Science, 28 (2): 135-145.

Bönick, J., G. Huschek, and H.M. Rawel. 2017. Determination of wheat, rye and spelt authenticity in bread by targeted peptide biomarkers. Journal of Food Composition and Analysis, 58: 82–91.

Caligiani, A., A. Palla., F. Bonzanini, A. Bianchi, and R. Bruni. 2011. A validated GC-MS method for the detection of tropane alkaloids in buckwheat (Fagopyron esculentum L.) fruits, flours and commercial foods. Food Chemistry, 127: 204-209.

Canadian total diet study in Quebec City and dietary intake estimates. Food Additives and. Contaminants, 28(6): 791–798.

Cao, X. L., C. Perez-Locas, G. Dufresne, G. Clement, S. Popovic, F. Beraldin, R.W. Dabeka, and M. Feeley. 2011. Concentrations of bisphenol A in the composite food samples from the 2008

Carloni, E., G. Amagliani, E. Omiccioli, V. Ceppetelli, M. Del Mastro, L. Rotundo, G. Brandi, and M. Magnani. 2016. Validation and application of a quantitative real-time PCR assay to detect common wheat adulteration of durum wheat for pasta production. Food Chemistry, 224: 86–91.

Casazza, A., C. Morcia, E. Ponzoni, F. Gavazzi, S. Benedettelli, and D. Breviario. 2012. A reliable assay for the detection of soft wheat adulteration in Italian pasta is based on the use of new DNA molecular markers capable of discriminating between Triticum aestivum and Triticum durum. Journal of Cereal Science, 56(3): 733–740.

Cavaliere, C., P. Foglia, E. Pastorini, R. Samperi, and A. Lagana. 2005. Development of a multiresidue method for analysis of major Fusarium mycotoxins in corn meal using liquid chromatography/tandem mass spectrometry. Rapid Communications in Mass Spectrometry, 19: 2085–2093.

Chala, A., W. Taye, A. Ayalew, R. Krska, M. Sulyok, and A. Logrieco. 2014. Multi mycotoxin analysis of sorghum (Sorghum bicolor L. Moench) and finger millet (Eleusine coracana L. Garten) from Ethiopia. Food Control, 45: 29-35.

Cheli, F., A. Campagnoli, V. Ventura, C. Brera, C. Berdini, E. Palmaccio, and V. Dell'Orto. 2010. Effects of industrial processing on the distribution of deoxynivalenol, cadmium and lead in durum wheat milling fractions. LW T - Food Science and Technology, 43: 1050-1057.

Chiang, P.O., C.-Y Chiu, M.K. Wang, and B.-T. Chen. 2011. Low-molecular-weight organic acids exuded by millet (Setaria italic (L.) Beauv.) roots and their effect on the remediation of cadmium-contaminated soil. Soil Science, 176(1): 33-38.

Choi, J.H., J.Y. Nah, M.J. Lee, J.Y. Jang, T. Lee, and J. Kim. 2021. Fusarium diversity and mycotoxin occurrence in proso millet in Korea. LWT – Food Science and Technology, 141: 110964.

Cocchi, M., C. Durante, G. Foca, A. Marchetti, L. Tassi, and A. Ulrici. 2006. Durum wheat adulteration detection by NIR spectroscopy multivariate calibration. Talanta, 68: 1505–1511.

Cocchi M., M. Corbellini, G. Foca, M. Lucisano, M.A. Pagani, L. Tassi, and A. Ulrici A. 2005. Classification of bread wheat flours in different quality categories by a wavelet-based feature selection/classification algorithm on NIR spectra. Analytica Chimicia Acta, 544(1-2): 100–107.

CODEX. 2021. https://www.fao.org/fao-who-codexalimentarius/meetings/en/. Accessed on: 21st November' 2021.

Contaminants in food and feed: inexpensive detection for control of exposure. 2008 CONFIDENCE CRAW Gembloux, Belgium. http://www.conffidence.eu. Accessed on: 20th November' 2021.

Correia, F.O., D.S. Silva, S.S.L. Costa, I.K.V. Silva, D.R.da Silva, H.A. José do Patrocínio, C.A. Garcia, T.D.A. Maranhao, E.A. Passos, and R.G. Araujo. 2017. Optimization of microwave digestion and inductively coupled plasma-based methods to characterize cassava, corn and wheat flours using chemometrics. Microchemical Journal., 135:190–198.

Cortés, V., J. Blasco, N. Aleixos, S. Cubero, and P. Talens. 2019. Monitoring strategies for quality control of agricultural products using visible and near-infrared spectroscopy: A review. Trends in Food Science & Technology, 85: 138-148.

Danezis, G.P., A.S. Tsagkaris, F. Camin, V. Brusic, and C.A. Georgiou. 2016. Food authentication: Techniques, trends & emerging approaches. TrAC Trends in Analytical Chemistry, 85: 123-132.

Defernez, M and R. H. Wilson. 1995. Mid-infrared spectroscopy and Chemometrics for determining the type of fruit used in jam. Journal of the Science of Food and Agriculture, 67(4): 461-467

De Girolamo, A., M.C. Arroyo, S. Cervellieri, M. Cortese, M. Pascale, A.F. Logrieco, and V. Lippolis. 2020. Detection of durum wheat pasta adulteration with common wheat by infrared spectroscopy and chemometrics: A case study. LWT –Food Science and Technology, 127: 109368.

De Girolamo, A., S. Cervellieri, M. Cortese, A.C.R. Porricelli, M. Pascale, F. Longobardi, C. von Holst, L. Ciaccheri, and V. Lippolis, V. 2019. F0urier transform near-infrared and midinfrared spectroscopy as efficient tools for rapid screening of deoxynivalenol contamination in wheat bran. Journal of the Science of Food and Agriculture,99(4): 1946-1953.

De la Guardia, M., and S. Garrigues. 2015. Handbook of Mineral Elements in Food. John Wiley & Sons.

Delmulle, B., S.D. Saeger, A. Adams, N.D. Kimpe, and C.V. Peteghem. 2006. Development of a liquid chromatography/tandem mass spectrometry method for the simultaneous determination of 16 mycotoxins on cellulose filters and in fungal cultures. Rapid Communications in Mass Spectrometry, 20: 771–776.

Delwiche, S.R., R.A. Graybosch, P. St Amand, and G. Bai. 2011. Starch waxiness in hexaploid wheat (Triticum aestivum L.) by NIR reflectance spectroscopy. Journal of Agricultural and Food Chemistry, 59: 4002-4008.

Delwiche, S.R. 2016. Advances in the identification of adulterated cereals and cereal products. In: Advances in Food Authenticity Testing, Woodhead Publishing, 491-518. .

Denery-Papini, S., M.F. Samson, and J.C. Autran. 2000. Anti-peptide antibodies directed against omega-glaidins for the detection of sequences from bread and durum wheats. Food and Agricultural Immunology, 12: 67 75.

De Noni, I., G. de Bernardi, and L. Pellegrino. 1994. Detection of common wheat (Triticum aestivum) flour in durum-wheat (Triticum durum) semolina by reverse-phase high-performance liquid chromatography (RP-HPLC) of specific albumins. Food Chemistry, 51: 325–329.

Dias-Martins, A.M., K.L.F. Pessanha, S. Pacheco, J.A.S. Rodrigues, and C.W.P. Carvalho. 2018. Potential use of pearl millet (Pennisetum galucum (L.) R. Br.) in Brazil: Food security, processing, health benefits and nutritional products. Food Research International, 109: 175-186.

DPR n. 187. 2001. Regolamento per la revisione della normativa sulla produzione e commercializzazione di sfarinati e paste alimentari, a norma dell'articolo 50 della legge 22 febbraio 1994, n. 146, Republic President, Italy.

Durazzo, A., J. Kiefer, M. Lucarini, E. Camilli, S. Marconi, P. Gabrielli, A. Augzzi, L. Gambelli, S. Lisciani, and L. Marletta. 2018. Qualitative analysis of traditional Italian dishes: FTIR Approach. Sustainability, 10: 4112.

EC. 2006 (EC 1881/2006). European Commission. 2006. Commission Regulation (EU) No. 1881/2006 of 19 December 2006 setting maximum levels for certain contaminants in foodstuffs. Official Journal of the European Communities L364, 4: 5–24.

EC Directive 91/173/EEC. https://op.europa.eu/en/publication-detail/-/publication/fc8f6d81-2963-4de8-81ff-b74e418b5cf4/language-en. Accessed on: 21st November' 2021.

EFSA. 2013. Scientific opinion on tropane alkaloids in food and feed. EFSA Journal 11: 3386. European Food Safety Authority (EFSA). 2009. Scientific opinion "cadmium in food"– scientific opinion of the panel on contaminants in the food chain. (Question No EFSA-Q-2007-138). Adopted on 30 January 2009. The EFSA Journal, 980: 1–139.

Ehling, S., S.H.O.I.P. Tefera, and I.P. Ho. 2007. High-performance liquid chromatographic method for the simultaneous detection of the adulteration of cereal flours with melamine and related triazine by-products ammeline, ammelide, and cyanuric acid. Food Additives and Contaminants, 24(12): 1319-1325.

European Commission (EC). 1992. Regulation 689/92 fixing the procedure and conditions for the taking-over of cereals by intervention agencies. Off J Eur Comm L074:18–22.

European Commission (EC). 2002. Directive 2002/32 on undesirable substances in animal feed. Off J Eur Comm L140:10–21.

European Commission (EC). 2004. Commission Directive 2004/19/EC of 1 March 2004 amending Directive 2002/72/EC relating to plastic materials and articles intended to come into contact with foodstuffs. Off J Eur Comm. L71:8–21

European Commission (EC). 2015. Commission Recommendation (EU) 2015/976 of 19 June 2015 on the monitoring of the presence of tropane alkaloids in food. Off. J. Eur. Union L157: 97-98.

European Commission (EC) 2016. Commission Regulation (EU) 2016/239 of 19 February 2016 amending Regulation (EC) No 1881/2006 as regards maximum levels of tropane alkaloids in certain cereal-based foods for infants and young children. Off. J. Eur. Union L45: 3-5.

European commission, Regulation (EC) No 396/2005 of the European Parliament and of the council of 23 February 2005 on maximum residue levels of pesticides in or on food and feed of plant and animal origin and amending Council Directive 91/414/EEC, [Online]. Available:

European Commission. 2008. Regulation 687/2008 establishing procedures for the taking-over of cereals by intervention agencies or paying agencies and laying down methods of analysis for determining the quality of cereals. Off J Eur Comm L192:20–48.

European Food Safety Authority (EFSA), 2008. Polyaromatic hydrocarbons in food. EFSA J. 724, 1e114.

European Food Safety Authority (EFSA). 2006. Opinion of the scientific panel on food additives, flavourings, processing aids and materials in contact with food (AFC) related to 2,2- bis(4-hydroxyphenyl) propane. EFSA J., 428: 1–6

Ezekiel, C.N., I.E. Udom, J.C. Frisvad, M.C. Adetunji, J. Houbraken, S.O. Fapohunda, R.A. Samson, O.O. Atanda, M.C. Agi-Otto, and O.A. Onashile. 2014. Assessment 27 of aflatoxigenic Aspergillus and other fungi in millet and sesame from Plateau State, Nigeria. Mycology, 5(1): 16-22.

Ezekiel, C.N., M. Sulyok, B. Warth, and R. Krska. 2012. Multi-microbial metabolites in fonio millet (acha) and sesame seeds in Plateau State, Nigeria. European Food Research Technology, 235(2): 285-293.

FAO & CIRAD. 2015. In: D. Bazile, D. Bertero, & C. Nieto (Eds.), State of the art report of quinoa in the world in 2013. Rome.

FAO, 2021. https://www.fao.org/3/cb2863en/cb2863en.pdf. Accessed on: 21st November' 2021.

FAO. 1986. Manuals of food quality control – 8. Food Analysis: quality, adulteration and tests of identity. FAO – Food and Nutrition Paper 14/8. 1986. Food and Agriculture Organization of the United Nations, Rome.

FAO. 2020. Brief. Available online at: https://www.fao.org/worldfoodsituation/csdb/en/. Accessed on: 20th November' 2021.

Ferreira, D. S., J.A.L. Pallone, and R.J. Poppi. 2015. Direct analysis of the main chemical constituents in Chenopodium quinoa grain using Fourier transform near-infrared spectroscopy. Food Control, 48: 91–95.

Filho, A. M. M., M.R. Pirozi, J.T.D.S. Borges, H.M. Pinheiro Sant'Ana, J.B.P. Chaves, and J.S.D.R. Coimbra. 2017. Quinoa: Nutritional, functional, and antinutritional aspects. Critical Reviews in Food Science and Nutrition, 57(8): 1618–1630.

FSSAI. 2011. https://www.fssai.gov.in/index.php?page=food-testing.php. Accessed on: 21st November' 2021.

FSSAI. 2020, https://www.fssai.gov.in/upload/uploadfiles/files/Compendium_Contaminants_Regulations_20_08_2020.pdf. Accessed on: 21st November' 2021.

Galleschi, L., M. Ingrosso, S. Ghiringhelli, L. Calucci, and R. Lorenzi. 2003. Isolation of wheat puroindoline-b from flour by preparative acid electrophoresis. Cereal chemistry, 80(1): 99.

Geng, P., J.M. Harnly, and P. Chen. 2016. Differentiation of bread made with whole grain and refined wheat (T. aestivum) flour using LC/MS-based chromatographic fingerprinting and chemometric approaches. Journal of Food Composition and Analysis, 47: 92–100.

Ghazouani, T., A. Atezi, W. Talbi, M.A. Fenu, C. Tuberoso, and S. Fattouch. 2021. Occurrence of acrylamide, hydroxymethylfurfural and furaldehyde as process contaminants in traditional breakfast cereals: "Bsissa". Food Control, 124: 107931.

Girolamo, A.D., M.C. Arroyo, V. Lippolis, S. Cervellieri, M. Cortese, M. Pascale, A.F. Logrieco, and C. von Holst. 2020. A simple design for the validation of a FT-NIR screening method: Application to the detection of durum wheat pasta adulteration. Food Chemistry, 333: 127449.

González-Muñoz, A., B. Montero, J. Enrione, and S. Matiacevich. 2016. Rapid prediction of moisture content of quinoa (Chenopodium quinoa Willd.) flour by Fourier transform infrared (FTIR) spectroscopy. Journal of Cereal Science, 71: 246–249.

GS.2021. General standards for contaminants and toxins in food and feed. https://www.fao.org/fao-who-codexalimentarius/shproxy/en/?lnk=1&url=https%253A%252F%252Fworkspace.fao.org%252Fsites%252Fcodex%2 52FStandards%252FCXS %2B193-1995 %252 FCXS_193e.pdf. Accessed on: 21st November' 2021.

Gyllenhammar, I., A. Glynn, P.O. Darnerud, S. Lignell, R. van Delft, and M. Aune. 2012. 4-Nonylphenol and bisphenol A in Swedish food and exposure in Swedish nursing women. Environment International, 43: 21–28.

Hadaruga, D.I., C.I. Costescu, L. Corpas N.G. Hadaruga, and H.D. Isengard HD. 2016. Differentiation of rye and wheat flour as well as mixtures by using the kinetics of Karl Fischer water titration. Food Chemistry, 195: 49–55.

Hajeb, P., j.j. Sloth, S. Shakibazadeh, N.A. Mahyudin, and L. Afsah-Hejri. 2014. Toxic Elements in Food: Occurrence, Binding, and Reduction Approaches. Comprehensive Reviews in Food Science and Food Safety, 13: 457–472.

Harcz, P., L. De Temmerman, S. De Voghel, N. Waegeneers, O. Wilmart, V. Vromman, J.-F. Schmit, E. Moons, C. Van Peteghem, S. De Saeger, Y.-J. Schneider, Y. Larondelle, and

Holland, P. M., R.D. Abramson, R. Watson, and D.H. Gelfand. 1991. Detection of specific polymerase chain reaction product by utilizing the 50-30 nuclease activity of Thermus aquaticus DNA polymerase. Proceedings of the National Academy of Sciences, 88(16): 7276–7280.

Hongxing, Z., and R. Yu-Kui. 2011. Determination of trace elements, heavy metals and rare earth elements in corn seeds from Beijing by ICP-MS simultaneously. E-Journal of Chemistry, 8(2): 782–786.

Horszwald, A., F.J. Morales, M.D. del Castillo, and H. Zieliński. 2010. Evaluation of antioxidant capacity and formation of processing contaminants during rye bread making. Journal of Food and Nutrition Research, 49(3): 149-159.

https://eur-lex.europa.eu/legal-content/EN/TXT/HTML/?uri=CELEX:32005R0396&from=DE, Accessed date: 20th November' 2021.

Hussain, N., D.-W. Sun, and H. Pu. 2019. Classical and emerging non-destructive technologies for safety and quality evaluation of cereals: A review of recent applications. Trends in Food Science and Technology, 91: 598-608.

Ibitoye, O.A., O.O. Olaniyi, C.O. Ogidi, and B.J. Akinyele. 2020. Lactic acid bacteria biodetoxified aflatoxins contaminated cereals, ameliorate toxicological effects and improve haemato-histological parameters in albino rats. Toxin Reviews,1-12.

Igrejas, G., H. Guedes-Pinto, V. Carnide, and G. Branlard. 1999. The high and low molecular weight glutenin subunits and ω-glaidin composition of bread and durum wheats commonly grown in Portugal. Plant Breeding, 118(4): 297-302.

International Agency for Research on Cancer (IARC). 1993. Some naturally occurring substances: food items and constituents' heterocyclic aromatic amines and mycotoxins. In: IARC monographs on the evaluation of carcinogenic risk to humans, 56: 359-444.

JECFA (the Joint FAO/WHO Expert Committee on Food Additives). 1974. Toxicological evaluation of certain food additives with a review of general principles and of specifications. 17th Report of the Joint FAO/WHO Expert Committee on Food Additives. World Health Organization Technical Report Series no. 539; Switzerland: FAO and WHO.

Jung, M.Y., J.H. Kang, H.J. Jung, and S.Y. Ma. 2018. Inorganic arsenic contents in ready-to-eat rice products and various Korean rice determined by a highly sensitive gas chromatography tandem mass spectrometry. Food chemistry, 240: 1179-1183.

Jurjevic, Z., J.P. Wilson, D.M. Wilson, and H.H. Casper. 2007. Changes in fungi and mycotoxins in pearl millet under controlled storage conditions. Mycopathologia, 164 (5): 229–239. Kamath, R. 2018. Food traceability in blockchain: Walmart's pork and mango pilots with IBM. The JBBA, 1(1): 47-53.

Kandpal, L.M., S. Lee, M.S. Kim, H. Bae, and B.-K. Cho. 2015. Short wave infrared (SWIR) hyperspectral imaging technique for examination of aflatoxin B 1 (AFB 1) on corn kernels. Food Control, 51:171-176.

Kara, G.N., F. Ozbey, and B. Kabak. 2015. Co-occurrence of aflatoxins and ochratoxin A in cereal flours commercialization in Turkey. Food Control, 54: 275-281.

Karoui, R., and C. Blecker. 2011. Fluorescence spectroscopy measurement for quality assessment of food systems—a review. Food and Bioprocess Technology, 4(3): 364–386.

Karoui, R., G. Cartaud, and E. Dufour. 2006. Front-face fluorescence spectroscopy as a rapid and nondestructive tool for differentiating various cereal products: a preliminary investigation. Journal of Agricultural and Food Chemistry, 54(6): 2027–34.

Kaur, M., D. Hüberli, and K.L. Bayliss. 2020. Cold Plasma: exploring the new option for management of postharvest fungal pathogens, mycotoxins and insect pests in Australian stored cereal grain. Crop & Pasture Science, 71: 715-724.

Kaur, S., P.S. Panesar, M.B. Bera, and V. Kaur. 2015. Simple sequence repeat markers in genetic divergence and marker-assisted selection of rice cultivars: a review. Critical Reviews in Food Science and Nutrition, 55: 41-49.

Khaniki, G.R.J., M. Yunesian, A.H. Mahvi, and S. Nazmara. 2005. Trace metal contaminants in Iranian Flat Breads. Journal of Agriculture and Social Sciences, 1(4): 301-303.

Kirincic, S., B. Škrjanc, N. Kos, B. Kozolc, N. Pirnat, and G. Tavčar-Kalcher. 2015. Mycotoxins in cereals and cereal products in Slovenia–Official control of foods in the years 2008–2012. Food Control, 50: 157–165.

Knödler, M., M. Most, A. Schieber, and R. Carle. 2010. A novel approach to authenticity control of whole grain durum wheat (*Triticum durum* Desf.) flour and pasta, based on analysis of alkylresorcinol composition. Food Chemistry, 118: 177-181.

Koenig A., K. Konitzer H. Wieser, and P. Koehler. 2015. Classification of spelt cultivars based on differences in storage protein compositions from wheat. Food Chemistry,168: 176–82.

Koerner, T.B., C. Cleroux, C. Poirier, I. Cantin, S.L. Vieille, S. Hayward, and S. Dubois. 2013. Gluten contamination of naturally gluten-free flours and starches used by Canadians with celiac disease. Food Additives and Contaminants, Part A, 30(12): 2017-2021.

Krenzelok, E.P. 2010. Aspects of Datura poisoning and treatment. Clinical Toxicology, 48: 104–110.

Kresse, M., H. Drinda, A. Romanotto, and K. Speer. 2019. Simultaneous determination of pesticides, mycotoxins, and metabolites as well as other contaminants in cereals by LC-LCMS/ MS. Journal of Chromatography B, 1117: 86-102.

Krska R and C. Crews. 2008. Significance, chemistry and determination of ergot alkaloids: A review. Food Additives and Contaminants, 25(6): 722–731.

L. Pussemier. 2007. Contaminants in organically and conventionally produced winter wheat (*Triticum aestivum*) in Belgium. Food Additives and Contaminants, 24 (7): 713– 720.

Lamacchia, C., A. Di Luccia, A. Baiano, G. Gambacorta, B. La Gatta, S. Pati, and E. La Notte. 2007. Changes in pasta proteins induced by drying cycles and their relationship to cooking behaviour. Journal of Cereal Science, 46: 58-63.

Laparra, J. M., and M. Haros. 2018. Inclusion of whole flour from Latin-American crops into bread formulations as substitute of wheat delays glucose release and uptake. Plant Foods for Human Nutrition, 73(1): 13-17.

Lasram, S., Z. Hamdi, and A. Ghorbel. 2020. Occurrence and exposure assessment of Aflatoxin B1 and Ochratoxin A in pearl millet (Pennisetum glaucum L.) from Tunisia. Journal of Food Quality and Hazards Control, 7: 157-161.

Laursen, K.H., J.K. Schjoerring, J.E. Olesen, M. Askegaard, U. Halekoh, and S. Husted. 2011. Multielemental fingerprinting as a tool for authentication of organic wheat, barley, faba bean, and potato. Journal of Agricultural and Food Chemistry, 59(9): 4385–4396.

Lee, B.J., Y. Zhou, J.S. Lee, B.K. Shin, J.A. Seo, D. Lee, Y.S. Kim, and H.K. Choi. 2018. Discrimination and prediction of the origin of Chinese and Korean soybeans using Fourier transform infrared spectrometry (FT-IR) with multivariate statistical analysis. PLoS ONE, 13(4): e0196315.

Lenhardt, L., I. Zeković, T. Dramićanin, B. Milićević, J. Burojević, and M.D. Dramićanin. 2017. Characterization of cereal flours by fluorescence spectroscopy coupled with PARAFAC. Food Chemistry, 229: 165–171.

Leslie, J. F., K.A. Zeller, S.C. Lamprecht, J.P. Rheeder, and W.F. Marasas. 2005. Toxicity, pathogenicity, and genetic differentiation of five species of Fusarium from sorghum and millet. Phytopathology, 95(3): 275–283.

Liao, C., F. Liu, and K. Kannan. 2013b. Occurrence of and dietary exposure to parabens in foodstuffs from the United States. Environmental Science and Technology, 47(8): 3918–3925.

Liao, C., L. Chen, and K. Kannan. 2013a. Occurrence of parabens in foodstuffs from China and its implications for human dietary exposure. Environment International, 57, 68–74.

Liu, R.H. 2007. Whole grain phytochemicals and health. Journal of Cereal Science, 46(3): 207–219.

Lohumi, S., H. Lee, M.S. Kim. L. Qin, and B.-K., Cho. 2019. Raman hyperspectral imaging and spectral similarity analysis for quantitative detection of multiple adulterants in wheat flour. Biosystems Engineering, 181: 103-113.

Massart, F., V. Meucci, G. Saggese, and G. Soldani. 2008. High growth rate of girls with precocious puberty exposed to estrogenic mycotoxins. Journal of Pediatrics, 152(5): 690-695.

Mendez, J., L. Mendoza, J.P. Cruz-Tirado, R. Quevedo, and R. Siche. 2019. Trends in application of NIR and hyperspectral imaging for food authentication. Scientia Agropecuaria, 10(1): 143-161.

Milani, J.M. 2013. Ecological conditions affecting mycotoxin production in cereals: a review. Veterinarni Medicina, 58(8): 405-411.

Mishra, N.K., and S.K. Daradhiyar. 1990. Mold flora and aflatoxin contamination of stored and cooked samples of pearl millet in the Paharia Tribal Belt of Santhal Pargana, Bihar, India. Applied and Environmental Biology, 57(4): 1223-1226.

Mishra, P., A. Herrero-Langreo, P. Barreiro, M. Roger Jean, B. Diezma, N. Gorretta, and L. Lleó. 2015. Detection and quantification of peanut traces in wheat flour by near infrared hyperspectral imaging spectroscopy using principal-component analysis. Journal of Near Infrared Spectroscopy, 23(1): 15–22.

Moon, J., J. Kim, and C. Jang. 2016. Development of multiplex PCR for species-specific identification of the Poaceae family based on chloroplast gene, rpoC2. Applied Biological Chemistry, 59(2): 201–207.

Mulder, P.P., D.P. Pereboom-de Fauw, R.L. Hoogenboom, J. de Stoppelaar, and M. de Nijs. 2015. Tropane and ergot alkaloids in grain-based products for infants and young children in the Netherlands in 2011–2014. Food Additives & Contaminants: Part B, 8(4): 284–290.

Mulder, P.P.J., M. de Nijs, M. Castellari, M. Hortos, S. MacDonald, C. Crews, C. Hajslova, and M. Stranska. 2016. Occurrence of tropane alkaloids in food. EFSA Supporting Publications, 13(12): 1140E.

Müller, J., and K. Wanke. 1998. Toxic psychoses from atropine and scopolamine. Fortschritte der neurologie Psychiatrie, 66: 289–295.

Nekvapil, F., I. Brezestean, D. Barchewitz, B. Glamuzina, V. Chiş,and S.C. Pinzaru. 2018. Citrus fruits freshness assessment using Raman spectroscopy. Food chemistry, 242: 560-

Odoemelam, S.A., and C.I. Osu. 2009. Aflatoxin B1 contamination of some edible grains marketed in Nigeria. E-Journal of Chemistry, 6(2): 308-314.

Paolini, M., L. Ziller, K.H. Laursen, S. Husted, and F. Camin. 2015. Compound-specific δ15N and δ13C analysis of amino acids for the potential discrimination between organically and conventionally grown wheat. Journal of Agricultural and Food Chemistry, 63(25): 5841-5850.

Pasqualone, A., C. Montemurro, A. Grinn-Gofron, G. Sonnante, and A. Blanco. 2007. Detection of soft wheat in semolina and durum wheat bread by analysis of DNA microsatellites. Journal of Agricultural and Food Chemistry, 55(9): 3312–3318.

Pasqualone, A. 2011. Authentication of durum wheat-based foods: Classical vs. innovative methods. In: Current Topics in Food Authenticity; M. B. P. P. Oliveira, I. Mafra, & J.S. Amaral (Eds.). Transworld Research Network (pp. 23–39).

Pastor, K, M. Acanski, Đ. Vujić, and A. Kondić-Špika. 2016c. Binary simple sugar profiling in corn and small grain flour authentication using GC/EI-qMS approach. Chromatographia, 79: 1553–9.

Pastor, K, M. Ačanski, D. Vujić, G. Bekavac, S. Milovac, and S. Kravić. 2016b. Rapid method for small grain and corn flour authentication using GC/EI–MS and multivariate analysis. Food Analytical Methods, 9(2): 443–50.

Pastor, K, M. Ačanski, Đ. Vujić, Đ. Jovanović, and S. Wienkoop. 2016a. Authentication of cereal flours by multivariate analysis of GC–MS data. Chromatographia, 79: 1387–93.

Pastor, K., L. Pezo, Đ. Vujić, Đ. Jovanović, and M. Ačanski. 2018. Discriminating cereal and pseudocereal species using binary system of GC/MS data: A pattern recognition approach. Journal of the Siberian Chemical Society, 83(3): 317-329.

Pastor, K., M. Ačanski, and D. Vujić. 2019. A review of adulteration versus authentication of flour. In: Flours and Breads and their Fortification in Health and Disease Prevention, 21-35.

Paśko, P., M. Sajewicz, S. Gorinstein, and Z. Zachwieja. 2008. Analysis of selected phenolic acids and flavonoids in Amaranthus cruentus and Chenopodium quinoa seeds and sprouts by HPLC. Acta Chromatographica, 20(4): 661–672.

Pegels, N., I. Gonzalez, T. Garcia, and R. Martin. 2015. Authenticity testing of wheat, barley, rye and oats in food and feed market samples by real-time PCR assays. LWT- Food Science and Technology, 60(2): 867–875.

Perharic, L. 2005. Mass tropane alkaloid poisoning due to buckwheat flour contamination. Clinical Toxicology, 43:413.

Peruchi, L.C., L.C. Nunes, G.G.A. de Carvalho, M.B.B. Guerra, E. de Almeida, I.A. Rufini, D. Santos Jr., and F.J. Krug. 2014. Determination of inorganic nutrients in wheat flour by laserinduced breakdown spectroscopy and energy dispersive X-ray fluorescence spectrometry. Spectrochimcia. Acta Part B: Atomic Spectroscopy, 100: 129–136.

Pitt, J., M.H. Taniwaki, and M. Cole. 2013. Mycotoxin production in major crops as influenced by growing, harvesting, storage and processing, with emphasis on the achievement of Food Safety Objectives. Food Control, 32(1): 205–215.

Psodorov, D., M. Acanski, D.D. Psodorov, D. Vujić, and K.A. Pastor. 2015. Determination of the content of buckwheat and wheat flours in bread using GC-MS and multivariate analysis. Journal of Food and Nutrition Research, 54(2): 179–83.

Pérez-Torrado, E., J. Blesa, J.C. Moltó, and G. Font. 2010. Pressurized liquid extraction followed by liquid chromatography –mass spectrometry for detection of zearalenone in cereal flours. Food Control, 21: 399-402.

Radiana, T.-B., P. Nicolae-Caprian, P. Stela, C. Stelica, C. Rodica, and T.-B. Suzana. 2010. Distribution of some toxic contaminants in the milling products, during the milling process. Romanian Biotechnological Letters, 15(3): 5281-5286.

Raghavender, C.R., B.N. Reddy, and G. Shobharani. 2007. Aflatoxin contamination of pearl millet during filed and storage conditions with reference to the stage of grain maturation and insect damage. Mycotoxin Research, 23(4): 199-209.

Rahmani, A., S. Jinap, and F. Soleimany. 2009. Qualitative and quantitative analysis of mycotoxins. Comprehensive Review in2Food Science and Food Safety, 8: 202–251.

Rammanee, K., and T. Hongpattarakere. 2011. Effects of tropical citrus essential oils on growth, aflatoxin production, and ultrastructure alterations of Aspergillus flavus and Aspergillus parasiticus. Food and Bioprocess Technology, 4(6): 1050–1059.

Rasmussen, P.H., F. Ghorbani, and T. Berg. 2003. Deoxynivalenol and other Fusarium toxins in the wheat and rye flour on the Danish market. Food Additives and Contaminants, 20(4): 396- 404.

Rausch, A.-K., R. Brockmeyer, and T. Schwerdtle. 2021. Development, validation, and application of multi-method for determination of mycotoxins, plant growth regulators, tropane alkaloids, and pesticides in cereals by two-dimensional liquid chromatography tandem mass spectrometry. Analytical and Bioanalytical Chemistry, 413: 3041-3054.

Ravikanth, L., C.B. Singh, D.S. Jayas, and N.D.G. White. 2015. Classification of contaminants from wheat using near-infrared hyperspectral imaging. Biosystems Engineering, 135: 73-86.

Resmini, P. 1968. Un nuovo metodo per identificare e dosare gli sfarinati di grano tenero presenti in quelli di grano duro e nelle paste alimentari. Tecnica Molitoria, 19: 145–167.

Ridgway, C., R. Davies., and J. Chambers. 1998. Imaging for the high-speed detection of pest insects and other contaminnats in cereal grain in transit. In: 2001 ASAE Annual Meeting (p.1). American Society of Agricultural and Biological Engineers.

Rodríguez, S.D., G. Rolandelli, and M.P. Buera. 2019. Detection of quinoa flour adulteration by means of FT-MIR spectroscopy combined with chemometric methods. Food chemistry, 274: 392-401.

Rosell, C.M., F. Barro, C. Sousa, and M.C. Mena. 2014. Cereals for developing gluten-free products and analytical tools for gluten detection. Journal of Cereal Science, 59: 354-364.

Russo, R., E. Cusano, A. Perissi, F. Ferron, V. Severino, A. Parente, and A. Chambery. 2014. Ultra-high performance liquid chromatography tandem mass spectrometry for the detection of durum wheat contamination or adulteration. Journal of Mass Spectrometry, 49: 1239-1246.

S.A. Lakho. 2017. FTIR spectroscopy combined with chemometric: A versatile tool for quality evaluation of fried vermicelli. Ukrainian Food Journal, 6: 61–76.

Saleh, B. 2012. Biochemical and genetic variation of some syrian wheat varieties using NIR, RAPD and AFLPs techniques. Journal of Plant Biology Research, 1(1):1–11. SCCS (Scientific Committee on Consumer Safety). 2011. Clarification on opinion SCCS/1348/10 in the light of the Danish clause of safeguard banning the use of parabens in cosmetic products intended for children under three years of age. Brussels, Belgium: European Commission; http://ec.europa.eu/health/scientific_committees/consumer_safety/docs/sccs_o_069.pdf. Accessed on: 21st November' 2021.

Scott, PM. 2005. Ergot alkaloids—analysis. In: Management of mycotoxins in food and feeds for improving public health IARC (International Agency for Research on Cancer), Lyon, France, publications 158.

Scott, PM. 2009. Ergot alkaloids: extent of human and animal exposure. World Mycotoxin J 2(2): 141–149 Shaikh, T.H., S.A. Mahesar, S.N. Shah, A.H. Kori, S.T.H. Sherazi, and

Shao, Y., G. Xuan, Z. Hu, and X. Gao. 2018. Identification of adulterated cooked millet flour with hyperspectral imaging analysis. IFAC-PapersOnLine, 51-17: 96-101.

Shao, Y., G. Xuan, Z. Hu, and Y. Wang. 2019. Detection of adulterants and authenticity discrimination of coarse grain flours using NIR hyperspectral imaging. Journal of Food Process Engineering, 42(7): e13265.

Silletti, S., L. Morello, F. Gavazzi, S. Gianì, L. Braglia, and D. Breviario. 2019. Untargeted DNA-based methods for the authentication of wheat species and related cereals in food products. Food Chemistry, 271: 410-418.

Sirma, A.J., D.M. Senerwa, D. Grace, K. Makita, N. Mtimet, E.K. Kang'ethe, and J.F. Lindahl. Aflatoxin B1 occurrence in millet, sorghum and maize from four agro-ecological zones in Kenya. African Journal of Food, Agriculture, Nutrition and Development, 16(3): 10991-11003.

Sirma, A.J., E.O. Ouko, G. Murithi, C. Mburugu, I. Mapeany, J.N. Ombui, E.K. Kang'ethe, and H. Korhonen. 2015. Prevalence of Aflatoxin contamination in cereals from Nandi County, Kenya. International Journal of Agricultural Sciences and Veterinary Medicine, 3(3): 55-63.

Sivam, A. S., D. Sun-Waterhouse, C.O. Perera, and G. I. N. Waterhouse. 2012. Exploring the interactions between black currant polyphenols, pectin and wheat biopolymers in model breads; a FTIR and HPLC investigation. Food Chemistry, 131(3): 802-810.

Sivam, A. S., D. Sun-Waterhouse, C.O. Perera, and G. I. N. Waterhouse. 2013. Application of FT-IR and Raman spectroscopy for the study of biopolymers in breads fortified with fibre and polyphenols. Food Research International, 50(2): 574-585.

Skendi, A., M. Papageorgiou, M. Irakli, and D. Katsantonis. 2019. Presence of mycotoxins, heavy metals and nitrate residues in organic commercial cereal-based foods sold in the Greek market. Journal of Consumer Protection and Food Safety, 15(2): 109-119.

Soleimany, F., S. Jinap, A. Faridah, and A. Khatib. 2012b. A UPLC-MS/MS for simultaneous determination of aflatoxins, ochratoxin A, zearalenone, DON, fumonisins, T-2 toxin and HT 2 toxin, in cereals. Food Control, 25: 647-653.

Soleimany, F., S. Jinap and F. Abas. 2012a. Determination of mycotoxins in cereals by liquid chromatography tadem mass spectrometry. Food Chemistry, 130: 1055-1060.

Sonnante, G., C. Montemurro, A. Morgese, W. Sabetta, A. Blanco, and A. Pasqualone. 2009. DNA microsatellite region for a reliable quantification of soft wheat adulteration in durum wheat-based foodstuffs by real-time PCR. Journal of Agricultural and Food Chemistry, 57(21):10199–10204.

Stevenson, A., P.K. McCarthy, and M., Griffin. 1994. Polyclonal antisera against unheated and heated common wheat specific gamma and omega gliadins for the detection of adulteration of durum wheat and durum wheat products with common wheats. Food and Agriculture Immunology, 6(4): 435–442.

Stikic, R., D. Glamoclija, M. Demin, B. Vucelic-Radovic, Z. Jovanovic, D. Milojkovic-Opsenica, S.-E. Jacobsen, and M. Milovanovic. 2012. Agronomical and nutritional evaluation of quinoa seeds (Chenopodium quinoa Willd.) as an ingredient in bread formulations. Journal of Cereal Science, 55(2): 132–138.

Størsrud, S., I.M. Yman, and R.A. Lenner. 2003. Gluten contamination in oat products and products naturally free from gluten. European Food Research and Technology, 217(6): 481-485.

Su, W.H., and D.W. Sun. 2016. Facilitated wavelength selection and model development for rapid determination of the purity of organic spelt (*Triticum spelta* L.) flour using spectral imaging. Talanta, 155: 347–57.

Su, W.H., and D.W. Sun. 2017. Evaluation of spectral imaging for inspection of adulterants in terms of common wheat flour, cassava flour and corn flour in organic Avatar wheat (Triticum spp.) flour. Journal of Food Engineering, 200: 59-69.

Sujka, K., P. Koczoń, A. Ceglińska, M. Reder, and H. Ciemniewska-Żytkiewicz. 2017. The Application of FT-IR spectroscopy for quality control of flours obtained from polish producers. Journal of Analytical Methods in Chemistry: 1–9.

Suzuki, Y., Y. Chikaraishi, N.O. Ogawa, N. Ohkouchi, T. Korenaga. 2008. Geographical origin of polished rice based on multiple element and stable isotope analyses. Food Chemistry, 109: 470-475.

Tahkapaa, S., R. Maijala, H. Korkeala, and M. Nevas. 2015. Patterns of food frauds and adulterations reported in the EU rapid alert system for food and feed and in Finland. Food Control, 47: 175–184.

Tanaka, H., M. Takino, Y. Sugita-Konishi, and T. Tanaka. 2006. Development of a liquid chromatography/time of the flight mass spectrometric method for simultaneous determination of trichotecenes, zearalenone and aflatoxins in foodstruffs. Rapid Communications in Mass Spectrometry, 20: 1422.

Tanaka, T., A. Yoneda, S. Inoue, Y. Sugiura, and Y. Ueno. 2000. Simultaneous determination of trichothecene mycotoxins and zearalenone in cereals by gas chromatography–mass spectrometry. Journal of Chromatography A, 882: 23-28.

Tao, F., L. Liu, C. Kucha, and M. Ngadi. 2021. Rapid and non-destructive detection of cassava flour adulterants in wheat flour using a handheld microNIR spectrometer. Biosystems Engineering, 203: 34-43.

Tautz, D. 1989. Hypervariability of simple sequences as a general source of polymorphic DNA markers. Nucleic Acids Research, 17(16): 6463–6471.

Terzi, V., B. Ferrari, F. Finocchiaro, N. Di Fonzo, A.M. Stanca, C. Lamacchia, J. Napier, P.R. Shewry, and P. Faccioli. 2002. TaqMan PCR for detection of genetically modified durum wheat. Journal of Cereal Science, 37(2): 157-163.

Terzi, V., C. Morcia, A. Gorrini, A. Stanca, P. Shewry, and P. Faccioli. 2005. DNA-based methods for identification and quantification of small grain cereal mixtures and fingerprinting of varieties. Journal of Cereal Science, 41(3): 213–220.

Terzi, V., M. Malnati, M. Barbanera, A.M. Stanca, and P. Faccioli. 2003. Development of analytical systems based on real-time PCR for Triticum species-specific detection and quantitation of bread wheat contamination in semolina and pasta. Journal of Cereal Science, 38:87-94.

Terzi, V., R. Pisani, N. Marziliano, G. Locatelli, and M. Malnati. 1999. Determinazione quantitativa della presenza di organismi geneticamente modificati in farine alimentari mediante real time detection PCR. XLIII Convegno SIGA, 47.

Thielecke, F., and A.P. Nugent. 2018. Contaminants in grain-A major risk for whole grain safety? Nutrients, 10: 1213.

Thompson, T., A.R. Lee, and T. Grace. 2010. Gluten contamination of grains, seeds, and flours in the United States: A pilot study. Journal of the American Dietetic Association, 110: 937-940.

Tichy, H.-V., A. Bruhs, and A. Palisch. 2020. Development of real-time polymerase chain reaction systems for the detection of so-called "superfoods" chia and quinoa in commercial food products. Journal of Agricultural and Food Chemistry, 68(49):14334–14342.

Tortoe, C., H. Ofori, P.T. Akonor, and H. Oduro-Obeng. 2018. Trace metal concentrations in three pastry products prepared from root and tuber and cereal crops composite flours. Congent Chemistry, 4: 1429157.

Ueno, Y., K. Umemori, E.C. Niimi, S.I. Tanuma, S.I. Nagata, M. Sugamata, T. Ihara, M. Sekijima, K.I. Kawai, I. Ueno, F. Tashiro. 1995. Introduction of apoptosis by T-2 toxin and other natural toxins in HL-60 human promyelotic leukemia cells. Natural Toxins, 3(3): 129-137.

US Department of Agriculture. 2009. Grain inspection, packers & stockyard administration. Grain, rice & pulses. Official US Standards. US Standards for Grains. US Department of Agriculture. https://www.ams.usda.gov/about-ams/programs-offices/federal-grain-inspection service. Accessed on: 20th November' 2021.

US Environmental Protection Agency (USEPA). 1993. Bisphenol A. (CASRN 80-05-7), integrated risk information system (IRIS). Washington, DC: USEPA.

Vaishnavi, A.S., G. Asvini, and S. Bhavani Ramya. 2015. Heavy metal contaminants in cereals used for the health mix preparation. Shanlax International Journal of Arts, Science and Humanities, 2(4): 140-143.

Van Raamsdonik, L. W. D., J. Vancutsem, and J.S. Jorgensen. 2009. A survey on the presence of undesirable botanical substances in feed in the European Union. Base, 13: 33–38.

Vasilachi, I.C., D.-M. Asiminicesei, M. Minut, M. Rosca, P. Cozma, C. Smaranda, and M. Gavrilescu. 2020. Human risks generated by the bioaccumulation of pollutants in cereals and their consumption. The 8th IEEE International Conference on E-Health and Bioengineering-EHB 2020.

Vendl, O., C. Crews, S. MacDonald, R. Krska, and F. Berthiller. 2010. Occurance of free and conjugated Fusarium mycotoxins in cereal-based food. Food additives and contaminants, 27(8): 1148-1152.

Verdú, S., F. Vásquez, R. Grau, E. Ivorra, A.J. Sánchez, and J.M. Barat. 2016. Detection of adulterations with different grains in wheat products based on the hyperspectral image technique: the specific cases of flour and bread. Food Control, 62: 373–380.

Vermeulen, P., M. Suman, J.A. Fernández Pierna, and V. Baeten. 2018. Discrimination between durum and common wheat kernels using near infrared hyperspectral imaging. Journal of Cereal Science, 84: 74–82.

Vermeulen, Ph., J.A. Fernández Pierna, H.P. van Egmond, J. Zegers, P. Dardenne, and V. Baeten. 2013. Validation and transferability study of a method based on near-infrared hyperspectral imaging for the detection and quantification of ergot bodies in cereals. Analytical and Bioanalytical Chemistry, 405(24): 7765-7772.

Vismer, H. F., G.S. Shephard, L. Van der Westhuizen, P. Mngqawa, V. Bushula-Njah, and J.F. Leslie. 2019. Mycotoxins produced by Fusarium proliferatum and F. pseudonygamai on maize, sorghum and pearl millet grains in vitro. International Journal of Food Microbiology, 296: 31– 36.

von Büren, M., M. Stadler, and J. Lüthy. 2001. Detection of wheat adulteration of spelt flour and products by PCR. European Food Research and Technology, 212: 234–239.

Vrček, I.V., D.V. Čepo, D. Rašić, M. Peraica, I. Žuntar, M. Bojić, M., G. Mendaš, and M. Medić-Šarić. 2014. A comparison of the nutritional value and food safety of organically and conventionally produced wheat flours. Food Chemistry, 143: 522-529.

Wagner, M., M.H. Morel, J. Bonicel, and B. Cuq. 2011. Mechanisms of heat-mediated aggregation of wheat gluten protein upon pasta processing. Journal of Agricultural and Food Chemistry, 59: 3146-3154.

Walker, R. 2002. Risk assessment of Ochratoxin: Current views of the European Scientific Committee on Food, The JECFA, and The CODEX Committee on food additives and contaminants. In: Mycotoxins and Food Safety, Eds. Trucksess *et al.*, Kluwer Academic/ Plenum Publishers, U.K.

Wilson, J.P., Z. Jurjevic, W.W. Hanna, D.M. Wilson, T.L. Potter, and A.E. Coy. 2006. Hostspecific variation in infection by toxigenic fungi and contamination by mycotoxins in pearl millet and corn. Mycopathologia, 161: 101-107.

Wu, X. and J.S. Smith. 2007. A Gas Chromatography– Flame Ionization Detection Method for Detection of Fusaproliferin in Corn. Journal of agricultural and food chemistry, 55(8): 3211-3216.

Xue, S.-S., J. Tan, J.-Y. Xie, and M.-F. Li. 2021. Rapid, simultaneous and non-destructive determination of maize flour and soybean flour adulterated in quinoa flour by front-face synchronous fluorescence spectroscopy. Food Control, 130: 108329.

Yapo-Crezoit, C.C.A., S.E. Kpan, A. Boni, I. Bonouman, J. Kalpy, J.D. N'guessan, A.J. Djaman, and M. Dosso. 2019. Evaluation of the allergenic potential of Anagobaka infant flours marketed in the markets of the district of Abidjan. Advances in Research, 19(2): 1-13.

Yaseen, T., D.W. Sun, and J.H. Cheng. 2017. Raman imaging for food quality and safety evaluation: Fundamentals and applications. Trends in Food Science & Technology, 62: 177-189.

Yu, Z., L. Zhang, D. Wu, and F. Liu. 2005. Anti-apoptotic action of zearalenone in MCF-7 cells. Ecotoxicology and Environmental Safety, 62: 441.

Zandomeneghi, M. 1999. Fluorescence of cereal flours. Journal of Agriculture and Food Chemistry, 47(3): 878–882.

Zeković, I., L. Lenhardt, T. Dramićanin, and M.D. Dramićanin. 2012. Classification of intact cereal flours by front-face synchronous fluorescence spectroscopy. Food Analytical Methods, 5(5): 1205–1213.

Ziegler, J.U., C.B. Steingass, C.F.H. Longin, T. W€urschum, R. Carle, R.M. Schweiggert. 2015. Alkylresorcinol composition allows the differentiation of Triticum spp. having different degrees of ploidy. Journal of Cereal Science, 65: 244–251.

Ziegler, J.U., M. Leitenberger, C.F.H. Longin, T. Wüurschum, R. Carle, and R.M. Schweiggert. 2016. Near-infrared reflectance spectroscopy for the rapid discrimination of kernels and flours of different wheat species. Journal of Food Composition and Analysis, 51:30–36.

Zinedine, A., J.M. Soriano, J.C. Moltó, and J. Mañes. 2007. Review on the toxicity, occurrence, detoxification, regulations and intake of zearalenone: An oestrogenic mycotoxin. Food Chemical Toxicology, 45, 1.

9

Emerging Trends in Detection of Adulteration in Oils and Fats An Overview

Rosy Bansal[1], Mukul Sain[2], Poornima[2] and Ajay Singh[3]

[1]*GSSDGS Khalsa College Patiala, Punjab, India*
[2]*ICAR-National Dairy Research Institute, Karnal, Haryana, India*
[3]*Mata Gujri College, Fatehgarh Sahib, Punjab, India*

Introduction

All the worldwide cuisines generally rely on plant-based edible oils as the main cooking ingredient due to their unique chemical composition, further presence of unsaturated fatty acids in plant-based oils makes it a healthy complement in the diet. Lipids are known to provide almost double calories than carbohydrates. For ages, vegetable oils have contributed maximally to our nutrition and economics, but their authenticity is still a considerable issue (Firestone and Reina, 1996) as for more profitability fraudulent blends lower grade oils to high-grade expensive oils. Adulteration of oils and fats poses serious health problems when toxic oils are being used for monetary benefits like admixing of argemone oil in mustard oil or non-edible rapeseed oil with edible rapeseed oil. Some of the adulterations also lead to fatal health effects which are a real cause of concern, so innovative techniques are in demand to spot the adulterants and the way old physical methods like melting point, refractive index, saponification number and iodine values are masking the realistic approach of adulteration (Azadmard *et al.*, 2015). Each oil has some specific major (triacylglycerols) or minor (carotenoids, sterols, tocopherols etc.) components that can be used as a tool for detection of adulteration e.g. presence of phytosterols in butter indicates the adulteration with vegetable oil fats. Food serves as fuel to our body and fats provide the extra mileage if compared to other available fuels of the body. The 5-gm oil gives 40 kcal of energy while the same quantity of sugar gives only 15 kcal. So, fat acts as a reserved energy source in the body that can be used at the time of starvation. Hence it is stipulated to come up with the most reliable methods to ascertain consumers regarding the purity of oils and fat.

1.1 Adulteration Practices of Oils and Fats

The adulteration is basically practised for economic benefits and to generate good revenue for the manufacturer. The main adulterations practised for oils and fats are of two types:

i) Adulteration of refined oil with cold press oil

ii) Admixture of cheap and low-grade oil with an expensive one

Mostly adulteration of fats and oils is done purposely for profits but it can be inadvertent also (Shukla *et al.*, 2005). The demand for specific oils due to their nutritional profile has alleviated and has captured a lot more attention. These oils are widely used in food formulation and diet in the form of salad dressings. These specific oils are expensive and some of these include olive oil, cocoa butter. Despite these oils, cold-pressed oils are also catching the market as these don't involve any sort of chemical refining. In fact, some nutritional components remain safe in cold press oil. Refining of oils results in the formation of trans fats and steradienes while these harmful components are not formed during cold press extraction of oil. The mixing with low grade oils like mineral oils, argemone or castor oil can have a negative impact on health, cause stomach, liver or heart problems, disruption in vision or even eyesight loss or in advanced cases can cause cancer (Sharma *et al.*, 2017). Research studies indicate that argemone oil admixture with mustard oil has negative repercussions on the human body including prevalent oedema, glaucoma and vision loss, while castor oil is accountable for gastritis, dysentery, ulcers and throat problems (Yadav, 2018). Recently Canola oil has been proved of many health benefits due to less saturated fat and adequate amount of ω-6 and ω -3 polyunsaturated fatty acids, thus is more liable to adulteration with cheaper oils resulting in a matter of concern and the health freaks (Anindita and Jain, 2018).

Recycled cooking oil is also a matter of concern due to adulterants in cooking oil and is not fit for human consumption on prolonged usage. The recycled cooking oil is reprocessed, bleach refined and deodorized to obtain adequate quality. This reprocessing is meant for non-edible and industrial purposes but unfortunately is used as an adulterant in cooking oil. Research studies indicate that fifty thousand reused cooking oil is produced in Malaysia per year and it has created land and water pollution due to inappropriate dumping. To prevent this pollution, oil is traded by vendors for industrial production of biodiesel, cattle feed, fertilizers candles and soaps. This reused oil is not allowed for human consumption but some corrupt dealers mix this oil with fresh oil and misled the consumers (Kou Y*et al.*, 2018). The adulteration of reused cooking oil in the range of more than one per cent was qualitatively analysed. The FTIR along with chemometric analysis is found to be an effective tool for detecting edible oil adulteration. Reused cooking oil can also be treated with magnesium trisilicate (a food additive) used as an absorbent to clarify the oil and filter the unwanted fat residues and other impurities (Lim *et al.*, 2018).

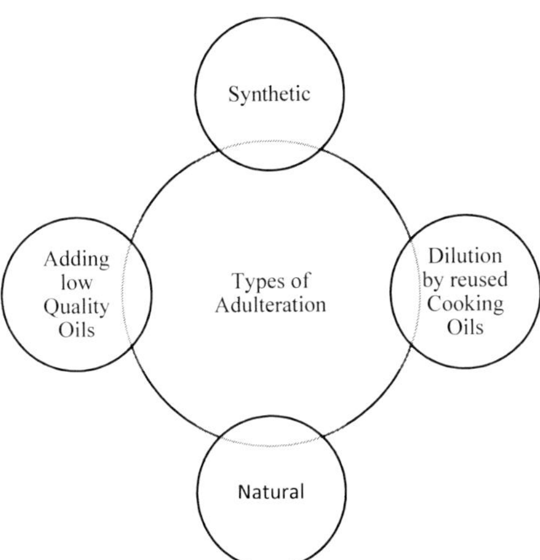

Fig. 1: Types of Fats and Oil Adulteration

Profitability Concerns

The expensive oils when replaced with cheaper ones provide access to profits from an economic standpoint. The oils high in cost are more likely to be trapped for adulteration. Olive oil (virgin) is more prone to adulteration as it is not necessary to combat all properties but can mask the quality to customers and offer economic gains to the manufacturer. This implies that low-grade oil is being sold at a higher price. Most oils are subjected to adulteration either for profits or for more consumer demands and inadequate availability (Vanstone *et al.*, 2018). The consumers are misled as they have the right to get natural, pure substances but adulteration leads to dilution of the plant-based oil. In general, high-income people are more aware of adulterated oils as compared to low-income. The research studies indicate that the difference in awareness level is due to the fact that high-income group people are more educated than the low-income group which substantially influences consumer awareness. Although the extent of the addition of low-grade oil does not vary in packaged or unpackaged oils like mustard and soyabean oil (Ayza and Belete, 2015).

The milk fat is more susceptible to being adulterated with cheaper vegetable or hydrogenated fats and mineral oils as milk is the most valued and costly food product across the world. The adulteration of milk fat has occupied a stained image of the dairy sector both in the national and international marketplace. Milk fat is not only a consumable item but it has religious importance in countries like India and so consumers are not only robbed of money but are potentially distressing their religious sentiments (Wheatley and Spink, 2013).

Fats and oils are now endorsed for the presence of ω-3 and ω-6 fatty acids (PUFA: Polyunsaturated fatty acids) wherein, ω-3, the first double bond is present on third carbon from the methyl end of the chain whereas it is on the sixth carbon in ω-6 fatty acid. Both fatty acids are beneficial from a health viewpoint but studies illustrate that excess of ω-6 FA is not required. Further, it has been shown that the replacement of saturated fats with ω-6 FA can be a greater risk factor for heart patients and can be lethal. The oils like sunflower, safflower and corn have the maximum amount of ω-6 FA (Fig. 2).

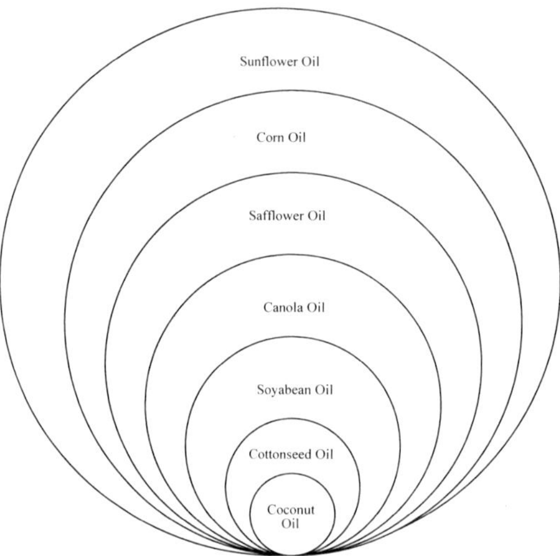

Fig. 2: ω 6 FA content present grams per tablespoon in different oils

Table 1: ω -6 and ω -3 Fatty Acids, Saturated and Monounsaturated Fatty Acid in grams per tablespoon in various Plant and Animal oils

S.No.	Source of Dietary lipid	ω -6 Fatty Acids	ω -3 Fatty Acids	Saturated Fatty Acid	Monounsaturated Fatty Acid
1.	Safflower Oil	2	-	0.8	10.2
2.	Canola Oil	2.8	1.3	1.0	8.2
3.	Flaxseed Oil	2.2	8.0	1.3	2.5
4.	Sunflower Oil	8.9	-	1.4	2.7
5.	Corn Oil	7.9	-	1.7	3.3
6.	Olive Oil	1.1		1.8	10
7.	Sesame Oil	5.6	-	1.9	5.4
8.	Soyabean Oil	6.9	0.9	2.0	3.2
9.	Peanut Oil	4.3	-	2.3	6.2
10.	Salmon Fat	-	4.8	2.7	3.8
11.	Cream Cheese	-	-	1.4	3.2
12.	Cottonseed Oil	7.0	-	3.5	2.4
13.	Chicken Fat	2.5	-	3.8	5.7
14.	Lard (Pork Fat)	1.3	-	5.0	5.8
15.	Beef Tallow	-	-	6.4	5.4
16.	Butter	-	-	6.4	5.4
17.	Cocoa butter	-	-	8.1	4.5
18.	Palm Kernel Oil	-	-	11.1	1.6
19.	Coconut Oil	-	-	11.8	0.8

Vegetable oils, on the subject to very high temperatures in the presence of a catalyst, get hydrogenated and converted into saturated fats, thus resulting in solidification of oil making it more stable towards heat. Partial hydrogenation causes the twisting of fatty acid carbon chains at double bonds leading to the formation of trans fats and so the liquid form of oil is transformed into a spreadable consistency. Due to resistance to rancidity, trans fats are mostly used for baking and deep-frying purposes thus enhancing the shelf life. The adulteration of edible oils is reported throughout the world. In India, coconut oil is most vulnerable to adulteration as it is produced from dry kernel of coconut and is cold-pressed oil. The manufacturing process is quite expensive and required a lot of labour, that is why coconut oil is mixed with cheap oils like palmolein. The purity of the oil is checked by the refractive index (RI) of oil which can be altered easily (Schonfeld and Wojtczak, 2016). The RI of oils upsurges with the surge in unsaturation and fatty acids chain length.

FSSAI Initiatives: An Indian Scenario

The exposure of fraudulent involved in ruining the quality of fats and oils by using cheaper versions for more profitability has become mandatory in recent times, thus FSSAI (Food Safety and Standards Authority India) has taken great initiatives for ensuring safe and quality produce distribution under some set of regulations for grading of fats and oils and its marketing. Under this set of rules and regulations, whosoever violates the legal limits will be accountable to punishment and penalty. The FSSAI personnel performs regular surveys, investigation, monitoring, random sampling of fats and oils available for sale in the market, procuring from manufacturers or through FBO's (Food Business operators) so as to ensure a healthy wholesome supply of the same stuffs kept in mind the betterment of consumer's health. The blending of two or more vegetable oils should be certified by Agmark as directed by FSSAI. The packaging of oils and fats should be carried out keeping in mind minimization of leaching of packaging constituents and cross contamination as well. The materials used for packaging fats and oils should also follow the BIS standards as declared by FSSAI regulations. The FSSAI provides an approved seal on pure oils as per Food Safety and Standards (Prohibition and Restrictions on Sales Regulations 2011) to aware consumers before buying. FSSAI has also formulated regulations that direct manufacturers not to sell or distribute, offer sales discounts, dispatch or deliver edible oils in loose form to anyone. Despite all these mandates sellers continue to sell the edible oils without any seal on packets and they are left unnoticed as consumers are lacking in awareness of the consequences of consumption of adulterated fats and oils.

Indian legislations, has been working on certification of unadulterated products and has set quality parameters for various features of ghee in different states and regions of the country. The lawbreakers will be legally liable to pay a heavy penalty. The establishment of purity and quality of milk and milk products is a complex process and milk fat adulteration is a serious threat to the consumer. The detection of milk fat adulteration has become a necessity and is a possible dimension especially in the domain of competition across the globe. Season, breed, species, stage of lactation etc. significantly affects the fatty acid configuration in milk and widen the complete range of fatty acids thus complicating the adulteration of milk. The fats of milk are unique and are different from other fats due to the presence of short-chain fatty acids while vegetable fats contain linoleic acids and animal fats are rich in oleic acids. Detection of adulteration in milk fats for the presence of soyabean oil can be done using Gas-Liquid Chromatography (Kumar *et al.*, 2015). In India, mixing of oil has been permitted by FSSAI under license as unauthorised and unempirical mixing can result in varied smoke points in oil mixtures thus causing health hazards.

Olive oil is most vulnerable to adulteration these days as it is consumed without an extensive refining process. Due to lower smoke points, olive oil is not used for cooking purposes but its remarkable sensory properties further increase its cost. Olive oil is adulterated deliberately in two ways: by combining high-grade olive oil with low-grade olive oil or by mixing it with other vegetable oils. So, standards are being set for refined and olive pomace oils by the International Olive Oil Council (IOOC) in 1993 and Codex Alimentarius Commission in 1993 to avoid adulteration.

Chemistry of Adulteration and Possible Adulterant Constituents

Fats and Oils can be obtained from both animals and plants sources. The basic nature of fats and oils is oleaginous and liberates long chains of fatty acids and glycerol upon hydrolysis in aqueous or alcoholic NaOH or KOH (Sodium or Potassium hydroxide). Oil is a combination of many fatty constituents, those available in milk and gums are liquid at room temperature shows sparingly solubility in organic solvents like alcohol. The Quality attributes of many processed foods rely on fats and oils.

The expensive oils are replaced by low-grade oils for monetary benefits and mostly oils with higher prices are having more potential for adulteration. like Olive oil obtained from fruits of the olive tree by controlled thermal conditions so as not to alter its nature and no chemical treatment is provided further. The virgin oil thus obtained is prone to be mixed with low price oils having similar fatty acid and sterol profiles (Bowadt and Aparicio, 2003). Composition variation among vegetable oils in terms of their fatty acid constituents can easily be detected at levels (< 5%) viz; peanut, canola, soyabean, mustard and rapeseed. In some cases, this quantification is little hard to detect owe to almost similar makeup as seen in olive oil if adulterated with hazelnut, perhaps difficult to detect (< 20%) (Christopoulou *et al.*, 2004). Adulteration of vegetable oil can be easily detected with a specific phytosterol group containing 4, 4 dimethyl sterols (triterpene alcohols, two methyl groups) as this group differs in vegetable oils as compared to the other two groups including 4-desmethyl sterols (without methyl group), 4-monomethyl sterols (one methyl group) (Azadmard and Torbati, 2015). Indian Ghee can be detected for the presence of goat body fat using Differential Thermal Analysis (DTA). There are many traditional tests that are being used to detect oil adulteration and are still used for the characterization of oils. Although many more sophisticated methods have been developed recently these are unaffordable to middle-class consumers.

For the same detection some qualitative tests for fats and oils are as follows;

i) Solubility test

ii) Translucent spot test
iii) Acrolein test
iv) Huble's test

i) Solubility Test

Organic solvents like chloroform, alcohol, etc. help in the solubilisation of fats. The formation of an oily layer by test sample over the surface of water indicates the presence of fat. Further adulteration with fat can be confirmed by its partial and fully solubility in alcohol and chloroform respectively. So miscibility of sample in chloroform and immiscibility in water confirms the presence of fats.

ii) Translucent Spot Test

To check the presence of fat, the sample can be tested by rubbing it between the folds of filter paper and observed for the appearance of a translucent spot.

1. Take the sample to be tested, press a little in the folds of the filter paper
2. The appearance of a greasy spot upon folding indicates the presence of oils or fats
3. Spot grows larger on heating and drying the filter paper

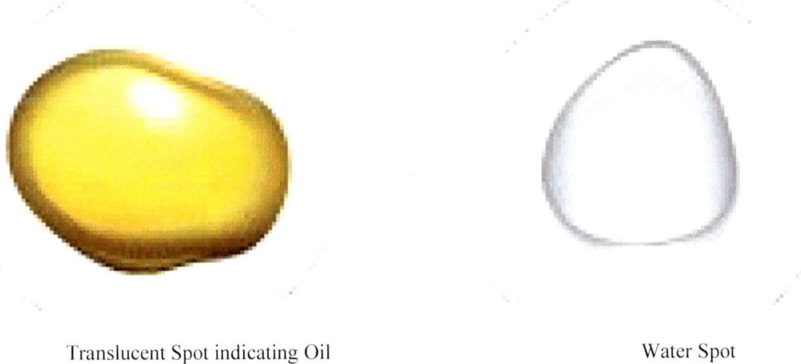

Translucent Spot indicating Oil　　　Water Spot

Fig. 3: Translucent test

iii) Acrolein Test

In this test, fats and oils are heated with potassium bisulphite ($KHSO_4$) crystals. The presence of fats or oils can be indicated by the formation of the pungent irritating aroma of acrolein.

$$\text{Oil/Fat} \xrightarrow{\text{Heating}} \text{Glycerol + fatty Acid}$$

$$\text{Glycerol} \xrightarrow{KHSO_4} \text{Acrolein + Water}$$

iv) Huble's Test

Comparison of different oils can be carried out on the basis of their degree of unsaturation which can be detected by Huble's test. The reaction of oils with Huble's reagent fads the violet colour of iodine indicates the presence of unsaturated fatty acids while if the colour of the iodine do not change then the given fat or oil have saturated fatty acids.

Huble's reagent – Equal volumes of 7% mercury chloride in alcohol with 5% iodine solution in 96% alcohol.

The commonly Used Test for the Identification of Fats and Oils Are

i) Halphen's test (Used to detect 0.1% of crude cottonseed oil)

Principle: if heated oil detected with intense red color within carbon disulphide solution, shows cottonseed oil in them. This test is also the basis for quality check in Hempseed oil, Kapokseed oil/oils as they are also provided with cyclopropenoid fatty constituents (sterculic and malvalic acid). Hydrogenation and deodorization obliterate the chromogens either wholly or partially and react with diminished intensity. The heating of oil at 250°C or above cannot result in a positive reaction. But on other hand, if animal fat (milk fat) fed on cottonseed meal (butter, lard) gives faint signs under the same test.

Reactions: Crude cottonseed oil contains malvalic and sterculic acid (XIVa & XIVb, respectively) cyclopropenic acids which are responsible for easy detection of adulteration. These compounds react with sulphur solution forming red coloured compounds (XVa & XVb, respectively) (Karleskind and WOLFF, 1996; Sukhla *et al.*, 2005)

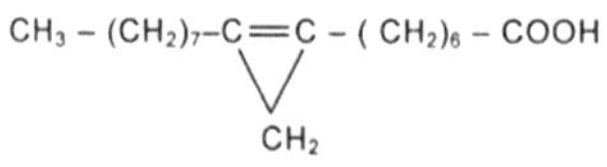

$CH_3-(CH_2)_7-C{=}C-(CH_2)_6-COOH$ (ring: CH_2)

(XIVa)

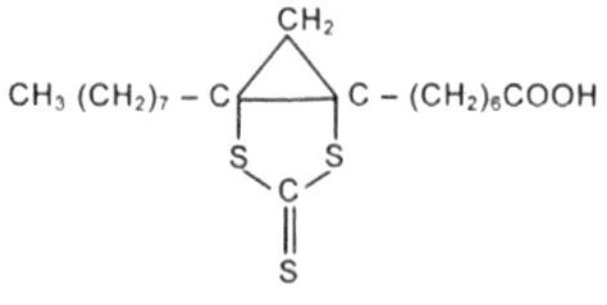

$CH_3-(CH_2)_7-C=C-(CH_2)_7-COOH$ (ring: CH_2)

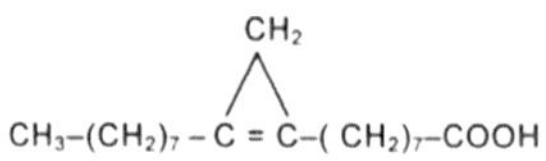

(XIVb)

$CH_3(CH_2)_7-C$ (CH_2) $C-(CH_2)_6COOH$; S, S, C, S

(XVa)

$CH_3(CH_2)_7-C$ (CH_2) $C-(CH_2)_6COOH$; HS, S, C=S, CH; $CH_3(CH_2)_7-C$ — $C-(CH_2)_6COOH$; S

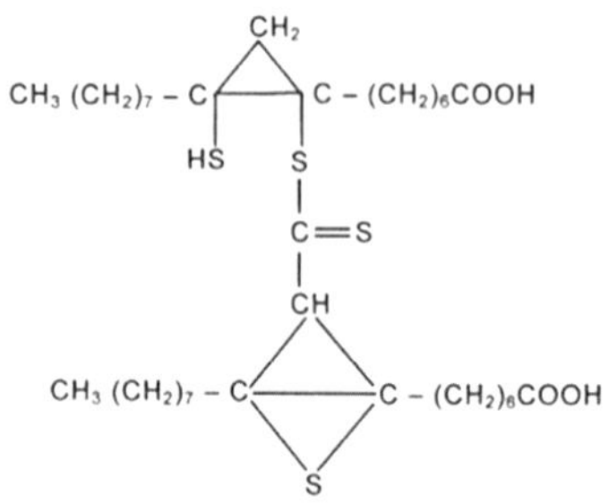

(XVb)

Apparatus

(a) Test tubes

(b) Water bath

(c) Oil bath or Brine bath (110°C– 115°C)

Reagent

1) Sulphur solution: Prepare 1% (w/v) solution of sulphur in carbon disulphide and then mix with an equal volume of amyl alcohol.

Procedure: To 5 ml of oil or melted fat, mix the contents thoroughly with carbon disulphide by shaking, ahead by heating (70 to 80 °C) with intermittent shaking in water bath for few minutes until sample stops foaming. Now place it in an oil bath and keep it for 2:30 h. development of intense red colour at the end of incubation period indicates adulteration with cottonseed oil. Sensitivity extent of given test (0.5%) is advantageous to detect cottonseed oil in other oils (I.S.I. Handbook of Food Analysis, pg 86).

ii) Baudouin test (Used to detect 1% sesame oil)

Principle: Sesame oil detection through development of pink color in furfural reagent in presence of hydrochloric acid (HCl) confirms its presence. This color appearance is because of sesamolin interaction with HCl, hence detected if added in other edible oils.

Reactions: Sesame oil contains two important antioxidants i.e. sesamol (Xc) and sesaminol (Xd) formed from sesamine (Xa) and sesamoline (Xb), respectively due to processing like refining. Confirmatory tests owe to pink color development among acid layer due to formation of XI as in presence of HCl, sesamol reacts with furfural, forming an addition compound (XI) as shown in reaction below (IS:548 (Part- II), (1976).

Apparatus

(a) Stoppered measuring cylinders

Reagents

1) Concentrated HCl (Sp. Gr. 1.19)
2) Furfural solution (2% furfural freshly distilled in ethyl alcohol)

Procedure: given oil (5 ml) added with 5 ml conc. HCl and 0.4 ml furfural solution. Shake it using glass stopper vigorously for about 2 minutes. Allow to stand for layers to separate. Lower layer detected with pink or red colour, indicate presence of sesame oil. Further, confirm the test by addition of 5 ml water and shake it again. If still colour persists, sesame oil is present. If colour disappears then it is absent. (As furfural gives violet tint with HCl, it is necessary to use the dilute solution specified).

Note: Test the sample for the presence of colouring substances that are cromogenic in presence of HCl. For this, add 5 ml of the sample in a 25ml glass stopper measuring cylinder and shake with an equal volume of conc. HCl. If there is no formation of pink or red colour in the aqueous layer apply the test as specified above. If the development of pink or red colour occurs in the aqueous layer, take out the lower red acid layer formed at the bottom and repeat the process until no further coloration occurs. After complete removal of the HCl layer perform the test as described above (I.S.I. Handbook of Food Analysis, Page 86 / A.O.A.C).

iii) Fitelson test (Used to indicate the presence of tea seed oil)

Principle: The test relies on the formation of red colour upon reaction with acetic anhydride in the presence of sulphuric acid when the oil in Chloroform is taken

Apparatus

a) Test tubes - 150mm x 15 mm

b) Pipette – 2 ml, graduated to 0.1 ml

c) Dropper (calibrated that 7 drops of oil weigh 0.22 gm)

d) Water bath (at 5 °C)

Reagents

1) Chloroform

2) Sulphuric acid (Conc.)

3) Acetic anhydride

4) Diethyl ether (anhydrous peroxide free, stored over Sodium)

Procedure

Pipette out 0.8 ml of acetic anhydride, 1.5 ml Chloroform and 0.2 ml Sulphuric acid in a test tube. Allow it to cool to 5 °C and subsequently add about 0.22gm (7 drops) of oil. If turbidity appears in the solution, add acetic anhydride dropwise with continuous shaking until it becomes clear. Keep the solution at 5 °C for 5 minutes. Add 10 ml Diethyl ether earlier chilled to 5°C. Insert stopper to test tube followed by mixing by inverting it twice. Keep the test tube in a water bath at 5 °C. An intense red colour formed about a minute after adding ether reaches the highest and fades away, showing the pure tea seed oil. A less intense colour indicates the presence of tea seed oil but care must be taken while interpretation of results in the presence of olive oil. In general, this test is applicable but some olive oils yield pink colour, therefore test is not reliable to detect less than 15 % of tea seed oil in olive oil (F.A.O Manuals of Food Quality Control 14 / 8, page 273 / A.O.A.C.),

iv) Evers's test (Used to detect 5-10% of groundnut oil)

v) Azo Dye Test (used to detect 2.5 % of rice bran oil)

Principle: The content obtained after the reaction between oil sample, sodium hydroxide and aniline which further reacts with benzene diazonium chloride solution in a mixture of aniline, dilute hydrochloric acid and sodium nitrate solution results in the formation of orange-red colour which indicates the positive result. In alkaline medium, when sodium nitrate solution and dilute HCl is present, diazotization of aniline occurs i.e. coupling reaction between VII (g-oryzanols of rice bran oil) and VIII compound resulting in the formation of phenyl azo-g-oryzanol or phenyl- azoferulic acid (IX) appearing as the orange-red colour within 10-20 seconds (Shukla *et al.*, 2004).

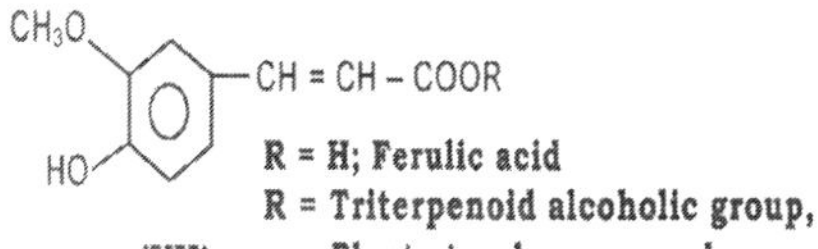

CH_3O, HO, $-CH=CH-COOR$

R = H; Ferulic acid
R = Triterpenoid alcoholic group, Plant sterol; γ-oryzanol

(VII)

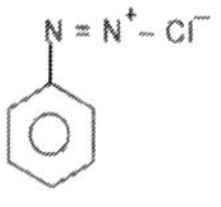

$N\equiv N^+-Cl^-$

(VIII) Bezenediazoniumchloride

CH_3O, HO, $CH=CH-COOR$, $N=N$

R = Triterpenoid alcoholic group, plant sterol; 5-phenylazo-γ-oryzanol
R = H; 5-phenylazoferulic acid

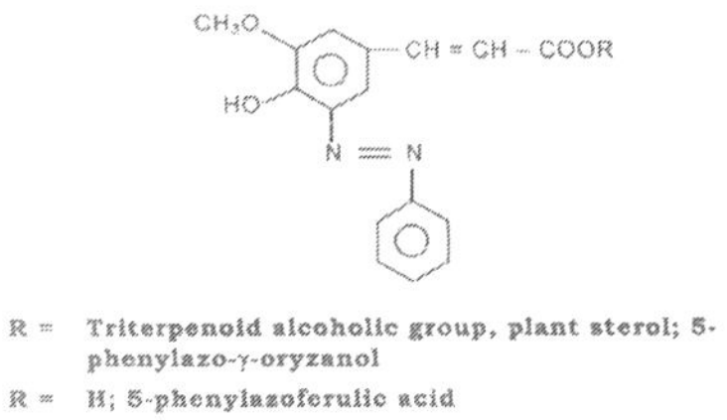

(IX)

Apparatus

a) Test tubes
b) Beakers
c) Ice water-bath

Reagents

1) Aniline
2) Sodium hydroxide (10%, w/v)
3) Dilute hydrochloric acid (dil. HCl)
4) Sodium nitrate solution (5 5, w/v)

Procedure: Take oil sample in a dry test tube and add 2-4 ml of sodium hydroxide and shake the content to form emulsion solution (a) for about 5-10 min. Take another dry test tube and add 1-2 drops of aniline and dissolve it in dil. HCl. Then, cool the content to 0-5°C and then add 2-3ml of sodium nitrate solution and shake vigorously. This results in the formation of benzene diazonium chloride solution (b). Now, mix the two solutions, a & b, shake content for a few seconds, an orange-red colour is developed within 10-20 sec indicating a positive result for the presence of rice bran oil in other oils.

vi) **Hexabromide Test** (used for the presence of 1 % Linseed Oil)

Principle: Development of hexabromide precipitates when a solution of oil in chloroform is reacted with bromine and then with alcohol and ether under chilled conditions shows the presence of linseed oil.

Reactions: Linseed oil contains linolenic acid (XII), when treated with bromine in chloroform, ether, alcohol it forms precipitates of hexabromide and further forms 9, 10, 12, 13, 15, 16- hexabromo octadecanoic acid (XIII) as red coloured precipitate as shown below. This test can also be used for oils and fats containing highly unsaturated fatty acids like fish oil, soyabean oil etc (IS:548, 1976, Dey *et al.*, 1956 and Felix, 1933).

$$CH_3\text{-}CH_2\text{-}CH{=}CH\text{-}CH_2\text{-}CH{=}CH\text{-}CH_2\text{-}CH{=}CH\text{-}(CH_2)_7\text{-}COOH$$

(XII)

$$CH_3\text{-}CH_2\text{-}\underset{Br}{CH}\text{-}\underset{Br}{CH}\text{-}CH_2\text{-}\underset{Br}{CH}\text{-}\underset{Br}{CH}\text{-}CH_2\text{-}\underset{Br}{CH}\text{-}\underset{Br}{CH}\text{-}(CH_2)_7\text{-}COOH$$

(XIII)

Apparatus

a) Boiling tubes (wide-mouth, 100 ml capacity)

b) Ice water-bath

Reagents

1) Chloroform
2) Liquid bromine
3) Ethyl alcohol
4) Diethyl ether

Procedure

Pipette out 1 ml of oil into a boiling tube following the addition of 5 ml chloroform and about 1ml bromine drop-wise till the solution becomes deep red. Cool the test tube in an ice water bath. Add about 1.5 ml of rectified spirit drop-wise while shaking the mixture until the precipitates that were first formed just dissolve and then add 10 ml of diethyl ether. Mix the contents and keep the tube in an ice water bath for 20 minutes. The appearance of precipitates indicates the presence of linseed oil.

Note

1. This test is not pertinent for the detection of linseed oil in mahua oil.
2. The use of safe and suitable pipettes i.e. Lunge-Ray pipette is recommended while handling and addition of bromine.
3. The test is also given by fish oils and fats containing highly unsaturated fatty acids. It has been observed that low erucic rapeseed oil and mahua oil having linolenic acid content higher than 12.0 % may also give a positive result. The results obtained in such cases have to be viewed with caution. Experiments conducted with these oils with or without added linseed oil have shown that, if linseed oil is present even at 1% level, hexabromide insoluble in the cold ether are formed within 20 min. Any hexabromides insoluble in cold ether formed after 20 min need not be taken for the presence of linseed oil. An explanation for

this behaviour of these oils can be given on the basis of glyceride structure. Oils and fats are mixed triacylglycerol i.e different fatty acids are present in each of the positions of the glycerol molecule. The exception to this rule is the oil /fat containing a particular fatty acid in amounts greater than 50 % where such a fatty acid may take all the three positions of the glycerol molecule giving rise to simple tri acyl glycerol. Linseed oil is such an example containing greater than 50 % Linolenic acid (Manual of Methods of Analysis for Adulterants and Contaminants in Foods, I.C.M.R (1990).

vii) **Solvent Partition method (used for detection of 2% Palmolein (palm oil))**

Principle: It is based on the observation of the treated sample for the presence or absence of greenish-yellow coloured fluorescence under UV-light where during solvent extraction contains pigments like lycopene, lycophyll, & licopinal (XVIIIa, XVIIIb, and XVIIIc, respectively) which are detectable (Moyoghare *et al.*, 1985).

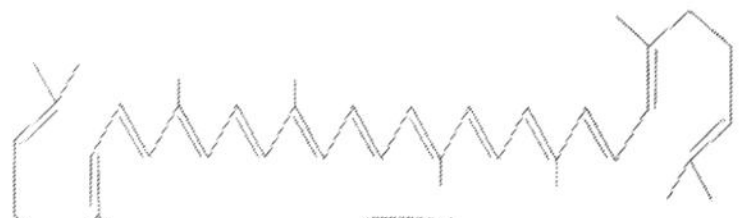

(XVIIIa)

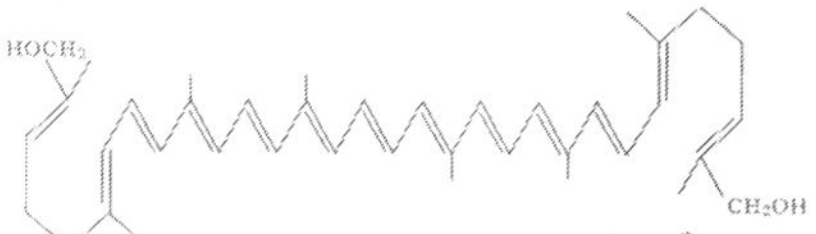

HOCH₂ CH₂OH

(XVIIIb)

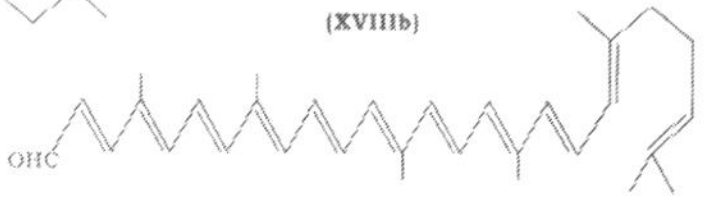

OHC

(XVIIIc)

Apparatus

a) Beakers
b) Separating funnel
c) Porcelain dish
d) Water bath
e) Test tubes
f) Ultraviolet (UV) light

Reagents

1) Hexane
2) Anhydrous sodium sulphate
3) Dimethylformamide (DMF)

Procedure: Take 5 ml oil sample in a test tube and add an equal volume of hexane. Pass the solution through anhydrous sodium sulphate while transferring it into a separating funnel. Add 3 ml DMF and shake the content for a minute and allow it to stand for a few minutes to form two separate layers. Discarded lower DMF layer. Repeat the washing if the DMF layer is deeply coloured. Now collect the hexane solution in a porcelain dish and place it on a water bath for evaporation. This treated sample is taken in a test tube and for control or reference, pure groundnut oil is treated the same as above and taken in another test tube. Then, the two treated samples are observed under UV light. Pure groundnut oil shows no greenish-yellow fluorescence whereas palmolein or palm oil adulterated ground oil shows greenish-yellow fluorescence under UV light.

Note: Palm oil can also be detected using phytosterol acetate test, thin layer and gas-liquid chromatography.

viii) **Polybromide test** (used for Mustard Oil)

Principle: Formation of precipitates upon treatment of an ethereal solution of fat or fatty acid with bromine, gives a qualitative indication of the presence of fatty acids with three or more non conjugated double bonds.

Reagents

a) Diethyl ether
b) Bromine

Apparatus

1) Conical Flask (100 ml capacity)
2) Burette with a finely drawn-out jet

Procedure: Dissolve approximately 3 gm of clear fat in 25 ml diethyl ether in the conical flask. Place the flask in a melting ice bath for 15 minutes and then slowly add 1 ml bromine dropwise from the burette with continuous swirling and cooling (the first half ml in 20 minutes and the remaining in 10 minutes). Cool the flask and keep it in the ice bath for further 3 hours. If a precipitate forms, the reaction is considered positive (Laboratory Handbook for Oil and Fat Analysis, Cocks and Reid, page 147-148).

ix) **Sodium azide test** (for artificial mustard oil; detection level 0.1% of added allyl isothiocyanate)

Principle: Allyl isothiocyanate reacts with sodium azide to form 1-allyl-2-tetrazoline-5-thione (**II**) which upon further treatment with neutral bismuth (III) nitrate solution forms a complex compound i.e. bis [(1-allyl-2-tetrazoline-5-thionato) (nitrato) bismuth (III)] (**III**) resulting in the development of a deep yellow colour (Shukla *et al.*, 2002).

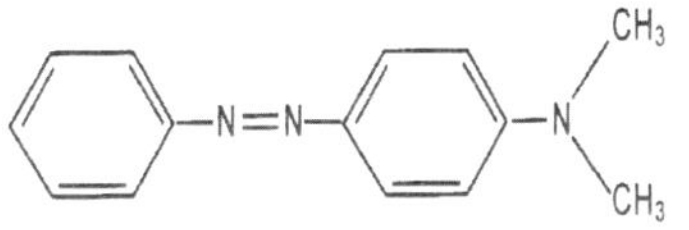

(p-Dimethylaminoazobenzene)
(Butter Yellow)
(I)

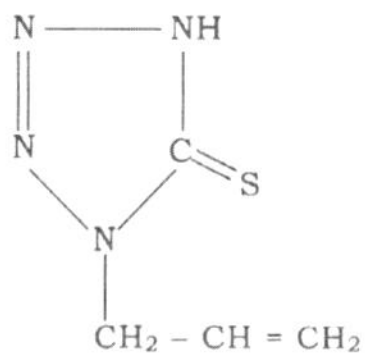

1- Allyl-2-tetrazoline-5-thione
(II)

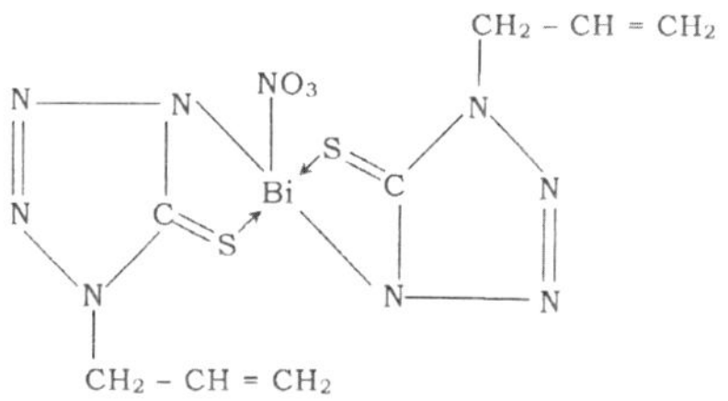

Bis (1-allyl-2-tetrazoline-5-thionato) (nitrato) bismuth (III)
(yellow)
(III)

Reagents

a) Sodium azide solution (2.0 gram per 100 mL)
b) Diethyl ether
c) Bismuth nitrate solution

Apparatus:

1) Hot plate
2) separating funnel (250 ml capacity)
3) Beakers
4) Test tubes

Procedure

Take 100ml sample and add 100ml sodium azide solution in a flask and heat the mixture directly or on a hot plate for around 3 hrs. Cool the content in a flask and transfer it to a separating funnel, and allow the formation of separate layers i.e. aqueous and oily layers. Transfer the lower aqueous layer into a beaker and discard the upper oily layer. Wash the aqueous layer with diethyl ether (twice -50ml each time) to remove a residual oily layer. After this, filter the content and concentrate the aqueous content to half of its volume. Now, take a test tube, add 1 ml of bismuth nitrate solution and 1 ml of concentrated solution and mix it thoroughly. Immediate development of deep yellow colour shows positive results indicating the presence of artificial or synthetic mustard oil in the given oil sample.

Health Impact

The adulterated fats and oils consumption has a negative impact on health and can lead to acute to chronic illnesses hence may prove to be fatal. The intake of adulterated fats and oils on regular basis can be allergic to some extent and also alleviates cholesterol levels thus harming cardiovascular health. The adulterated fats and oils can incite the diseases like cancer, paralysis, liver damage and CVD (Cardiovascular disorders). The common adulteration of mustard oil with argemone oil and butter yellow has been found to be the main reason for gall bladder cancer and also lead to other ailments pertaining to eyes like glaucoma, epidemic dropsy and even loss of eyesight (Alessandra *et al.*, 2021). The fats and oils should be monitored regularly for adulteration as even a small amount of adulteration can lead to allergies and irritation which further can cause severe skin problems in long run. The quality of oils and fats equally matters as of others. Consumer organization has conducted the latest study on the adulteration of fats and oils and stated that 85% of edible fats and oils are subject to adulteration across fifteen Indian states. According to the study, the main oils targeted for adulteration are mainly mustard, coconut, sunflower, soya bean, sesame, palm olein, groundnut and cottonseed. People buy loose oil for a cheaper price but it has more adverse effects on health and can be fatal also. Unfortunately, the quality check for adulterated oils is not performed which motivates the profit-seekers to play with the health of consumers. The quality should be monitored regularly and is directly linked to the health status of the consumers.

Developed Technological Measures to Test Adulteration of Oil and Fats

The lower price oils are usually admixed with expensive oils from an economic point of view. There are many techniques being developed to date to check the purity of the oil. Many conventional methods have been still used to check the adulteration of olive oil with vegetable oil. Many tests are based on the detection of a difference in fatty acid and glycerol compositions in olive oil and other vegetable oils. Hazelnut oil is generally used as an adulterant in olive oil due to its analogous chemical composition. The admixture of hazelnut oil below 20% is difficult to detect by conventional tests. Gas chromatographic techniques can be employed to analyse olive oil adulteration with almond and hazelnut oil at a level of less than 5% (Fang *et al.*, 2013).

The costly essential oils are extended with almond oil or coconut oil and are usually diluted by putting alcohol, surfactants and emulsifiers to make these less expensive. The presence of these adulterants can easily be detected by the greasy feel on hands as pure essential oil does not leave any greasiness on the skin. The purity of essential oil can also be determined by putting a drop on paper, pure essential oil does not show any greasy marks on paper upon drying. The addition of emulsifiers can be tested by putting a drop of oil in water, the pure oil will float on the surface while emulsified will be liquefied in water and make it turbid. Many diluents like Benzyl benzoate, Di propylene Glycol, Carbitol, Hercolyn D, Isopar, Isopropyl Myristate, Triacetin and Diethyl Phthalate are added as adulterants and do not contribute any peculiar odour to essential oil and sometimes these remained unnoticed at a level of 15%. These adulterants can be tested with alcohol solubility tests and can be further identified by means of various Gas chromatographic techniques.

The colorimetric techniques have also been used to check adulteration in coloured oils. For example, sunflower oil is light yellow in colour and if it is admixed with some other oil, then the absorption of light at a given wavelength will change. Similarly, PCA (Principal component analysis) identifies the lowest level of adulteration. The mixing of argemone oil in mustard oil is very common and it can be detected by the HPTLC technique. The mineral oil can also be estimated in various vegetable oils by HPTLC fingerprinting technique. The component variation is an amazing tool for the detection of adulteration as oils has a specific composition, for example, brassica sterol is solely present in canola oil, sesamin is wholly present in sesame oil, stigmasterol is present in sunflower oil and camp sterol is the exclusive feature of soyabean oil (Reid *et al.*, 2006).

Emerging Techniques

The latest trends for detecting adulteration in olive oil and butter are TLC (Thin Layer Chromatography), NIR (Near Infra-Red Spectroscopy), DSC (Differential Scanning Colorimetry). There has been a complete range of diagnostic procedures that are approved to detect adulteration in fats and oils including chromatographic techniques like GC (Gas Chromatography), HPLC (High-Performance Liquid Chromatography), and SFC (Supercritical Fluid Chromatography). Further ELSD (Evaporative Light Scattering Detection), Spectroscopic techniques like NIR (Near Infra-Red), FTIR (Fourier Transform Infrared), NMR (Nuclear Magnetic Spectroscopy), Mass Spectroscopy including electrospray ionization mass spectrometry and many more techniques have been developed to test the purity of fats and oils (Yang *et al.*, 2005). Margarine is often used as an adulterant in butter which is a cause of concern for consumers and so the latest analytical techniques are required to confirm the quality of products. Research studies indicate that Raman spectroscopy could be efficiently used to measure butter contamination with margarine. The Raman spectra have shown a difference in peak intensity wherein the more margarine addition has given more peak intensity (Nedeljkovic *et al.*, 2016).

The analytical instruments commonly used to detect adulteration in oils are GC (Gas Chromatography), GC-FID (Gas Chromatography-Flame Ionization Detector), GC-MS (Gas Chromatography-Mass Spectrometry), GC-IRMS (Gas Chromatography – Isotope-Ratio Mass Spectrometry), SNIF-NMR (Site-Specific Natural Isotope Fractionation NMR), chiral GC-MS. The adulterant may be artificial or natural, but these do have some marker or impurity and identification of these exceptional markers for each detailed adulterant is a more operative technique of sensing oil adulteration. GC-IRMS measures the isotopic ratio and has some limitations like some compounds based on natural precursors are not detected, thus not a reliable technique (Ilya Strashnov *et al.*, 2019). Similarly, Chiral GC-MS also has some limitations due to its applicability on chiral molecules only and the enantioselective ratio of chiral compounds differs. The SNIF-NMR is applied for monoterpenes but essential oils are composed of sesquiterpenes and diterpenes also, so the technique needs pure isolated compounds (Abbas and Baeten, 2016).

Table 2: Advanced techniques for detection of adulterants in oils and fats

Type of product	Type of adulterant	Method of analysis	Detection limit	Reference
Argan oil	Vegetable oil	UV-visible, FTIR spectroscopic, UPLC-MS, selected ion mass spectrometry, Visible and near infra-red spectroscopy	1 %	Farres *et al.* (2019), Kharbach *et al.* (2017), Kharbach *et al.* (2018)
Camellia oil	Peanut, rapeseed, & soybean oil or mixture thereof	NIR spectroscopy and sub-window permutation analysis	-	Sun *et al.* (2015)
	Corn, rapeseed, & sunflower oil	Proton nuclear magnetic resonance spectroscopy	-	Shi *et al.* (2018)
	Peanut, rapeseed, & soybean	LC-MS/MS	-	Dou *et al.* (2018)
	Cheap vegetable oils	Excitation-emission matrix fluorescence spectroscopy combined with chemometrics	-	Wang *et al.* (2019)
	Canola, corn, sesame, sunflower, & peanut oil	DSC/gas chromatography	-	Li *et al.* (2016)
	Cheap vegetable oils like peanut oil, & sunflower oil	TRES, and Immunoassays	-	Chen *et al.* (2018); Guillaume *et al.* (2021)
Canola, Palm and Palm kernel oil	Lard	FTIR, Liquid chromatography with Mass spectrum	2%	Marikkar *et al.* (2001); Marikkar *et al.* (2005)
Olive oil	Vegetable oil like corn, canola, parilla, sesame and soyabean oil	Dielectric Spectroscopy (101 Hz–1 MHz)	0.5%	Li *et al.* (2018)
RBD palm oil	Lard	E-Nose	1%	Nurjuliana *et al.* (2011)

Conclusion

Oils and Fats adulteration is a thoughtful and serious safety issue, as it can create havoc on consumers. Toxic Oil Syndrome of 1991 in which rapeseed oil on denaturation with aniline was sold as Olive oil and melamine mixing in baby formulas are some of the examples of adulterations. Adulterations not only result in economic loss to consumers but also affect their safety. Fats and oils adulteration is a subject of increasing attention worldwide and there is a requirement of highly sophisticated techniques for the detection of adulterants in essential oils. Fats and oils have always been an essential commodity of our meals and amongst animals, human has included these in their diet and rest animals depend on raw food for nourishment. A full proof approach is required to solve this problem and with the advancements in technology and analytical methods, detection has become easy. Some conventional tests based on colour change reactions are safe at pocket and can be used for detecting synthetic oils and physically refined oils while sophisticated instrumentation techniques based on chromatography and spectroscopy are very expensive and out of reach for many consumers. An extensive study is needed to be developed for analysing individual oils based on their composition and scientific evidence are required to prove the superiority of oil over the other. The awareness and prevention of adulteration is the superlative mode to avoid health problems. Contamination whether intentional or unintentional should be a matter of concern and everyone should be educated enough and take practical interest to eradicate the health hazard.

References

Abbas, V. Baeten,"Advances in the Identification of Adulterated Vegetable Oils",Advances in Food Authenticity Testing,Woodhead Publishing,2016,Pages 519-542,ISBN 9780081002209,https://doi.org/10.1016/B978-0-08-100220-9.00019-9.

Alessandra Mazzocchi, Valentina De Cosmi, PatriziaRisé, Gregorio Paolo Milani, Stefano Turolo, Marie-Louise Syrén, Angelo Sala, Carlo Agostoni, "Bioactive Compounds in Edible Oils and Their Role in Oxidative Stress and Inflammation", Frontiers in Physiology, 10.3389/fphys. 2021; 659551, 12.

Anindita D. P., Jain Arpita "Adulteration in Commonly Used Cooking Oils of Kolkata: Evaluation of Consumer Perception and Detection of Adulterants" International Journal of Health Sciences and Research. 2018; 8 (12): 30-37

AOAC (2000) Official Method 920.118-1920, Butter. Examination of fat. In: Official methods of analysis of AOAC International (17th ed.). Gaithersburg, MD: AOAC International

Ayza A, Belete E. "Food adulteration: Its challenges and impacts. Food Science and Quality Management". 2015; 41: 50-56.

Bøwadt S., Aparicio R. "The detection of the adulteration of olive oil with hazelnut oil: a challenge for the chemist". Inform. 2003;14: 342-344

Chen H, Chen B, Lu D. A novel method for detection of camellia oil adulteration based on time-resolved emission fluorescence. Scientific reports. 2018 Sep 13;8(1):1-6.

Christopoulou E., Lazaraki M., Komaitis M., Kaselimis K. "Effectiveness of determinations of fatty acids and triglycerides for the detection of adulteration of olive oils with vegetable oils". Food Chemistry. 2004; 84: 463-474

Codex Alimentarius Commission. 1993. Report of the Fourteenth Session of the Codex Committee on Fats and Oils, London, 27 September - 1 October 1993, Alinorm 95/17. Rome, FAO/WHO.

Dey DC, Mitra SN, Sanyal AK. 1956. A Modified Hexabromide Test for the Detection of Linseed and Other Highly Unsaturated Oils in Mustard Oil. Current Science. 1;25(7):227-.

Dou X, Mao J, Zhang L, Xie H, Chen L, Yu L, Ma F, Wang X, Zhang Q, Li P. Multispecies adulteration detection of camellia oil by chemical markers. Molecules. 2018 Feb;23(2):241.

Evers, J. & Wightman, Leah & Crawford, Robert &Contarini, Giovanna & Coors, Ursula & Farrington, Derek &Molkentin, Joachim & Nicolas, Marina.. "A precise method to measure the total fat content of spreadable fats". International Dairy Journal.2000: 10. 815-827. 10.1016/S0958-6946(01)00025-5.

F.A.O Manuals of Food Quality Control 14 / 8, page 273 / A.O.A.C 17th edn , 2000, Official Method 936.12 Oil (Teaseed) in olive oil)

Fang, G.; Goh, J.Y.; Tay, M.; Lau, H.F.; Li, S.F.Y. Characterization of oils and fats by ^{1}H NMR and GC/MS fingerprinting: Classification, prediction and detection of adulteration. Food Chem.**2013**, 138, 1461–1469.

Farres S, Srata L, Fethi F, Kadaoui A. Argan oil authentication using visible/near infrared spectroscopy combined to chemometrics tools. Vibrational Spectroscopy. 2019 May 1;102:79-84.

Felix F. Hexabromide determination in linseed oil. Chem.,-Zig. 1933;57:596

Firestone D., Reina R.J. (1996) Authenticity of vegetable oils. In: Ashurst P.R., Dennis M.J. (eds) Food Authentication. Springer, Boston, MA. https://doi.org/10.1007/978-1-4613-1119-5_7

Guillaume C, Miller P, Cantrill R. Fraud in fats and oils. InFood Fraud 2021 Jan 1 (pp. 151-175). Academic Press.

I.S.I. Handbook of Food Analysis (Part XIII)-1984 Page 86 / A.O.A.C 17th edn,2000, Official method 893.01-Oil (sesame) in Oils and Fats Modified Villavecchia Test)

I.S.I. Handbook of Food Analysis of (Part XIII)-1984 Page 86 / A.O.A.C 17th edn, 2000, Official method 197.02-oil (cottonseed) in oils and fats /F.A.O. Manuals of Food Quality Control 14 / 8 Page 271)

Ilya Strashnov, Jamie D. Gilmour, Andrew Cannavan, Gang Chen, Champa Dissanayake, Biyagamage Ruchika Fernando, Simon Kelly, Syahidah Akmal Muhammad, Dileep Kumar Singh, Zhenzhen Xu. Atmospheric pressure chemical ionisation (APCI) and photoionisation (APPI) mass spectrometry for detection of unsaturated fatty acids: potential for rapid detection of adulteration of vegetable oils. Analytical Methods**2019,** *11* (30) , 3819-3828. https://doi.org/10.1039/C9AY00977A

International Olive Oil Council (IOOC). 1993a. International trade standard applying to olive oils and olive-pomace oils. COI/T. 15/NC, No. 1/Rev. 6, Madrid.

IS:548 (Part- II), Methods of Sampling and Test for Oils and Fats, (1976).

Kharbach M, Kamal R, Marmouzi I, Barra I, Cherrah Y, Alaoui K, Vander Heyden Y, Bouklouze A. Fatty-acid profiling vs UV-Visible fingerprints for geographical classification of Moroccan Argan oils. Food control. 2019 Jan 1;95:95-105.

Kharbach M, Viaene J, Vercammen J, Bouklouze A, Vander Heyden Y. POSTER: Geographic origin discrimination of Moroccan Argan oils. InChimiométrie 2017 2017.

Kou Y, Li Q, Liu X, Zhang R, Yu X. Efficient Detection of Edible Oils Adulterated with Used Frying Oils through PE-film-based FTIR Spectroscopy Combined with DA and PLS. J Oleo Sci. 2018 Sep 1;67(9):1083-1089. doi: 10.5650/jos.ess18029. Epub 2018 Aug 14. PMID: 30111676.

Kumar A, Upadhyay N, Padghan PV, Gandhi K, Lal D, Sharma V, "Detection of vegetable oil and animal depot fat adulteration in anhydrous milk fat (Ghee) using fatty acid composition" MOJ Food Processing & Technology. 2015; 1(3): 46-53

Laboratory Handbook for Oil and Fat Analysis, Cocks and Reid, page 147- 148)

Li R, Huang J, Huang L, Teng J, Xia N, Wei B, Zhao M. Comparison of GC and DSC monitoring the adulteration of camellia oil with selected vegetable oils. Journal of Thermal Analysis and Calorimetry. 2016 Dec;126(3):1735-46.

Lim SY, Mutalib MSA, Khazaai H, Chang SK "Detection of fresh palm oil adulteration with recycled cooking oil using fatty acid composition and FTIR spectral analysis",International Journal of Food Properties,2018: 21:1,2428-2451.

Manual of Methods of Analysis for Adulterants and Contaminants in Foods, I.C.M.R (1990.) Page 5 / I.S.I. Handbook of Food Anaysis Part (XIII) – 1984 page 86)

Marikkar JM, Ghazali HM, Man YC, Peiris TS, Lai OM. Distinguishing lard from other animal fats in admixtures of some vegetable oils using liquid chromatographic data coupled with multivariate data analysis. Food chemistry. 2005 Jun 1;91(1):5-14.

Marikkar JM, Lai OM, Ghazali HM, Che Man YB. Detection of lard and randomized lard as adulterants in refined-bleached-deodorized palm oil by differential scanning calorimetry. Journal of the American Oil Chemists' Society. 2001 Nov;78(11):1113-9.

Motghare M, Sarma NV, Nagaraja CV, Grover MR, Kamath KS. A simple method for detection of palmolein oil in groundnut oil. Journal of food science and technology (Mysore). 1985;22(4):287-8.

Nedeljković, A., Rösch, P., Popp, J. *et al.* "Raman Spectroscopy as a Rapid Tool for Quantitative Analysis of Butter Adulterated with Margarine". *Food Anal. Methods*, 2016; **9,** 1315–1320. https://doi.org/10.1007/s12161-015-0317-1

Nick Vanstone, Andrew Moore,Perry Martos and Suresh Neethirajan "Detection of the adulteration of extra virgin olive oil by near-infrared spectroscopy and chemometric techniques" J of Food Quality and Safety, 2018; 2, 189–198 doi:10.1093/fqsafe/fyy018 Advance Access publication

Nurjuliana M, Che Man YB, Mat Hashim D. Analysis of lard's aroma by an electronic nose for rapid halal authentication. Journal of the American Oil Chemists' Society. 2011 Jan;88(1):75-82.

Peter Schönfeld,Lech Wojtczak. "Short- and medium-chain fatty acids in energy metabolism: the cellular perspective".Journal of Lipid Research 2016, 57 .

Reid, L.M.; O'Donnell, C.P.; Downey, G. Recent technological advances for the determination of food authenticity. Trends Food Sci. Technol.**2006**, 17, 344–353.

S. Azadmard-Damirchi1, M. "Torbati "Adulterations in Some Edible Oils and Fats and Their Detection Methods" Journal of Food Quality and Hazards Control 2015; 2, 38-44

Sharma A, Batra N, Garg A, *et al.* "Food adulteration: A review". International J. for Research in Applied Science and Engineering Technology. 2017; 5(3): 686- 689.

Shi T, Zhu M, Chen Y, Yan X, Chen Q, Wu X, Lin J, Xie M. 1H NMR combined with chemometrics for the rapid detection of adulteration in camellia oils. Food chemistry. 2018 Mar 1;242:308-15.

Shukla AK, Dixit AK, Singh RP. Detection of adulteration in edible oils. Journal of oleo science. 2005;54(6):317-24.

Shukla AK, Johar SS, Dixit AK, Singh RP. Identification of Physically Refined Rice Bran Oil and Its Simple Detection in Other Oils. Journal of Oleo Science. 2004;53(8):413-5.

Sun T, Wu Y, Li X. Discrimination of camellia oil adulteration by NIR spectra and subwindow permutation analysis. Acta Optica Sinica. 2015;35(6):350â.

Wang T, Wu HL, Long WJ, Hu Y, Cheng L, Chen AQ, Yu RQ. Rapid identification and quantification of cheaper vegetable oil adulteration in camellia oil by using excitation-emission matrix fluorescence spectroscopy combined with chemometrics. Food chemistry. 2019 Sep 30;293:348-57.

Wheatley, V.; Spink, J. Defining the public health threat of dietary supplement fraud. Compr. Rev. Food Sci. Food Saf.2013, 12, 599–613.

Yadav S. "Edible oil adulterations: Current issues, detection techniques, and health hazards". International Journal of Chemical Studies. 2018; 6(2): 1393-1397

Yang H, Irudayaraj J, Paradkar MM "Discriminant analysis of edible oils and fats by FTIR, FT-NIR and FT-Raman spectroscopy" 2005; Food Chem 93(1):25–32

10

Adulteration of Fruits, Vegetables and Their Products: Impact on Healthy Living

Janifer Raj Xavier

Defence Food Research Laboratory, Defence Research and Development Organization, Mysore, Karnataka, India

Introduction

Fruits and vegetables are an important part of healthy human diet. Consumption of vegetables is widely promoted by Governmental health agencies since it provide consumers essential nutrients such as vitamins, minerals, dietary fiber and phytochemical compounds, at a relatively low calorie density. Furthermore, the consumption of vegetables has been strongly associated with reduced chronic and degenerative diseases such as cancer and heart related ailments. The United States Department of Agriculture recommends five to nine servings of fruits and vegetables daily (Berger *et al* 2010). In recent years, fresh-cut produce has become more popular due to their convenience, freshness, taste and assumed "Healthiness". Consumers around the globe demand for fresh fruits and vegetables with high-quality and extended shelf life. Almost every fruit and vegetable commercially available in modern markets contains chemicals that are not natural. Fresh produce are transported over long distances and pre washes and coating are used as post-harvest procedures to shield surfaces of fruits and vegetables from and microbes and insects. Approved wax coatings include beeswax, candelilla wax, carnauba wax and shellac while paraffins, montanic acid esters and polyethylene wax oxides are not permitted for coating of fresh produce. Non permitted additives such as morpholine, a precursor for carcinogenic nitrosamines are used along with permitted wax coatings for ease in water based application over the fresh produce (Ritter *et al* 2001). Extensive pesticide and fertilizer application and using polluted water for irrigation could lead to residues of harmful chemicals such as toxic heavy metals being found in fruit and vegetables and these residues pose risks to human health (Zhao *et al* 2019). Heavy metal contamination of food items is one of the most important aspects of food quality assurance. International and national regulations on food quality have lowered the maximum permissible levels of toxic metals due to an increased awareness of the risk, these metals pose to food chain (FAO/WHO, 2007; FSSAI, 2016).

Food colouring agents are the major chemicals among food additives in many processed foods. Colouring dyes are mostly azo compounds which are added to different items to impart new color and making them more attractive and palatable (Forgacs *et al* 2004). Thousands of azo dyes are used in many sectors like textile firms, paper production industries, food industries, cosmetics and pharmaceutical companies. It has been found that many kinds of azo dyes produce carcinogenicity and induce genetic disorders in human beings (Lau *et al* 2006). Synthetic food colours are a major source of food intoxication and lead to severe health problems such as low haemoglobin concentration, allergic reactions, mutations, cancers, irritability, restlessness, sleeping disturbances, effects on the liver, kidney, and intestine, hyperactive effects on children, ear infections, asthma, and eczemas (Bachalla, 2016; Nath *et al* 2015). Permitted synthetic colours for food use up to 100 ppm are carmoisine, erythrosine, ponceau, tartrazine, sunset yellow FCF, indigo carmine, brilliant blue FCF and fast green FCF. Apart from the above mentioned permitted synthetic food colours, non-permitted synthetic colours such as fast red, rhodamine b, alizarin, metanil yellow, bromocresol purple, sudan 1, sudan 2, sudan 3 and sudan 4 are being used in various types of processed foods. According to a survey conducted in India, 72.3 percent of samples were adulterated using permitted colours exceeding their prescribed limits while 28.7 percent samples contained non-permitted colours (Dixit *et al* 2013).

Detection of such harmful contaminants and coatings are essential to ensure the food quality of the processing materials. Presently, azo dye based kits are used for colorimetric detection of heavy metals in foods and drinks with testing times as less as 5 to 8 minutes. Rapid test kits are essential for onsite detection of harmful residues and toxic metals for fresh produce. Medina *et al* (2019) reported barcoding in form of fingerprinting of foods as a tool for food authentication for prevention of food fraud. Food metabolites of molecular weight less than 1200 Daltons for use as biomarkers were compiled from recent studies and their potential has been reviewed. Patra *et al* (2019) elaborated quality control and safety of beverages using rapid adulteration detection methods. Substances such as urea, formaldehyde, melamine, toxic heavy metals commonly affect food safety of beverages. Graphene based sensors for rapid detection of food adulterants up to trace quantities was described. Nano graphene derivatives could find use in fabrication of portable and rapid sensors based on thermal, chemical as well as electro chemical methods for various foods (Raghavan *et al* 2021). This chapter provides a brief review of the current knowledge and summarizes data on adulteration of fresh fruits and vegetables and their processed products and food safety aspects related to non-permitted and over used additives such as food colours, harmful coatings etc. Methods of recent and rapid detection of adulterants, Indian and global safe limits for additives, food recalls are also discussed.

Common Adulterants of Fruits and Vegetables and Their Products

i. Non-permitted Food Colouring Substances

Colour of food is considered as an essential sensory parameter which influences acceptability and appearance of the same. Natural and synthetic colouring agents are being used in various types of foods to impart desirable sensory acceptability (Corradini, 2019). Due to paradigm shift in consumer awareness for health foods, natural colours are preferred in lieu of synthetic food colours. However, usage of permitted and non-permitted synthetic colours in excess for economical gain has been reported which raises health concerns for human kind. Chao *et al* (2018) reported a proof-of-concept methodology for using thermal desorption electro spray ionization/ mass spectrometry for detection of basic colorants in processed vegetable based foods. Homogeneous processed foods could be tested for adulterants during food safety applications. Bilberry (*Vaccinium myrtillus*) is rich in anthocyanins and used as herbal product. Amaranth naphthylazo food dye was used to impart the colour to resemble anthocyanin content of bilberry. Spectrophotometric assays routinely used for detection of anthocyanin content may not be sufficient to detect presence of adulterants as reported. High-performance liquid chromatography based methods are essential for detection of intentional addition of low cost dyes for gaining economic benefits (Penman *et al* 2006).

Ashok *et al* (2014) described a simple micellar liquid chromatography method for quick detection of malachite green (4-[(4-dimethylaminophenyl) phenyl-methyl]-N,N- dimethylaniline), a non-permitted color in foods such as green pea and selected ready-to-eat products. Isocratic elution of malachite green using reverse phase C_{18} column with mobile phase consisting of sodium dodecyl sulphate (0.15 M) and pentanol (6.0%) at pH 5 was standardised. Detection at wavelength of 620 nm and elution in 9.2 min was observed while limit of detection (LOD) was 0.10 ppm and limit of quantification (LOQ) was 0.25 ppm for the tested hazardous dye. Metanil yellow and malachite green were detected using Surface Enhanced Raman Spectroscopy (SERS) based paper swab technique for quality testing of yellow splits of red gram, green peas and green chillies. Thin flexible films based on nano-structured-silver were manufactured to obtain paper-based SERS inkjet-printed swabs for quick detection of non-permitted colourants. Combination of portable Raman spectrometer was used for signal collection during analysis. Furthermore, washes and ozoniser used for removing contaminants in form of non-permitted colours were also tested in the study (Kumar and Santhanam, 2019). Ntrallou *et al* (2020) reviewed in detail the advancements in sample preparation techniques for analytical applications for determination of food colourants in complex matrices such as food.

ii. Non-permitted Coatings and Additives for Shelf life Extension

Duration until food maintains its acceptability in terms of microbial safety and sensory acceptability is termed as shelf life of that particular food. Physical and chemical factors play a major role in determining shelf stability of foods (Kilcast and Subramaniam, 2011). Traditional food preservation processes included reduction of water activity and pH, low temperature storage and use of preservatives to extend shelf life of foods. Combination processing methods using chemical, physical, biological and their combinations are growing due to latest technological interventions in the field of food science and technology. Recent food preservation methods also rely on use of chemicals in form of additives to enhance shelf stability of foods. Usage of chemicals based on permitted limits by food regulatory organizations are barely followed by food processing units for economical benefits which may result in health hazards of the consumer. Therefore, adulteration detection of non-permitted chemicals and additives in foods has become essential to ensure safety of foods. Medina *et al* (2019) reported barcoding in form of fingerprinting of foods as a tool for food authentication for prevention of food fraud. Food metabolites of molecular weight less than 1200 Daltons for use as biomarkers were compiled from recent studies and their potential has been reviewed. Graphene based sensors for rapid detection of food adulterants up to trace quantities was described. Nano graphene derivatives could find use in fabrication of portable and rapid sensors based on thermal, chemical as well as electro chemical methods for various foods (Raghavan *et al* 2021). Tattoo paper based on gold island film capable of transferring plasmonic patterns were screened using surface enhanced Raman spectroscopy (SERS) for food toxins was reported. Sputter coäting on fruit surface of orange for *in situ* detection of Thiabendazole using gold tattoo paper as well as *ex situ* molecular detection of Di 2-ethylhexyl phthalate in commercial sports drinks using copper tattoo paper was described (Wang *et al* 2018). Richardson *et al* (2019) developed a method for rapid detection and quantification of adulterants in processed coconut water developed using young coconuts obtained from Costa Rica. Commonly used diluents water, single and mixed sugar syrups as well as high-fructose corn syrup were detected using Raman spectroscopy at wavelength of 785 nm. Chemometric tools were used to analyse the obtained spectra and derived successful models for detection of adulterants in the range of 1.9 and 2.6 percent for single sugars, 9.8 percent for sugar mixtures and high-fructose corn syrup up to 7.1 percent in processed coconut water.

Patra *et al* (2019) elaborated quality control and safety of beverages using rapid adulteration detection methods. Substances such as urea, formaldehyde, melamine, toxic heavy metals as well as nano particles commonly affect food safety of beverages. Novel analytical methods for detection the above listed contaminants were described. Wu *et al* (2021) reported a method for authentication of fresh fruit and vegetable based juices based on detection of undeclared addition of diluents and sugars used in Not-from-concentrate juices to gain economically. Sugar specific isotope analysis was used to identify various simple sugars such as glucose, fructose and sucrose as well as complex plant sugars based on number of carbon atoms. Fruit nectars obtained from fruits such as orange, grape, passion and peach adulterated with common diluents and adulterants namely juices and syrups were detected using Attenuated total reflectance Fourier transform mid infrared (ATR-FTIR) method. Models were developed using partial least squares method and validated for detection of adulteration of juices (Miaw *et al* 2018). Iridium quadruple - disk electrode was used for detection of glucose - fructose syrup added to apple juice using deep neural networks models to profiling commercially available apple juices (Wójcik and Jakubowska, 2021). High-performance liquid chromatography derived amino acid profile of fruit juices were subjected to chemometric analysis using tools for multivariate analysis viz. partial least squares and principal component analysis. Presence of amino acids such as glutamic acid and valine indicated presence of blood orange juice while adulteration with other orange juices indicated presence of phenylalanine and glycine. Adulteration of approximately 20% blond orange juice was observed in the reported study (Wistaff *et al* 2021).

iii. Non-permitted Chemicals for Washing/Cleaning/Artificial Ripening

Post harvest losses account to major constraint in fruits and vegetable marketing. To extend the shelf life of the harvested fresh produce till it reaches the retailer, fruits and vegetables are subjected to treatments such as washing, coating etc. Moreover, harvesting of mature fruits before ripening and degreening prior to retailing is being practiced to reduce losses due to over ripening, lack of appeal due to colour and textural changes during transport and handling of ripened fruits. Usage of non-permitted chemicals for the above mentioned purposes has drawn attention in consideration due to awareness of health hazards associated with them. Chlorinated water has been used time immemorial for drenching fruits and vegetables as surface decontaminating agent at packing houses. Non-permitted chemicals such as formalin are being used for pre washing or sanitization of fresh produce such as mushrooms and cauliflower. Zhang *et al* (2022) has developed a low-cost, hand-held detector with a screen-printed electrode amperometric sensor and an open-sourced potentiostat for quick detection of formaldehyde upto levels of 100 μmol/L. Six types of vegetables such as cabbage (regular, Chinese and purple), lettuce (whole and leaf) and rapeseed were tested and validated for adulteration of formaldehyde.

Calcium carbide is widely used for induction of ripening of fruits such as mango, banana etc., due to release of acetylene gas when moisture contacts the calcium carbide powder. Sreeraj *et al* (2020) reported development of Artificial intelligence (AI) based device (CLadron®) for identification of fruits ripened through artificial ripening. The machine could identify fruits such as mango, guava, papaya, banana and apple when ripened using calcium carbide with sensitivity, specificity and precision at levels of 0.9216, 0.6842 and 0.8868, respectively. Haar Filters for detection of fruits ripened using calcium carbide involving AI technology was developed (Bagyalakshmi *et al* 2019). Gold nano particle based colorimetric method was also reported for quick detection of calcium carbide ripened mangoes at household level (Lakade *et al* 2018). Lakade *et al* (2019) also reported a near infrared spectroscopy (wavelength 600 to 1100 nm) based method for detection of artificial ripening using the chemical calcium carbide in mangoes. Hassan *et al* (2019) reported increased levels of elements such as Fe, Co and heavy metals such as As, and Pb using inductively coupled plasma-mass spectrometry (ICP-MS) in mango peels on treatment of mangoes using commercial grade calcium carbide

Safe Limits/Tolerable Limits for Food Additives

Food and Agricultural Organization (FAO)

WHO guidelines (https://www.who.int/news-room/fact-sheets/detail/food-additives 31 January 2018) on food safety are as below:

- Food additives are substances added to food to maintain or improve its safety, freshness, taste, texture, or appearance.
- Food additives need to be checked for potential harmful effects on human health before they can be used.
- The Joint FAO/WHO Expert Committee on Food Additives (JECFA), is the international body responsible for evaluating the safety of food additives.
- Only food additives that have been evaluated and deemed safe by JECFA, on the basis of which maximum use levels have been established by the Codex Alimentarius Commission, can be used in foods that are traded internationally.

A manual on principles and methods for the risk assessment of chemicals in food and environmental health criteria has been published in 2009 to reflect significant advances in chemical analysis, toxicological assessment, and risk assessment procedures, this detailed monograph on the methods and principles for the risk assessment of chemicals in food (https://www.who.int/publications/i/item/9789241572408). Safety evaluation of certain food additives including flavouring agents was prepared by the Eighty-second meeting of the Joint FAO/WHO Expert Committee on Food Additives (JECFA). Geneva: World Health Organization; 2017 (WHO Food Additives Series, No. 73). General standard for food additives is based on CODEX STAN 192-1995 Adopted in 1995. Revision 1997, 1999, 2001, 2003, 2004, 2005, 2006, 2007, 2008, 2009, 2010, 2011, 2012, 2013, 2014, 2015, 2016, 2017, 2018, 2019.

(https://www.fao.org/fao-who-codexalimentarius)

Food Safety and Standards Authority of India (FSSAI)

Eight synthetic dyes which are allowed to be used as red, yellow, blue and green colors in food are Brilliant Blue FCF (Blue 1), Indigo Carmine (Blue 2), Fast Green FCF (Green 3), Tartrazine (Yellow 5), Sunset Yellow FCF (Yellow 6), Erythrosine (Red 3), Carmoisine (Red 10) and Ponceau 4R (Red 18) according to the Indian FSSAI (FSSAI, 2011). (https://www.fssai.gov.in/dam/jcr:99067191-c774-4c81-b9c8-708b9e72b770/Food_ Additives _ Regulations.pdf)

Detection methods (lab level & household rapid kits) /chemistry/principle

Detection of adulteration of foods has an imperative role to guarantee safe, healthy and quality food. Food fraud accounts to 40 billion US dollars per annum and gains obtained by fraudsters are considered no less than profits of drug trafficking (PwC, 2016; Esteki *et al* 2019). Traditional analytical methods in food science and technology has taken a new model, by using contemporary approach by integration of various latest technologies for food quality testing, safety, traceability, interaction of food components as well as their implications on human health and wellness. Traditional methods to ensure food safety are reliable and precise however, has limitations in terms of consumption of long time to obtain precise results while cutting edge technologies are quick and achieve objectives which are impossible by other analytical methods. In this context, detection methods has emerged as a relevant platform with integration of advanced technologies such as genomics, proteomics, transcriptomics, metabolomics and chemometrics to achieve the goal of food safety by rapid and precise detection of food adulteration (Table 1).

Table 1: Technologies available for detection of adulteration in selected fruits and vegetables and their products

Sl.No.	Fruits and vegetables	Adulterant	Technique for analysis	Principle	Reference
I. Household methods					
1	Bitter gourd, green chilies	Malachite green	Method -1 Take a cotton soaked in vegetable oil or water and gently rub on the surface of the vegetable. If cotton turns green, then the vegetable is adultered with malachite green. Method-2 Take a small portion of the sample of vegetables and place it on moistened blotting paper. If the blotting paper turns green, the vegetable is adultered with malachite green.	Absorption	FSSAI DART
2	Green peas	Artificial color	Mix a few grams of green peas with water in a transparent container. Wait for half an hour. Clear separation of water will indicate the presence of artificial color.	Dissolving of colour into solvent	FSSAI DART
3	Sweet potato	Rhodamine B	Take cotton soaked in vegetable oil or water and gently rub on the surface of the vegetable. If cotton absorbs color, the vegetable is adultered with Rhodamine B.	Absorption	FSSAI DART

Sl.No.	Fruits and vegetables	Adulterant	Technique for analysis	Principle	Reference
II. Laboratory methods					
A. Vegetables					
4	Vegetables (green peas and green chillies)	Non-permitted colourant (Malachite Green)	Paper swab based SERS detection	Paper-based SERS (P-SERS) swabs macroscopically uniform thin-films in combination with a portable Raman spectrometer equipped with orbital raster scanning (ORS™) technology for signal collection	Kumar and Santhanam (2019)
5	Bitter gourd, okra, pointed gourd, yellow, green peas and red chilies	Copper sulfate, malachite green, and Sudan red	^{1}H-nuclear magnetic resonance (NMR) relaxometry	NMR relaxometry with nonlinear regression-based trend modeling	Shomaji *et al* (2021)
6	Vegetable samples	Plant growth regulators	Polyacrylonitrile nanofibers mats	Functionalisation of using sulfhydryl groups as adsorbent for solid phase extraction	Cao *et al* (2020)
7	Apple	Soluble solid content	Near-infrared spectroscopy	Nondestructive determination	Schulz and Baranska (2009)

Sl.No.	Fruits and vegetables	Adulterant	Technique for analysis	Principle	Reference
B. Fruits					
8	Strawberry achene	Changes in cell wall	Fourier transform infrared spectroscopy (FTIR) spectroscopy	Variations in degree of esterification	Schulz and Baranska (2009)
9	Orange	Thiabendazole	Surface enhanced Raman spectroscopy (SERS)	*In situ* molecular detection using gold island film (GIF) coated tattoo paper to transfer plasmonic patterns to SERS	Wang *et al* (2018
10	*Vaccinium myrtillus* (Bilberry fruit)	Naphthylazo dye (3-hydroxy-4-[(4-sulfo-1-naphthalenyl)azo]-2,7-naphthalenedisulfonic acid trisodium salt) from amaranth	High-performance liquid chromatography (HPLC)	Accurate determination of anthocyanin source	Penman *et al* (2006)
11	Small berry fruits	Differentiation from other fruits	Sanger sequencing-based DNA barcoding	mini-barcodes (*rbcL* and *ITS*) medium-barcode (*psbA-trnH*)	Wu *et al* (2018)
12	Wax apple, Lemon, Tomato, Red Pepper and Carrot	Freshness indicator	Graphene enhanced Raman spectroscopy (GERS)	Gold/silver nanoparticles seeded graphene sheets	Gopal *et al* (2016)
C. Processed foods/Beverages					

Sl.No.	Fruits and vegetables	Adulterant	Technique for analysis	Principle	Reference
13	Green pea and some ready-to-eat foodstuffs	Malachite green (4-[(4-dimethylaminophenyl) phenyl-methyl]-N,N-dimethylaniline)	Micellar liquid chromatography	RP C18 column with isocratic elution. Mobile phase (0.15 M sodium dodecyl sulfate and, 6.0% pentanol at pH 5). Detection at 620 nm	Vipin *et al* (2014)
14	Processed vegetables	Basic colorants	Thermal desorption electrospray ionization/ mass spectrometry (TD-ESI-MS)	Direct qualitative screening using a quickly interchangeable ionization source	Chao *et al* (2018)
15	Grape, orange, peach and passion fruit nectars	Apple and cashew juices as adulterant	Attenuated total reflectance Fourier transform mid infrared (ATR-FTIR) spectroscopy	Predictive multivariate calibration models	Miaw *et al* (2018)
16	Fresh fruit and vegetable juices	Undeclared addition of water and sugar	HPLC-co-IRMS based Sugar-specific isotope analysis (SSIA) technology	Carbon and oxygen stable isotope ratios (δ13C and δ18O values) to detect C4 plant sugars	Wu *et al* (2021)
17	Fruits, alcoholic and non-alcoholic beverages	Food genuineness	Combinatorial peptide ligand libraries	Proteomic fingerprint	Fasoli and Righetti (2015)
18	Beverages	Organic, inorganic, nanoparticles, and microorganisms	Laboratory to the Point-of-Care Detection	Quality control for safety	Grumezescu and Holban (2019)
19	Commercial juices	Saccharin adulterant	Fourier Transform Near-Infrared Spectroscopy (FT-NIRS) combined with partial least square (PLS)		Mabood *et al* (2018)

Sl.No.	Fruits and vegetables	Adulterant	Technique for analysis	Principle	Reference
20	Apple juice	Adulteration with cane sugar	Fourier Transform Infrared Spectroscopy based chemometric model	Specific peak in the fingerprint region	Dhaulaniya *et al* (2019)
21	Apple juice	Maltose as an indicator for adulteration or use of low-quality material	Gas chromatography (GC-FID)	Detection of presence of silylated oxime derivatives	Woźniak *et al* (2020)
22	Apple juice	Glucose - fructose syrup added as an adulterant	Universal model based on Deep neural networks	Voltammetric signals measured using iridium quadruple - disk electrode	Wójcik and Jakubowska (2021)
23	Blood orange juice	Blond orange juice as adulterant	High-performance liquid chromatography (HPLC)	Whole amino acid profiling	Wiśtaff *et al* (2021)
24	Orange juice	Grape fruit juice	Raman spectroscopy combined with multivariate analysis	Relative ratio of glucose, fructose and sucrose in orange juice (1 : 1 : 2and grapefruit juice (1 : 1 : 1).	Varnasseri *et al* (2022)
25	Pomegranate fruit juice	Apple juice as adulterant	UV–visible spectroscopy with chemometrics	Differentiation between apple and pomegranate juice based on spectra	Pappalardo (2022)
26	Lemon juice	Identification and authentication	Targeted screening (LC-UV-MS) and untargeted screening (UHPLC-QTOF/MS)	Vicenin 2, Lucenin 2-4′ - methyl ether, Eriocitrin, Scoparin, Orientin 4′ - methyl ether and Hesperidin	Lyu *et al* (2022)
27	Coconut water	Water, single sugars, mixtures of sugars, and high-fructose corn syrup as adulterants	Raman spectroscopy at 785 nm excitation	Raman spectroscopy with chemometrics	Richardson *et al* (2019)

Sl.No.	Fruits and vegetables	Adulterant	Technique for analysis	Principle	Reference
28	Raw banana flour	Low cost flours	Microscopy	Starch granule	Sard´a *et al.* (2016)
29	Dried horticultural powders	Detect adulteration	Recent methods	Spectroscopy, DNA and proteome barcoding.	Ndlovu *et al* (2022)
D. Foods					
30	Foods	Sensing preservatives/ adulterants	Nanobiosensors	Nanomaterial-based fluorescence techniques	Patel *et al* (2020)
31	Foods	Heavy metal–sensing	Nanobiosensors	Metallic and carbon nanomaterials	Patel *et al* (2020)
32	Foods	Authenticity, and adulteration	Liquid chromatography coupled to mass spectrometry (LC–MS)	Separation	Antunes *et al* (2021)
33	Foods	Adulterants, contaminants, toxins, pathogens and allergens	Smartphone-based biosensors	Fusion of smart phones with different kinds of sensitive and selective biosensors	Franca and Oliveira (2021)
34	Foods	Food authenticity and safety	Biomarkers	Metabolites (<1200 Da)	Medina *et al* (2019)
35	Foods	Adulterants and toxicants	Graphene-based nanosensors	Optical, electrical (field-effect) and electrochemical sensing	Raghavan *et al* (2021)
36	Foods	Food authentication	GC-MS, LC- q-TOF-MS and NMR followed by PCA and PLS-DA	Chemical fingerprints	Medina *et al* (2019)
37	Food and agro-products	Identification and authentication	Deep learning	Spectral analysis based predictive modeling	Zhang *et al* (2021)

Health Impact of Contaminated Fruits, Vegetables and their Products

Toxicity of additives which impart artificial colour to foods has been studied in detail in recent years due to growing awareness on health and wellness. Usage of synthetic non-permitted colours and usage of permitted colours in excess are known to be hazardous to health. Ramesh and Muthuraman (2018) reviewed in detail on the health risks caused by flavouring and colouring agents used widely in foods. Reduction of azo dyes causes release of carcinogenic metabolites on action by micro biota residing in intestine (Silva *et al* 2022). Synthetic colours such as allura red, azorubine, brilliant black, brilliant blue FCF, brown HT, erythrosine, green S, indigo carmine, patent blue, ponceau 4R, quinoline yellow, sunset yellow and tartrazine and awareness of their toxicity profiles were reviewed in detail based on the European regulation (EC) No. 1333/2008 (Amchova *et al* 2015). Tartrazine, a synthetic lemon-yellow azo dye (E102), has been reported to cause the most allergic reactions (Tuormaa, 1997) as well as childhood hyper activity disorders (McCann *et al* 2007). Sunset yellow FCF (E110), a petroleum-derived orange azo dye imparts yellow-orange colour to food products. Health issues such as allergy, asthma, hyperkinesia, DNA damage, increased incidence of cancer, stunted growth as well as weight loss were reported when consumed (Inetianbor *et al* 2015). Green S (E142), a synthetic coal tartriaryl methane derived food colour widely used was reported to leave carcinogenic metabolites after digestion in human body. Brilliant black PN (E151), Brown HT (E155) and Litholrubine BK (E180) were reported to trigger allergic reactions such as asthma as well as concerns related to genotoxicity due to harmful metabolites generated after intestinal digestion of these azo dyes (Voss, 2011). McCann *et al* (2007) reported that artificial colours and permitted preservative namely sodium benzoate added to beverages increased the cases of hyperactivity in children aged 3, 8 or 9 in the tested population.

Application of industrial grade calcium carbide on fruits for degreening application leaves a toxic residue of arsenic and phosphorus hydride which are harmful to human health. Reports on ill effects of calcium carbide indicate carcinogenic, reproductive and neurological disorders as well as damage to intestinal lining causing irritation. Sojinu *et al* (2021) studied the impact of artificial ripening on changes in mineral composition of plantain (*Musa paradisiaca*). Increased levels of calcium and decreased iron content was found in produce ripened using the ripening agent calcium carbide. Ubuoh *et al* (2022) reported effect of artificial ripening of banana using calcium carbide on nutritional quality pertaining to calorific value, proximate composition, mineral, and vitamin content as well as toxicity profile based on hazard quotient and index. Minerals and vitamin content reduced while heavy metals such as mercury and cadmium, were found to be higher than FAO permitted safe limits in artificially ripened banana in comparison to natural ripening. Okeke *et al* (2022) reviewed in detail toxicity and health effects of artificial ripening of fruits using calcium carbide in current scenario. De *et al* (2021) reported increased cell distress followed by apoptosis in mammalian fibroblast L929 cells on exposure to calcium carbide for 24 h duration at 0.2 μg/μl and recommends stoppage of its use for degreening of fruits.

Food preservatives such as sorbates and benzoates are reported to be intrinsically free of carcinogenic property. Nevertheless, conversion of these safe compounds into mutagenic compounds was reported in various animal studies. Genotoxic effects and allergic reactions as well as hyperactivity disorders were reported for sorbates (Mamur *et al* 2012) and benzoates (Piper *et al* 2017), respectively. Chaleshtori *et al* (2018) investigated effect of preservatives such as sodium benzoate and potassium sorbate in 103 food products collected from commercial outlets of Iran. Most of the samples tested contained excess preservatives in comparison to prescribed safe limits (mg/kg body weight/day) such as 0.0–5.0 and 0.0–25.0 for benzoate and sorbate salts, respectively (WHO,2016). Risk assessment studies indicated non-carcinogenic risks associated with consumption of the studied products. Shelf life extension of fruits and vegetables by intentional usage of formaldehyde, a Group I carcinogen to humans was reported. Serious health issues ranging from coma, vomiting and pain especially in children are a matter of concern (Uddin *et al* 2011).

Food recalls by producer of the produce/product or by government agencies are aimed at ensuring the health of consumers by recalling the produce/product from the market to avoid consumption. Majority of food recalls happen due to presence of food borne pathogens, foreign body such as broken pieces of glass, detection or undeclared allergens etc. FDA Recalls, Outbreaks & Emergencies maintains a list of recalls of food and agricultural produce.

References

Antunes, M., Rosado, T., Simão, A. Y., Gonçalves, J., Soares, S., Barroso, M., & Gallardo, E. (2021). Liquid chromatography–mass spectrometry as a tool to identify adulteration in different food industries. In Food Toxicology and Forensics (pp. 123-180). Academic Press.

Ashok, V., Agrawal, N., Durgbanshi, A., Esteve-Romero, J., & Bose, D. 2014. Determination of adulteration of malachite green in green pea and some prepared foodstuffs by micellar liquid chromatography. Journal of AOAC International, 97(5), 1387-1392.

Bagyalakshmi, N., Gokul, M., Sankaraiya, R., 2019. Detection of calcium carbide based ripeness in mango fruit. Int. J. Adv. Res. Electr. Electron. Instrum. Eng. 8, 17–21.

Cao, J., Li, R., Liang, S., Li, J., Xu, Q., & Wang, C. (2020). Simultaneous extraction of four plant growth regulators residues in vegetable samples using solid phase extraction based on thiol-functionalized nanofibers mat. Food Chemistry, 310, 125859.

Chaleshtori, F. S., Arian, A., & Chaleshtori, R. S. (2018). Assessment of sodium benzoate and potassium sorbate preservatives in some products in Kashan, Iran with estimation of human health risk. Food and Chemical Toxicology, 120, 634-638.

De, I., S, R., Kour, A., Wani, H., Sharma, P., Panda, J. J., & Singh, M. (2021). Exposure of calcium carbide induces apoptosis in mammalian fibroblast L929 cells. Toxicology mechanisms and methods, 31(3), 159–168. https://doi.org/10.1080/15376516.2020.1849484

Dhaulaniya et al (2019) https://doi.org/10.1080/19440049.2020.1718774

Fasoli, E., & Righetti, P. G. (2015). Proteomics of fruits and beverages. Current Opinion in Food Science, 4, 76-85.

Franca, A. S., & Oliveira, L. S. (2021). Applications of smartphones in food analysis. In Smartphone-Based Detection Devices (pp. 249-268). Elsevier.

Gopal, J., Abdelhamid, H. N., Huang, J. H., & Wu, H. F. (2016). Nondestructive detection of the freshness of fruits and vegetables using gold and silver nanoparticle mediated graphene enhanced Raman spectroscopy. Sensors and Actuators B: Chemical, 224, 413-424.

Hassan, S., Mazhar, W., Farooq, S., Ali, A., & Musharraf, S. G. (2019). Assessment of heavy metals in calcium carbide treated mangoes by inductively coupled plasma-mass spectrometry (ICP-MS). Food additives & contaminants. Part A, Chemistry, analysis, control, exposure & risk assessment, 36(12), 1769–1776. https://doi.org/10.1080/19440049.2019.1671990

Inetianbor, J.E.; Ykubu, J.M.; Ezeonu, S.C. Effects of food additives and preservatives on man—A review. Asian J. Sci. Tech. 2015, 6, 1118–1135.

Kumar, A., & Santhanam, V. (2019). Paper swab based SERS detection of non-permitted colourants from dals and vegetables using a portable spectrometer. Analytica chimica acta, 1090, 106-113.

Lakade, A. J., Sundar, K., & Shetty, P. H. (2018). Gold nanoparticle-based method for detection of calcium carbide in artificially ripened mangoes (Magnifera indica). Food additives & contaminants. Part A, Chemistry, analysis, control, exposure & risk assessment, 35(6), 1078–1084. https://doi.org/10.1080/19440049.2018.1449969

Lakade, A. J., V, V., Ramasamy, R., & Shetty, P. H. (2019). NIR spectroscopic method for the detection of calcium carbide in artificial ripening of mangoes (Magnifera indica). Food additives & contaminants. Part A, Chemistry, analysis, control, exposure & risk assessment, 36(7), 989–995. https://doi.org/10.1080/19440049.2019.1605206

McCann, D.; Barrett, A.; Cooper, A.; Crumpler, D.; Dalen, L.; Grimshaw, K.; Kitchin, E.; Lok, K.; Porteous, L.; Prince, E.; et al. Food additives and hyperactive behavior in 3-year-old and 8/9-year-old children in the community: A randomized, double-blinded, placebo-controlled trial. Lancet 2007, 370, 1560–1567. McCann et al (2007) https://doi.org/10.1016/S0140-6736(07)61306-3

Lyu et al (2022) org/10.1016/j.foodchem.2021.131424.

Medina, S., Pereira, J. A., Silva, P., Perestrelo, R., & Câmara, J. S. (2019). Food fingerprints–A valuable tool to monitor food authenticity and safety. Food Chemistry, 278, 144-162.

Mabood et al (2018) https://doi.org/10.1080/19440049.2018.1457802

Ntrallou, K., Gika, H., & Tsochatzis, E. (2020). Analytical and sample preparation techniques for the determination of food colorants in food matrices. Foods, 9(1), 58.

Okeke, E. S., Okagu, I. U., Okoye, C. O., & Ezeorba, T. (2022). The use of calcium carbide in food and fruit ripening: Potential mechanisms of toxicity to humans and future prospects. Toxicology, 468, 153112. https://doi.org/10.1016/j.tox.2022.153112

Patel, G., Pillai, V., Bhatt, P., & Mohammad, S. (2020). Application of nanosensors in the food industry. In Nanosensors for Smart Cities (pp. 355-368). Elsevier.

Patra, S., Choudhary, R., Madhuri, R., & Sharma, P. K. (2019). Quality control of beverages for health safety: Starting from laboratory to the point-of-care detection techniques. In Quality Control in the Beverage Industry (pp. 39-83). Academic Press.

Patra, S., Choudhary, R., Madhuri, R., & Sharma, P. K. (2019). Quality control of beverages for health safety: Starting from laboratory to the point-of-care detection techniques. In Quality Control in the Beverage Industry (pp. 39-83). Academic Press.

Piper, J.D., Piper, P.W., 2017. Benzoate and Sorbate Salts: a systematic review of the potential hazards of these invaluable preservatives and the expanding spectrum of clinical uses for sodium benzoate. Compr. Rev. Food Sci. Food Saf. 16, 868–880.

Pappalardo (2022) https:// doi. org/10. 1038/ s41598- 022- 07979-7.

Raghavan, V. S., O'Driscoll, B., Bloor, J. M., Li, B., Katare, P., Sethi, J., S.S Gorthi... & Jenkins, D. (2021). Emerging graphene-based sensors for the detection of food adulterants and toxicants–A review. Food Chemistry, 355, 129547.

Ramesh, M., & Muthuraman, A. (2018). Flavoring and coloring agents: Health risks and potential problems. In Natural and artificial flavoring agents and food dyes (pp. 1-28). Academic Press.

Richardson, P. I., Muhamadali, H., Ellis, D. I., & Goodacre, R. (2019). Rapid quantification of the adulteration of fresh coconut water by dilution and sugars using Raman spectroscopy and chemometrics. Food Chemistry, 272, 157-164.

Schulz, H., & Baranska, M. (2009). Fruits and vegetables. In Infrared spectroscopy for food quality analysis and control (pp. 321-353).

Silva, M.M.; Reboredo, F.H.;Lidon, F.C. Food Colour Additives: A Synoptical Overview on Their Chemical Properties, Applications in Food Products, and Health Side Effects. Foods 2022, 11, 379. https:// doi.org/10.3390/foods11030379

Sard´a, F. A. H., de Lima, F. N., Lopes, N. T., Santos, A. D. O., Tobaruela, E. D. C., Kato, E. T., & Menezes, E. W. (2016). Identification of carbohydrate parameters in commercial unripe banana flour. Food Research International, 81, 203–209.

Shomaji, S.; Masna, N.V.R.; Ariando, D.; Deb Paul, S.; Horace-Herron, K.; Forte, D.; Mandal, S.; Bhunia, S. Detecting Dye-Contaminated Vegetables Using Low-Field NMR Relaxometry. Foods 2021, 10, 2232. https://doi.org/ 10.3390/foods10092232

Tuormaa, T.E. The adverse effects of food additives on health: A review of the literature with special emphasis on childhood hyperactivity. J. Orthomol. Med. 1994, 9, 225–243.

Uddin, R.; Wahid, M. I.; Jasmeen, T.; Huda, N. H.; Sutradhar, K. B. Detection of Formalin in Fish Samples Collected from Dhaka City, Bangladesh. Stamford J. Pharm. Sci. 2011, 4, 49–52.

Varnasseri et al (2022) DOI: 10.1039/d2ay00219a rsc.li/methods

Voss, C. Veneno no Seu Prato? Utilidades e Riscos Dos Aditivos Alimentares, 3rd ed.; EDIDECO—Editores para a Defesa do Consumidor Lda.: Lisboa, Portugal, 2011

Wang, C. M., Roy, P. K., Juluri, B. K., & Chattopadhyay, S. (2018). A SERS tattoo for in situ, ex situ, and multiplexed detection of toxic food additives. Sensors and Actuators B: Chemical, 261, 218-225.

Whei Miaw, C. S., Assis, C., Carolino Sales Silva, A. R., Cunha, M. L., Sena, M. M., & Carvalho de Souza, S. V. (2018). Determination of main fruits in adulterated nectars by ATR-FTIR spectroscopy combined with multivariate calibration and variable selection methods. Food Chemistry, 254, 272-280.

Wistaff, E. A., Beller, S., Schmid, A., Neville, J. J., & Nietner, T. (2021). Chemometric analysis of amino acid profiles for detection of fruit juice adulterations–Application to verify authenticity of blood orange juice. Food Chemistry, 343, 128452.

Wójcik, S., & Jakubowska, M. (2021). Deep neural networks in profiling of apple juice adulteration based on voltammetric signal of the iridium quadruple-disk electrode. Chemometrics and Intelligent Laboratory Systems, 209, 104246.

Woźniak, Ł., Szczepańska, J., Roszko, M., & Skąpska, S. (2020). Occurrence of maltose in apple juices: Improved method of analysis, typical levels, and factors affecting it. LWT, 124, 109154. Wu et al (2018) doi: 10.1111/1750-3841.14177

Wu, H., Zhou, X., Chen, H., Jin, B., Yan, Z., Xie, L., ... & Lin, G. (2022). A comparative authentication study of fresh fruit and vegetable juices using whole juice and sugar-specific stable isotopes. Food Chemistry, 373, 131535.

Wu, N., Gao, W., Zhou, L., Lian, Y., Li, F., & Han, W. (2015). Identifying potential sources of Sudan I contamination in Capsicum fruits over its growth period. Food Chemistry, 173, 99-104.

Zhang, H., Wu, Z., Zhi, Z., Gao, W., Sun, W., Hua, Z., & Wu, Y. (2021). Practical and Efficient: A Pocket-Sized Device Enabling Detection of Formaldehyde Adulteration in Vegetables. ACS omega, 7(1), 160-167.

Zhang, X., Yang, J., Lin, T., & Ying, Y. (2021). Food and agro-product quality evaluation based on spectroscopy and deep learning: A review. Trends in Food Science & Technology, 112, 431-441.

11

Contamination in Your Cup of Tea? Recent Advances to Reveal Adulterants

K.G. Nelum P. Piyasena[1] and Mahasen A.B. Ranatunga[2]

[1]*Biochemistry Division, Tea Research Institute of Sri Lanka, Talawakelle Sri Lanka*

[2]*Plant Breeding Division, Tea Research Institute of Sri Lanka, Talawakelle Sri Lanka*

Tea as a Beverage

Tea is the most popular beverage consumed in the world after water. All types of teas including black. green, oolong, white and pu-erh tea are mainly made from the buds and young leaves of tea plants, *Camellia sinensis* (L.) O. Kuntze belongs to the family Theaceae [1]. Depending on the different processing techniques of tea leaves, tea types are categorized as "Green Tea" (unfermented), "Black Tea" (fully fermented), "Oolong Tea" (partially fermented), "Pu-ErhTea" (post fermented) and white tea (unfermented) [2]. The manufacturing process for these different types of teas mainly differ in the degree of enzymatic oxidation (fermentation) process [1]. For the manufacturing of green tea the polyphenol oxidase enzymes are deactivated by heat before rolling [3]. Black tea is extensively oxidized before being processed and is fully fermented form of tea, which has two main process types; orthodox and cut, tear, and curl (CTC) [4]. Oolong tea is semi fermented and partial oxidation is arrested by polyphenol oxidase enzyme deactivation during the process. Terminal bud and/or first leaves only are plucked for the production of white tea and dried with minimal processing [3].

ISO Definitions of Tea

Black Tea

As per the ISO 3720:2011, black tea is defined as "Tea derived solely and exclusively, and produced by acceptable process, notably withering, leaf maceration, aeration, and drying, from the tender shoots of varieties of the species *Camellia sinensis* (L.) O. Kuntze, known to be suitable for making tea for consumption as a beverage" [5]. Consequently, manufacturing of black tea is all about manipulating the biochemical and chemical reactions in an optimized way to ensure high quality end product [6].

Therefore, tea shall have no taint and shall be reasonably free from extraneous matter and is considered as 100% natural, safe product, as per the definition no extraneous matter can be added during tea manufacturing process [6].

Green Tea

According to ISO 11287:2011 green tea is defined as tea derived solely and exclusively, and produced by acceptable processes, notably enzyme inactivation and commonly rolling or comminution, followed by drying, from the tender leaves, buds, and shoots of varieties of the species *Camellia sinensis* (L.) O. Kuntze, known to be suitable for making tea for consumption as a beverage [[7]].

Oolong Tea

According to BS ISO 20716 oolong tea is defined as tea derived solely and exclusively, and produced by acceptable processes, notably withering, tumbling, light/medium/ high aeration, enzyme inactivation, rolling / shaping and drying, the tender shoots of varieties of the species *Camellia sinensis* (L.) O. Kuntze, known to be suitable for making tea for consumption as a beverage [[8]].

White Tea

According to ISO/DIS 20716 or ISO/TR 12591:2013(en) white tea is defined as tea derived solely and exclusively, and produced by acceptable processes, by harvesting and a single withering/drying stage of the bud or bud and tender shoots (one to three leaves) of varieties of the species *Camellia sinensis (L.) O. Kuntze* known to be suitable for making tea for consumption as a beverage [[8]]. Therefore, black, green, oolong and white tea shall have no taint and shall be reasonably free from extraneous matter [6,7,8].

Adulteration of Food

Food adulteration is one of the serious problems faced worldwide, it is raised considerable economic issues and increased health risks for consumers [[9]]. In addition, food fraud leads about adverse economic allegations for consumers and traders in food business. Consequently, it leads to collapse in consumer assurance, loss of sales and of tax from sales, implement stringent regulatory guidelines and increased costs for regulatory bodies. Consumers and stakeholders have been becoming gradually aware of food fraud, and researchers have suggested various approaches for preventing or identifying food fraud [[9]].

Adulteration of Tea

The questions related to tea adulteration are one of the serious issues emerging globally. Adulteration of teas has been recorded since 1800 to earn unreasonable profits through mixing with extraneous matters. An adulterant can be considered as any substances that has been added thereto or mixed or packed therewith so as to increase its bulk or weight, reduce its quality or strength, or make it appear better or of greater value than it is [[10]].

Trends in Adulteration of Tea

1. **Adulteration of tea by plant derived materials:** Adulteration of tea by plant derived products dated back to 1800. Tea was mixed with leaves from other plants including bilberry, willow, elder, sloe, hawthorn, and beech leaves [[11]] and leaves of Acer negundo, Albizia amara, Chenopodium ambrosioides, Epilobium angustifolium and Dendrocalamus hamiltonii bamboo [[12]]. Also, adulteration of tea with spent tea leaves (tea leaves that had been already brewed and dried) has been a common practice [[13]]. In some Asian countries, reports are available on adulteration of tea products with cashew nut husk, which is the cover surrounding the cashew kernel, is a very common and profitable practice that created global issue of food safety due to the detrimental effects to human health [[14]]. Also, reports are available mixing of tea with Acacia arabica powdered bark foreign vegetables, saw dust, husk of pulses, chicory and cereal starch [[15]].
2. **Adulteration of tea by non- plant derived materials:** Tea is also adulterated by colouring substances that increases the astringence such as catch, borax and sodium carbonate. Other adulterants of tea are iron, coal tar, leather flakes, sand powdered animal hide and fabric dyes [[16]].
3. **Adulteration of tea by sugar components *viz* sucrose, glucose and sugar syrups :** In recent years, it is evident that further to satisfy the customer's sensory needs and earn unreasonable profits, low-cost sugar and glucose syrup has been allegedly added to tea in the manufacturing process to improve the lustrousness and taste of tea, tighten the leaf strip and simultaneously achieve the weight gains. It is reported that there were a number of tea factories raided due to adding sucrose illegally these incidents have occurred frequently and it has become an open secret in the tea industry [[17,18]]. Tea with added sucrose and glucose is difficult to preserve, due to microbial contaminations and it affects the quality of tea and leads to health issues, etc. Moreover, sugar-added tea with low cost and quality will harm the interests of legitimate businessmen as well as consumers [18].
4. **Adulteration of tea with coloring agents:** The color of black tea infusion is a very important factor in determining the black tea quality, which depends on the concentration of theaflavins and thearubigins formed during the black tea manufacturing process [[1]]. Nevertheless, some unscrupulous tea traders deceitfully improve the appearance of black tea and black tea infusion by illegally adding artificial colorants or dyes to black tea [[19]]. It is revealed that the artificial colorant, carmine commonly used in the food industry is used by fraudulent tea producers to adulterate black tea, for the of earning illegal profits [[19]]. Further, black tea occasionally contains extraneous coloring material including bismark brown, potassium blue, turmeric, indigo, ferrous sulphate, and plumbago. These coloring materials are added to black tea to enhance its appearance [[20]]. Further, sodium bicarbonate is added to the black tea manufacturing process to improve the appearance of the end product [[21]].

Occasionally the resulting color of the black tea was not sufficient, accordingly anything from sheep's dung to poisonous copper carbonate was added to make the appearance of black tea more authentic [[19]].

5. **Adulteration of tea with different origins or inferior quality tea:** The chemical constituents of tea, is highly affected by geographical features and climate condition, such as characteristics of soil, exposure to sunlight, rainfall distribution and temperature, besides tea cultivars, harvesting time and processing methods [[22]]. All these parameters determine the quality characteristics of black, green and oolong tea and parameters are varied with geographical area of production [[23], [24], [25]]. As a result, tea produced from specific geographical origins possess unique quality characteristics and fetches high prices in the market. "Darjeelin", a unique tea produced in West Bengal, India which is certified as a Geographic Indicator (GI) since 2004 and gained a reputation for its distinctiveness and quality. Initially, produce Darjeelin black tea and currently expanded to produce green, white and oolong tea from some selected estates. In Sri Lanka, orthodox black tea produced in Western slope of the country popularly known as "Uva tea" has gained wide acceptance among global tea consumers for its unique tase and aroma during the flavour season. Similarly, Shandong green tea produced in Northern part of China has characteristic penetrating fragrance, thick mellow taste and long brewing time due to the day-night temperature difference and special geographic location than southern green tea. Tieguanyin is an oolong tea produced in China which contains the highest amino acids, vitamins, minerals, tea polyphenols, and alkaloids, which has the most nutritional and pharmacodynamic ingredients compared with other oolong teas, and its price is higher than other oolong teas [[26]]. The high prices gained by this unique tea tempted some manufacturers to mix them with inferior quality tea or tea from other origins [26]. This could be considered as an adulteration or fraudulent labelling of high value tea and many cases have been reported in the market.

Adulteration Detection in Tea

1 Detection of plant derived adulterants

1.1 Visual Techniques

Using microscopic and macroscopic techniques, tea adulteration with other herbs could be detected easily. Initial attempts on identification of adulteration of tea by parts from other plants using microscopic observations dated back to 1889. Taylor et al (1889) discussed that the presence of stone-cells or idioblasts in tea-leaves is unique to *Camellia sinensis* and their absence in all other plant-leaves could be used to detect adulterants of tea [[27]].

1.2 Biochemical Analysis

As tea leaves that had been already brewed are not contained the required amount of total polyphenols, fiber and ash content according to the ISO standards, tea adulteration with tea which have been already brewed could be identified by quantifying individual catechins, polyphenols, fiber, ash contents from tea extracts using ISO standards, it is possible to confirm if the tea has been adulterated [8].

1.3 DNA based Techniques

Molecular biological techniques or DNA based techniques have become popular for identification of plants and plant-based products due to the precise detection that can be made even with low quantity of material available. Lagiotis et al (2020) reported use of universal plant DNA barcoding marker coupled with High Resolution Melting (Bar-HRM) analysis to authenticate tea products from cashew ground nut. In the study, cashew-specific markers combination with HRM technology were used to detect adulterated tea samples with cashew DNA subsequently to quantify level of adulteration. The authors proves that the methodology can reliably detect admixtures as low as 1% v/v cashew in commercial tea products. Hence, HRM technology can be considered as a strong molecular approach in tea authentication, capable of detecting very low adulterations in DNA admixtures (Lagiotis et al 2020). In another study, Dhiman and Singh (2003) developed a simple PCR-based method to detect cashew husk adulteration in tea. They developed species-specific PCR primers from intergenic spacer regions of 5S ribosomal RNA genes and used successfully in the detection of cashew husk in tea market samples. Dhiman and Singh (2003) further discussed the potential of application of similar approaches to detect adulteration by other materials of plant or animal origin in tea and discuss the potential to develop a comprehensive technology for automation and simultaneous authentication of commercial teas [28, 29].

2 Detection of Non-plant Derived Adulterants

2.1 Physical Analysis

Tea adulterated with non- plant-derived materials could be identified by performing simple physical tests. The bulk density of tea defines as the ratio of the mass of the powder to its bulk volume that varies when mixed with solid particles. Therefore, by analyzing the bulk density of tea samples it could be judged whether the tea is adulterated. The value is the bulk volume of the tea sample and bulk density could be determined [11].

Iron particles adulterated with tea can be easily separated using a magnet [30]. In order to identify if tea is adulterated with leather flakes, a small quantity of tea is burnt to observe characteristic odour emitted from leather [30]. When shaking tea in water, adulterated sand particles get sedimented [30].

2.2 Chemical Analysis

As explained by (Mishra and Mishra 2017), when coal tar is present in tea, adulterated tea sample can be melted in a test tube and once 5 ml of dilute sulphuric acid or concentrated Hydrochloric acid is added, pink or crimson colour could be observed [30]. To identify catechu in tea, tea brew can be added to a porcelain dish washed with dilute solution of ferric chloride and green colour can be observed [30].

3 Detection of Coloring Agents

Though it is very difficult to identify color adulteration in tea with the naked eye, simple screening tests could be useful in detecting color in tea.

3.1 Visual Techniques

The light microscope could be used to visualize the color in a magnified manner under microscope and the coloring matter is appeared in small dots.

3.2 Physical techniques

Artificially colored tea could be detected by simply rubbing a small quantity of tea between thumb and fore finger. Artificially colored tea leaves a bright stain when rubbed. Another simple way to detect coloring in tea is to fill a clear glass with cold water and put some tea on the surface. If the color of water changes immediately, it is confirmed that tea has been dyed with water soluble colors [19, 20, 21].

3.3 Chemical Techniques

Sodium bicarbonate adulterated black tea showed high alkalinity beyond the ISO 3720. Black tea adulterated by ferrous sulphate resulted in a reduction of total polyphenol content which is beyond the ISO 3720:2011(Black tea-Definition and basic requirements) [20]. Detection of coloring agents in tea (which is either soluble or insoluble in water) could be identified by using chromatographic methods such as derivative spectrophotometry, HPLC (high-performance liquid chromatography) alone, or HPLC/MS, raman spectrometry, GC-MS (gas chromatography-mass spectrometry), as well as capillary electrophoresis and immunoassays [19, 20]. It is revealed that carmine in black tea was identified by using HPLC combined with a predictive model [19].

4 Detection of Sugars

4.1 High Performance Liquid Chromatography

The sugar components in the made tea sample were quantified according to Anan's method with a few alterations. A finely ground tea sample (2.5 g) was spiked with an internal standard, sorbitol (0.1 g), and extracted with hot water (50 mL) at 80 °C for 15 min. The supernatant (2 mL) was passed through pre-equilibrated ion-exchange resins after centrifugation. The eluate was filtered through a filter disk (0.45-*m*m) and sucrose, glucose, and fructose concentrations in the filtrate were quantified by using HPLC [31].

4.2 liquid Chromatography with Tandem Mass Spectrometry (LC-MS/MS)

A solid phase extraction-liquid chromatography-tandem mass spectrometry (SPE-LC–MS/MS) method was established for the simultaneous determination of fructose, glucose, and sucrose in tea. Made tea samples were ultrasonically extracted with ultrapure water and cleaned up by Oasis PRIME HLB SPE cartridge. The separation of the analyte was performed on X Bridge BEH Amide column (2.1 × 150 mm, 5 μm) with isocratic elution. Detection of the target compounds was carried out under the multiple reaction monitoring (MRM) mode with negative electrospray ionization. The qualitative screening was achieved with retention time and qualitative ion, then confirmation analysis with peak area and quantitative ion. The limit of detection (LODs, S/N ≥ 3) of fructose, glucose, and sucrose was 0.02 mg/L, 0.03 mg/L, and 0.08 mg/L, respectively. The average recoveries varied from 92.1 %–108.5 % for three levels. Linearity was obtained with the r (correlation coefficient) greater than 0.992 for sugar components. The average recovery of fructose, glucose, and sucrose was 98.80 % - 106.27 % [[10]].

Detection of Sugar Adulteration in Tea- Case Study

The inquiries on the adulteration of made tea with sucrose and glucose is one of the serious problems emerging frequently in tea industry. In recent years, it is evident that further to satisfy the customer's sensory requests and earn unreasonable profits, low-cost sugar has been allegedly added to made tea in the tea manufacturing process to improve the lustrousness as well as simultaneously achieve weight gains [[18]]. Moreover, it is reported that some enterprises make illegal profits by adding sugar and glucose syrup to improve the taste, tighten the tea leaf strip and reduce costs in roasted green tea [[17]]. Therefore, the method developed to detect sucrose and glucose-adulterated made teas is a crucial issue. As tea is a natural beverage derived from fresh tender leaves of the tea plant, the existence of sugar components in made tea is unavoidable.

Previous research studies revealed that the presence of sucrose, glucose and fructose content in tender tea shoot tips have been reported [[32]]. In addition to that, in tea extract solids, around 4% polysaccharides, 0.15% pectin, and 6.5% sugars including fructose, glucose, sucrose, *m* -inositol along with small amounts of maltose and raffinose have been determined [[33]]. Consequently, as made tea also contains inherent fructose, glucose and sucrose, prior to the detection, the establishment of baseline values on the inherent sugar components in made tea is an essential requirement. Once establishing the maximum permissible glucose, sucrose, and fructose levels in made tea, it is directly used as the detection of the adulteration of made tea with sucrose and glucose. The analytical methods should be rapid, simple and accurate which applicable for batch determination of fructose, glucose and sucrose concentrations in made tea. Therefore, a rapid and sensitive method was developed for the quantitative estimation of fructose, glucose and sucrose in black tea brew using high performance liquid chromatographic (HPLC)[[31]].

HPLC methodology was developed for the simultaneous quantification of fructose, glucose, and sucrose in made tea. The made tea (4 g) was ground through 0.5 mm mesh and homogenized with distilled water (50 mL) for 30 mins at room temperature, in a magnetic shaker, and centrifuged. The extraction (1 ML) was diluted with 4 mL of 100% Acetonitrile and filtered through 0.45 μM Nylon filters prior to the HPLC injection [35]. The dry matter content was measured by the loss in mass at 103 °C [36]. HPLC instrumentation with separation module coupled to Refractive Index Detector (RID) system, a built-in column oven, and a vacuum degasser. NH2 column (150 x 4.6 mm (5 μm)) was used. The column oven temperature was 33 ºC in RID. Injection volume and flow rate were 10 μL and 1.5 mL min-1 for all the standards and the samples. The method utilized an isocratic system with a mobile phase comprised of 83% Acetonitrile: 17 % water. The concentration values of 35 ppm and 50 ppm were the limits of detection and limit of quantification values respectively for glucose, sucrose, and fructose [35].

5 Detection of other Origin and Inferior Quality Tea

The isotope ratio, the mineral element content (elemental fingerprinting), the presence of volatile and non-volatile substances and other biochemical components used to determine origin of tea varied among climatic conditions, geographical locations and cultivation methods of the crop [37, 38,39]. Also, recently developed technologies, such as near infrared (NIR) spectroscopy, Raman spectroscopy [40] X-ray electronic tongue (E-tongue), electronic nose (E-nose), computer vision, integration of multiple sensors, Mobile technology are becoming very popular for origin verifications of tea and adulteration detection [41]. Combine use of instrumentation and chemometrics analyses with relevant data processing algorithms facilitate discrimination of suspected adulterations in tea. The recent attempts made on origin verification and adulteration detection in tea using spectroscopic techniques are summarized in Table 1.

Table 1: Summary of spectroscopic application on origin verification of tea

Tea type	Sample description	Methodology	Main conclusions	Reference
Black tea	140 black tea samples from different geographical origins	Fourier transform (FT) NIR spectroscopy	Discrimination of broken black tea originating in geographical origins (e.g. India, Kenya, Sri Lanka, and Yunnan) was not successful from that grown in other locations.	42
Black & green tea	31 green teas and 51 black tea of 05 brands of each type	NIR spectroscopy	All tea samples were discriminated into 5 differing tea classes (Argentinean green tea; Brazilian green tea; Argentinean black tea; Brazilian black tea; and Sri Lankan black tea) using 12 wave numbers alone with a developed SPA-LDA model. However, 100 % classification accuracy was not achieved when the full-spectrum was used with PLS-DA and SIMCA models.	43
Black & Herbal tea	07 varieties of tea samples (herbal Tea) black tea and earl grey tea	Raman spectroscopy	SERS spectral data can be used to detect any adulterant that in the content of tea samples. SERS in combination with PCA resulted better discrimination of different kinds of samples.	44
Black tea	38 widely differing black teas representing seven growing areas (China, Northern India, Southern India, Bangladesh, Sri Lanka, East Africa, Central Africa)	NIR spectroscopy	Reported a reasonable ability of successful discrimination of black teas of differing sensory profiles using wavelengths corresponding to absorption bands in the spectra of teas.	45
Black tea	Nine different tea samples are collected from the general stores	NIR spectroscopy	Nine varieties of commercial Indian teas were successfully discriminated.	46
Green tea	37 green teas were collected from five production zones in China	NIR spectroscopy	Green teas were classified according to the geographical origin using pretreated NIR spectra of tea powder and water extracted tea brew.	47
Green tea	240 samples of eight typical kinds of tea	NIR spectroscopy	The relationship between the reflectance spectra and tea varieties was established.	48

Tea type	Sample description	Methodology	Main conclusions	Reference
Green tea	All tea samples of four varieties came from different provinces in China	NIR spectroscopy	NIR spectroscopy technique has high potential to identify the tea varieties and determine simultaneously some chemical compositions in tea.	49
Green tea	50 green tea samples collected from local markets	NIR spectroscopy & UV-Vis spectroscopy	The resulting RF model based on the fusion-data had better potential to successfully discriminate 5 varieties of green tea.	50
Green tea	Six varieties from different provinces in China.	NIR spectroscopy	Quality of different tea was identified using NIR spectroscopy combined with CART.	51
Oolong & green tea	Seven varieties from five different provinces in China	Fourier transform (FT) NIR spectroscopy.	The FT-NIR spectroscopy technique coupled with the pattern recognition has a high potential to identify the tea varieties.	52
Oolong tea	450 samples collected from 30 main Tieguanyin-producing areas of Anxi County and 120 NATT samples from Yongchun, Huaan, Xiandu,	Fourier transform (FT) NIR spectroscopy.	Results demonstrated the feasibility of combining NIR spectroscopy and PLSDA for discriminating the geographical origin of Tieguanyin tea.	53

6. Analytical Methods Used to Test for Authenticity

Sensory assessment and tea tasting are the initial quality control methods used in the tea industry and accepted in internationally. Sensorial, visual and analytical controls of the tea are done both by the vender and purchaser (Table 2) [[11]].

Table 2: Analytical methods used to test for authenticity

Analytical method	Test for authenticity
Macroscopic analyses [[11]]	Identification of undeclared products or unwanted substances including non-vegetable materials, non-tea materials and molds
Sensory testing [[11]]	Judge the overall tea quality by considering the size and shape of leaf, colour and flavour. Organoleptically determine the market value of the final product and used throughout the whole tea supply chain.No ISO standard for tea sensory analyses
Determination of compounds [[11]]	Tea catechins, purine alkaloids, theanine and gallic acid provide information about the variety, category and geographical origin of the tea. ISO 14502-1 & ISO 14502-2

With respect to rapidly responding to tea adulteration, fraud, and unauthentic tea products, different techniques are used for instantly determining chemical composition. The methods for authenticity testing are summarized in table 3.

Table 3: Overview of methods for authenticity testing [[11]]

Analytical techniques	Indicative data or analyte	Authenticity issue/ information
Sensory analysis	Visual, olfactory and taste scores	Tea grades, adulterations
Microscopic observation	Visual assessment	Adulterations
Sieving	Foreign bodies	Adulterations
HPLC	Catechins and theaflavins	Geographical origin
HPTLC	Catechin compounds	Geographical origin
CE	Catechins, caffeine, theanine and other amino acids	Geographical origin
IRMS	Catechins and theaflavins	Geographical origin
NIR	Catechin compounds	Tea categories, geographical origin, verities, age
NMR	Amino acids, organic acids, caffeine and catechins	Geographical origin and quality
SSR fingerprinting	Genome	Varietal identification
SNP fingerprinting	Genome EST transcriptome	Varietal identification
E-Nose	Volatile organic compounds	Tea grades, types, varieties, categories, geographical origin, storage times

Analytical techniques	Indicative data or analyte	Authenticity issue/ information
E-Tongue	Catechins, amino acids and caffeine	Tea varieties, geographical origin, grades, fermentation degree
Multi-spectral imaging	Image analysis	Tea categories, grades, brands
Hyperspectral Imaging	Image analysis	Tea grades, quality
Normal Camera Imaging	Image analysis	Tea grades, varieties, colour

Safe limits/Tolerable Limits for Contaminants in Tea: International Regulations/Laws

ISO is a worldwide federation of national standards bodies (ISO member bodies) which has produced several standards relating to tea. The standards are produced to provide world-class specifications for tea products, services and systems, to ensure quality, safety and efficiency. The ISO methods produced mainly compromise of definition and basic requirements of black, green, white and instant tea. ISO has also developed a range of standard methods of determining quantities of important chemical parameters of tea which include: total ash, water-soluble ash and water-insoluble ash, acid-insoluble ash, alkalinity of water-soluble ash, total polyphenols, catechins, theanine and crude fiber. There is no specific regulation on tea and tea products in the European Union. All these products fall under general food regulations. Tea producing countries are established minimum quality standards for tea representing foreign matters (completely free), detectable metals (acceptable levels for iron copper, lead, zinc, cadmium and sodium), pesticides residue and non-pesticides residue.

Furthermore, made tea adulterated with sugars would harm the interests of genuine tea producers as well as consumers. Presently, international standards for the contents of sucrose, fructose and glucose in made tea are not declared. However, several research studies revealed the concentration range of sucrose, fructose, and glucose in made tea. Also, made tea adulterated with sugars would harm the interests of genuine tea producers as well as consumers. Presently, international standards for the contents of sucrose, fructose and glucose in made tea are not declared. However, several research studies revealed the concentration range of sucrose, fructose, and glucose in made tea. It is reported that, currently, regulatory bodies in tea-producing countries have a law enforcement basis for the detected tea adulterants. These regulatory bodies are developed methods for on-site evidence collection, treatment in the production field, and analytical tools used to detect tea adulterants. It is not only difficult to enforce the law, but also open opportunities for unscrupulous traders to replace high-quality made tea with shoddy tea.

Tea Adulteration: Impact on Health and Wellness

Tea adulterated with sucrose and glucose leads to microbial growth and is difficult to store. It would affect the quality of made tea and occur health issues for the consumers [18]. Adulterated tea with non-plant-derived agents ie coloring agents and other plant derived agents, may be toxic and affect health and cause intoxication or problems such as allergy in sensitized individuals [11].

Future Prospects

Chromatographic methods including HPLC (high-performance Liquid chromatography), liquid chromatography with tandem mass spectrometry (LC-MS/MS), and gas chromatography-mass spectrometry (GC-MS) have been utilized to determine the adulterants in made tea. However, these hi-tech instruments are highly sophisticated, expensive, laborious, and time-consuming. And also these analytical methods require a complex method for sample preparation, skilled people for operating the instruments, and/or large amounts of solvents used for detection. Therefore, these hi-tech instruments are not suitable for quality assurance during tea processing/or made tea. Hence, rapid analysis or online monitoring techniques, as well as reliable and inexpensive techniques, is required for the convenient identification of adulterants used in tea.

References

1 Engelhardt U. H. (2010). Chemistry of tea. Elsevier Ltd, 1000-1027.

2 Carvalho, E., Punyasiri, P. N., Somasiri, H. S., Abeysinghe, I. S. B., & Martens, S. (2014). Quantification of γ-aminobutyric acid in Sri Lankan tea by means of Ultra Performance Tandem Mass Spectrometry. Natural Product Communications, 9(4), 1934578X1400900426.

3 Hilal, Y., & Engelhardt, U. (2007). Characterisation of white tea–Comparison to green and black tea. Journal für Verbraucherschutz und Lebensmittelsicherheit, 2(4), 414-421.

4 Wickremasinghe, R. L. (1978). Monograph on tea production in Sri Lanka No: 07. Tea Research Institute of Sri Lanka, Talawakelle, Sri Lanka, 16-17.

5 ISO 3720 (2011) Black Tea-Definition and basic requirements.

6 Robertson, A. (1992). The chemistry and biochemistry of black tea production—the non-volatiles. In Tea (pp. 555-601). Springer, Dordrecht.

7 ISO 11287 (2011) Green Tea-Definition and basic requirements.

8 https://www.iso.org

9 Spink, J. (2014). Safety of food and beverages: risks of food adulteration.

10 Forte, W. E. (1965). The Food and Drug Administration and the Economic Adulteration of Foods. Ind. LJ, 41, 346.

11 Lees, M., & Morin, J. F. (Eds.). (2018). FoodIntegrity Handbook: A Guide to Food Authenticity Issues and Analytical Solutions. Eurofins Analytics France.

12 Garg, S. (1992). Substitute and adulterant plants. Periodical Exports Book Agency.

13 Sudershan, R. V., & Bhat, R. V. (1995). Changing profile of food adulteration: perception of food analysts. Journal of Food Science and Technology, Mysore, 32(5), 368-372.

14 Breiteneder, H., & Radauer, C. (2004). A classification of plant food allergens. Journal of allergy and clinical immunology, 113(5), 821-830.

15 Israel, A. H., & Issar, R. K. (1973). Adulterated drugs and their standardization. Ind J Pharm, 35, 208-9.

16 Dhiman, B., & Singh, M. (2003). Molecular detection of cashew husk (Anacardium occidentale) adulteration in market samples of dry tea (Camellia sinensis). Planta medica, 69(09), 882-884.

17 Luqing, L., Lingdong, W., Jingming, N., & Zhengzhu, Z. (2015). Detection and quantification of sugar and glucose syrup in roasted green tea using near infrared spectroscopy. Journal of Near Infrared Spectroscopy, 23(5), 317-325.

18 Wang, H., Hu, L., Zhou, P., Ouyang, L., Chen, B., Li, Y., ... & Zhou, J. (2021). Simultaneous determination of fructose, glucose and sucrose by solid phase extraction-liquid chromatography-tandem mass spectrometry and its application to source and adulteration analysis of sucrose in tea. Journal of Food Composition and Analysis, 96, 103730.

19 Wei, L., Yang, Y., & Sun, D. (2020). Rapid detection of carmine in black tea with spectrophotometry coupled predictive modelling. Food Chemistry, 329, 127177.

20 http://www.teaboard.gov.in

21 http://www.erepo.lib.uwu.ac.lk/handle/123456789/6995

22 Jayasekera, S., Kaur, L., Molan, A. L., Garg, M. L., & Moughan, P. J. (2014). Effects of season and plantation on phenolic content of unfermented and fermented Sri Lankan tea. Food Chemistry, 152, 546-551.

23 Aisaka, H., Kosuge, M., & Yamanishi, T. (1978). Comparison of the flavors of Chinese 'Keemun'black tea and Ceylon black tea. Agricultural and Biological Chemistry, 42(11), 2157-2159.

24 Owuor, P. O., Obanda, M., Nyirenda, H. E., & Mandala, W. L. (2008). Influence of region of production on clonal black tea chemical characteristics. Food Chemistry, 108(1), 263-271.

25 Wickremasinghe, R. L. (1974). The mechanism of operation of climatic factors in the biogenesis of tea flavour. Phytochemistry, 13(10), 2057-2063.

26 Hu, Y., & Kang, Z. (2022). The Rapid Non-Destructive Detection of Adulteration and Its Degree of Tieguanyin by Fluorescence Hyperspectral Technology. Molecules, 27(4), 1196.

27 Taylor, T. (1889, January). Microscopic investigations relating to tea and its adulterations. In Proceedings of the American Society of Microscopists (Vol. 11, pp. 46-52). American Microscopical Society, Wiley.

28 Lagiotis, G., Stavridou, E., Bosmali, I., Osathanunkul, M., Haider, N., & Madesis, P. (2020). Detection and quantification of cashew in commercial tea products using High Resolution Melting (HRM) analysis. Journal of food science, 85(6), 1629-1634.

29 Dhiman, B., & Singh, M. (2003). Molecular detection of cashew husk (Anacardium occidentale) adulteration in market samples of dry tea (Camellia sinensis). Planta medica, 69(09), 882-884.

30 Rakesh Kumar Mishra, M. K. Mishra (2017) Analysis of tea powder for adulterant IOSR Journal of Pharmacy and Biological Sciences Volume 12, Issue 4 Ver. VI (Jul – Aug 2017), PP 37-42 .

31 Mizukami, Y., Kohata, K., Yamaguchi, Y., Hayashi, N., Sawai, Y., Chuda, Y., ... & Yoshida, M. (2006). Analysis of acrylamide in green tea by gas chromatography– mass spectrometry. Journal of Agricultural and food Chemistry, 54(19), 7370-7377.

32 Piyasena, K. N. P., Hettiarachchi, L. S. K., Jayawardhane, S. A. D. P. S., Edirisinghe, E. N. U., & Jayasinghe, W. S. (2022). Evaluation of inherent fructose, glucose and sucrose concentrations in tea leaves (Camellia sinensis L.) and in black tea. Applied Food Research, 2(1), 100100.

33 Mander, L., & Liu, H. W. (2010). Comprehensive natural products II: chemistry and biology (Vol. 1). Elsevier.

34 Jayawardhane, S. A. D. P. S., Gunarathna, N. A. J. C., Lakshman, P. L. N., Jinendra, B. M., Perera, G. A. A. R., Mewan, K. M., ... & Fernando, M. S. C. (2016). Development of a Method for The Detection of Sugar Adulteration in Black Tea. In Proceedings of the 6 th plantation crop research symposium, Held at BMICH, in Novermber.

35 Jayawardhane, M. H. S. A. D. P. S., Edirisinghe, E. N. U., Fernando, M. S. C., & Tharangika, H. B. (2015). Development and validation of a hplc based analytical method towards the determination of sugar concentration in processed black tea. Biochemistry DivisionTea Research Institute of Sri Lanka-Thalawakele, 23(5), 317-325.

36 ISO-1573. (1980). Tea. Determination of loss in mass at 103°C.

37 Kelly, S., Heaton, K., & Hoogewerff, J. (2005). Tracing the geographical origin of food: The application of multi-element and multi-isotope analysis. Trends in Food Science & Technology, 16(12), 555-567.

38 Zhao, H., & Yang, Q. (2019). The suitability of rare earth elements for geographical traceability of tea leaves. Journal of the Science of Food and Agriculture, 99(14), 6509-6514.

39 Wang, Y., & Ho, C. T. (2009). Polyphenolic chemistry of tea and coffee: a century of progress. Journal of agricultural and food chemistry, 57(18), 8109-8114.

40 Buyukgoz, G. G., Soforoglu, M., Basaran Akgul, N., & Boyaci, I. H. (2016). Spectroscopic fingerprint of tea varieties by surface enhanced Raman spectroscopy. Journal of food science and technology, 53(3), 1709-1716.

41 Chen, Q., Zhang, D., Pan, W., Ouyang, Q., Li, H., Urmila, K., & Zhao, J. (2015). Recent developments of green analytical techniques in analysis of tea's quality and nutrition. Trends in Food Science & Technology, 43(1), 63-82.

43Ren, G., Wang, S., Ning, J., Xu, R., Wang, Y., Xing, Z., & Zhang, Z. (2013). Quantitative analysis and geographical traceability of black tea using Fourier transform near-infrared spectroscopy (FT-NIRS). Food Research International, 53(2), 822-826.

44 Diniz, P. H. G. D., Gomes, A. A., Pistonesi, M. F., Band, B. S. F., & de Araújo, M. C. U. (2014). Simultaneous classification of teas according to their varieties and geographical origins by using NIR spectroscopy and SPA-LDA. Food Analytical Methods, 7(8), 1712-1718.

45 Buyukgoz, G. G., Soforoglu, M., Basaran Akgul, N., & Boyaci, I. H. (2016). Spectroscopic fingerprint of tea varieties by surface enhanced Raman spectroscopy. Journal of food science and technology, 53(3), 1709-1716.

46 Osborne, B. G., & Fearn, T. (1988). Discriminant analysis of black tea by near infrared reflectance spectroscopy. Food chemistry, 29(3), 233-238.

47 Kumar, S., Panchariya, P. C., Prasad, B., & Sharma, A. L. (2013). Discrimination of Indian tea varieties using UV-VIS-NIR spectrophotometer and pattern recognition techniques. Int. J. of Comp. Sci. and Comm., 2, 15-9.

48 Long Zhang,aJiarong Pan,bCheng Zhu (2015) The Use of Near Infrared Spectroscopy for Geographical Origin Traceability of Chinese Green Teas. Journal of Food Engineering and Technology 4:1.

49 He, Y., Li, X., & Deng, X. (2007). Discrimination of varieties of tea using near infrared spectroscopy by principal component analysis and BP model. Journal of food Engineering, 79(4), 1238-1242.

50 CheN, Q., Zhao, J., Liu, M., & Cai, J. (2008). Nondestructive identification of tea (*Camellia sinensis* L.) varieties using FT-NIR spectroscopy and pattern recognition. Czech J. Food Sci, 26(5), 360-367.

51 Wang, X., Huang, J., Fan, W., & Lu, H. (2015). Identification of green tea varieties and fast quantification of total polyphenols by near-infrared spectroscopy and ultraviolet-visible spectroscopy with chemometric algorithms. Analytical Methods, 7(2), 787-792.
52 Tan, S. M., Luo, R. M., Zhou, Y. P., Gong, H., & Tan, Z. (2012). Rapid and non-destructive discrimination of tea varieties by near infrared diffuse reflection spectroscopy coupled with classification and regression trees. African Journal of Biotechnology, 11(9), 2303-2312.
53 Chen, Q., Zhao, J., Zhang, H., & Wang, X. (2006). Feasibility study on qualitative and quantitative analysis in tea by near infrared spectroscopy with multivariate calibration. Analytica chimica acta, 572(1), 77-84.
54 Yan, S. M., Liu, J. P., Xu, L., Fu, X. S., Cui, H. F., Yun, Z. Y., & Ye, Z. H. (2014). Rapid discrimination of the geographical origins of an Oolong tea (Anxi-Tieguanyin) by near-infrared spectroscopy and partial least squares discriminant analysis. Journal of Analytical Methods in Chemistry.

12

Contamination of Your Cup of Coffee? Recent Methods to Reveal Adulteration

Pal Murugan, Alan Jerish J., Abhishek B.S., Santosh Pal Wadikar D.D. and A.D. Semwal

Defence Food Research Laboratory, Defence Research and Development Organization, Mysore, Karnataka, India

Introduction

Coffee is a beverage made from the roasted and ground seeds of the African tropical evergreen coffee plant. Coffee is one of the world's three most popular beverages (together with water and tea), as well as one of the most profitable international commodities, and it is the world's second most marketed product after petroleum (Myhrvold, 2021).Coffee is a popular beverage that is enjoyed around the world for its taste, health benefits, and ability to enhance physical performance (Simkin, 2008).

Coffee plants, which normally bear fruit for up to 20 years, are kept short to best conserve their energy and facilitate harvesting. Each tree is characterized by two waxy, green leaves growing opposite each other in pairs. It takes around 12 months for a coffee cherry to fully mature after it first blooms, and approximately 5 years to reach maximum fruit production. Although these trees have the potential to live for up to 100 years, they are most productive between the ages of 7 and 20, and can yield up to 10 pounds of coffee cherry annually, or 2 pounds of green beans (www.ncausa.org).

Fig. 1: Coffee Cherry (Image source: www.pbs.org)

Classification of Coffee Plants

Roasted seeds harvested from trees in the Rubiaceae botanical family, namely the species Coffea, are the source of the beloved coffee that millions of people enjoy across the world (A.P. Davis. 2006). Carolus Linneaus, a Swedish botanist, originally described the genus in the 18th century. In 1753, he also described *Coffea arabica* in his book Species Plantarum. Since coffee plants may vary greatly, botanists have argued on the proper categorization ever since. They can range in size from little shrubs to large trees, and their leaves can range in colour from purple or yellow to the more common dark green (www.ncausa.org).The coffee tree belongs to the Angiospermae subkingdom of plants, which means that it reproduces by way of seeds that are housed in a box-like compartment called the ovary at the base of the flower. They are shrubs that naturally grow in the shade of African forests, such as those found on the islands of Madagascar and Mauritius (the latter of which was formerly referred to as a member of the Mascarene Islands), and they are also grown in tropical regions like equatorial Africa, Java, Sumatra, and other Dutch East Indies, West Indies, India, Arabia, the islands of the Pacific, Mexico, and Central and South America (Davis AP. 2007).

There are between 25 and 100 different types of coffee plants, with *Coffea arabica* and *Coffea canephorais* being the two most often utilised commercially. In addition to Arabica and Robusta coffees, other species including Libercia (Coffea liberica) and Excelsa (Coffea excelsa (or) Coffea liberica var dewevre), which provide just 1 percent of the world's total supply and are mostly farmed in Africa, (Rubayiza, 2005)

Since all other species in the genus are diploid and typically incompatible with one another, *C. arabica* is the only tetraploid species (2n = 4x = 44) that is self-fertile. The original coffee trees were found in Ethiopia, and *Coffea arabica* is a descendant of those. These trees account for around 70 percent of the world's coffee output and yielda good, mellow, fragrant coffee (Charrier & Berthaud, 1985). Due to its organoleptic features, which tend to convey more fruity notes, a buttery odour, and a roastyscent, Arabica coffee is thought to be superior to Robusta and is, therefore, more costly(Sanz, C 2002). The taste of robusta coffee has been described as bland, under whelming, and occasionally having a strong, noticeable bitterness (Alonso-Salces & R.M, 2009). Arabica coffee is more expensive than Robusta because of better taste profiles and other agronomical advantages (such as growing at higher altitude sand being more susceptible to illnesses) (Valdenebro, M.S. 1999).

In order to meet the need for plants with greater output, disease resistance, and improved cup quality, a number of novel cultivars have been developed along the way

From chosen variety, and new wild kinds have also been identified throughout time. The genus Coffea now has approximately 100 species included in the catalogue (A. P. Davis, 2006).

The Structure of the Coffee Cherry

Coffee bean, parchment, silverskin, pulp and skin all make up a coffee cherry. Ripeness causes the pericarp, often known as the skin, to turn crimson below the skin, there is the yellowish, fibrous and sweet pulp, the outer mesocarp. This covers a thin layer of mucilage, the so-called pectin layer. The next component is the parchment or endocarp, which is followed by the silverskin. In the centre of the fruit, there is the coffee bean, the endosperm of the fruit (Esquivel, P., & Jimenez, V. M. 2012).

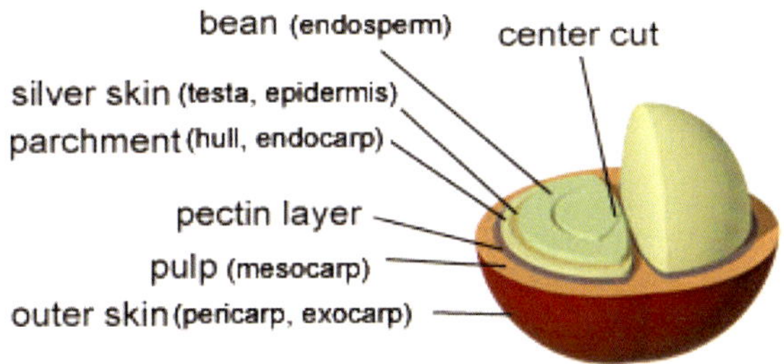

Fig. 2: Coffee cherry anatomy (Image source: www.ncausa.org)

Two beans are positioned side by side inside the parchment and are each individually protected by a further thin membrane. Although this seed skin's scientific term is the spermoderm, it is more often referred to as silver skin in the coffee industry. One bean is found inside the cherry in just 5 percent of the coffee produced worldwide. It is a natural mutation known as a peaberry (also known as a caracol, or "snail" in Spanish). Peaberries, according to some, are really sweeter and tastier than regular beans (www.ncausa.org).

Processing of Coffee Bean

The two main methods used in the initial processing of coffee are dry processing and wet processing. In the dry processing technique, the ripe cherry is left to dry in the sun and the silverskin, parchment, and pulp are removed by mechanical means. In the wet processing technique, the ripe cherry is soaked in water and the silverskin and parchment are removed, leaving the green bean and the pulp. The green bean is then dried and ready for roasting. In both techniques, only 50-55 percent of the dry mass of the ripe cherry consists of the commercially valuable green coffee (Vincent. 1987). In addition, the wet processing method is more labour-intensive and requires more water than the dry processing technique. Furthermore, wet processed coffee is typically more acidic and flavourful than dry processed coffee. The process flowchart is given below

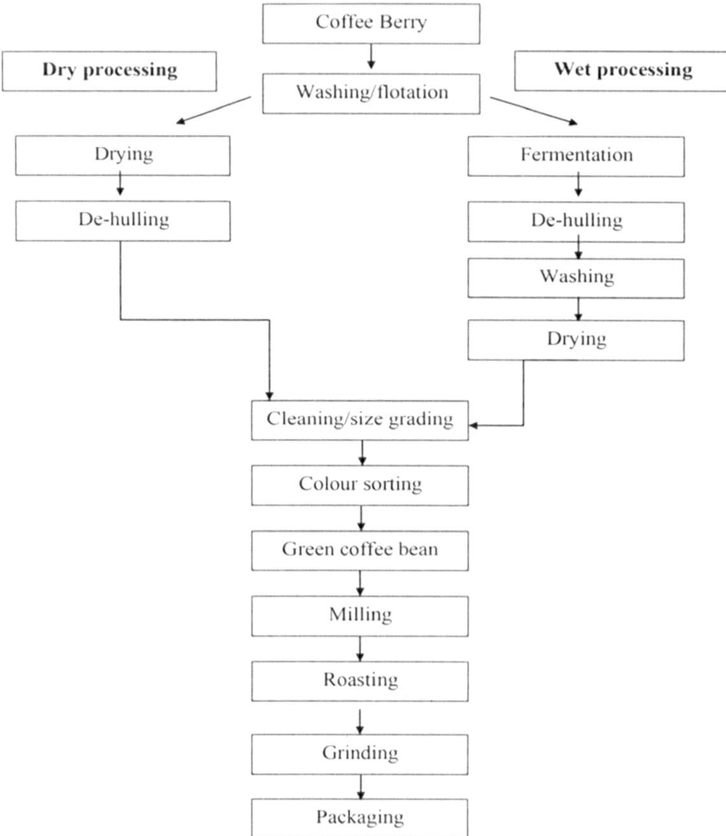

Fig. 3: Coffee Processing Flowchart

Dry Processing is the easiest method for processing coffee cherries is dry processing. Thecoffeecherries are dried to a moisture percentage of 10-11 percent after harvest. After that, coffee beans are separated by having the dehulling machine remove the beans' outerskin, pulp, parchment and silverskin. Silverskin and coffee husks (outer skin, pulp and parchment) are the names for the generated solid leftovers. This method of processingis used in Brazil for both Robusta and the vast majority of *Arabica coffees*. This can be ascribed to the method's simplicity and the availability of vast spaces for sun-drying coffee cherries. Additionally, this procedure is practical for picking fruits using the stripping method, which concurrently selects ripe, overripe and unripe fruits (Vincent, 1987). Drying can be done "naturally" or "artificially," depending on the situation. In big farms, the approach that is most frequently used is natural or solar drying.It has the benefit of not requiring energy expenditures or equipment investments.But it needs extensive drying zones (usually concrete surfaces). The fruit are often spread out on a thin layer throughout the lengthy, three to four week process to prevent fermentation. Additionally, regular raking is necessary to maintain uniform drying conditions and prevent the growth of mould (Sivetz & Desrozier, 1979).

It is possible to use artificial drying in addition to or as a replacement for natural drying. The process can be continuous or done in batches, and a variety of equipment is used, including static, rotary, horizontal and vertical dryers (Borém, 2008).

Wet Processing

The cherries do not need to be dried for this technique. In this method of processing, the outer skin and pulp are first removed mechanically, leaving the solid residue known as coffee pulp. The beans can either go through a fermentation process to remove any remaining pulp material, in which case the finished product is known as pulped coffee (café despolpado), or they can go straight through drying, in which case the name of the finished product is dehulled cherry coffee (café cerejadescascado). The beans are once again dehulled to remove the parchment in both situations after drying to a moisture level of around 12 percent. The resultant solid trash, consisting of parchment and silverskin, is referred to as parchment husks.

Following the removal of the coffee husks/pulp, a process known as curing is carried out that involves washing, size grading, density and colorimetric sorting, and storing of the green coffee beans. The main method used to separate faulty and non-defective coffee beans is electronic colour sorting (Franca & Oliveira, 2008). Coffee beans travel via an electronic eye or camera system in the electronic sorters one at a time. Depending on wavelength measurements, the bean is either permitted to pass or it is fired with a puff of air into a reject pile.

Roasting

The taste and aroma that are trapped inside green coffee beans are released during roasting. Beans are stored green, a state in which they can be kept without loss of quality or taste. As the beans are quickly heated to extremely high temperatures, such as 180 to 250 °C for 7 to 20 minutes, roasting results in chemical changes. When they reach the peak of perfection, they are quickly cooled to stop the process. Roasted beans smell like coffee, and weigh less because the moisture has been roasted out. They are crunchy to the bite, ready to be ground and brewed (*www.ncausa.org*). During roasting a huge number of changes occur on the coffee beans. Their volume becomes twice the green beans, while the weight decreases of 15-20 percent mainly for water evaporation. Also, chemical composition significantly modifies. Caffeine content decreasesduetosublimation and dragging with vapor, while several thousands of chemicalcompounds are obtained improving the sensorial profile of the beans and enriching their antioxidant activity (Severini, 2018).

Grinding

It is the closest step to making a coffee cup that hardly changes its quality. Being all the brewing methods based on solid-liquid extraction, the particle size of ground coffee has a great influence on the sensorial and healthy properties of coffee. Among those commonly used around the world several differences exist (Ephraim, 2016.)

Instant Coffee

Roasted and ground coffee beans are used to create instant, or soluble, coffee. The coffee flavour and fragrance are then recovered by extracting the ground beans with hot water. The procedure is comparable to making coffee at home in a percolator. Next, one of two methods for drying the coffee extract is used.

Spray-drying

The coffee extract is blasted into a heated air stream at the top of a tall cylindrical tower during spray drying. The droplets dry as they fall, and by the time they reach the bottom, they are a fine powder. To make dosing and dissolving easier, the powder may then be texturized into granules. Thanks to the extremely quick drying that takes place during this procedure, the purity of the fragrance and flavour is kept. The most popular method of drying is spraying.

Freeze-drying

The coffee extract is chopped into granules and frozen to a temperature of around -40°C for freeze-drying. The frozen granules are then vacuum-and low-temperature dried. The extremely low temperature and delicate drying procedures preserve the purity of the fragrance and flavour.

Coffee Adulterants: Scope of Detection and Procedures

The adulteration of ground coffee, particularly Arabica and Robusta varieties, is a major concern for the food industry, regulators and retailers due to potential economic and health risks for consumers. This issue is of particular importance due to the high value of the product and the high levels of consumption. As such, detecting fraudulent or accidental adulteration is critical. Additionally, it is important to note that coffee adulteration is not only a financial issue, but can have serious health implications as well. The commonly used adulterants in the case of coffee are as follows

Chicory

Chicory *(Chichorium intybus)* also known as blueweed, coffee-weed, blue sailors, bunk, or blue daisy, is familiar to many of us as an erect weed along roadsides or in fields, displaying small, startlingly bright blue flowers (Mitich, 1993).It is a typical adulterant present in coffee powder that is made from chicory plant roots. Trade in coffee has grown to be a very successful and wealthy industry. Only behind crude oil, raw coffee is now the second most valuable commodity traded internationally. Chicory, which has the same flavour and scent as coffee, was also added more, which results in adulteration.

Caramel

Caramel is a medium to dark-orange confectionery product made by heating a variety of sugars. It can be used as flavouring and colouring agent. Additionally, it serves as an emulsifier and a colouring ingredient in coffee adulterants.

Date Seeds

Dates (*Phoenix dactylifera*), seeds of dates were commonly used as an adulterant in coffee. They look similar to the coffee bean and they also have no toxic effect. So it is roasted along with the coffee beans and grinded into coffee powder. It is also known as a non-caffeine source of coffee adulteration.

Clay

Fine clay from mud is grinded coarsely into a powder and they are added in the coffee to increase the quantity of coffee and to reduce the market price. Usually these kinds of adulterants are seen in brewing cups.

Coffee Husk and Stem

The stem and coffee husk are components of the coffee plant. Additionally, they are frequently utilised as adulterants in coffee powder. Along with the coffee beans, the plant components were dried and ground into powder.

Maize and Barley

Barley (*(Hordeum vulgare)* and maize (*Zea mays L.*), which are also untraceable, have qualities comparable to those of coffee. To make more cups, roasted corn and barley were also ground with the coffee. According to the report, roasted corn and barley made up 5-20 percent of the final coffee powder produced by coffee manufacturers.

Wheat Middling

Wheat middlings (also known as mill feed, wheat mill run, or wheat midds) are the by-product of wheat flour production obtained by the milling of wheat grains (*Triticum aestivum L.*) which is roasted and added with the coffee as a non-caffeine adulterant.

Brown sugar

The inclusion of molasses gives brown sugar, a sucrose sugar product, its unique brown hue. It is either an unprocessed or partially refined soft sugar made up of sugar crystals that still include some molasses. Coffee adulteration frequently involves the use of this sugar. It increases coffee's consistency and flavour and is undetectable.

Soybean

A type of legume native to East Asia, the soybean, soy bean, or soya bean (*Glycine max*) is widely cultivated for its edible bean, which has a variety of applications. They may be used to generate a beverage that tastes quite similar to coffee after being roasted and roughly ground. Because soy beans are similar to coffee and less expensive, they are increasingly being adulterated in coffee.

Rye and Triticale

Rye (*Secale cereale*), a grass that is widely used as a grain, and Triticale, a wheat and rye hybrid used as a cover crop and grazing crop. It belongs to the Triticeae family of

Plants, which includes wheat and barley, and is closely related to both. Rye grain is utilized to make animal feed, flour, bread, beer, crisp bread, certain whiskeys and some vodka. Rye is powdered and added after being roasted to lower the price of the coffee.

Acai

The acai berry is an inch-long, reddish-purple fruit. It comes from the acai palm tree, which is native to Central and South America. They are also used in adulteration of coffee as a recent trend.

These are generally mixed with the coffee beans before grinding. Once the mixture is ground, it is tough to find out the presence of any of these just by way of visual inspection. However, all these adulterants have different properties from coffee beans, and one can rely on these properties to determine the quality of the coffee.

Testing Methods

The techniques and methods for testing for adulterants in coffee can be done in a lab or at home and described below.

Household Tests

Pinch test	Take some coffee and pinch it between your fingers, it should form asolid, cake-like texture if it is adulterated with chicory.This is becauseCoffee grains are hard and large; they do not stick together. The grains ofchicory, however, are small and soft and tend to stick to each other. So, ifthe grains in your coffee like to stick together, it's probably not coffee.
Water droplet test	Put some coffee powder on a dish in a pinch. Add some water to it.Chicory is present if it rapidly absorbs the water and turns soft. It iscoffee if it takes some time to absorb water and stays firm.
Tumbler test	Pick up a clear glass. Add water to it. Gently sprinkle some coffeepowder over it. Coffee is the powder if it floats for a while beforesinking. The powder is chicory or another seed if it sinks rapidly. Itincludes caramel or chicory if it quickly emits a brownish or yellowishcolour.
Test for coffee	A clear glass of water should include 1/2 teaspoon of coffee powder.After one minute of stirring, let it aside for five minutes. Look at thebottom of the glass. No clay flakes will settle to the bottom when usingpure coffee powder. Clay particles will accumulate at the bottom of acoffee powder mixture if it is tainted.

(www.citygreens.in); (www.fssai.gov.in)

Lab-based Tests

Microscopic Method

The foundation of optical microscopy is the employment of ocular and objective lenses to focus on the object being investigated. Methods based on optical microscopy for the identification of roasted and ground coffee that has been tampered with were presented in the 1950s. These methods need sample processing such as sifting, drying, And degreasing using an organic solvent (Menezes & Bicud, 1958). The results of the studies rely on how closely the properties of the unknown material and roasted coffee particles correspond. By lighting the object with radiation that has a shorter wavelength than visible light, resolution may be increased. Since the depth of field and magnification are inversely related, a flawless smooth surface under observation is desirable but impractical for coffee analysis.In this instance, chemical reagents are used to produce the microscope slides, and the measurement of contaminants is done by contrasting the sample's aqueous extract percent with that of pure coffee. The morphologyof a solid sample may be quickly determined using Scanning Electron Microscopy (SEM), which can also be used to determine the chemical components that arepresent.In place of the photons used in traditional optical microscopy, it employs an electron beam.Instead of using photons to explore the surface of the sample, SEM worksonthe principle of using small-diameter electron beams that travel in successive lines, exploring the sample point by point. The detector signal is then sent to a cathodicscreen whose scan rate is perfectly synchronised with the incident beam (Lopez, 1983).

Infrared Spectroscopy

The development of direct injection techniques has replaced the separation of analytes in the past and provides a perfect digital representation of the sample. Previously, spectrometric techniques were only employed to detect the analytes. One of the physical methods for identifying adulterants in coffee is infrared (IR) spectroscopy. It is renowned for requiring less time for the analysis, requiring fewer samples processing, and producing less trash after the analysis. There are difficult to analyse because of the intense peak overlapping caused by the compounds in the roasted coffee. Coffee adulteration has been frequently detected using FT-IR techniques (Silva, 2011).

Multispectral Imaging

This method of detecting coffee tainted with maize and barley uses reflectance technology. Given that various organic components included in ground coffees have diverse spectral signatures, the approach was based on image analysis. A charge coupled device (CCD) camera and a glass magnifier were used to create multispectral pictures of coffee samples. Red, Green and Blue (RGB) photos of the sample were taken by the camera and afterwards processed by a digital image processing application to determine the relative proportions of the sample's constituent parts. The photos were quickly captured, but sample preparation procedures including cleaning, drying, and homogenization were necessary (Assad, 2002).

Mass Spectrometry

A new development in the investigation of coffee bean adulteration is mass spectrometry. It is often used to identify and distinguish between unroasted, green, and pure coffee beans. According to bean qualities like green, ripe, and over ripe, as well as post-harvesting processes like wet, dry, and semi-wet, direct infusion electrospray ionisation mass spectrometry has been utilised to differentiate between pure green and Roasted *arabica coffees*. The most crucial factors in the identification of faulty beans were chlorogenic acids and short-chain organic acids produced from sugars (Amorim, 2009).

NMR Spectroscopy

This method reveals the spin state information of carbon and hydrogen nuclei, providing specific information regarding the number of magnetically distinct atoms of the compounds being investigated. With ^{1}H-NMR, one can determine the number of distinctive hydrogen nuclei present, as well as the nature of their surrounding environment. This characteristic technique allows analysis of brew samples without chemical derivatization and separation (Pavia,2001). This technique provides precise information on the number of magnetically unique atoms in the compounds under investigation by revealing the spin state information of the hydrogen and carbon nuclei. One may detect the number of different hydrogen nuclei present as well as the characteristics of their surroundings using 1H-NMR. Brew samples may be analysed using this distinctive method without the need for chemical derivatization and separation (Pavia, 2001).

Photo acoustic Method (PA)

The main reason this spectroscopic method works so well is that only the light that is absorbed is transformed into pressure changes in the gas cell. The solid in the cell is cycled heated by the absorption of pulsed light, and this cycled heating causes a periodic heat transfer from the solid to the surrounding gas, which is the main source of the acoustic signal in the cell. The periodic heat transfer from the solid to the surrounding gas is considered to only affect a relatively small layer of gas next to the solid's surface thermally.The gas boundary layer then functions like a piston, producing the audible sound picked up by the microphone. There is a strong correlation between the strength of the acoustic signal and the quantity of light absorbed by the solid because the size of the periodic pressure fluctuations in the cell is proportionate to the amount of heat radiating from the solid absorber. This method's viability for identifying coffee adulteration was established. The sole need for the approach was that the samples have a reasonably homogenous particle size distribution. Excellent mathematical models for maize and husks were created, and a linear link between the PA signal and the bulk concentrations of all the adulterants was discovered. The mathematical model for barley, however, did not perform as well; possibly as a result of the selection of a response wavelength that was extremely similar to that of pure coffee (Cesar, 1984).Another technique to identify the presence of maize in brewed coffee is thermal lens spectrometry. The method entails illuminating a sample with a modulated light beam, followed by the measurement of the temperature change brought on by non-radiative de-excitation processes within the material. These spectroscopic approaches should be useful for research using powdered materials since the photo thermal signal is exclusively obtained from the absorbed light and does not significantly depend on the effects of scattered light (Fontes, 2001).

Chromatography Methods

One of the most effective methods for spotting coffee fraud is chromatography. Coffee sample mixtures of chemical species can be separated using gas or liquid chromatography in the presence of adequate detectors with the use of ideal columns and mobile phases.

High-performance Liquid Chromatography (HPLC)

HPLC or high-performance liquid chromatography is an amazing analytical technique for chemical compounds including biopolymers, small molecules, and polymers. In this method, a sample is first dissolved to make a solution. Then, a "column" containing resin that will interact with the sample is injected with this solution. As a result, the sample will move more slowly along the "column," and it will be easier to identify when it exits on the other side. This enables us to understand both the moment the sample is released and the level of sample intensity (www.scigine.com). HPLC may be used to examine coffee for adulterants. The study was undertaken to analyse the carbohydrate levels in coffee and disclose the presence of adulterants on it. The proof of HPLC analysis over coffee was discovered in 1989. (Blanc, 1989).One significant chemical component of coffee that may be examined using HPLC is caffeine (Martn, 1998). Coffee husk and maize were also found using high-performance anion exchange chromatography with pulsed amperometric detection (HPAEC-PAD). Non-derivatized analytes are identified using this type of detection by varying the potentials applied to the working electrode over a predetermined amount of time. On the electrode surface, the hydroxyl groups in carbohydrates are oxidised, and the resultant current is measured. One of the most sensitive detection techniques for the identification of adulterants in coffee, the sensitivity of the PAD method may be obtained up to picomole to femtomole levels (Garcia, 2009).For improved resolution and sensitivity of the analyte, coupled procedures like HPLC-HPAEC-PAD,

(RP)-HPLC followed by methylation and analysis using GC-MS and HPLC-UV-Vis can also is utilised. This raises the minimal threshold of detection for the adulterants in coffee.

Gas Chromatography

A Gas Chromatography system consists of two phases: an immiscible stationary phase that is put in a column and an inert gas termed the mobile phase. During the procedure, analytes are divided between the stationary and mobile phases according to their relative affinity for both. Higher affinity molecules are therefore maintained longer and elute later than other compounds. Even though complex mixes of volatile substances can be directly studied by GC, derivatization of nonvolatile polar molecules or compounds with limited thermal stability is necessary for their analysis by this approach (Ruiz-Matute, 2018).The technique was first applied in 1995 for the detection of adulterants in coffee. They estimated the levels of furfural and 5-methylfurfural, which are present in maize, and 2-methylbutanal, 3-methylbutanal, and furan, which are present in barley, using the dynamic degassing method for the Analysis of roasted maize and barley that is adulterated in coffee (Ruiz. 1995). By assessing their sterol concentrations, the proportion of arabica coffee in blends of roasted arabica/robusta coffee was calculated using GC (Valdenebro, 1999). By utilising GC to measure the fatty acid methyl ester (FAME) compositions of the various varietals, the degree of adulteration in roasted coffee with maize may be evaluated. For clarifying the complex mixes contained in the contaminated sample, additional techniques including two-dimensional gas chromatography (GC-GC), GC-MS, and SPME-GC-MS might be applied (Toci, 2008).

Capillary Electrophoresis

A separation method called capillary electrophoresis (CE) divides molecules into groups based on their size and charge in an electric field. CE is carried out in an electrolyte solution-filled capillary, a tiny glass tube used for the procedure. Due to changes in electrophoretic mobility, which varies with charge, solvent viscosity, and size, analytes are separated. The voltage that may be used in traditional gel electrophoresis is constrained because Joule heating effects would destroy the gel and the separation. Capillaries disperse heat more effectively because they have a high surface area to volume ratio. As a result, a capillary electrophoresis experiment often uses voltages of 10,000-20,000 V. (JoVE Science Education Database, 2021). When investigating low-molecular-mass organic acids in complicated aqueous samples, CE has proven to be an effective method with minimal sample pre-treatment (Saavedra, 2000). Based on the carefully regulated acid hydrolysis of the polysaccharides xylan and starch, a technique for separating plant-based adulterants from coffee was developed (Nogueira, 2009). For determining the carbohydrate profiles of adulterants like coffee husks and maize grains, capillary electrophoresis has been found to be just as effective as liquid chromatography. One of the key elements for determining the quality of coffee is its acidity, which can be determined by using CE (Galli, 2004).

Polymerised Chain Reaction (PCR)

Another effective method for specifically identifying adulterants in coffee is DNA detection. Small quantities of certain nucleic acid sequences contained in each distinct coffee species, as well as other foreign biological components, may be found and quantified using this method. Real-time PCR is a method that keeps track of the amplifying of a specific DNA region that displays genetic differences across coffee species. Real-time PCR is reportedly used to identify and measure cereals that are adulterants in ground, roasted, and soluble coffees (Ferreira, 2016).High resolution melting (HRM) analysis uses the melting profile of real-time PCR results to enable genotyping by differentiating single nucleotide polymorphisms (SNPs). By using HRM analysis on chloroplast genome SNP sequences, the identification of the Arabica and Robusta coffee species and measurement of their mix ratios were carried out. The detection thresholds of admixture varied from 1 to 5 percent of Robusta in Arabica for green beans, showing that adulteration could be detected and quantified for green coffeebeans using this DNA approach (Combes, 2018). The sample preparation and sampling procedure used for the detection of adulterants in coffee is shown in the Table 1

Table 1: Sample preparation and sampling procedure

Method	Sample	Extraction Procedure	Reference
Microscopic method	Coffee power	Boil about 1 gm of sample with 50 ml of 2% sodium hydroxide for about 2 - 3 minutes. Dilute and filter and wash the residue with water till the filtrate is free of alkali. Theprocess is repeated till the residue gives no colour with water. A drop of residue materialin lycerine on a clear microscopic slide a slip cover is placed over the sample andexamined under microscope	FSSAI lab manual no. 4,
Multispectral imaging	Coffee bean	10-g total mass of sample were obtained they were then passed through finely meshedsieves to secure approximately the same small spread distribution of grain sizes inall samples. The data were acquired by exposing the samples to Ar/Kr ion lasers with beam power typically set at about 80 mW with an otherwise standard gas microphonePA detection set up operating at a 300-H~ chopping frequency. Pill-shaped compressed(l-ton) coffee samples were made and mounted in the PA cell fitted with a Knowlesmicrophone	Cesar, 1984
FT-IR	Roasted coffee powder	1 g of the roasted and ground coffee was placed in the sampling accessory and pressed;the background material was obtained from readings of the accessory without anysample. All spectra were recorded within a range of 1800-1600 cm−1. The resolutionused was 4 cm−1 with 20 scans and submitted to background subtraction; all sampleswere analyzed in triplicate.	Monje, 2018
NMR	Coffee powder	The coffee samples were pulverised, and 150 mg of the extract were vortexed for one minute and centrifuged for ten minutes using one mL of deuterated water. A 600 MHzNMR spectrometer was then used to analyse the supernatant after being transferred to an NMR tube. The run with the pure coffee samples was used to compare the results fromthe sample runs.The extraction process can also make use of methanol.	de Moura Ribeiro, 2017 Arana, 2015
HPLC	Coffee powder	Two grams each of ground coffee beans and adulterated coffee samples were brewed with 25 mL water for 10 min. Each sample was centrifuged at 1000 × g for 5 min and thenfiltered with No. 1 filter paper. The filtrate was passed through a 0.2-μm PTFE filter and Then transferred into a 1.5 mL vial for analysis.	Cheah, W. L., 2020
GC-SPME	Coffee powder	5 mL sealed vials containing 3 grams of sample were heated for 10 minutes at 70 °C. The Fibre was then exposed to the headspace above the sample for 40 minutes at 70 C afterthe SPME needle had been put into the vial. The fibre was thermally desorbed in the GCinjection port for 10 minutes at 250 C after sample.	Oliveira, 2009

Method	Sample	Extraction Procedure	Reference
Capillary Electrophoresis	Coffee powder	For the extraction procedure, 10 ml of filtered water and 1 g of coffee with a particlesize under 0.4 mm were mechanically shaken for 10 minutes. After that, samples were centrifuged for 5 minutes at 2700 rpm. Prior to analysis, all sample solutions were filtered using 0.22 m nylon filters and stored at 20 °C.	Galli, 2004
PCR	Coffee beans	**DNA isolation** Ground beans (1 g) were placed in a sterile 15 mL tube with 3 mL of freshly stirredCTAB extraction buffer (1.5 M NaCl, 0.12 M sorbitol, 1% PVPP, 0.08% CTAB, 1%N-lauroylsarcosine, 5 mM EDTA, pH 8.0). After incubation at 65°C, the tube was thencooled to room temperature. Afterwards, 1 mL of 5 M potassium acetate was added andthe mixture was incubated overnight at -20°C. The sample was then thawed under ice,and centrifuged at 4000 g for 15 min. A 1 mL aliquot of supernatant was then transferredto a sterile 2 mL micro centrifuge tube. Subsequently, 5 μL of 1 U/μL RNase A was added and the solution was incubated at 37°C for 15 min. After extraction with 600 μL of chloroform/isoamylic alcohol (24:1), the sample was centrifuged at 4300g for 15 min.The aqueous phase was then transferred to a 1.5 mL micro centrifuge tube, and 800 μ **Sampling** 10 L of the final eluate/resuspended pellets were electrophoresed on a 0.8% agarose gel containing 0.3 mg/mL ethidium bromide to determine the DNA concentration and purity(EtBr).	Martellossi, 2005

Regulations of Coffee

FSSAI-Approved Decaffeinated Roasted and Ground Coffee

Decaffeinated coffee is made from dried seeds of *Coffea arabica, Coffea liberica, Coffea excelsa, or Coffea canephora (Robusta)* that have had their husks (mesocarp and endocarp) removed and decaffeinated to remove virtually all of the caffeine from the beans. Before the beans are roasted, they are decaffeinated.

Roasted decaffeinated coffee is defined as well cleaned green coffee that has been decaffeinated, roasted to a brown colour, and developed its distinctive scent.

Ground decaffeinated coffee refers to powdered items made from roasted decaffeinated coffee that are devoid of husk.

It must be free of artificial colouring, flavouring, facing, extraneous materials, or glazing ingredients, and it must be dry and fresh, with no rancid or offensive flavours. On a dry weight basis, it must meet the following requirements: -

Sl.No.	Parameter	Limits
1	Moisture, percent by mass, Max	5.0
2	Total Ash, percent by mass, Max	3.0 to 6.0
3	Acid insoluble ash, percent by mass, max	0.1
4	Water soluble ash, percent by mass, min	65.0
5	Alkalinity of soluble ash in ml of 0.1 N hydrochloric acid per gram of material, percent by mass, ml	3.5 to 5.0
6	Aqueous extracts, percent by mass	26.0 to 35.0
7	Caffeine (anhydrous) percent by mass, max	0.1

Decaffeinated Soluble Coffee Powder - as per FSSAI

Decaffeinated soluble coffee powder is coffee powder made from freshly roasted and ground pure coffee beans that has had the majority of the caffeine removed. The product must be in the form of a free flowing powder or granules with the colour, taste, and flavour of coffee. It must be devoid of contaminants and include no chicory or other additives.

On a dry weight basis, Decaffeinated Soluble Coffee powder or granules must meet the following requirements:

Sl.No.	Parameter	Limits
1	Moisture, percent by mass, Max	4.0
2	Total Ash, percent by mass	12.0
3	Caffeine (Anhydrous,) percent by mass, Max	0.3
4	Solubility in boiling water	Dissolves in 30 seconds with moderate stirring
5	Solubility in cold water at 16+/- 2 C	Dissolve in 3 minutes with moderate stirring

Roasted Chicory Powder

With or without the inclusion of edible oils and fats or sugar (such as glucose, sucrose) in a proportion that does not exceed 2.0 percent by weight in total

Total ash (on dry basis) m/m	not less than 3.5 % and not more than 8.0 %
Acid insoluble ash (on dry basis) m/m in diluted HCl	not more than 2.5 %
Aqueous extracts (on dry basis) m/m	not less than 55 %

Coffee-chicory Mixture

It is prepared by combining roasted and ground coffee with roasted and ground chicory.

- It needs to be sound, dry, and dust-free, with no rotten flavour.
- It needs to be in the form of a free-flowing powder with the colour, taste, and flavour of coffee - chicory powder.
- It needs to be devoid of contaminants and have no additional added ingredients.
- The coffee component of the combination must be at least 51 percent by mass.

Moisture	not more than 5.0 %
Total ash on dry basis	not more than 7.50 %
Acid insoluble ash on dry basis	not more than 0.6 %
Caffeine content on dry basis	not less than 0.6 %
Aqueous extracts	not more than 50 %

Instant-coffee Chicory Mixture

- A product made with roasted and ground coffee as well as roasted and ground chicory.
- It must be clean, dry, and dust-free, with no rotten or offensive odours.
- It must be a free-flowing powder or agglomerated (granules) powder with the colour, taste, and flavour characteristics of coffee chicory powder.
- It must be devoid of contaminants and include no other additional substances.
- The mixture's coffee content must be at least 51% of the total mass on a dry basis.

Moisture	not more than 4.0 %
Total ash on dry basis	not more than 10.0 %
Acid insoluble ash on dry basis	not more than 0.6 %
Caffeine (anhydrous) content on dry basis	not less than 1.4 %
Solubility in boiling	dissolve readily in water (30 seconds) with moderate stirring
Solubility in cold at 16 ± 2 ºC	soluble in water (3 minutes) with moderate stirring

Copper and arsenic levels in dried or roasted chicory shouldn't be greater than 30.0 and 4.0 parts per million by weight, respectively (FSSR, 2011c)

International standards - A Speciality Coffee Association has framed some international protocols for coffee (www.sca.coffee)

Lighting for Green Grading

When grading green coffee, the light level on the tabletop and grading mat shall be full spectrum and at least 4000 Kelvin (K) /1200 Lux (lx) /120 foot-candles (fc).

Surface for Grading

Green grading shall be done on a table at least two (2) feet long and two (2) feet wide (0.6096 meters by 0.6096 meters), on a black grading mat of at least the same size.

Water Activity Standard

Specialty grade green coffee beans shall have a water activity measurement lower than 0.70 aw.

Golden Cup Standard

Coffee shall exhibit brew strength, measured in Total Dissolved Solids, of 11.5 to 13.5 grams per litre, corresponding to 1.15 to 1.35 "percent" on the SCA Brewing Control Chart, resulting from a soluble extraction yield of 18 to 22 percent.

Coffee and its Adverse Impact in Human Health

Food adulteration affects people's health in several ways. Anemia, an upset stomach, miscarriage, paralysis, and a growth in tumours and lesions in important organs are some of the health risks (Thapa, 2016). A study was done on roasted coffee, instant coffee, and coffee replacements in 2013 by Mojska, H., and Gielecinska. Acrylamide concentrations of 818 pg/kg were found in coffee replacements, 358 µg/kg in instant coffee and 179 µg/kg in roasted coffee, according to the LC-MS/MS analysis of a few chosen samples. The average amount of acrylamide in a single cup of coffee (160 ml) ranged from 0.45 µg in roasted coffee to 3.21 µg in coffee substitutes.The use of adulterants and other processing techniques during coffee processing result in the presence of acrylamide, which has been linked to the development of cancers in humans. Acute coronary syndrome is diagnosed in those who drink more than 600 mL of coffee daily, according to studies on coffee intake (Panagiotakos*et al.*, 2003). A known risk factor for CHD and stroke is hypertension. It is widely known that both normotensive and hypertensive people's blood pressure increases with acute caffeine ingestion at dietary levels. In normotensive people, a 200-250 mg dosage of caffeine, or around 2-3 cups of coffee, has been shown to elevate systolic blood pressure by 3-14 mm Hg and diastolic blood pressure by 4-13 mg Hg (Nurminen*et al.*, 1999). In those with insufficient calcium intakes, coffee use results in a minor negative calcium balance, according to controlled calcium balance studies in humans. Due to a modest decrease in the effectiveness of calcium absorption, which causes osteoporosis and hip fracture, the negative shift in calcium balance is estimated to be 4-6 mg of calcium per cup of coffee (Barger-Lux & Heaney, 1995). A high coffee intake can also cause iron and zinc absorption to be impeded, which can result.

In mineral deficiencies. Additionally, dietary inhibitors of zinc absorption, notably phytate, increase the risk of zinc shortage. The phenols in coffee can bond with the non-heme group of iron and delay its absorption in the small intestine (Fairweather-Tait, 2004; Institute of Medicine and Food and Nutrition Board, 2001). Overall, moderate coffee consumption among adults is associated with few health risks and some health benefits; however, the evidence that is currently available suggests that pregnant women, lactating women, and women who are trying to get pregnant should limit their caffeine intake to no more than three cups per day (300 mg/d) (Nawrot*et al.*, 2003).

Case Studies in Coffee

South Korea, Japan ban Import of Kenyan Coffee for 3 Years

Due to claims of high levels of chemical contamination in Japan and South Korea, the Kenyan coffee, one of the most sought-after coffees in the world for its powerful taste, substantial body, and pleasant scent, may lose its attractiveness on a worldwide scale. According to reports by Nation Kenya, the chemical levels of Ochratoxin, a naturally occurring food borne mycotoxin occurring in a wide variety of agricultural commodities, was found to have exceeded the allowable minimum. Due to this, the two countries have banned the commodity for three years with local stakeholders now raising concerns that if this is not reversed, then the produce could face total ban and could have a negative effect on other importing countries as well. The flagging of Kenyan coffee by the key markets due to high levels of contamination does not augur well for the sector. The government has to move fast in addressing this challenge. However, Kenyan coffee production has been declining from a high of 130,000 tonnes realised in the late 80s to the current 40,000 tonnes. Majority of the smallholder coffee farmers are currently producing less than two kilograms (kg) per tree per year against an annual potential of 35kg per tree. The fall in production is blamed on poor management of farmer cooperatives and infiltration of the sector by cartels that have seen growers' earnings fall every other year. To revive the sector, the government announced that coffee trading will be regulated by the Capital Markets Authority in a move aimed at taming cartels blamed for diminishing earnings by farmers (www.chinagoabroad.com).

Conclusion

Coffee is a widely consumed food item that has a large economic impact on the nations that produce and export it. Coffee provides several health benefits for people since it is an excellent source of critical vitamins and minerals including riboflavin (vitamin B2), pantothenic acid (vitamin B5), manganese, potassium, magnesium, and niacin (vitamin B3). Coffee is a rich source of caffeine, which will increase the neurotransmitters norepinephrine and dopamine. These neurotransmitters have accelerated neuronal activity, which keeps us sharper by boosting our energy levels. Additionally, it is a healthy beverage that enhances cognitive performance. Caffeine, which is a component of coffee, is also utilised to burn body fat.

According to studies on coffee's health advantages, the phytochemicals found there may lower the chance of developing type 2 diabetes, Alzheimer's disease, dementia and Parkinson's disease. Additionally, it has the power to fend against sadness and safeguard the liver from cirrhosis. Even though it is expensive and has many health advantages, it is prone to adulteration. The likelihood of coffee adulteration is substantial, despite the fact that there are several procedures available for its detection. The creation of a broadly applicable, sensitive technique that can address the numerous facets of coffee adulteration is still required. This includes the ability to distinguish between different species, find beans that aren't right, and recognise the presence of outside agents. Consumers may be protected from food adulteration by recent developments including increased proactivity, high levels of openness, and improved governmental monitoring over farmers, suppliers, and vendors.

References

A. P. Davis, R. Govaerts , M. D. Bridson and P. Stoffelen , Bot. J. Linn. Soc., 2006, 152 , 65 Alonso-Salces, R. M., Serra, F., Reniero, F., &HEberger, K. (2009). Botanical and Geographical Characterization of Green Coffee (Coffeaarabica and Coffeacanephora): Chemometric Evaluation of Phenolic and Methylxanthine Contents. Journal of Agricultural and Food Chemistry, 57(10), 4224-4235.

Amorim, A. C. L.; Hovell, A. M. C.; Pinto, A. C.; Eberlin, M. N.; Arruda, N. P.; Pereira, E. J.; Bizzo, H. R.; Catharino, R. R.; MoraisFilho, Z. B.; Rezende, C. M. Green and Roasted Arabica Coffees Differentiated by Ripeness, Process and Cup Quality via Electrospray Ionization Mass Spectrometry Fingerprinting. J. Braz. Chem. Soc. 2009, 20(2), 313-321. Arana, V. A., Medina, J., Alarcon, R., Moreno, E., Heintz, L., Schäfer, H., &Wist, J. (2015).Coffee's country of origin determined by NMR: The Colombian case. Food Chemistry, 175, 500-506.

Assad, E. D.; Sano, E. E.; Cunha, S. A. R.; Correa, T. B. S.; Rodrigues, H. R. Identifying Adulteration in Coffee Powders by Spectral Signatures and Digital Analyses. Pesqui. Agropecu.Bras. 2002, 37(2), 211-216.

Barger-Lux, M. J., & Heaney, R. P. (1995). Caffeine and the calcium economy revisited. Osteoporosis International, 5(2), 97-102.

Blanc, M. B.; Davis, G. E.; Parchet, J. M.; Viani, R. Chromatographic Profile of Carbohydrates in Commercial Soluble Coffees. J. Agric. Food Chem. 1989, 37, 926-930. Borém, F.M. (2008) Pós-Colheita do Café (Coffee Post Processing), Lavras, Brazil: Editora UFLA.

Cesar, C. L., Vargas, H., Lima, C. A., Mendes Filho, J., & Miranda, L. C. (1984). On the use of photoacoustic spectroscopy for investigating adulterated or altered powdered coffee samples. Journal of Agricultural and Food Chemistry, 32(6), 1355-1358.

Cesar, C. L.; Vargas, H.; Lima, C. A. S.; Mendes Filho, J.; Miranda, L. C. M. On the Use of Photoacoustic Spectroscopy for Investigating Adulter- ated or Altered Powdered Coffee Samples. J. Agric. Food Chem. 1984, 32, 1355-1358.

Charrier A, Berthaud J (1985) Botanical classification of coffee. In: Clifford MN, Wilson KC (Eds), Coffee: botany, biochemistry and production of beans and beverage. Croom Helm, London, pp 13-47

Cheah, W. L., & Fang, M. (2020).HPLC-based chemometric analysis for coffee adulteration. Foods, 9(7), 880. coffee Ephraim D. 2016.

Combes, M. C., Joët, T., &Lashermes, P. (2018). Development of a rapid and efficient DNA-based method to detect and quantify adulterations in coffee (Arabica versus Robusta). Food Control, 88, 198-206.

Davis AP, Chester M, Maurin O, Fay MF.Searching for the relatives of Coffea (rubiaceae, ixoroideae): the circumscription and phylogeny of Coffeeae based on plastid sequence data and morphology. Am J Bot 2007;94(3):313-29.

deMouraRibeiro, M. V., Boralle, N., Pezza, H. R., Pezza, L., &Toci, A. T. (2017). Authenticity of roasted coffee using1H NMR spectroscopy. Journal of Food Composition and Analysis, 57, 24-30.

Esquivel, P., & Jimenez, V. M. (2012).Functional properties of coffee and coffee by-products. Food Research International, 46(2), 488-495.

Fairweather-Tait, S. J. (2004). Iron nutrition in the UK: getting the balance right. Proceedings of the Nutrition Society, 63(4), 519-528.

Ferreira, T., Farah, A., Oliveira, T. C., Lima, I. S., Vitório, F., & Oliveira, E. M. (2016).Using Real-Time PCR as a tool for monitoring the authenticity of commercial coffees. Food Chemistry, 199, 433-438.

Fontes, A. S.; Bento, A. C.; Miranda, L. C. M.; Baesso, M. L. Thermal Lens Evaluation of the Presence of Adulterants in Brewed Coffee. Anal. Sci. 2001, 17, 526-529. Franca, A.S., Oliveira, L.S., Ferreira, M.E. (2008) Kinetics and equilibrium studies of methylene blue adsorption by spent coffee grounds, Desalination, in press.

FSSAI lab manual no. 4, Manual of methods of analysis of foods

Galli, V., &Barbas, C. (2004).Capillary electrophoresis for the analysis of short-chain organic acids in coffee. Journal of Chromatography A, 1032(1-2), 299-304.

Garcia, L. M. Z.; Pauli, E. D.; Cristiano, V.; Camara, C. A. P.; Scarminio, I. S.; Nixdorf, S. L. Profile in Coffee Due to Maize and Husk by Determining Carbohydrates Using HPAEC-PAD. J. Chromatogr. Sci. 2009, 47, 825-832.

Garza. (2010) Starbucks Coffee from: http://www.chinagoabroad.com/en/article/south-korea-japan-ban-importation-of-kenyan-coffee-for-3-years (Accessed on 06 October 2021).

https://fssai.gov.in/dart/

https://sca.coffee/research/coffee-standards

https://scigine.com/blog/hplc-analysis-method-guide/ https://www.britannica.com/topic/coffee

https://www.citygreens.in/blogs/post/check-adulteration-coffee-powder

https://www.ncausa.org/about-coffee/coffee-roasts-guide#:~:text=Roasting%20brings%20out%20the%20aroma,loss%20of%20quality%20or%20taste.&text=Roasting%20causes%20chemical%20changes%20to,brought%20to%20very%20high%20temperatures.

https://www.ncausa.org/About-Coffee/What-is-Coffee (Accessed on 08-09-2021). https://www.pbs.org/food/the-history-kitchen/history-coffee/

https://www.royalcupcoffee.com/blog/articles/case-study-improve-hotel-coffee-improve-guest-satisfaction

https://www.uhu.es/45122/temas/P&SC/Theme1_StarbucksCoffe_CaseStudy.pdf (Accessed on06 October 2021).

Institute of Medicine and Food and Nutrition Board. 2001. Zinc. In: Dietary reference intakes for vitamin A, vitamin K, boron, chromium, copper, iodine, iron, manganese, molybdenum, nickel, silicon, vanadium, and zinc. pp. 442-501. National Academy Press, Washington D.C.

JoVE Science Education Database.<em>Analytical Chemistry.</em> Capillary Electrophoresis (CE).JoVE, Cambridge, MA, (2021).

Lopez, F. C. (1983). Determinacaoquantitativa das principaissubstanciasutilizadasparafraudaro cafe torrado e moido. Rev. Inst. Adolfo Lutz, 3-8.

Martellossi, C., Taylor, E. J., Lee, D., Graziosi, G., &Donini, P. (2005).DNA Extraction and Analysis from Processed Coffee Beans. Journal of Agricultural and Food Chemistry, 53(22), 8432-8436.

Martín, M. J.,Pablos, F.,González, A. G. (1998). Discrimination between arabica and robusta green coffee varieties according to their chemical composition. Talanta, 46: 1259-1264. Medina, S., Perestrelo, R., Silva, P., Pereira, J.A.M., &Caˆmara, J.S. (2019) Trends Food Sci. Technol. 85, 163-176

Menezes, Jr., J. B. F.; Bicudo, B. A. A. Sobre um MetodoMicroscopicoparaContagem de Cascas no Cafe em Po.Rev. Inst. Adolfo Lutz 1958, 11, 13.

Mitich, L. W. (1993). Chicory. Weed Technology, 7(1), 274-277.

Monje, A. F. B., Parrado, L. X., & Gutiérrez-Guzmán, N. (2018). ATR-FTIR for discrimination of espresso and americano Coffee pods.

Myhrvold, N. (2021, August 23). coffee. Encyclopedia Britannica.

Nawrot, P., Jordan, S., Eastwood, J., Rotstein, J., Hugenholtz, A., &Feeley, M. (2003). Effects of caffeine on human health. Food Additives & Contaminants, 20(1), 1-30.

Nogueira, T.; Lago, C. L. Detection of Adulterations in Processed Coffee with Cereals and Coffee Husks Using Capillary Zone Electrophoresis. J. Sep. Sci. 2009, 32, 3507-3511 Nurminen ML, Niittynen L, Korpela R, Vapaatalo H. Coffee, caffeine and blood pressure: a critical review. Eur J ClinNutr. 1999 Nov;53(11):831-9. doi: 10.1038/sj.ejcn.1600899. PMID: 10556993.

Oliveira, R. C. S., Oliveira, L. S., Franca, A. S., &Augusti, R. (2009). Evaluation of the potential of SPME-GC-MS and chemometrics to detect adulteration of ground roasted coffee with roasted barley. Journal of Food Composition and Analysis, 22(3), 257-261.

Panagiotakos, D. B., Pitsavos, C., Chrysohoou, C., Kokkinos, P., Toutouzas, P., &Stefanadis, C. (2003). The J-shaped effect of coffee consumption on the risk of developing acute coronary syndromes: the CARDIO2000 case-control study. The journal of nutrition, 133(10), 3228-3232.

Pavia, D.L., Lampmanm, G.M., &Kriz, G.S. (2001) Introduction to Spectroscopy, 3rd Ed

Rubayiza, A.B.; Meurens, M. Chemical discrimination of arabica and robusta coffees by fourier transform raman spectroscopy. J. Agric. Food Chem. 2005, 53, 4654-4659. Ruiz, A. D.; Suarez, M.; Moreno, E.; Bautista, E. Volatiles of Roasted Maize (Zea mays), Barley (Hordeumvulgare) and Coffee (Coffeaarabica).Effect of Addition of These Cereals on the Aroma of Coffee. Rev. Colombiana Quim. 1995, 24, 13-23.

Ruiz-Matute, A. I., Rodríguez-Sánchez, S., Sanz, M. L., &Soria, A. C. (2018). Chromatographic Technique: Gas Chromatography (GC) Modern Techniques for Food Authentication, 415-458.

Saavedra, L., Garcıa, A., &Barbas, C. (2000).Development and validation of a capillary electrophoresis method for direct measurement of isocitric, citric, tartaric and malic acids as adulteration markers in orange juice. Journal of Chromatography A, 881(1-2), 395-401.

Sanz, C., Maeztu, L., Zapelena, M.J., Bello, J., & Cid, C. (2002) J. Sci. Food Agric. 82, 840-847 Severini, C., Derossi, A., Ricci, I., Caporizzi, R., & Fiore, A. (2018) Roasting conditions, grinding level and brewing method highly affect the healthy benefits of a coffee cup. Int J ClinNutr Diet, 4(1), 127.

Silva, L., Oliveira, L., & Franca, A. (2011). Fourier transform infra-red (FTIR) spectroscopy applied to the detection of roasted coffee adulteration with spent coffee grounds. In Pure and applied chemistry international conference (food safety) p219-221.

Simkin, A. J., Moreau, H., Kuntz, M., Pagny, G., Lin, C., Tanksley, S., & McCarthy, J. (2008). An investigation of carotenoid biosynthesis in Coffeacanephora and Coffeaarabica. Journal of plant physiology, 165(10), 1087-1106.

Sivetz, M. and Desrosier, N.W. (1979).Coffee Technology. Westport, CO: Avi Publishing Co. Soares, M., Christen, P., Pandey, A., Soccol, C.R. (2000) Fruity flavour production by Ceratocystisfimbriata grown on coffee husk in solid-state fermentation, Process Biochemistry, 35, 857-861

Song, H. Y., Jang, H. W., Debnath, T., & Lee, K. G. (2019). Analytical method to detect adulteration of ground roasted coffee. International Journal of Food Science & Technology, 54(1), 256-262.

Thapa, M. B. Food Adulteration And Its Effects On Human Health. 2016

Toci, A. T.; Farah, A. Volatile Compounds as Potential Defective Coffee Beans' Markers. Food Chem. 2008, 108, 1133-1141.

Valdenebro, M.S., Leon-Camacho, M., Pablos, F., Gonzalez, A.G., & Martin, M.J. (1999) Analyst 124, 999-1002

Valdenebro, M. S.; Leon-Camacho,M.; Pablos, F.; Gonzalez, A.G.;Martın, M. J. Determination of the Arabica/Robusta Composition of Roasted Coffee According to Their SterolicContent. Analyst 1999, 124, 999-1002.

Vincent, J.C. (1987) Green coffee processing. In: R.J. Clarke and R. Macrae, (Eds.), Coffee Technology, New York: Elsevier Applied Science Publishers Ltd., pp. 1-33. y syndrome. (2013). Studies of acrylamide level in coffee and coffee substitutes: influence of raw material and manufacturing conditions. RocznikiPaństwowegoZakładuHigieny, 64(3).

13

Beyond Sweetness: A Comprehensive Guide to Unmask Adulterants in Cocoa Products

Pal Murugan, Abhishek B S, Kavanashree N, Wadikar, D D and Semwal A D

Defence Food Research Laboratory, Defence Research and Development Organization, Mysore, Karnataka, India

Introduction

Cocoa beans are extracted from the pod of a tree crop *Theobroma cacao*. It is mainly cultivated in tropical and subtropical countries. In terms of world production, Ivory Coast and Ghana contribute about 65 percent of world production. Cocoa beans are the main ingredient in chocolate. (Teye, E *et al 2014*). One of the earliest examples of chocolate consumption comes from the Mayans. The Maya drink was prepared with cocoa and water and often included cinnamon and black pepper. It was also known as the "Food of the Gods" and offered to Emperor Moctezuma II by the Aztecs, (Verna, R. 2013). Christopher Columbus was the first European to encounter cocoa in 1502. He captured a canoe that contained cocoa beans, which he identified as "mysterious-looking almonds" and considered a form of currency in Mesoamerica. Cocoa also appeared in Europe in 1528 when Hernán Cortés brought samples to King Charles of Spain, where it spread throughout the region. The cocoa plant that we now know as *Theobroma cacao* was first discovered in 1753 and named by Swedish scientist Carl Linnaeus. This name is a mix of Latin (Theobroma) and Aztec (xocolatl).

Cocoa is among the top commercially cultivated cash crops for many tropical and subtropical countries. More importantly, cocoa represents a major exportable agricultural commodity for several countries in West Africa (Teye, E *et al.*, 2020). Cocoa beans are widely produced in developing countries and are consumed worldwide by people of all races and cultures. There was a time when cocoa was used for medicinal, ritual and nutritional purposes because of its unique taste and aroma; today it's valued exclusively as a delicious treat. Cocoa is a crop which can be grown by smallholder farmers in the lowland tropics, including parts of Latin America, West Africa, and Indonesia (Franzen P, Mulder M. 2007). Cocoa is one of the most important agricultural crops and it makes up about 80 percent of all cash crops (Assoumou 1977). In 1900 Africa had only around a 17 percent share in the world cocoa production. Nearly 20 years later (1996), by far the four largest producers accounted for 65 percent of world output(ICCO 1997).

Compared to other agricultural activities, cocoa has been shown to be akey subsector in economic growth and development in those counties. However, since the late 1980s this sector was subjected to some major institutional and organizationalchanges which led to new frameworks.

In recent years, a number of biological properties and potential health benefits of consuming cocoa-derived products have been investigated, e.g., antioxidant, anti-inflammatory, anticarcinogenic, and antifungal properties. Much research has focused on the cocoa phenolic compounds, such as flavonoids (procyanidins, anthocyanins, flavonols, etc.), as a potential health promoting compounds due to its antioxidant capacity and abundance in the cocoa beans.

Scientific Classification and Anatomy of Cocoa Bean

The scientific classification of cocoa plant is listed in Table.1

Table 1: Scientific classification of cocoa Plant

Scientific Classification	
Kingdom	Plantae
Division	Magnoliophyta
Class	Magnolopsida
Order	Malvales
Family	Sterculiaceae
Genus	Theobroma
Species	Cacao

Cocoa beans are the seeds found within the large pods of the cocoa tree. Each pod contains numerous seeds embedded in a sweet, mucilaginous pulp. Cocoa pods are large, elongated fruits with a tough, thick rind. The pods vary in color, including yellow, orange, red, or purple, depending on the cocoa variety. The seeds are surrounded by a sweet, mucilaginous pulp that provides nutrients for the developing seeds. This pulp is essential for fermentation, a crucial step in cocoa bean processing. Each cocoa bean consists of a seed surrounded by a protective seed coat or shell.

Fig. 1: Anatomy of Cocoa Beans (*Source*: www.pngitem.com)

The seed contains the cotyledon, which is the embryonic plant that will eventually sprout and grow into a cocoa tree if planted. The seed is covered by a protective seed coat or shell. This outer layer helps protect the internal components of the seed. The cotyledons are the fleshy, internal structures of the seed. These are the parts that are processed to produce cocoa products. The cotyledons contain the cocoa solids and cocoa butter. Within the cotyledons, there is an embryo, which is the early stage of the cocoa plant. This part is vital for the germination process if the seed is planted. When cocoa beans are processed, the cotyledons are ground into small pieces known as cocoa nibs. Cocoa nibs contain both cocoa solids and cocoa butter.

Types of Cocoa Beans

There are four main types of Cocoa beans namely:

A. Criollo

B. Forastero

C. Trinitario

D. Nacional

A. Criollo is characterized by pods which are rough surfaced and somewhat pointed at the ends. Beans of freshly opened pods are non-pigmented and need onlya relatively short fermentation period and mild roasting condition to develop chocolate flavor. Historically, the flavor of criollo beans has been recognized as especially desirable. Criollo's native habitat is west and north of the Amazon to Mexico. 5 percent of the world Cocoa pods are made from Criollo tree which belonged to central and south America and the Caribbean islands.

Fig. 2: Criollo Cocoa pod
(*Source*: http://cocoaskiss.blogspot.com)

B. Forastero is a native to the Amazon and Orinoco River basins, and its smoother and rounder Criollo-like pods produce purple beans. As opposed to Criollo beans, roasting allows for the best taste development, which calls for alonger fermentation time (Oracz, J., & Zyzelewicz, D. 2019).

Fig 3. Forastero Cocoa pod (*Source*: https://www.montosogardens.com/)

C. Trinitario is a natural hybrid biological class resulting from cross-pollination. Legend recounts that it first came into existence on the Island of Trinidad,after a hurricane nearly completely destroyed the local Criollo crops in 1727. Assuming all the trees were dead, the plantations were replanted with Forastero,but spontaneous hybrids appeared. The hardiness and great yield of Forastero and the sophisticated flavour of Criollo are combined in Trinitario, which is the better of the two other major kinds. The grade of the cocoa fluctuates between normal and exceptional. It dominates the exquisite flavour of cocoa. Because the parents have such diverse personalities, Trinitario populations typically have varying pod and bean characteristics. All of the nations where Criollo cocoa was formerly grown, including Mexico, the Caribbean islands, Colombia, Venezuela, and several regions of Southeast Asia, now host them.

Fig 4. Trinitario Cocoa pod
(*Source*: chocolateclass.wordpress.com)

D. Arriba, a premium-flavor chocolate, is made using Nacional cocoa from Ecuador. The exact origin of the Nacional type is not known, but it is generally considered as native to Ecuador. Recent studies using Restriction Fragment Length Polymorphism (RFLP) probes indicated that Nacional cocoa genotypes are genetically different from the Forastero, Criollo or Trinitario groups. The well-known Ecuadorian cocoas, derived from the Nacional variety, are considered to be fine-flavoured. The production of this specific Arriba flavour not only depends on genetic factors, but also on the post-harvest fermentation and the processing in the factory (Crouzillat, D *et al* 2000).

Fig 5. Nacional Cocoa pod (*Source*: https://www.snacq.com/)

Processing of cocoa beans

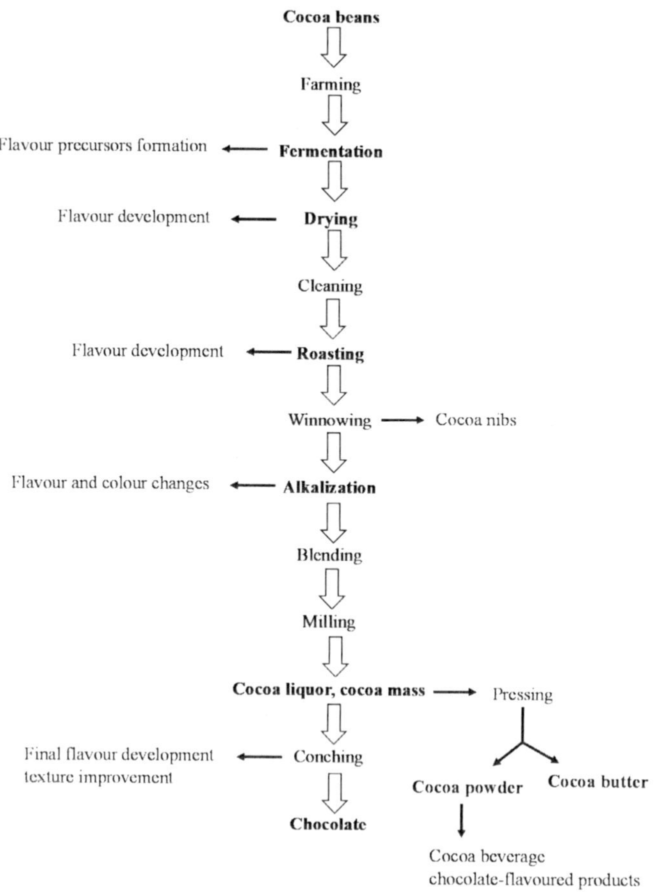

Fig. 6: Steps involved in Cocoa Processing (*Source*: Aprotosoaie, A. C et.al, 2016)

Processing steps of Cocoa Beans

1. **Harvesting/Cleaning:** The cocoa is usually harvested with a major peak and a minor peak of pod ripeness. Careful removal of the pods from the trees with a knife is required to avoid damage to the flower cushions. After inspecting and conducting the necessary tests to ensure compliance with international standards and regulatory requirements, the cocoa beans cleaning process takes place.

   The inside of an undamaged cocoa bean is clean in all respects (removal of foreign material ranging from machete blades to shotgun shells), and will remain so as long as it is stored properly and is not mixed with contaminated materials not completely removed during cleaning or subsequent steps downstream.

2. **Fermentation:** Chocolate flavour is developed in two parts: the first is on the farm through correct fermentation of the wet beans by the grower, and the

second by the processor in the factory at the roasting step. Good chocolate flavour cannot be produced by adhering to only one of these stages.

Between 36 and 72 hours, the beans go through the earliest phases of fermentation, during which most of the pulp drains away. Despite recent improvements achieved via the application of specialized analytical and sensory assessment (flavour profiling) tools, the mechanisms of flavour production are complicated and yet largely unknown.

3. **Drying:** After fermentation, the moisture level of the beans must be lowered from 55 percent to 7.5 percent - an acceptable moisture content for stable storage of cocoa for a few months in the tropics. Small holders lay the wet beans on raised bamboo mats or, less satisfactorily (for hygiene considerations), on concrete platforms on the ground in the villages. The duration of the drying step depends on the weather, but it is unusual for sun-drying in West Africa, for example to be completed in less than a week.

4. **Roasting:** Roasting cocoa beans can be described as an individual process. While all manufacturers have a similar goal of making products efficiently, the flavour objectives for cocoa liquors usually differ from company to company and from country to country. As a result, the flavour plays an important role in deciding the type and mixof cocoa beans to be processed, whether to roast whole beans or nibs, the type of roasting equipment used, and the roasting parameters used.

   In a nutshell, there is not a right or wrong roast level nor is there a correct or incorrect way to obtain the target roast level. The correct way to roast and the proper roast level is the process that provides nibs efficiently and cost-effectively with the chosen flavour system and yields products meeting consumer needs in a specified market.

5. **Winnowing:** Winnowing, cracking, fanning, and hulling are among of the terminology and phrases used to describe the separation of the shell (hull) and flesh of the bean (nib). A clean separation of the two components is achieved by a process that is driven by economics, product integrity, and, in many nations, government regulation.

6. **Nibs grinding:** Nibs grinding has seen many advances in the last half of the 20th century. A mill, one method of grinding nibs was mixing nibs with granulated sugar and placing the mixture in a mélangeur. This process yielded a material with a consistency ranging from a paste to a fluid.

7. **Alkalization:** The alkalizing procedure can be used to change the flavour and colour of chocolate liquors and cocoa powders. It is sometimes referred to as Dutch processing. The selected cocoa material (cocoa cake, nibs, or cocoa liquid) is mixed with an aqueous solution of a specified alkaline chemical at raised temperatures and perhaps enhanced pressures. Potassium carbonate, calcium carbonate, and sodium hydroxide are examples of common alkaline chemicals. The finished product ranges in colour from pale crimson to charcoal black.

8. **Liquor pressing:** Typically, the separation or pressing process begins by pumping hot cocoa liquor (200 degrees Celsius) into a horizontal hydraulic press with operating pressures of up to 550 bars. The cocoa cake formed from this pressing operation is then broken and milled to specified particle sizes as determined by end use.

9. **Cocoa grinding:** A succession of chilling stages must be completed before the pressed cocoa cake may be ground. The industry's most frequent mills are the hammer mill and disc mill, and a classifier is generally incorporated as part of the system to optimize grinding efficiency.

   In reality, the term grinding is misleading. The particle size of the finished powder is dependent upon the particle size of the cocoa liquor pressed. Commonly used mills do not reduce the particle size of the cake; instead, they tend to break up agglomerates. It must also be mentioned that cocoa powder must be tempered. This is a sort of controlled cooling operation during which the powder is held at specified temperatures for predetermined times to allow for the cocoa butter to form into a stable crystal configuration. The powder should then be stored properly if it is to maintain its colour and remain soft and in a flow able state.

10. **Cocoa butter:** Cocoa butter is another byproduct of pressing cocoa liquor. This is the costliest of the primary components in a chocolate recipe. Cocoa butter colour can be an excellent indicator of potential problems. For example, shipments that are dark brown rather than golden in colour warrant further testing. It can indicate improper handling during processing and storage.

11. **Chocolate manufacturing:** Cocoa liquor, cocoa butter and sugar are the three major ingredients required to produce the various types of chocolate found throughout the world.

Overall, the manufacturing of chocolate is divided into four areas

1. **Batching**, which is the combining of cocoa liquor, sugar or sweeteners, milk powder (if a milk chocolate), cocoa butter and non-volatile flavouring materials as specified by a recipe or formula.

2. **The particle reduction** of chocolate components is essential to generate the smooth result that most customers anticipate.

3. **Conching** evenly distributes Cocoa butter in chocolate acting as particle polisher in the manufacturing process. Conching is a complicated process that has been and continues to be studiedin great depth in order to fully understand the changes that occur in chocolate during this process (especially in the flavour of chocolate).

4. **Standardization** which is the step where the viscosity of the chocolate is adjusted with the addition of fats, usually cocoa butter, and emulsifiers. Volatileflavouring materials may also add at this point. (https://www.icco.org/processing-cocoa/)

Importance of Cocoa Beans

Cocoa has an essential role in a variety of disciplines, including the economy and health care. Cocoa bean is important to the economy since it is the primary raw material used in the creation of chocolate and other goods. Cocoa is highly essential economically for small growers. As a cash crop, it may offer critical money for food purchases, which is especially significant in places where food security has been an issue. In southern Cameroon, for example, cocoa appears to be the major source of revenue. In a sample of 21 communities in southern Cameroon, farming accounts for two-thirds of total income in each village, while Gockowski and Ndoumbe' indicate that cocoa provides around one-third of revenue in homes where it is a significant source of income (Franzen, M. and Mulder M. B, 2007). Cocoa is a valuable commodity all around the world. The majority of its production originates from West African nations (mostly the Ivory Coast and Ghana, which account for around 60 percent of the world's total cocoa), however it is generally processed in the European Union (EU; 1.3 million tonnes or 40 percent of the worldwide processing market in 2014). Apart from the EU, considerable amounts of cocoa beans are processed in Indonesia, the United States, and Brazil. (QuelalVásconez, M *et al.*, 2020)

Cocoa and chocolate have recently received significant scientific attention, owing to their intriguing phytochemical makeup. They are high in flavonoids, which are found in red wine and green tea. These substances may be beneficial to one's health. Cocoa and chocolate are plant foods made from cacao beans, which are *Theobroma cacao* seeds. Fermentation, drying, roasting, and grinding of the beans result in an intermediate product known as "cocoa mass." To manufacture chocolate, sugar is added to cocoa mass, which is then processed. The bulk of the fat is removed during the manufacture of cocoa powder. The manner in which cocoa beans are processed has a significant impact on the content of the final goods. Fermentation and alkali treatment, in particular, lower the flavonoid concentration of the final product.

The general qualitative characteristics of cocoa beans cocoa powder and chocolate (CAOBISCO-ECA-FCC, 2015; ADM Cocoa Manual, 2006) are listed Table .2.

A cocoa bean in general contains proteins, amino acids, lipids, triglycerides, phospholipids, carbohydrates, polyphenol, organic acids, purinic alkaloids and biogenic amines. Many enzymes may be found in raw cocoa beans, however the majority of them are inactivated as a result of the manufacturing process. They include β-glucosidase,β -fructosidase, α- amylase, β-gactosidase, proteinase, alkaline and acid phosphatases, lipase, polyphenol oxidase, pectinesterase, catalase, and peroxidase. (Bertazzo, A *et al., 2013*). The major components of cocoa powder (ICCO, 2012; Krahmer *et al.*, 2015; Lacueva *et al., 2008* are listed in Table .3

Tabel 2: Qualitative characteristics of cocoa beans

Quality attributes	Details	Observation
Cocoa beans		
Size (beans /100g)	≤100	Standard beans
	101 to 110	Medium beans
	> 120	Very small beans
Uniformity	Beans of varying sizes are moredifficult to break and deshell.	
Fermentation	5% slaty,5% defectiveness	Good fermented
	10% slaty, 10% defectiveness	Fair fermented
Moisture	<8%	Acceptable
Chemical residues	According to Authority Regulation	Under limits
Cocoa powder		
Solubility	95%	Good solubility
Shell content	<5% in fat free- dry cocoa.	
Chocolate		
Fat quality	Low in free fatty acids, withmelting and solidification properties.	
Aroma and flavour	Without mouldy off-flavors, smoky taints, acidic off-flavors,or closeness to other strong-smelling items.	Characteristic
Colour		Characteristic

Tabel 3: Major Components of Cocoa powder

Components	Major components (%)
Fat	11
Moisture	3.0
Total nitrogen	4.3
Nitrogen (corrected for alkaloids)	3.4
Protein	20
Nitrogen corrected for alkaloids x 6.25%	21.2
Ash	5.5
Water soluble ash	2.2
Phosphate (as P2O5)	1.9
Ash insoluble in 50% HCL	0.08
Flavanols	**Minor compounds (mg/g)**
Catechin	0.6
Epicatechin	5.7
Methylxanthines	
Caffeine	6
Theobromine	28
Other compounds	
Total procyanidins	22
Total amino acids	3.4
Total sugars	8.9

Common Adulterants Found in Cocoa beans and Cocoa Powder

1. Chestnut shell Adulteration

Chestnut shell adulteration in cocoa is a concern in the chocolate industry. It involves adding ground or powdered chestnut shells to cocoa products. This adulteration is considered fraudulent and unethical, as it aims to increase the weight of the cocoa product without adding any value in terms of cocoa content. Chestnut shells are cheaper than cocoa, so this adulteration can be a way for unscrupulous individuals or companies to increase profits at the expense of product quality.

Chestnuts are the seeds of chestnut trees, belonging to the genus Castanea. They are characterized by a hard outer shell, or husk, and a smooth, shiny, brown seed inside. The chestnut tree produces large, toothed leaves and, in the fall, the tree drops its spiky husks, revealing the nuts inside. The chestnut shell, also known as the husk or burr, is the protective outer covering of the chestnut fruit, which encases the chestnut seed. Chestnut shells are typically brown and have a rough, textured surface due to the spiky projections. The size of the husk can vary depending on the chestnut species. The size of the husk can vary depending on the chestnut species.

2. Peanut shell & Peanut Flour Adulteration

Peanut shell adulteration in cocoa powder refers to the unethical practice of adding ground or powdered peanut shells to cocoa products. This type of adulteration is considered fraudulent and poses serious risks, especially for individuals with peanut allergies. Peanut flour also added in the cocoa powder to enrich the richness of cocoa powder. Peanut allergies can be life-threatening for some individuals, leading to anaphylaxis. Adding peanut shells to cocoa without proper labelling poses a significant risk to consumers who may unknowingly consume the adulterated product. Apart from the health risks, peanut shell adulteration can affect the quality, flavor, and overall sensory characteristics of the cocoa product. Cocoa is prized for its unique taste and aroma, and the addition of peanut shells can alter these attributes negatively. The peanut shell is a thin and papery covering that encases the edible seeds or peanuts. Its primary function is to protect the seeds from environmental factors, pests, and damage. The shell has a rough and fibrous texture, providing a protective layer for the seeds. Peanut shells are typically light brown to tan in color.

Peanut flour is a finely ground powder made from roasted and defatted peanuts, resulting in a versatile ingredient with a concentrated peanut flavor. It is commonly used in both savory and sweet culinary applications, such as baking, cooking, and smoothies, to add nutty richness and protein.

3. Carob Adulteration

Adulteration of cocoa powder with ground or powdered carob involves the unethical practice of incorporating carob, derived from the pods of a leguminous tree, as a potential chocolate substitute due to its naturally sweet flavor. Such adulteration is considered fraudulent, leading to potential consequences for the quality and authenticity of the cocoa product.

Carob refers to the edible pod from the carob tree (*Ceratonia siliqua*), a leguminous evergreen tree native to the Mediterranean region. The carob pod contains seeds, and after processing, it can be ground into a powder commonly used as a substitute for cocoa or chocolate. Carob is known for its naturally sweet flavor, which comes from sugars like sucrose and fructose, and it is often utilized in various culinary applications as a healthier alternative to cocoa due to its lower fat content. Additionally, carob is rich in fiber, antioxidants, and certain minerals, contributing to its nutritional appeal.

4. Lard Adulteration

Adulterating cocoa products with lard is deceptive and goes against ethical food standards. It can compromise the integrity of the product and mislead consumers who may expect a vegetarian or vegan-friendly item. Many cultures and religions have dietary restrictions, and the inclusion of lard in cocoa products may offend or violate these beliefs. Proper labelling and transparency are crucial to respecting diverse dietary practices. Lard has a distinct flavor and texture that can alter the taste and quality of cocoa products. Cocoa is appreciated for its unique characteristics, and adulteration may negatively affect its sensory profile.

Lard is a type of fat that is obtained from the fatty tissues of pigs, specifically the visceral fat that surrounds the kidneys and loin. It is a semi-solid at room temperature and has been traditionally used in cooking and food preparation for centuries. Lard is primarily composed of triglycerides, with a high proportion of saturated and monounsaturated fatty acids. It contains oleic acid, the same monounsaturated fatty acid found in olive oil. At room temperature, lard is semi-solid, but it becomes liquid when heated. This characteristic makes it a useful ingredient in both baking and frying.

5. Starch Adulteration in Cocoa Powder

Starch, being a carbohydrate, lacks the rich flavor and characteristics of cocoa. The addition of starch can dilute the cocoa flavor and result in an inferior product in terms of taste and quality. Detecting starch adulteration in cocoa powder may be challenging, especially if the starch is finely ground and well-blended. Advanced analytical techniques, such as chromatography or spectroscopy, may be needed for accurate detection. Dextrin is a common starch added to cocoa powder. This addition can serve various purposes, including modifying the texture, enhancing solubility, or serving as a thickening agent. However, it's essential to distinguish between intentional and unintentional additions, as the latter may be considered adulteration. Dextrin can enhance the solubility of cocoa powder, making it easier to mix with liquids and resulting in a smoother beverage.

6. Gelatin Adulteration in Cocoa Powder

Gelatin is derived from animal collagen, usually from the skin, bones, and connective tissues of pigs or cows. Its use in cocoa products makes those products unsuitable for individuals following vegetarian or vegan diets. Gelatin is commonly used as a gelling agent, stabilizer, or thickener in various food applications. However, the addition of gelatin to cocoa powder can have specific implications, both in terms of intentional use and potential concerns for certain dietary preferences or restrictions. Gelatin is known for its ability to form a gel-like structure when dissolved in liquid. In certain culinary applications, gelatin is intentionally added to cocoa powder to provide specific textures, such as in the production of chocolate-flavored desserts or mousse. The addition of gelatin can contribute to the improvement of texture and mouthfeel in cocoa-containing products, providing a smooth and creamy consistency.

Adulteration of cocoa can have various negative consequences, including compromising the flavor, aroma, and overall quality of the cocoa product. Additionally, it may mislead consumers who expect pure cocoa but end up with a product that contains lower-quality or unrelated ingredients.

To combat such practices, food regulatory authorities and industry organizations often establish quality standards and conduct regular inspections to ensure compliance. Manufacturers and producers are expected to adhere to these standards to maintain the integrity of their products and protect consumer interests.

Consumers can play a role in preventing adulteration by purchasing cocoa products from reputable sources, checking for quality certifications, and being aware of common signs of adulteration. Additionally, staying informed about industry regulations and standards can contribute to making informed choices when buying cocoa products.

Scope of Detection of Adulterants

The increase in cocoa demand, volatile prices, and uncertain global cocoa production, which is at risk due to climatic change, can lead to cases of cocoa adulteration. The addition of natural or synthetic compounds changes the chemical composition and sensory qualities, with the goal of replacing certain valued components with less expensive ones and gaining unjustified advantages. Use of additional components in higher than recommended dosages. To avoid these or to ensure that the product we are consuming is safe for us, detection of adulteration is required.

Testing Methods for Detecting of Adulteration

The tests that are used to detect the adulterants in cocoa beans, powder are

- Cut test
- High Performance Liquid Chromatography (HPLC)
- Gas chromatography (GC)
- Near Infrared Spectroscopy (NIR)

Cut Test

Cut tests are performed throughout the fermentation process. Farmers will evaluate a small amount each day of fermentation to monitor the progress and to spot any health issues or infestation early on. Farmers may conduct further cut tests once the beans have been taken from their fermenting boxes and left out to dry. At the conclusion of processing, they frequently perform a big cut test on 50 or more beans (occasionally using a guillotine). Thisis done to determine the total level of fermentation for each batch. Throughout the process, if a large number of samples in a batch have mold, are slaty (not grooved, but slick and greyish), or have germinated beans (the germ and/or cotyledon are still present), the batch is deemed useless and destroyed

To perform a cut test, you simply cut fermented cacao beans in half to evaluate their centers. They are usually performed on at least ten beans to have a representative sample of each batch. At least 70 percent of the beans must be fermented to the correct level to be considered successful.

(https://perfectdailygrind.com/2019/08/what-is-a-cacao- cut-test-when-is-it-useful/)

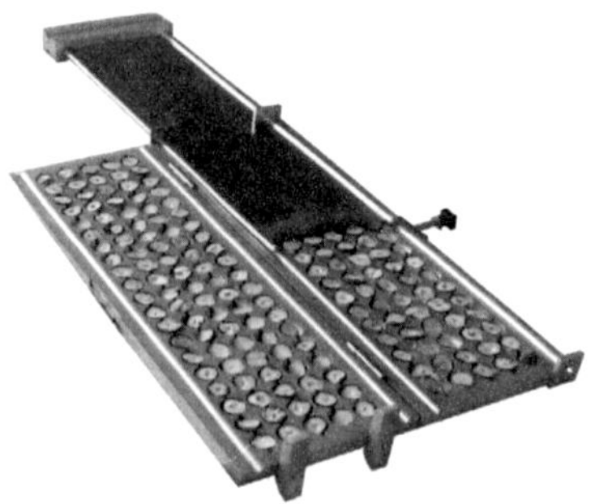

Fig. 6. Instrument used for Cut test (*Source*: https://innovas.en.ec21.com/)

High Performance Liquid Chromatography (HPLC)

HPLC is a chromatographic method used in phytochemical and analytical chemistry to identify, measure, and purify the various components of a mixture. The analyte (sample) is distributed between a mobile phase (eluent) and a stationary phase in HPLC (packing material of the column). The molecules are slowed while passing through the stationary phase depending on the chemical structure of the analyte. The unique intermolecular interactions of a sample's molecules with the packing material dictate their duration "on-column." As a result, various elements of a sample elute at varying rates. As a result, the sample components are separated.

Yang *et al.* (2015) developed a novel analysis approach for detecting adulterated cocoa powder utilizing high-performance liquid chromatography (HPLC) fingerprint profiling in conjunction with principal component analysis (PCA). In their study, polysaccharides were extracted from 11 batches of cocoa powder samples using hot water under ultrasound-assisted conditions. Subsequently, enzymatic removal of proteins and complete hydrolysis with sulfuric acid were performed on the extracted polysaccharides. The resulting hydrolysates underwent HPLC analysis, employing pre-column 1-phenyl-3-methyl-pyrazolone derivatization to generate fingerprints of cocoa powder polysaccharides. By processing the relative areas of 12 characteristic peaks in the fingerprints and corresponding chromatogram data of adulterated samples through PCA, the study demonstrated the capability to identify adulterated cocoa powder samples mixed with 15% or more cocoa shell powder and 10% or more exogenous plant material, such as chestnut shell powder, longan shell powder, peanut shell powder, starch, wheat powder, or pumpkin powder, under the same chromatographic conditions.

Gas Chromatography

The sample solution injected into the instrument enters a gas stream that takes the sample into a separation tube known as the "column." (As the so-called carrier gas, helium or nitrogen is employed.) Within the column, the various components are segregated. The detector counts the number of components that escape the column. To measure a sample with an unknown concentration, a known concentration standard sample is introduced into the device. To compute the concentration, the standard sample peak retention time (appearance time) and area are compared to the test sample. (www.shimadzu.com)

In 2019, Barbosa-Pereira *et al.* conducted a comprehensive study aimed at characterizing and classifying cocoa bean shells (CBS) derived from various cultivars and geographical origins based on their volatile profiles. The researchers employed headspace solid-phase microextraction gas chromatography-mass spectrometry (HS-SPME/GC-qMS) and electronic nose (E-nose) techniques, coupled with principal component analysis (PCA). This pioneering study provided a detailed fingerprint and semi-quantitative data for over 100 volatile organic compounds (VOCs), encompassing aldehydes, ketones, pyrazines, alcohols, and acids. Through PCA, a distinct separation of the Criollo cultivar from other cultivars was achieved using both GC-qMS and E-nose methods, attributed to the elevated content of key-aroma VOCs. Various biomarkers identified by GC-qMS, such as 2-hepanol, 2-methylpropanoic acid, and 2,3,5-trimethylpyrazine, recognized as crucial aroma compounds for cocoa beans, proved effective for classifying CBSs based on their quality and origin. The study indicated a high correlation between GC-qMS and E-nose, suggesting their suitability as analytical approaches for CBS classification. The volatile fingerprint and classification of CBSs hold potential for selecting samples with specific flavor profiles tailored to different food applications, presenting an intriguing avenue for valorizing this by-product as a valuable food ingredient.

Near Infrared Spectroscopy (NIR)

NIR spectroscopy is used to analyze the composition, functionality, and sensory properties of materials, intermediates, and finished products. It is used in the food and feed, agricultural, dairy, pharmaceutical, and chemical sectors, which are under continual pressure to produce goods that fulfil client demands while boosting plant productivity and profitability. NIR can be used for quantitative analysis (determining substance concentrations), qualitative analysis (identification of raw materials, intermediate and final products), and process control. It can offer information on moisture, protein, fat, and carbohydrate composition. NIR applications vary by sector and are individually built to fit distinct firms and their particular goods and demands.

The absorption of electromagnetic (EM) radiation at wavelengths ranging from 780 to

2,500 nm is the basis for near infrared (NIR) spectroscopy. The detector measures the sample's transmittance and absorbance after the light interacts with it. B. G. Osborne (2006)

Quelal-Vásconez *et al.* (2018) employed rapid techniques, specifically Near-Infrared (NIR) technology coupled with multivariate analysis, to detect the presence of carob flours in cocoa powder. The diffuse reflectance spectra of samples ranging from 1100 to 2500 nm were captured using a Foss NIR spectrophotometer. Both qualitative and quantitative analyses were conducted. In the qualitative analysis, Principal Component Analysis (PCA) and Partial Least Squares Discriminant Analysis (PLS-DA) were employed, resulting in excellent outcomes with 100% classification accuracy. This success suggests the capability to differentiate between pure cocoa powders and adulterated samples. For quantitative analysis, Partial Least Squares (PLS) regression analysis was utilized. The most robust PLS prediction model, achieved with a single factor (LV), demonstrated a coefficient of determination for prediction (RP^2) of 0.974 and a root mean square error of prediction (RMSEP) of 3.2% for the external set. These findings lead to the conclusion that the combination of NIR technology and multivariate analysis facilitates the identification and quantification of natural cocoa powder in mixtures adulterated with carob flour.

Extraction Method of Cocoa for Different Testing Methods

Table 4: Sample Preparation and sampling procedure

Method	Sample	Extraction Procedure	References
HPLC	Cococa Powder	Weigh, to the nearest 0.1 g, a test portion of 5 ginto a high-speed blender. Add 100mL of aqueous solution of sodium hydrogen carbonate (0.1 percent) and polyethylene glycol (0.3 percent). Stopper the jar and blend samples at high speed for 2 min; or (b) weigh, to the nearest 0.1 g, a test portion of 5 g intoa 300mL Erlenmeyer flask. Add 100mL of aqueous solution of sodium hydrogen carbonate (0.1 percent) and polyethylene glycol (0.3 percent). Stopper the flask and shake samples at medium speed for 30–45 min in wrist action shaker. Filter the extract through folded filter paper. Draw 20mL of filtrate and dilute with 20 mL of PBS. Mix thoroughly. Apply the diluted sample to the IAC, as described underneath. If the sample should be slightly cloudy, re-filter the extract through glass microfiber filters.	Brera, C *et al*, 2003

Method	Sample	Extraction Procedure	References
NIR	Cocoabeans	Multiple beans were collected from different regions then sealed in air-tight bag and transported to the lab. All the samples were ground separately for 15 s in multi-purpose grinder container was allowed to cool each sample to avoid the loss of volatile compounds. Each ground sample was sieved with a 500 μm meshbefore further analysis. The adulterated samples were prepared by accurately weighing portions of 5–40 wt/wt. percent content of UFC in FC content and shaken vigorously to ensure total homogenization. ADC was prepared to cover as much the perceived range as possible that pertains in the market. The samples were immediately analyzed	Teye, E *et al*, 2014
Cut Test	CocoaBeans	In order to reveal the maximal cut surface of the cotyledons, 300 beans were sliced lengthwise through the centre. In broad daylight, both parts were evaluated and classified as one of the following: plum, light plum, brown, slaty, germinated, andmouldy	Guehi, T. S *et al*, 2007
Gas chro-matography	CocoaBeans	A 50 mL screw cap centrifuge tube was filled with 10 g of dried cocoa beans. The mixture was then aggressively agitated manually for 1 minute andagain using a vortex mixer for another 1 minute. The mixture was then centrifuged at 12,000 rpm for5 minutes at 4° C. In a scintillation vial, 2 mL of thesupernatant was combined with 4 mL of n-hexane. Before analysis, the vial was vortexed for 30 seconds and an aliquot of 1 mL acetonitrile layer was filtered via a 0.2 mL PVDF filter into the auto sampler vial.	Zainudin, B. H *et al*, 2015

Standards for Cocoa Products

Cocoa Powder Description

(a) Cocoa powder is the substance formed by mechanically transforming cocoa press cake or cocoa mass into powder form after the partial removal of fat from the ground nib of well-fermented sound roasted beans of Theobroma cacao L.

(b) It must be in powder form and have a distinct taste and flavour.

(c) It must be devoid of dirt, filth, harmful compounds, adulterants, and added colouring matter, as well as rancidity, off-flavor, mould development, and insect infestation.

(d) It must also meet the following specifications:

Table 5: FSSAI Standard for Cocoa Powder

		Requirements (in per cent.)		
S.No.	**Characteristics**	**Cocoa Powder**	**Fat reduced cocoa powder (Medium fat)**	**Highly Fat reduced cocoa powder (Low fat)**
1	Moisture content byweight	Not more than7.0	Not more than 7.0	Not more than7.0
2	Cocoa butter by weight(on dry basis)	Not less than20.0	Not less than10.0- Not more than 20.0	Not more than10.0
3	Acid insoluble ash by weight (on moisture andfat-free basis)	Not more than1.0	Not more than 1.0	Not more than1.0
4	Alkalinity of total Ashas K2O by weight (onmoisture and fat-free basis)	Not more than12.0	Not more than 12.0	Not more than12.0
5	Crude fibre by weight (on moisture and fat-freebasis)	Not more than7.0	Not more than 7.0	Not more than7.0

(Food safety and standards (Food Products Standards and Food Additives) Fourteenth Amendment Regulations, 2018)

Low- Fat and High-Fat Cocoa Powder means the powder which is the partially defattedproduct derived from the cocoa bean the seed of Theobroma cocoa L. During the manufacturing process, it may be treated with alkali and/or magnesium carbonate, bicarbonate, and tartaric, citric, or phosphoric acids. It must not have rancidity, dirt, filth, insects or insect fragments, or fungal infections. The product may include Appendix A-approved food additives. It must meet the following requirements:

Table 6: FSSAI specification for Low- Fat and High-Fat Cocoa Powder

Total ash	Not more than 14.0 per cent (on moisture and fat free basis).
Ash insoluble in dilute HCl	Not more than 1.0 per cent (on moisture and fat free basis).
Alkalinity of total ash	Not more than 6.0 per cent as K2O (on moisture and fat free basis)
Cocoa butter (i) for low fat (ii) for high fat	Not less than 10.0 percent (on moisture free basis)Not less than 20.0 percent (on moisture free basis)

(Food safety and standards (food products standards and food additives) regulations, 2011)

Cocoa Liquor or Cocoa Mass, and Cocoa Cake

1. Description

(a) Cocoa (Cacao) Mass or Cocoa/Chocolate Liquor is a product made from cocoa beans, the seeds of Theobroma cacao, by appropriate grinding of clean, almost shell-free nibs (cotyledons) with or without roasting, and with or without removal or addition of any of its ingredients.

(b) Cocoa Cake is a product made by removing some or all of the fat from cocoa nibs or cocoa powder.

(c) The items must have their own distinctive color, odour, and flavour, and they must be devoid of any additional colouring, flavouring, or fats other than cocoa butter.

(d) It also has to meet the following requirements:

Table 7: FSSAI Specification for Cocoa mass & Cocoa cake

			Requirements (in per cent.)	
S.No.	**Characteristics**		**Cocoa Mass**	**Cocoa Cake**
1	Moisture content by weight		Not more than 10.0	Not more than 10.0
2	Cocoa Shell and Germ % by weight	Calculated on thefat free drymatte	Not more than 5.0	Not more than 5.0
		Calculated on an alkali free basis (forcocoa shell only)	Not more than 1.75	Not more than 4.5
3	Cocoa Butter % by weight		47.0 -60.0	Not applicable
4	Alkalinity of total Ash as K2O % by weight (on moisture and fat-free basis)		Not more than 12.0	Not more than 12.0
5	Acid insoluble ash % (on moisture and fat-free basis)		Not more than 1.0	Not more than 1.0
6	Crude fibre % by weight (on moistureand fat-free basis)		Not more than 7.0	Not more than 7.0

(Food Safety and Standards (Food Products Standards and Food Additives) Fourteenth Amendment Regulations, 2018.)

International Regulations/Laws

According to the Food Safety and Standards (Food Products Standards and Food Additives) Regulations 2011, chocolate defined as a homogeneous product obtained through an appropriate manufacturing process from a mixture of one or more of the ingredients, namely, cocoa beans, cocoa nib, cocoa mass, cocoa press cake, and cocoa dust (cocoa fines/powder), including fat reduced cocoa powder with or without sugar addition, cocoa butter, milk solids including milk fat The chocolates must not include any vegetable fat other than cocoa butter.

The material must be devoid of rancidity or off-odors, bug and fungal infestations, dirt, adulterants, and any other dangerous or hurtful materials.The chocolates must be of the following varieties:

Milk chocolates are made with one or more of the following ingredients: cocoa nib, cocoa mass, cocoa press cake, cocoa powder, including low-fat cocoa powder with sugar, and milk solids, which would include milk fat and cocoa butter.

Plain chocolate is made using one or more of the following ingredients: cocoa nib, cocoa mass, cocoa press cake, cocoa powder (including low-fat cocoa powder with sugar), and cocoa butter.

Plain Covering Chocolate- similar to plain chocolate, but designed for covering.

Blended Chocolate is a combination of milk and plain chocolates in various amounts.

White chocolate is prepared by cocoa butter, milk solids (containing milk fat and sugar), and sugar.

Filled Chocolates are items with an external covering of chocolate and a centre that is obviously distinguishable from the external coating by its composition, but do not contain flour confectionery pastry and biscuit products. The coating should be of chocolate that fits the standards of one or more of the chocolate varieties indicated above. The chocolate component of the coating must account for at least 25 percent of the overall mass of the completed product.

A composite chocolate product consists of at least 60 percent chocolate by weight plus edible healthy components such as fruits and nuts. It must contain one or more edible wholesome components that account for at least 10 percent of the total mass of the completed product. Provided that it may contain artificial sweeteners as provided in regulation 3.1.3 and label declaration as provided under Regulation 2.4.5 (24, 25, 26, 28 & 29) of Food Safety and Standards (Packaging and Labelling) Regulations, 2011. Furthermore, in addition to the components listed above, the chocolate may contain one or more of the chemicals listed under distinct varieties of chocolates below.

(a) Edible salts.

(b) Spices and condiments.

(c) Permitted emulsifying and stabilizing agents.

(d) Permitted sequestering and buffering agents.

Impact on Health and Wellness

Cocoa is most commonly used for heart disease and high blood pressure.

- According to certain studies, consuming chocolate reduces the risk of heart disease and mortality. Cocoa may have this impact through decreasing blood pressure and enhancing the function of blood arteries.
- According to the majority studies, eating dark chocolate or cocoa goods for 2-8 weeks can reduce the top number in a blood pressure reading (systolic blood pressure) by 4 mmHg. In persons with high blood pressure, the bottom number (diastolic blood pressure) is likewise reduced by 2 mmHg. Eating cocoa may also assist persons with normal blood pressure avoid high blood pressure, but only if the cocoa comes from simple chocolate. Consuming chocolate in desserts may raise the chance of developing high blood pressure.
- Some research shows that taking cocoa extract alone or in combination with other ingredients might improve skin wrinkles, elasticity, and roughness.

- According to a recent study, using cocoa extract alone or in combination with other substances may improve skin wrinkles, suppleness, and roughness.
- Taking a large amount of cocoa on a daily basis can help persons with chronic fatigue syndrome minimize exhaustion, anxiety, and sadness, as well as improve their general capacity to operate.
- Cocoa may aid some individuals to boost their memory and reasoning skills. However, not all study agrees. The advantage, if any, may be dependent on the person's age, length of cocoa consumption, cocoa dose, and the complexity of the cognitive tests. Memory and reasoning abilities decline in elderly persons that is greater than average for their age. Early study suggests that cocoa may boost memory and reasoning skills in persons with moderate cognitive impairment.

Table 8: FSSAI Standards for Chocolates

Characteristics	**Requirements for**					
	Milk Chocolate	**Milk Covering Chocolate**	**Plain Chocolate**	**Plain Covering chocolate**	**White chocolate**	**Blended chocolate**
Total fat (on dry basis) per cent by weight. Not less than	25	25	25	25	25	25
Milk fat (on dry basis) Per cent by weight. Not less than	2	2	-	-	2	-
Cocoa solids (on Moisture-free and fat-freebasis) per cent by weight. Not less than	2.5	2.5	12	12	-	3.0
Milk Solids (on Moisture-free and fat-free basis) per cent by weight. A) Not less than B) Not more	10.5 -	10.5 -	- -	- -	10.5 -	1 9
Acid-insoluble ash (on moisture fat and sugar-free basis) per cent by weight. Not more than	0.2	0.2	0.2	0.2	0.2	0.2

Source: Food Safety and Standards (Food Products Standards and Food Additives) Regulations, 2011

- When consumed in moderation or at proportions usually found in meals, cocoa may be safe during pregnancy and breast-feeding. However, keep an eye on your consumption. Because of the caffeine content, consuming more cocoa during pregnancy may be hazardous. Cocoa contains caffeine, which crosses the placenta. Although contentious, some research shows that increased caffeine intake during pregnancy may be linked to early birth, low birth weight, and miscarriage. Caffeine use should be limited to less than 300 mg per day during pregnancy, according to some specialists. Take into account that chocolate products contain between 2 and 25 mg of caffeine per serving, whereas a cup of hot chocolate has about 10 mg. As a result, these goods are probably not a major risk. Unsweetened, dry cocoa powder, on the other hand, can contain up to 200 mg of caffeine per cup. Caffeine is also an issue with breast-feeding. Caffeine concentrations in breast milk are considered to be around half those of the mother's blood. Because of the caffeine, if the mother consumes too much chocolate (16 oz per day), the nursing newborn may become fussy and have frequent bowel movements. (https://www.webmd.com)

Over the last decade, studies on the health benefits of cocoa and cocoa products have been done, with a particular emphasis on degenerative disorders. These advantages might be attributed to their high levels of flavonoid monomers (catechin and epicatechin) up to tetradecamers. The majority of studies have focused on the flavonoids found in cocoa and cocoa products and their contributions to health benefits, but it should be noted that cocoa and cocoa products are also high in methylxanthines, such as caffeine, theobromine, and theophylline, and studies have shown that methylxanthines can have both positive and negative health effects. Caffeine consumption, for example, has been linked to poor reproductive health. Caffeine, on the other hand, increased net hepatic glucose absorption via increasing glucose-6-phosphate synthesis in the liver (Ying, W and Waller, D. P. 1994). Caffeine, on the other hand, increased net hepatic glucose absorption via increasing glucose-6-phosphate synthesis in the liver.

Cocoa is very high in micronutrients, and micronutrients like copper contained in cocoa might considerably contribute to human nutritional consumption. Because cocoa includes a variety of bioactive components, it is conceivable to speculate that there may be direct or indirect synergism between these components in terms of providing their health benefits. Cooper *et al.* proposed that if the biological effects are attributable to cocoa flavonoids rather than other components, chocolate or cocoa products containing other than flavonoids would be the ideal control. Factors that influence the quality and amount of cocoa and cocoa-based products during production and manufacture are critical in providing the optimal health effects. These factors have the potential to drastically diminish the polyphenol content of the chosen goods. When evaluating their contribution to the biological system, physiological aspects such as bioavailability and antioxidant qualities should be evaluated. Cooper and colleagues (2008)

Case Studies

Obesity and Lipid Metabolism

Recent research has looked on the preventive and therapeutic benefits of cocoa and cocoa ingredients on obesity and metabolic syndrome. Cocoa consumption reduced the amount of visceral adipose tissue in rats. The DNA study of the liver and mesenteric fat tissue yielded intriguing results. In the study, there was a decrease in the expression of several genes involved in fatty acid transport and synthesis in the liver and mesenteric fat, as well as an increase in the expression of genes involved in thermogenesis. Smelling dark chocolate was used in a clinical investigation to test hunger response. Chocolate causes a satiation reaction and reduces hunger, therefore it may help prevent weight gain. Furthermore, flavonoids can cause metabolic processes that reduce lipogenesis, enhance lipolysis, and boost adiponectin production; these events lower lipid deposition and insulin resistance, hence reducing obesity. Research found that participants who consumed chocolate on a regular basis gained much more weight over time. However, no information was supplied on the recruited individuals' consumer profiles or the sort of chocolate ingested (in particular, the specific amount of dark chocolate). (Gu *et al.* 2014, Matsui *et al.* 2005, Massolt *et al.* 2010, Magrone *et al.* 2017)

Psychological Aspects

The psychological and social backdrop of daily living influences metabolic health, emotions, and moods, and can influence food choices. Chocolate eating has been linked to a kind of depression known as hysteroid dysphoria in some circumstances. This disorder is characterized by frequent periods of despair in reaction to feelings of inadequacy or social rejection, culminating in real bulimic attacks for candy and chocolate. A real chocolate addiction (being a chocoholic) is similar to alcoholism and nicotine addiction; it affects 40 percent of females and 15 percent of males in Western countries. The symptoms include sensitivity to medicines that increase serotonin transmission; this shows that chocolate eating may involve central serotonin pathways. The availabilityof serotonin may explain why sugar and sweets are highly sought during chocolate bulimic episodes. Carbohydrate consumption (e.g., bread and chocolate) increases therelationship between plasma tryptophan and other neutral amino acids; as a result, tryptophan transport across the blood-brain barrier is activated, resulting in an increasein cerebral serotonin synthesis and a sense of energy and pleasure. (Alonso, C. *et al.* 2008; Martin, F. P. J *et al.* 2012)

Impact on Skin

Cocoa's skin benefits may extend beyond its usage as a topical treatment. Cocoa butter is found in commercially accessible skin care products. Many studies have found that cocoa flavanols protect skin from UV radiation damage, which is thought to be mediated by processes such as the formation of ROS and inflammatory markerssuch as prostagladins, NO, leukotrienes, and histamine. In one research, 12 weeks of high-flavanol cocoa intake reduced erythema generated by UV light by 25 percent, while in another, specifically made high-flavanol chocolate (>600 mg flavanols and >10 000 ORAC units per 20 g chunk) more than quadrupled the dosage of UV radiation required to create erythema. However, commercially available chocolate did not have this impact. Cocoa flavanols' antioxidant properties are one probable method for skin protection.

Improvement in Coronary Circulation

Shiina *et al.* found improvement in coronary circulation related with dark chocolate consumption in 2009 research. For two weeks, 39 healthy males were randomly allocated to consume either 45 g of flavonoid-rich dark chocolate (550 mg polyphenols) or 35 g of flavonoid-free white chocolate. At the beginning and conclusion of the trial, echocardiography was used to determine CFVR, blood pressure, serum lipids, and oxidative stress markers. Dark chocolate, but not white chocolate, enhanced CFVR considerably, regardless of other covariates.

Conclusion

Chocolate is made mostly from cocoa beans. Cocoa beans are cultivated in tropical zones surrounding the Equator, which is the best place for growing cocoa beans. Approximately 70 percent of the world's cocoa beans are produced in four West African countries: Ivory Coast, Ghana, Nigeria, and Cameroon. When taken in moderation, chocolate provides several health advantages. Dark chocolate is the healthiest option since it is the least processed, containing the maximum amount of flavonoid-rich cocoa bean. Cocoa is being utilized as a treatment for conditions such as diarrhea, asthma, and bronchitis. Cocoa has the potential to induce adverse skin responses, constipation, and migraine headaches. Chocolate includes some iron, calcium, magnesium, potassium, and vitamin E and B vitamins. Chocolate contains both fat and sugar. Cocoa/chocolate has various advantages and negatives, hence it should be used in moderation for optimal health. It is still crucial to develop a broadly applicableand sensitive technique that can handle the many features of cocoa adulteration. This involves species classification, detecting faulty beans, and detecting the presence of exogenous chemicals. Recent developments such as increased proactivity, high levelsof openness, and improved government monitoring of suppliers and vendors may safeguard consumers against food adulteration.

References

Alonso, C., Guilarte, M., Vicario, M., Ramos, L., Ramadan, Z., Antolín, M., & Malagelada, J. R. (2008). Maladaptive intestinal epithelial responses to life stress may predispose healthy women to gut mucosal inflammation. Gastroenterology, 135(1), 163-172.

Aprotosoaie, A. C., Luca, S. V., & Miron, A. (2016). Flavor chemistry of cocoa and cocoa products—an overview. Comprehensive Reviews in Food Science and Food Safety, 15(1), 73-91.

Barbosa-Pereira, L., Rojo-Poveda, O., Ferrocino, I., Giordano, M., & Zeppa, G. (2019). Assessment of volatile fingerprint by HS-SPME/GC-qMS and E-nose for the classification of cocoa bean shells using chemometrics. Food research international (Ottawa, Ont.), 123, 684–696. https://doi.org/10.1016/j.foodres.2019.05.041

Bertazzo, A., Comai, S., Mangiarini, F., & Chen, S. (2013). Composition of cacao beans. In Chocolate in health and nutrition (pp. 105-117). Humana Press, Totowa, NJ.

Brera, C., Grossi, S., de Santis, B., & Miraglia, M. (2003). High performance liquid chromatographic method for the determination of ochratoxin a in cocoa powder. Journal of Liquid Chromatography & Related Technologies, 26(4), 585-598.

CAOBISCO-ECA-FCC. (2015). Cocoa beans: Chocolate and cocoa industry quality requirements. Retrieved from http://www.cocoa quality.eu/

Chocolate, B. (2015). Confectionery of Europe (CAOBISCO). European Cocoa Association, (ECA), and Federation of Cocoa Commerce (FCC), Cocoa in grain: quality requirements for the chocolate and cocoa industry. [Online], End, MJ & Dand, R., Eds. Europe. September.

Coe, S.D.; Coe, M.D. The True History of Chocolate, 1st ed.; Thames and Hudson: London, UK, 1996.

Cooper, K. A., Donovan, J. L., Waterhouse, A. L., & Williamson, G. (2008). Cocoa and health: a decade of research. British Journal of Nutrition, 99(1), 1-11.

Crouzillat, D., Bellanger, L., Rigoreau, M., Bucheli, P., & Pétiard, V. (2000). Genetic structure, characterisation and selection of Nacional cocoa compared to other genetic groups. In international workshop on new Technologies and Cocoa Breeding (p. 30).

Feasibility study on the use of Fourier transform near-infrared spectroscopy together with chemometrics to discriminate and quantify adulteration in cocoa beans Ernest Teye, Xing-yi Huang, Wu Lei, Huang Dai.

Franzen, M., & Mulder, M. B. (2007). Ecological, economic and social perspectives on cocoa production worldwide. Biodiversity and Conservation, 16(13), 3835-3849.

Gu, Y., Yu, S., & Lambert, J. D. (2014). Dietary cocoa ameliorates obesity-related inflammationin high fat-fed mice. European journal of nutrition, 53(1), 149-158.

Guehi, T. S., Konan, Y. M., Koffi-Nevry, R., N'dri, D. Y., & Manizan, N. P. (2007). Enumeration and identification of main fungal isolates and evaluation of fermentation's degree of Ivorian raw cocoa beans. Aust. J. Basic Appl. Sci, 1(4), 479-486.

Hannum, S. M., & Erdman Jr, J. W. (2000). Emerging health benefits from cocoa and chocolate. ICCO. (2012). Physical and chemical information on cocoa beans, butter, mass and powder.Retrived from http://www.icco.org/faq/ 61-physical-and-chemical-information-on cocoa/106 physical-and-chemical–information-on-cocoa-beans-butter-mass–and-power.html beans,

Katz, D. L., Doughty, K., & Ali, A. (2011). Cocoa and chocolate in human health and disease. Antioxidants & Redox Signaling, 15(10), 2779-2811.

Lippi, D. (2013). Chocolate in history: food, medicine, medi-food. Vlaeminck, P., Vandoren, J., Vranken, L., Squicciarini, M., & Swinnen, J. (2016). Consumers' willingness to pay for fair trade chocolate. The Economics of Chocolate, 180.

Magrone, T., Russo, M. A., & Jirillo, E. (2017). Cocoa and dark chocolate polyphenols: from biology to clinical applications. Frontiers in Immunology, 8, 677.

Martin, F. P. J., Antille, N., Rezzi, S., & Kochhar, S. (2012). Everyday eating experiences of chocolate and non-chocolate snacks impact postprandial anxiety, energy and emotional states. Nutrients, 4(6), 554-567.

Massolt, E. T., van Haard, P. M., Rehfeld, J. F., Posthuma, E. F., van der Veer, E., & Schweitzer, D. H. (2010). Appetite suppression through smelling of dark chocolate correlates with changes in ghrelin in young women. Regulatory Peptides, 161(1-3), 81-86.

Matsui, N., Ito, R., Nishimura, E., Yoshikawa, M., Kato, M., Kamei, M., ... & Hashizume, S. (2005). Ingested cocoa can prevent high-fat diet-induced obesity by regulating the expression of genes for fatty acid metabolism. Nutrition, 21(5), 594-601.

Oracz, J., & Zyzelewicz, D. (2019). In vitro antioxidant activity and ftir characterization of high-molecular weight melanoidin fractions from different types of cocoa beans. Antioxidants, 8(11), 560

Osborne, B. G. (2006). Near-infrared spectroscopy in food analysis. Encyclopedia of analytical chemistry: applications, theory and instrumentation.

Quelal-Vásconez, M. A., Lerma-García, M. J., Pérez-Esteve, É., Talens, P., & Barat, J. M. (2020). Roadmap of cocoa quality and authenticity control in the industry: A review of conventional and alternative methods. Comprehensive Reviews in Food Science and Food Safety, 19(2), 448-478.

Quelal-Vásconez, M. A., Pérez-Esteve, É, Arnau-Bonachera, A., Barat, J. M., & Talens, P. (2018). Rapid fraud detection of cocoa powder with carob flour using near infrared spectroscopy. Food Control, 92, 183-189

Teye, E., Anyidoho, E., Agbemafle, R., Sam-Amoah, L. K., & Elliott, C. (2020). Cocoa bean and cocoa bean products quality evaluation by NIR spectroscopy and chemometrics: A review. Infrared Physics & Technology, 104, 103127.

Teye, E., Huang, X. Y., Lei, W., & Dai, H. (2014). Feasibility study on the use of Fourier transform near-infrared spectroscopy together with chemometrics to discriminate and quantify adulteration in cocoa beans. Food Research International, 55, 288-293.

Verna, R. (2013). The history and science of chocolate. The Malaysian journal of pathology, 35(2), 111.

Vlaeminck, P., Vandoren, J., Vranken, L., Squicciarini, M., & Swinnen, J. (2016). Consumers' willingness to pay for fair trade chocolate. The economics of chocolate, 180.

Yang, W. L., Hu, M. H., Chen, S. W., Wang, Q., Zhu, S., Dai, J., & Li, X. Z. (2015). Identification of adulterated cocoa powder using chromatographic fingerprints of polysaccharides coupled with principal component analysis. Food analytical methods, 8(9), 2360-2367.

Ying, W., & Waller, D. P. (1994). Theobromine toxicity on Sertoli cells and comparison with cocoa extract in male rats. Toxicology Letters, 70(2), 155-164.

Zainudin, B. H., Salleh, S., Mohamed, R., Yap, K. C., & Muhamad, H. (2015). Development, validation and determination of multiclass pesticide residues in cocoa beans using gas chromatography and liquid chromatography tandem mass spectrometry. Food Chemistry, 172, 585-595.

14

Innovations in Detection of Adulterants in Spices and Condiments

Snigdha Mohandas[1], C Sarathambal[1], K Anees[1], E Jayashree[1] and Janifer Raj Xavier[2]

[1]ICAR-Indian Institute of Spices Research Marikunnu (PO) Kozhikode, India
[2]Defence Food Research Laboratory (DFRL), Defence Research and Development Organisation (DRDO), Mysore, Karnataka, India

Introduction

Spices are botanical products of tropical and subtropical plants, mainly fruits, seeds, flowers, roots, stems, leaves, bark, or whole plant (Golob & Nations, 1999). The dominant spices of trade including cardamom, cinnamon, cloves, ginger and pepper (A. Osman *et al.*, 2019)including aroma, flavor, and color, to foods and beverages in an effort to enhance their palatability. Herbs and spices exhibit a plethora of medicinal properties including antimicrobial and antioxidant properties. Spices serve as natural preservatives to delay food spoilage and extend shelf life. As spice trading has exponentially increased globally, herbs and spices have become prone to adulteration, which can be deliberate or unintentional. Deliberate adulteration is usually economically motivated, aimed at maximizing profit margins, whereas unintentional adulteration is often attributed to improper harvesting or processing of the plant material or collecting/substituting closely related species. Adulteration of herbs or spices with toxic ingredients such as Sudan dyes or metal salts (e.g., lead chromate. Spices used in culinary, industrial cooking and pharmaceuticals can be rhizome (ginger, turmeric), leaf (curry leaf), berries (black pepper), buds (clove), seeds (cumin), bark (cinnamon), or even stigma of flower (saffron). Although spices and herbs have been used since prehistoric times, they are playing a new and important role in modern food preparation. They not only provide distinct flavors to our cuisine, but they also add colour and diversity. Certain spices and herbs, whether used alone or in combinations, can replace or reduce the amount of salt and sugar in dishes.. It can be used fresh, dry, whole, sliced or crushed in the preparation of foods and drinks. Spices are an important part of human nutrition and contain proteins, fibre, sugars, essential oils, minerals, pigments besides bioactive compounds such as phenolic acids, sterols and pungent principles like piperine, capsaicin etc. Based on Indian Brand Equity Foundation, India has exported more than 1 billion kg spices of various types with a total export market value of $3.11 billion dollars during 2017-2018 to different countries of the world. Standard ISO 676: has listed 109 spices, out of which, spices board of Indian has listed 52 spices as Indian. Spices are usually used as flavoring, coloring , preservatives and also as medicine (A. Osman *et al.*, 2019)including aroma, flavor, and color, to foods and beverages in an effort to enhance their palatability. Herbs and spices exhibit a plethora of medicinal properties including antimicrobial and antioxidant properties. Spices serve as natural preservatives to delay food spoilage and extend shelf life. As spice trading has exponentially increased globally, herbs and spices have become prone to adulteration, which can be deliberate or unintentional. Deliberate adulteration is usually economically motivated, aimed at maximizing profit margins, whereas unintentional adulteration is often attributed to improper harvesting or processing of the plant material or collecting/substituting closely related species. Adulteration of herbs or spices with toxic ingredients such as Sudan dyes or metal salts (e.g., lead chromate).

Adulteration has been reported in all foods including spices. Adulteration usually refers to the mixing of inferior and sometimes harmful quality substances with food or drink intended to be sold. Food or drink becomes tainted and unsuitable for human consumption as a result of adulteration. India provides a wide range of spices, which are essential in every Indian cooking.

If branded and packaged spices are not available, most people will buy loose spices from their local grocery shop. This raises the possibility of ingesting adulterated spices., for example, The most expensive spice in the world Saffron is adulterated by colored dried tendrils of maize cob (Kumar *et al.*, 2014). Unlike this particular adulterant which is not unhealthy (but illegal), most of the adulterants are unhealthy and can cause serious and fatal damage to our system. Spices are more prone to adulteration due its increased price and high demand (Hart, 1952).

Some unscrupulous traders are irresponsible and perform adulteration. Moreover, some spices are easily misused unconsciously due to their similar appearance. Papaya seeds, black berries are the common adulterant used in black pepper as they are almost similar in size but tasteless (Orrillo *et al.*, 2019). Spices are often processed into powdered products for convenient usage. For such powdered spices, it is rather difficult to distinguish the genuine products from adulterated ones with naked eyes. Generally, some merchants mixed large proportions of crop plant-based adulterants such as rice, corn or wheat flour into powdered products without notification. Such practices damage the economic interests of consumers and pose risks to their health and safety (A. Osman *et al.*, 2019) including aroma, flavor, and color, to foods and beverages in an effort to enhance their palatability. Herbs and spices exhibit a plethora of medicinal properties including antimicrobial and antioxidant properties. Spices serve as natural preservatives to delay food spoilage and extend shelf life. As spice trading has exponentially increased globally, herbs and spices have become prone to adulteration, which can be deliberate or unintentional. Deliberate adulteration is usually economically motivated, aimed at maximizing profit margins, whereas unintentional adulteration is often attributed to improper harvesting or processing of the plant material or collecting/substituting closely related species. Adulteration of herbs or spices with toxic ingredients such as Sudan dyes or metal salts (e.g., lead chromate. Therefore, the identification and quality control of natural spices is very important for protecting the health, rights, and interests of consumers. The Indian standard approved by Agricultural and Food Products Division Council of India regulates sensory methods for the spice's determination, which rely upon the color, smell, and taste of the products. European Union legislation in EU has its own control measures for preventing contamination in trade which is carried out through RASFF – the Rapid Alert System for Food and Feed. To increase weight and improve look, ground spices may be adulterated with fake colours, flour, chalk powder, and other ingredients. For commercial benefit, high-value ground spices are routinely contaminated.

Adulteration is difficult to identify by visual and sensory inputs alone. The EU has one of the best food safety standards in the world, owing in large part to a robust set of EU regulations in place that assures food safety for consumers. Rapid alert system for food and feed is an essential tool for ensuring the flow of information and enabling quick response when threats to public health are discovered in the food chain (RASSF).

Table 1: Annual reports of rejected spices from India due to contamination

Hazard	No. of Rejections
Mycotoxin	574
Pesticide Residues	202
Pathogenic Micro-organisms	794
Poor or Insufficient Controls	3
Foreign Bodies	108
Adulteration/Fraud	8
Food Additives and Flavourings	20
Metals	13

Source: Annual report 2019 (RASSF)

Indian standard uses morphological, microscopic, and physical and chemical identification methods general identification spices (Dhanya & Sasikumar, 2010). For example in case of mustard seeds, argemone seed is the common adulterant used, which is identified using magnifying glass (Sanyal, 1950). Argemone seeds are grainy rough and black. Although the morphology-based method is effective for the black pepper from its adulterant, it cannot be applied the powdered products. Another powerful method is chemical method using spectrometric instruments which needs a standard laboratory and professional analysts (Vaclavik *et al.*, 2014)2014. With the further development of analytical techniques based on molecular biology, DNA barcoding technology is gradually being applied for the molecular identification of medicinal plant species (Zhang *et al.*, 2019). DNA barcoding is a new biological identification method that uses relatively short genomic DNA fragments as markers for species. There are several nondestructive methods of detection of adulteration which has been recently adapted by several researchers. This chapter is the summary of adulteration in spices and the various techniques for it detection.

Common Adulterants/Contaminants in Spices

Common Intentional Adulterants in Spices

Spice or its powders are adulterated intentionally to due to its high price and demand. Presence of adulterant in spices are there from several centuries. However a proper detection and investigation of them were investigated since 1902 (A. G. Osman *et al*., 2019). Several countries around the world have different associations such as American Spice Trade Association (ASTA) and Indian Standards Institution (ISI), for identification, standardization and quantification of adulterants in food commodities including spices. (Krug, 1902; ASTA, 2004; ISI, 2005; IISR, 2007).

Table 2: Available adulterants of spices as described by ASTA and ISI, Delhi

Product	Adulterants (ASTA)	Adulterants (ISI, DELHI)
Asafoetida (Heeng)		Soap stone, other earthy matter and chalk
Black pepper (ground)	Buckwheat and millet seed	Papaya seeds and light berries
Capsicums	Tomato skin, dextrose, monosaccharides, disaccharides, Sudan red and related dyes	
Chilies		Brick powder grit, sand, dirt and filth
Cinnamon	Coffee husks	low quality cinnamon
Coriander powder		Dung powder and common salt
Cumin seed, poppy seed, black pepper		Artificially colored foreign seeds, Grass seeds colored with charcoal
Ground black and white pepper	Buckwheat and millet seed	Powdered bran and saw dust
Ground spices	Spent spices (defatted), starch, grains, hulls, oleoresins	Powdered bran and saw dust
Mustard seeds		Argenome seeds
Nutmeg	Coffee husks	coffee husks
Oregano leaves	Olive leaves, sumac leaves, thyme	
Saffron	artificial colour and dried waste of flower	beet, pomegranate fibers, red-dyed silk fibers, safflower and marigold to red stigma of saffron
Turmeric		Lead chromate and metanil yellow
Turmeric powder		Starch of maize, wheat, tapioca, rice

Source: (Galvin-King *et al*., 2018; September, 2011; TREMlovÁ, 2001)

Common Incidental Contaminants in Spices

Spice powders can be adulterated two either unintentionally or intentionally. Spices are contaminated unintentionally with toxins by the infection of microbes. Mycotoxins produced by mycotoxigenic fungi are one of the most common contaminants in spices. Mycotoxins such as aflatoxin , ochratoxin , fumonisins etc. are found to be present in spices (Thanushree *et al.*, 2019). AFB1 was detected in chilli, cumin ginger, nutmeg etc. (Llewellyn *et al.*, 1992). Mycotoxins are a group of toxic secondary metabolites produced by specific group of fungus. Under favourable conditions of temperature and humidity, these fungi grow on certain foods and feeds, resulting in the production of mycotoxins. According to the FAO, approximately 25% of the world's agricultural products is contaminated with mycotoxins and these toxins can be formed either before or after harvesting the produce and levels may increase during postharvest handling and storage. The consequences of food-borne mycotoxins are severe, with signs of severe disease developing shortly after consuming contaminated food products. Significant mycotoxins in spices include aflatoxins, ochratoxins, zearalenone, fumonisins, and trichothecenes. Among the mycotoxins, aflatoxins are highly carcinogenic compounds, produced by A. flavus and A. parasiticus, are the leading toxins present in black pepper, nutmeg, ginger and cardamom. They play an important role in the national economy of several of the producing, exporting and importing countries. Presence of ochratoxin was found in cardamom coriander, and black pepper. The common reason behind this is improper storage conditions (Ozbey & Kabak, 2012). High temperatures, heavy rains, and humidity encourage fungal growth, resulting in a rise in the presence of mycotoxins in spices. Furthermore, in India, the lack of good farming techniques, good manufacturing technique, and good hygienic behavior is a major source of concern (Pickova *et al.*, 2020). International standards and codes of practice to limit exposure to mycotoxins from certain foods are established by the Codex Alimentarius Commission. Particularly in America, Europe and Japan, regulatory control of imported spices has become stricter, with extremely low mycotoxins and also for microbiological contamination (Table 3).

Table 3: Limit of Aflatoxin level for export of spices in different countries

Country	**Total Aflatoxin (µg/kg)**	**Aflatoxin B_1(µg/kg)**
European Union	10	5
European country	10	5
South Africa	10	5
Malaysia	10	5
Japan	10	
Australia	15	
Canada	15	
Newzland	15	
USA	20	
Middle East	30	
others	30	

Another form of adulteration comes when synthetic chemicals are added to increase the shelf life. For example chemical preservatives—including benzoate, propionate, sorbate, nitrate, nitrite, and sulfites—are used to avoid fungal spoilage increase the shelf life of several spices such as chilli and cardamom (Anand & Sati, 2013). Use of ethylene oxide is a common form of preservation in spices such as all spice and pepper (Sádecká, 2008). However, the flavor and volatile compounds are deteriorated due to its usage. Other forms of adulteration/contamination in food application of a systemic pesticides. Plants have the capacity to uptake heavy metal, pesticides, herbicides and insecticides through roots and leaves (Hwang *et al.*, 2017). These chemicals may finally accumulate in the edible fruit which are used for human consumption. The insecticide nicotine is found to be accumulated peppermint plants due to its uptake and accumulation after soil application (Selmar *et al.*, 2015). In agriculture, pesticides, insecticides, and herbicides are commonly used to protect plants against insects, pests, and fungal infections well as to increase the quantity and quality of harvest. Spices being high value, export-oriented produce and mostly consumed in its raw form, it is of paramount importance that they are free of pesticide residues. Moreover, stringent standards set by the importing countries often result in rejection of the produce leading to severe losses to producers and traders. Hence, identification of safer alternatives for the management of spice crop pests is of paramount importance. Biorational insecticides have emerged as a viable alternative form of pest control to meet this demand. It is also important that biorational strategies are integrated with other ecofriendly strategies such as cultural methods, use of resistant lines, biocontrol agents and safer insecticides in a holistic manner to improve crop health and productivity (Devasahayam *et al.*, 2020). However, the accumulation of pesticide residues in plants may have harmful and allergic consequences for human health if contaminated spices and herbs are consumed. Their residues are deposited in the soil and transferred throughout the food chain. Pesticides such as carbendazim, diazinon, malathion, metalaxyl, Acetamiprid; cymoxanil, cyproconazole, difenoconazole, dimethoate,; flusilazole, metalaxyl, pirimiphos-methyl, propamocarb, tebuconazole, tetraconazole, thiacloprid thiamethoxam etc are found to be present in black pepper, basil, oregano and thyme (Reinholds *et al.*, 2017). Cadmium chromium lead etc were found in ginger, black pepper, nutmeg , clove etc (Mubeen *et al.*, 2009). Admittedly, the extent of contamination of the spices with heavy metal varies from one plant to another. Several researchers reported that that the tendency to accumulate Cd is species dependent while Pb uptake occurs rather by chance (Chizzola *et al.*, 2003). In recent years, there has been a lot of discussion about the toxicity of trace metals on human health and the environment. Plants are the primary pathway for heavy metals to be transferred from polluted soil to humans. Heavy metals can build up in the food chain. Heavy metals have low excretion rates via the kidney, which means that even at extremely low quantities, they can be harmful to humans. Metals such as zinc, copper, iron, manganese, and chromium are necessary nutrients; they are required for the human body's physiological and biological processes. However, exceeding certain allowed limitations in their consumption might be hazardous (Korfali *et al.*, 2013). A reduction in immunological defences, heart malfunction, foetal deformity, poor psychosocial and neurological behaviour, gastrointestinal cancer, and many other health issues have been related to high dietary heavy metal absorption (Singh *et al.*, 2011). Heavy metal contamination of herbal treatments has already been observed in numerous Asian, South American, and African herbal items from various nations (Rahimi *et al.*, 2012).

Adulteration of Spices: Impact on Health and Wellness

Even though we use spices are used as a whole, most popular form is powder. This make it more susceptible to adulteration. Consumption of contaminated spices can result in a variety of illnesses such as skin allergies, liver disorders, and so on. Due to high probability of adulteration in ground spices, FSSAI has banned the sale of ground spices.

Table 4: Toxic elements in noxious addition of spices with possible outcomes

Contaminants	Spice type	Possible outcome
Mycotoxins	Nutmeg, chilli, ginger, black pepper	Hepatotoxic, carcinogenic, immuno-suppressive
Coloring agents chrome, tartrazine, and erythrosine	Mixed spices	Cancer in kidney, liver, skin, prostate, and lungs
Coal tar and industrial dyes	Powdered spices	Carcinogenic
Brick dust	Chili powder	Respiratory problem
Metanil yellow aniline dyes	Turmeric powder	Carcinogenic
Yellow and Sudan red colors	Chili powder	Tumors in liver and bladder and finally for cancer
Lead (II) chromate (PbCrO4)	Turmeric powder	Lead poisoning may cause neurotoxicity, nephrotoxicity, anemia, reduced male fertility. Chromium (VI) is involved in the process of carcinogenesis.
Sudan red (typically used to color oils and waxes, including shoe polish)	Paprika, chili powders	Carcinogens classified as Class III because of their teratogenicity, genotoxicity, and carcinogenicity to humans.
Cassia	Cinnamon	Coumarin is abundant in Cassia cinnamon, but not in Ceylon.
Argemone seeds	Mustard seeds.	Vomiting, diarrhoea, nausea, limb swelling, erythema, pitting edoema, and shortness of breath were all symptoms of argemone oil poisoning.
Used pods	Green cardamom pods	Newer pods will be plump and full, whereas worn ones will be wrinkled and empty, with no health advantages.
Papaya seeds, Piperine, expired green and red chili	Black pepper	The presence of papain in the body can trigger natural uterine contractions, which can lead to abortions.

(Hore *et al.*, 2019; Lungarini *et al.*, 2008; Rahim *et al.*, 2012; Shah, 2017; Woolf & Woolf, 2005)

Certain consumer products are often implicated. Between 2008 and 2017, the New York City Department of Health and Mental Hygiene tested more than 3000 samples of consumer products during lead poisoning case investigations and surveys of local stores, and of these, spices were the most frequently tested (almost 40% of the samples spices in loose form. Moreover, these adulterated spices can be a cause of diseases such as cancer, high BP, allergy, intestinal problems etc. The addition of brick dust or lead oxide to chili powder to mimic it color is a typical example of intentional adulteration (Hore *et al.*, 2019)respectively, these are not the only possible lead sources. Certain consumer products are often implicated. Between 2008 and 2017, the New York City Department of Health and Mental Hygiene tested more than 3000 samples of consumer products during lead poisoning case investigations and surveys of local stores, and of these, spices were the most frequently tested (almost 40% of the samples. Cinnamon is often adulterated with a look alike bark cassia, which is toxic (He *et al.*, 2005)the dried stem bark of Cinnamomum cassia Presl. (Lauraceae. Cassia cinnamon (C. cassia) is reported to possess a strong and spicy taste whereas true cinnamon is sweeter. Many reports reveal that 85% of the ground cinnamon in our markets is Chinese cassia. Consumers are warned about coumarin content of cassia as it can affect liver and also tend to cause cancer. Cassia is reported to contain up to 1% 'coumarin' (2H-chromen-2-one), while true cinnamon has only trace (0.004%), which is a major factor that differentiates cassia and cinnamon. Although coumarin was suggested to be a genotoxic and carcinogenic, recent study suggest adverse effects in humans after coumarin exposure is rare and if any is reported, is only at high dosage and for long duration during clinical therapies. Due to the potential hepatotoxic effect in humans, Codex Alimentarius commission suggested a Tolerable Daily Intake (TDI) limit for the consumption of coumarin in food. According to this, the amount of coumarin in food and beverages should be limited to 2 mg kg-1 food/day. Jose *et al.* (2019) reported the availability of C. cassia trees with low coumarin and high levels of cinnamaldehyde in India. The tested samples Variability in quality of Indian spices 18 contained low coumarin Mustard seeds are adulterated with argemone seeds , which is also not good for consumption (Sanyal, 1950). Another adulteration which cause serious health hazard is of saw dust with cumin (A. G. Osman *et al.*, 2019). All these adulteration causes serious long-term health hazard and has to be identified and prevented at the earliest. Detailed list of Table 5.

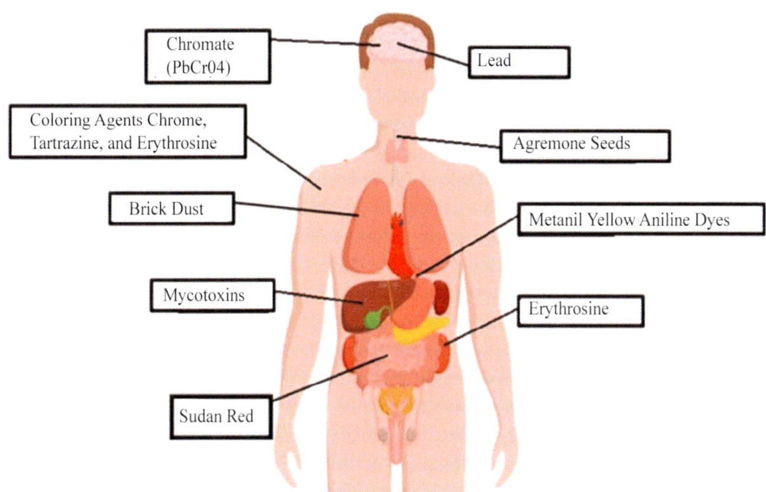

Fig. 1: Effect of various adulterants on different human body parts

Tolerable Limits for Contaminants in Spices

The safe limit of any containment in food and feed is based on Codex Alimentarius Commission (Thompson & Darwish, 2019). It is a 13-volume compilation of Food Standards, Codes of Practice, Guidelines and Recommendations, with World's authoritative reference on Food Standards. It is a joint FAO/WHO organisation that is used by the National Food Inspection System, the World Trade Organization, the food industry, and consumer advocates.

In India to protect public health and safety, the Government of India imposes regulatory restrictions on the manufacturing and sale of products and services, including those imported from other countries (Shukla *et al.*, 2014)stricter sanitary and phyto-sanitary requirements and improved global product norms trigger the necessity to improve legislative network of a country based on codex Hazard Analysis Critical Control Point (HACCP. Regulations are incorporated in several Acts, Rules, and Regulations established by various Ministries of the Government of India. The enactment of Acts, Rules, and Regulations is done in accordance with the relevant WTO Agreements. Quality Regulation Bodies in India and World Wide.

There are several bodies worldwide for the for regulation of quality and maintain standard for various commodities. ISO is the most significant of these. It defines spices and condiments as vegetable products or mixtures of vegetable products free of foreign matter used for flavouring, seasoning, and imparting aroma to foods; the products may be in whole or ground form. The Indian government also has several quality regulations for the safe consumption and trading of food commodities, including spices. The American Spice Trade Association represents the interests of the U.S. spice industry (ASTA). ASTA guidance of "Clean, Safe Spices" was published in March 2011 as guidance for the spice industry on food safety. Companies are encouraged to implement the recommendations contained in the guidance and distribute the guidance to their suppliers and customers. The recommendation was to reduce the danger of adulterant being introduced into the supply chain. During processing and storage, avoid environmental contamination, cross-contamination, and post-processing contamination. Use microbial reduction approaches that have been proven to work. Perform post-treatment testing to ensure that the product is safe. Verify that the manufacturing environment is clean and safe. The Food and Drug Administration is the oldest comprehensive consumer protection agency in the U. S. federal government. FDA is responsible for the monitoring and supervision of food safety in order to safeguard and promote public health.

Food safety, tobacco products, dietary supplements, prescription and over-the-counter pharmaceutical drugs (medications), vaccines, biopharmaceuticals, blood transfusions, and medical deviants are all under the FDA's jurisdiction to protect and promote public health. FSSAI in accordance with Food Safety and Standards Act, 2006 formulated Rules, Regulations and Standards of identity and purity for food products which are science based. Regulation 2.9 of FSSR (FPS and FA) 2011 prescribes standards for 29 spices and spice products

Three types of standards:

1. Cleanliness standards,
2. Quality standards
3. Acceptable contamination standards

The Bureau of Indian Standards (BIS) was established on April 1, 1987, after the Parliament enacted the Bureau of Indian Standards Act in 1986. In India, the ISI mark is used to identify industrial items. The mark verifies that a product complies with the Indian Standard (IS), which is set forth by the Indian Standard Institute. BIS is the new name for ISI (Bureau of Indian Standards). Spices are defined by IS 1877.

Table 5: The list of regulations to control adulteration in trade in India

Regulations to regulate trade in India	**Agencies**
Prevention of Food Adulteration Act 1954 (on the pattern of USFDA)	Directorate General of Health Services
Plant Quarantine (Regulation of Import into India) Order, 2003	Department of Agriculture & Cooperation
Insecticides Act	Dept. Of Agriculture and co-operation
Fruit Products Order, 1955	Ministry of Food Processing Industries
Standards on Weight and Measurement Act 1976	Department of Legal Metrology
Bureau of Indian Standards act, 1986	Department of Consumer Affairs, Govt. of India
AGMARK Act 1937	Directorate of Marketing and Inspection

(Dhanya and Sasikumar, 2010; A. G. Osman *et al.*, 2019; Shukla *et al.*, 2014)

Detection Mechanisms of Adulterated Spices

To understand the various techniques for identification and quantification of contaminants several research has been carried out to on different times (Dhanya & Sasikumar, 2010; Tripathi *et al.*, 2007). In 1980 a detailed study on adulteration in spices of past 20 years was carried out which explains the adulteration of more than 10 spices (Lupien, 1980) The details are presented in Table 2 This survey made evident the requirement for basic, fast, dependable and economical insightful strategies for the recognition of various adulterants in spices. Later in 2010 another review was published which covered the adulteration till that dat.) reviewed spice adulteration studies completed prior to 2009; a selection of which will be briefly discussed (Dhanya & Sasikumar, 2010).

Various techniques are adopted from olden days to detect and identify various adulterants in food including spices. Generally best and simple way to detect the presence of any adulterant is identifying the presence of any foreign material. Those include physical, chemical or biochemical, molecular tools and antibody-based tools. The macroscopic and microscopic detection of adulterant comes under the physical methods. It also include analysis of texture, solubility etc. thin layer chromatography, high performance liquid chromatography, gas chromatography , mass spectroscopy, electronic nose and tongue etc. comes under chemical or biochemical detection methods (Galvin-King *et al.*, 2018). Capillary electrophoresis and polyacrylamide gel electrophoresis (PAGE) also comes under this (Dhanya & Sasikumar, 2010). Antibody methods such as ELISA is also widely used. DNA based molecular tools are suitable for identification of adulterants which are plant derived materials which are similar in morphology and chemically (Zhang *et al.*, 2019).

Physical and Chemical Method of Detection

The utilization of artificial colors in food stuffs has been reviewed in detail (Tripathi *et al.*, 2007). They reported that Sudan I and Sudan III for chilli; Orange II and Metanil for turmeric. This study states that adulteration with chemical colors occurred in several items in both provincial and metropolitan market areas despite the fact that administrative observation was set up (Tripathi *et al.*, 2007).

Several techniques are adopted to identify adulteration using dyes or artificial colours. Simple chemical analysis is available to detect artificial colours in chilli. To detect the presence of red lead salt , diluted nitic acid along with potassium iodide was added to check the formation of yellow coloured precipitate which indicates red lead salts (Sen *et al.*, 2017). Several other techniques adopted are UV visible spectroscopy, thin layer chromatography and HPLC.

Table 6: List of physical and chemical methods used to detect adulteration in spices

Spice	Adulterant	Technique for analysis	Means of confirmation
Vanilla extract	Foreign plant material	Paper chromatography	Foreign spots in chromatogram
Capsicum spices	Dehydrated red beets	Paper chromatography UV-Vis spectrophotometry Microscopic examination	Red to pinkish-red chromatogram Difference in absorbance Red particles (parenchyma)
Poppy seeds (Papaver somniferum)	Rajeera (Amaranthus paniculatas) seeds	Analysis of fat and squalene contents (AOAC method) Evaluation of puffing	Decrease in fat content Increase in squalene content Flavoured product
Chilli and curry powder	Synthetic colours/dyes (metanil yellow, sudan I, sudan III and orange II)	Paper chromatography, UV –light (365 nm)	Fluorescent spots
Turmeric powder	Curcuma zedoariapowder		
Whole black pepper	Light berries	Starch determination	Variation in starch content
Whole black pepper	Papaya seeds	Flotation test Microscopic examination	Floating papaya seeds (density difference) Visual confirmation of floaters
Whole black pepper	Papaya seed	GC	Presence of benzyl glucosinolate
Black pepper powder	Ground papaya seed	TLC	Fluorescent bands observed at 366 nm at R_f 0.172 and R_f 0.943

These methods were mainly used to detect sudan I and sudan III for chilli, orange II and metanil for turmeric and various dyes in saffron etc. The other main method of adulteration is addition of similar low-quality plant-based product which are similar in appearance. Black pepper seeds used to get adulterated with dried papaya (TREMlovÁ, 2001). This can be easily detected by floatation method. In order to detect ground papaya seed in pepper powder thin layer chromatography was used. An ethylene dichloride extract give a on bluish green fluorescence at 366 nm at Rf 0.943. This method claim to identify adulteration using papaya at a level of 20 g/ kg (Shah, 2017). In another work super critical carbon dioxide extracts seems to be more efficient than ethylene dichloride extract when used along with TLC.

Adulterant Detection using DNA Based Methods

DNA fingerprints are bar-code-like patterns created by amplification of an individual's chromosomal DNA that can be used to differentiate one person from another. DNA typing is also known as genetic fingerprinting, DNA profiling, and DNA typing. There are a variety of applications and uses for DNA fingerprinting in plants including detection of adulteration in spices. To assess DNA polymorphism, a variety of DNA-based approaches are available. Hybridization-based methods, polymerase chain reaction (PCR)-based methods, and sequencing-based methods are the three types. Hybridization based techniques are Restriction Fragment Length Polymorphism (RFLP), Variable Number Tandem Repeat (VNTR), Probe hybridization with Micro and Minisatellite, Random Genomic Clone and cDNA Clone. PCR based technique is simple, sensitive and specific detection method. Recently the rapid processing time and low cost of PCR made it more convenient that other techniques. There are several PCR based methods used mainly are RAPD, DAF, RFLP,ISSR, AFLP, DAMD, SCAR, ARMS, SSR etc (Parvathy *et al.*, 2015)turmeric [Curcuma longa L. (Zingiberaceae. The most widely used among them is RAPD with some changes. When RAPD markers are converted to SCAR markers, they become more specific and sensitive. There are several techniques to quantify the amount of adulterants in spices such ad quantitative or real time PCR and quantitative competitive PCR (Zhang *et al.*, 2019). Sequence based techniques are Simple Sequence Repeats (SSR), Sequence Characterized Amplified Region (SCAR), Cleaved Amplified Polymorphic Sequence (CAPS) and Single Nucleotide Polymorphism (SNP). Among DNA based methods there are techniques which are sequencing based. The detect the variation genome of two related species thus identifies adulteration. However prior in species specific the variations in the species-specific region of the genome (amplified rRNA genes, mitochondrial genes or chloroplast genes) due to transversions, transitions, insertions or deletions present are commonly identified. Anyhow information of sequence is required for primer designing and amplifying particular gene (Zhang *et al.*, 2019).

Table 7: List of DNA based methods used to detect adulteration in spices

Adulterant to be tested	**Technique used**
Detection of cashew husk (Anacardium occidentale L.) adulteration in tea [Camellia sinensis (L.) samples	Species-specific PCR
Differentiation of 'Arabica' and 'Robusta' coffeebeans	PCR-RFLP
Detection of allergenic mustard (Sinapis alba, Brassica juncea, Brassica nigra) in food.	Real-Time PCR
Detection of allergenic sesame (Sesamum indicum) in food	Real-Time PCR
Detection of adulterant in traded turmeric powder	RAPD
Detection of adulterant in traded chilli powder	RAPD
Detection of adulterant in traded black pepper powder	SCAR
Detection of adulterant in traded oregano	RAPD
Identification of cinnamon (Cinnamomum cinnamomum) from its adulterants (Cinnamomum cassia, C. zeylanicum, C. burmannii and C. sieboldii).	Sequencing; SSCP
Identification of ginger (Zingiber officinale) from crude drugs and multicomponent formulations.	SCAR

Nondestructive Adulteration Detection Techniques

Spice adulteration is an overall concerned issue with regards to quality and wellbeing, and it is hard to be identified. There are several techniques available in the present scenario to identify adulterants in spices. Techniques such as chromatography (HPTLC, HPLC, GCMS) can not only identify adulterant but also quantify them. However, all these have a complex sample preparation where sample is destroyed completely and cannot be reused. The need for rapid, nondestructive and accurate adulterant detection systems should be available for a better adulteration free future. There are several nondestructive techniques available now which can be used in future for the detection of adulteration in spices.

Computer Vision

In order to accurately estimate the color, shape, texture and size of computer simulation of biological vision is carried out using computers. The pictures obtained will be analyzed to estimate the wanted information. Computer vison was used to identify adulterated saffron based on their color (Kiani *et al*., 2017). This can be used as successful nondestructive method of detection of adulteration.

Infrared Spectroscopy

Infrared spectroscopy analyses the absorption spectra of sample after irradiating with infrared light. They mainly include NIR spectroscopy and MIR spectroscopy. Fourier transform infrared spectroscopy (FTIR) is one of the forms in infrared spectroscopy. FTIR identifies the fundamental vibration It can monitor the vibration and rotational stretching modes of the molecule, resulting in a chemical profile of the sample. FTIR was successful in detecting adulteration using metanil yellow up to 1% in turmeric power (Kaavya *et al*., 2020).

Raman Spectroscopy and Raman Imaging

Raman spectroscopy is an advanced nondestructive technique. It gives the chemical as well as structural information of adulterants both qualitative and quantitative. Raman imaging technique later gives the distribution image of relative substance. There are several reports of using Raman spectroscopy for identification of adulterant in spices. In one study adulteration of saffron was easily detected 5 contaminates in saffron from 5-25% of total concentration (Oliveira *et al.*, 2019).

Terahertz Spectroscopy and Terahertz Imaging

In this technique object is irradiated with terahertz spectral radiation. The refection or transmission of object is collected and it reconstruct eh geometric image of the object. Moreover, it can understand the chemical composition of the object through this. Adulteration of turmeric with yellow chalk powder was easily detected by this technique than conventional methods (Nallappan *et al.*, 2013).

Electronic Nose and Electronic Tongue

These devices mimic the human nose and tongue electronically and detect the compounds. There are several adulterants detected using them which are difficult with traditional methods. Barley is used as an adulterant in coffee which can be easily detected electronic tongue. Adulteration of saffron is also easily detected using this technique (Baldwin *et al.*, 2011).

Commercial Rapid Adulteration Detection Systems for Spices

Currently, it is evident that food suppliers work with food manufacturers for seeking rapid, user-friendly methods to detect food adulteration. In addition to being reproducible and robust, these technologies should be portable (ideally a handheld system) so that they can be taken to or be positioned on/at-line points of vulnerability along food supply networks while requiring a minimum amount of training (Kucharska-Ambrożej & Karpinska, 2020). It is always necessary that we should be able to detect adulteration in spices in a rapid and user-friendly methods. These techniques should be portable, reproducible and also robust, where person with minimum training can use it. There are several rapid systems such as booklets and kits which are commercially available to check the adulteration of spices.

Detect Adulteration with Rapid Test (DART) is a booklet which contains the common simple test for detecting food contamination including spices. The test included in these booklets are rapid, simple and can be identify adulteration without the help of any instrument or addition of any reagent. Another booklet commercially available is Food Safety Magic Box Companion guidebook. This contains tests which can be performed without or without any reagents. More than 100 tests are included here. Another manual which better that DART and Food Safety Magic Box Companion Book is Manual of simple methods for testing common adulterants In food (Food Safety on Wheels). This manual has more than 100 tests for identification of adulterant in food including spices. There are several commercially available adulteration kits available in market which are chemical based, which can be purchased and used by common man. In India we have company named Veripure, Codon biotech which has rapid kits for detecting adulteration in spices powder.

Table 8: List of public funded labs for quality analysis of spices

Name of the lab	Address
Quality Evaluation Lab, Cochin	Quality Evaluation Laboratory Spices Board (Ministry of Commerce, Govt. of India) Sugandha Bhavan, NH Bye Pass, Palarivattom PO, Kochi 688025 sbqel@indianspices.com
Quality Evaluation Lab, Mumbai	Quality Evaluation Laboratory, Spices Board (Ministry of Commerce and Industry,Govt. of India), EL-184, Electronic Zone TTC Industrial Area, M I D C Navi Mumbai-400710 Email: sbqelmumbai @ gmail.com
Quality Evaluation Lab, Guntur	Quality Evaluation Laboratory, Spices Board, Ministry of Commerce Industry, (govt of India), Chuttugunta Centre, GT Road, Guntur-522004. Email: sbzognt@gmail.com
Quality Evaluation Lab, Tuticorin	Quality Evaluation Laboratory, Spices Board, World Trade Avenue, Harbour Estate, New Port, TUTICORIN-628004 Email-sbqeltuticorin@gmail.com
Quality Evaluation Lab, Narela	Quality Evaluation Laboratory Spices Board, Near Satyawadi Raja Harishchandra Hospital, Behind Indian Oil petrol pump, Narela, New Delhi, India sbqelnarela@gmail.com
Quality Evaluation Lab, Chennai	Spices Board, Quality Evaluation Laboratory, Plot No. R-11, SIPCOT Industrial Complex, North End Road, Gummidipoondi-601 201 E-mail: sbqelchennai@gmail.com
Quality Evaluation lab, Kantla	Spices Board, Quality Evaluation Laboratory, Plot No. 22A, Sector 8 (Govt office Area), Tagore Road, Gandhidham, Kutch, Gujarat 370201 sbqelkandla@gmail.com
Quality Evaluation lab, Kolkata	Quality Evaluation Laboratory Kolkata Spices Board of India, PO - Bamangachi, PS-Baruipur, Dist. South 24 PGS, Landmark- DRDO & West Bengal Tele Academy Kolkata – 700145 Email: sbqelkolkata@gmail.com
ICAR IISR, Kozhikode	ICAR-Indian Institute of Spices Research, Kozhikode, Marikkunnu P.O, Vellimadukunnu, Kerala 673012 Email: director.spices@icar.gov.in

Name of the lab	Address
Food Safety and Quality Testing Laboratory, IICPT	Indian Institute of Crop Processing Technology, Food Safety and Quality Testing Laboratory, Pudukkottai Road, Thanjavur – 613005, Tamil Nadu
ICAR-NRC for Grapes	P.B. No. No. 3, P.O. Manjri Farm, Solapur Road, Pune – 412307, Maharashtra, India
Food Safety & Analytical Quality Control Laboratory, CFTRI, Mysore	Food Safety & Analytical Quality Control Laboratory, C/o Central Food Technological Research Institute, Mysore-570013

Conclusions and Future Prospects

The chapter explains various adulteration happening in spices and their detection methods. in recent times, various new techniques have been investigated for the identification of adulteration in spices. Moreover, the studies revealed that there has been a lot advancement in the area of detection of adulteration with regard to efficacy and data processing. However, several challenges need to be solved in order to get the improved assessment to confirm the adulteration level of classification up to 100%. To detect adulterants in spices, strong sensors with high repeatability are required. These sensors must be dependable and mass-produced in order to handle bulk screening of ever-increasing sample volumes.

There is still plenty to be done to increase the accuracy food adulteration detecting system. It is now been revealed deep learning strategies for recognizing food adulteration are in high demand. There is also a need to develop online tools for detecting food adulteration. The future research should be focused on these needs. The research should be done to design a low-cost smartphone-system for identifying adulteration, which can help end-users with their quality satisfaction.

Nondestructive methods are recently gaining attention. However, compare to conventional methods these techniques are expensive. Altogether, the various nondestructive systems combined with proper chemometric tools has proved to be a, reliable, and rapid technique in the detection of adulteration of spices. Continuous research and development in various techniques are essential for numerous spice adulteration issues.

References

Anand, S. P., & Sati, N. (2013). Artificial preservatives and their harmful effects: Looking toward nature for safer alternatives. International Journal of Pharmaceutical Sciences and Research, 4(7), 2496.

Baldwin, E. A., Bai, J., Plotto, A., & Dea, S. (2011). Electronic Noses and Tongues: Applications for the Food and Pharmaceutical Industries. Sensors (Basel, Switzerland), 11(5), 4744–4766. https://doi.org/10.3390/s110504744

Chizzola, R., Michitsch, H., & Franz, C. (2003). Monitoring of metallic micronutrients and heavy metals in herbs, spices and medicinal plants from Austria. European Food Research and Technology, 216(5), 407–411. https://doi.org/10.1007/s00217-003-0675-6

Devasahayam, S., Jacob, T. K., Kumar, C. M. S., & Rajkumar, M. B. (2020). Biorational strategies for the management of insect pests of spice crops. Journal of Spices and Aromatic Crops, 21–37. https://doi.org/10.25081/josac.2020.v29.i1.6330

Dhanya, K., & Sasikumar, B. (2010). Molecular Marker Based Adulteration Detection in Traded Food andAgricultural Commodities of Plant Origin with Special Reference to Spices. Current Trends in Biotechnology and Pharmacy, 4(1), 454–489.

Galvin-King, P., Haughey, S. A., & Elliott, C. T. (2018). Herb and spice fraud; the drivers, challenges and detection. Food Control, 88, 85–97.

Golob, P., & Nations, F. and A. O. of the U. (1999). The Use of Spices and Medicinals as Bioactive Protectants for Grains. Food & Agriculture Org.

Hart, F. L. (1952). A history of the adulteration of food before 1906. Food Drug Cosm. LJ, 7, 5.

He, Z.-D., Qiao, C.-F., Han, Q.-B., Cheng, C.-L., Xu, H.-X., Jiang, R.-W., But, P. P.-H., & Shaw, P.-C. (2005). Authentication and Quantitative Analysis on the Chemical Profile of Cassia Bark (Cortex Cinnamomi) by High-Pressure Liquid Chromatography. Journal of Agricultural and Food Chemistry, 53(7), 2424–2428. https://doi.org/10.1021/jf048116s

Hore, P., Alex-Oni, K., Sedlar, S., & Nagin, D. (2019). A Spoonful of Lead: A 10-Year Look at Spices as a Potential Source of Lead Exposure. Journal of Public Health Management and Practice, 25, S63. https://doi.org/10.1097/PHH.0000000000000876

Hwang, J.-I., Lee, S.-E., & Kim, J.-E. (2017). Comparison of theoretical and experimental values for plant uptake of pesticide from soil. PLOS ONE, 12(2), e0172254. https://doi.org/10.1371/journal.pone.0172254

Kaavya, R., Pandiselvam, R., Mohammed, M., Dakshayani, R., Kothakota, A., Ramesh, S. V., Cozzolino, D., & Ashokkumar, C. (2020). Application of infrared spectroscopy techniques for the assessment of quality and safety in spices: A review. Applied Spectroscopy Reviews, 55(7), 593–611. https://doi.org/10.1080/05704928.2020.1713801

Kiani, S., Minaei, S., & Ghasemi-Varnamkhasti, M. (2017). Integration of computer vision and electronic nose as non-destructive systems for saffron adulteration detection. Computers and Electronics in Agriculture, 141, 46–53. https://doi.org/10.1016/j.compag.2017.06.018

Korfali, S., Mroueh, M., Al-Zein, M., & Salem, R. (2013). Metal Concentration in Commonly Used Medicinal Herbs and Infusion by Lebanese Population: Health Impact. Journal of Food Research, 2(2), p70. https://doi.org/10.5539/jfr.v2n2p70

Kucharska-Ambrożej, K., & Karpinska, J. (2020). The application of spectroscopic techniques in combination with chemometrics for detection adulteration of some herbs and spices. Microchemical Journal, 153, 104278. https://doi.org/10.1016/j.microc.2019.104278

Kumar, C. R., Nageswara Rao, D., & Srinivasa Rao, G. (2014). Consumer protection need of the hour. History, 4(11), 15–19.

Llewellyn, G. C., Mooney, R. L., Cheatle, T. F., & Flannigan, B. (1992). Mycotoxin contamination of spices—An update. International Biodeterioration & Biodegradation, 29(2), 111–121.

Lungarini, S., Aureli, F., & Coni, E. (2008). Coumarin and Cinnamaldehyde in Cinnamon Marketed in Italy: A Natural Chemical Hazard? Food Additives & Contaminants. Part A, Chemistry, Analysis, Control, Exposure & Risk Assessment, 25, 1297–1305. https://doi.org/10.1080/02652030802105274

Lupien, J. R. (1980). Food Problems and Consumer Protection. In W. Santos, N. Lopes, J. J. Barbosa, D. Chaves, & J. C. Valente (Eds.), Nutrition and Food Science: Present Knowledge and Utilization: Volume 1 Food and Nutrition Policies and Programs (pp. 739–744). Springer US. https://doi.org/10.1007/978-1-4684-1030-3_67

Mubeen, H., Naeem, I., Taskeen, A., & Saddiqe, Z. (2009). Investigations Of Heavy Metals In Commercial Spices Brands. New York Science Journal, 2.

Nallappan, K., Dash, J., Ray, S., & Pesala, B. (2013). Identification of adulterants in turmeric powder using terahertz spectroscopy. In International Conference on Infrared, Millimeter, and Terahertz Waves, IRMMW-THz (p. 2). https://doi.org/10.1109/IRMMW-THz.2013.6665688

Oliveira, M. M., Cruz-Tirado, J. P., & Barbin, D. F. (2019). Nontargeted Analytical Methods as a Powerful Tool for the Authentication of Spices and Herbs: A Review. Comprehensive Reviews in Food Science and Food Safety, 18(3), 670–689. https://doi.org/10.1111/1541-4337.12436

Orrillo, I., Cruz-Tirado, J. P., Cardenas, A., Oruna, M., Carnero, A., Barbin, D. F., & Siche, R. (2019). Hyperspectral imaging as a powerful tool for identification of papaya seeds in black pepper. Food Control, 101, 45–52.

Osman, A. G., Raman, V., Haider, S., Ali, Z., Chittiboyina, A. G., & Khan, I. A. (2019). Overview of analytical tools for the identification of adulterants in commonly traded herbs and spices. Journal of AOAC International, 102(2), 376–385.

Osman, A., Raman, V., Ali, Z., Chittiboyina, A., & Khan, I. (2019). Overview of Analytical Tools for the Identification of Adulterants in Commonly Traded Herbs and Spices. Journal of AOAC International. https://doi.org/10.5740/jaoacint.18-0389

Ozbey, F., & Kabak, B. (2012). Natural co-occurrence of aflatoxins and ochratoxin A in spices. Food Control, 28(2), 354–361.

Parvathy, V. A., Swetha, V. P., Sheeja, T. E., & Sasikumar, B. (2015). Detection of plant-based adulterants in turmeric powder using DNA barcoding. Pharmaceutical Biology, 53(12), 1774–1779. https://doi.org/10.3109/13880209.2015.1005756

Pickova, D., Ostry, V., Malir, J., Toman, J., & Malir, F. (2020). A Review on Mycotoxins and Microfungi in Spices in the Light of the Last Five Years. Toxins, 12(12), 789. https://doi.org/10.3390/toxins12120789

Rahim, S. A., Idris, W. M. R., Rahman, Z. A., Lihan, T., Omar, R., & Yan, L. K. (2012). Heavy Metal Content in Selected Flavouring Plants and in Ultra-Basic Soil of Felda Rokan Barat, Negeri Sembilan, Malaysia. Sains Malaysiana, 41(1), 11–21.

Rahimi, M., Farhadi, R., & Balashahri, M. S. (2012). Effects of heavy metals on the medicinal plant. International Journal of Agronomy and Plant Production, 3(4), 154–158.

Reinholds, I., Pugajeva, I., Bavrins, K., Kuckovska, G., & Bartkevics, V. (2017). Mycotoxins, pesticides and toxic metals in commercial spices and herbs. Food Additives & Contaminants: Part B, 10(1), 5–14. https://doi.org/10.1080/19393210.2016.1210244

Sádecká, J. (2008). Irradiation of spices – a review. Czech Journal of Food Sciences, 25(No. 5), 231–242. https://doi.org/10.17221/684-CJFS

Sanyal, P. K. (1950). Argemone and mustard seeds. The Indian Medical Gazette, 85(11), 498.

Selmar, D., Engelhardt, U. H., Hänsel, S., Thräne, C., Nowak, M., & Kleinwächter, M. (2015). Nicotine uptake by peppermint plants as a possible source of nicotine in plant-derived products. Agronomy for Sustainable Development, 35(3), 1185–1190. https://doi.org/10.1007/s13593-015-0298-x

Sen, S., Mohanty, P., & Vuppu, S. (2017). Detection of Food Adulterants in Chilli, Turmeric and Coriander Powders by Physical and Chemical Methods. Research Journal of Pharmacy and Technology, 10, 3057–3060. https://doi.org/10.5958/0974-360X.2017.00542.X

September, D. J. F. (2011). Detection and quantification of spice adulteration by near infrared hyperspectral imaging [PhD Thesis]. Stellenbosch: University of Stellenbosch.

Shah, R. (2017). Identification and estimation of non-permitted food colours (metanil yellow and aniline dyes) in turmeric powder by rapid color test and thin layer chromatography. https://doi.org/10.20959/WJPPS20178-9867

Shukla, S., Shankar, R., & Singh, S. P. (2014). Food safety regulatory model in India. Food Control, 37, 401–413. https://doi.org/10.1016/j.foodcont.2013.08.015

Singh, R., Gautam, N., Mishra, A., & Gupta, R. (2011). Heavy metals and living systems: An overview. Indian Journal of Pharmacology, 43(3), 246–253. https://doi.org/10.4103/0253-7613.81505

Thanushree, M. P., Sailendri, D., Yoha, K. S., Moses, J. A., & Anandharamakrishnan, C. (2019). Mycotoxin contamination in food: An exposition on spices. Trends in Food Science & Technology, 93, 69–80.

Thompson, L. A., & Darwish, W. S. (2019). Environmental Chemical Contaminants in Food: Review of a Global Problem. Journal of Toxicology, 2019, e2345283. https://doi.org/10.1155/2019/2345283

TREMlovÁ, B. (2001). Evidence of spice black pepper adulteration. Czech Journal of Food Sciences, 19(6), 235–238.

Tripathi, M., Khanna, S. K., & Das, M. (2007). Surveillance on use of synthetic colours in eatables vis a vis Prevention of Food Adulteration Act of India. Food Control, 18(3), 211–219. https://doi.org/10.1016/j.foodcont.2005.09.016

Vaclavik, L., Krynitsky, A. J., & Rader, J. I. (2014). Mass spectrometric analysis of pharmaceutical adulterants in products labeled as botanical dietary supplements or herbal remedies: A review. Analytical and Bioanalytical Chemistry, 406(27), 6767–6790. https://doi.org/10.1007/s00216-014-8159-z

Woolf, A. D., & Woolf, N. T. (2005). Childhood lead poisoning in 2 families associated with spices used in food preparation. Pediatrics, 116(2), e314-318. https://doi.org/10.1542/peds.2004-2884

Zhang, M., Shi, Y., Sun, W., Wu, L., Xiong, C., Zhu, Z., Zhao, H., Zhang, B., Wang, C., & Liu, X. (2019). An efficient DNA barcoding based method for the authentication and adulteration detection of the powdered natural spices. Food Control, 106, 106745.

15

An Overview of Adulteration and Detection of Milk and Milk Products

Elizabeth Thomas and Narender Raju Panjagari

Dairy Technology Division, ICAR-National Dairy Research Institute Karnal, Haryana, India

Introduction

Milk is regarded as the "ideal food" as it contains nutrients necessary for both newborns and adults (Azad & Ahmed, 2016). It is high in protein, fat, carbohydrates, vitamins, and minerals. Unfortunately, milk is easily adulterated all over the world for several reasons, including a gap between supply and demand, the inherent nature of milk which can imbibe many adulterants that are difficult to identify, the low purchasing power of consumers, an insufficient share of the organized sector in the dairy business, and, most importantly, a lack of efficient and rapid methods for detecting adulterants (Ellis *et al.*, 2012; Moore *et al.*, 2012). The incentive for food fraud is purely economic, but the consequences include serious public health issues, technological difficulties, and economic loss to processing industries as well as to consumers (Ellis *et al.*, 2012; Singh & Gandhi, 2015). Adulteration of milk is the purposeful reduction of its quality when offered for sale, either by mixing, the substitution of cheaper components, or the removal of a valuable constituent. It can happen accidentally through natural processes, such as the naturally occurring entry of antibiotics into milk from cattle treated for mastitis, or dust particles or other extraneous objects that may have entered the milk during processing. It can also happen intentionally by adding extraneous water, nondairy proteins, melamine, urea, animal fat, reconstituted milk, or synthetic milk in order to increase profits (Poonia *et al.*, 2017).

Food adulteration is the process of lowering food quality by adding poorer-grade material or extracting valuable ingredients. This method completely or partially substitutes the original product, and the added extra compounds may be harmful to one's health (Spink & Moyer, 2011). Milk adulteration is described as the removal or substitution of milk components with inferior ones without the knowledge of the customer (Santos *et al.*, 2013a). In the Indian scenario, adulteration of milk is mainly done for two different purposes: (1) To avoid loss from curdling of milk due to higher ambient temperature, inferior microbiological quality of the initial milk, improper cleanliness of vessels used for collection, storage, and transport and absence of cold chain facility. The use of neutralizers like sodium carbonate, sodium bicarbonate, or sodium hydroxide, preservatives like formalin and hydrogen peroxide, and antibiotics usually prevents milk from curdling (Karima *et al.*, 2015), (2) To make a profit through unethical means from the dairy business which includes replacement of fat in milk with vegetable oil in which detergent is commonly added as an emulsifier (Azad & Ahmed, 2016), and the inclusion of water which reduces solids-not-fat (SNF) content thereby necessitating the incorporation of other materials to maintain SNF. Many chemicals are used for maintaining SNF such as maltodextrin, starch, glucose, sucrose, urea, and sodium chloride (Francis *et al.*, 2020; Singh & Gandhi, 2015). According to several review studies conducted by Moore *et al.* (2012), after olive oil, milk is the food item most likely to be contaminated.

Food fraud, including the better-defined subcategory of economically motivated adulteration (EMA), is a food danger that is getting attention and concern since it poses a hazard to public health on a global scale (Spink & Moyer, 2011). The Food and Drug Administration has adopted a working definition of EMA, defining it as the "fraudulent, intentional substitution or addition of a substance in a product for the purpose of increasing the apparent value of the product or lowering the cost of its production, that is for economic gain" (Poonia *et al.*, 2017). FSSAI (2011) performed a survey intending to identify prevalent milk adulterants in rural and urban regions and common adulterants in loose and packaged milk. According to the findings, about 32% of the samples examined complied with Indian food rules while the remaining were unsuccessful and the most prevalent adulterant in milk is water followed by skim milk powder (SMP). Water depletes milk's nutritional content and may spread water-borne diseases (Poonia *et al.*, 2017). Kumar *et al.* (2015) did another investigation in Varanasi, India, to analyze the amounts of several adulterants in milk. The examinations revealed that a high number of samples did not meet the regulatory requirements established by FSSAI. The findings indicated that the majority of the milk samples were adulterated. The extent of adulteration varied significantly, with starch being the least adulterant (20%) and sodium chloride being the highest adulterant (80%), while neutralizers, urea, formalin, hydrogen peroxide, and detergents were found to be 28%, 60%, 30%, 36%, and 44%, respectively. Similarly, Maurya *et al.* (2017) looked into frequent milk, paneer, and khoa adulterants in Lucknow. They reported formalin as the most common milk adulterant (60%), while paneer (50%) and khoa (80%) are mostly adulterated with starch. Adulteration with urea, melamine, and other compounds in dairy products has recently increased, posing serious health hazards. Furthermore, processing might change the composition of minor constituents, complicating detection. This book chapter explains in-depth several methods used to identify different adulterants in milk and dairy products in this respect.

Adulterants in Milk and Related Health Complications

Carbohydrates

Different carbohydrates are commonly added to falsely elevate the total solids and to mask the addition of water.

Glucose and Sucrose

To disguise the addition of additional water in milk, glucose and/or sucrose are added. Furthermore, sugar enhances the specific gravity and therefore lactometer reading of the milk that has been diluted with water, so that the detection of added water by the lactometer is concealed (Singh & Gandhi, 2015).

Starch and Cereal Flours

To boost the viscosity and SNF content of milk, starches, dextrins, wheat flour, rice flour, arrowroot, and other ingredients can be added. The most frequent of these adulterants is starch. These components improve density and prevent the identification of additional water. Consumption of starch-adulterated milk, on the other hand, may cause diarrhoea in humans and its accumulation in the body may be lethal for diabetic people (Singuluri & Sukumaran, 2015). To detect starch, an iodine solution is added, which leads to the development of bluish-black due to starch–iodo complex. Iodine tincture or solution is used to detect starch adulteration in milk. The appearance of blue implies the presence of starch (Chauhan *et al.*, 2019).

Maltodextrin

Maltodextrin is a polysaccharide that is devoid of gluten and can be produced by partially hydrolyzing a number of starches, such as potato, wheat, maize or rice starch. It is a common ingredient in many processed foods. To increase the total solids content of raw milk, milk suppliers may add maltodextrin. Also, it works effectively as a fat substitute in milk due to its bulking properties (Singh & Gandhi, 2015). When maltodextrin is present, adding 2 mL of a dilute (0.05 N) iodine solution to 5 mL of milk results in a chocolate red-brown color. Flatulence and stomach cramps might result from an overabundance of fermentable carbohydrates in the large intestine (Marchal *et al.*, 1999).

Saccharin

Saccharin [1,2-benzisothiazol-3(2H)-on-1,1-dioxide] is an artificial sweetener derived from the term "saccharine," meaning "similar to sugar." According to the Codex Alimentarius Commission, the maximum acceptable quantity of saccharin in dairy-based beverages, cheese substitutes, and dairy-based desserts is 80–100 mg/kg, although it is not permitted in milk(Singh & Gandhi, 2015). However, as an adulterant, it is used to disguise the poor quality of dairy products. Long-term use of saccharin has been found in studies to increase the risk of obesity, diabetes, hepatic dysfunction, and renal impairment (Azeez *et al.*, 2019).

Salts and Fertilizers

Milk and its products are usually adulterated with urea, ammonium sulphate, sodium chloride, nitrates, and other additives to conceal the impact of additional water and enhance total solids.

Urea

It is a natural component of milk and the main component of milk's nonprotein nitrogen. The urea concentration in natural milk ranges from 200 to 700 ppm. Urea level beyond 700 ppm, on the other hand, implies "added urea." Urea is incorporated in milk to enhance whiteness, boost consistency, increase nonprotein nitrogen content, and level the quantities of solids-not-fat (Gandhi *et al.*, 2020). It is also utilized in synthetic milk production. Acidity, indigestion, ulcers, and cancer are all potential side effects. Urea is damaging to the heart, liver, and kidneys, as the kidneys have to work harder to eliminate urea from the body. Milk containing ammonia causes regression, loss of acquired speech, and sensory problems (Maheshwara Reddy *et al.*, 2017).

Ammonium Sulphate

It is a food additive that is GRAS (generally recognized as safe) by the USFDA. It is a water-soluble nitrogenous chemical that is used to disguise the effects of water dilution and to raise the apparent protein content of milk (Goswami & Gupta, 2008). It is an irritant, and its consumption can cause gastrointestinal discomfort, including nausea, vomiting, and diarrhoea. Intoxication from ammonium sulphate may have negative consequences on the respiratory, gastrointestinal, and cutaneous systems (Singh & Gandhi, 2015).

Sodium Chloride and Chlorine

From several years, sodium chloride has been used as a culinary ingredient. It is used to improve the specific gravity of milk and as a thickening agent in the production of synthetic milk. The addition of salt increases the freezing point depression value greatly, disguising the identification of added water. A qualitative test with silver nitrate and potassium chromate can identify sodium chloride in milk (Singh & Gandhi, 2015). Chlorine is added to compensate for the density of the diluted milk after it has been watered down. Chlorinated milk can induce artery blockage and cardiac problems. Chloride in milk disrupts the body's acid-base equilibrium as well as blood pH (Maheshwara Reddy *et al.*, 2017).

Neutralizers

The intention of adding neutralizers to milk, such as sodium hydroxide, is to disguise the acidity of damaged milk, which is detectable by titratable acidity. Milk neutralization inevitably raises the ash content and overall alkalinity of a certain quantity of ash (Singh & Gandhi, 2015).

Sodium Hydroxide

To counteract the acidic impact, sodium hydroxide/caustic soda is frequently added to synthetic milk. Synthetic milk, made by combining urea, caustic soda, refined oil, and ordinary detergents, is a frequent problem in India. The Codex Alimentarius Commission states that under good manufacturing practices, sodium hydroxide may be used in fermented milk and other dairy products but not in raw milk (Singh & Gandhi, 2015). Caustic soda includes sodium, which works as a gradual poison for those with hypertension and cardiac problems. Caustic soda prevents the body from using lysine, an important amino acid found in milk and needed by newborns (Maheshwara Reddy *et al.*, 2017). Additionally, it is an extremely corrosive substance that is also a strong irritant. If consumed orally, it can cause severe burns to the tongue, lips, oral mucosa, and oesophagus as well as vomiting, collapse, and esophageal cancer (Singh & Gandhi, 2015).

Sodium Carbonate and Bicarbonate

Foods may include up to 2% of the GRAS food additive sodium bicarbonate. All dairy products may contain sodium carbonate in compliance with Good Manufacturing Practices, according to the Codex Alimentarius Commission. Ash alkalinity testing or the rosalic acid test can be used to check milk for sodium carbonates and bicarbonates. On the other hand, ingesting milk that has carbonates or bicarbonates can cause electrolyte imbalance, stomach ulcers, diarrhea, and colon ulcers (Barham *et al.*, 2014).

Preservatives

The growth of microorganisms ruins the milk, and spoilt milk is harmful to one's health. Boric acid, salicylic acid, formalin, sodium carbonate, sodium bicarbonate, benzoic acid, and sodium azides may preserve milk but are deadly and can cause death. It develops stomach discomfort, diarrhoea, vomiting, and other poison-related symptoms (Maheshwara Reddy *et al.*, 2017).

Hydrogen Peroxide

Milk is treated briefly with hydrogen peroxide to lessen the bacterial load and maintain the quality over the long term. Additionally, it is included to render the aflatoxin in contaminated food inactive. The concentration and quantity of hydrogen peroxide consumed determine its adverse effects on people. In addition to symptoms including sinus tachycardia, lethargy, disorientation, convulsions, subepiglottic constriction, and cardiorespiratory arrest, the gastrointestinal system is also inflamed (Watt *et al.*, 2012).

Formaldehyde/Formalin

Due to antibacterial properties, a 40% solution of formaldehyde is purposely and illegally added to raw milk to serve as a preservative and extend shelf life (Moosavy *et al.*, 2019). The International Agency for Research on Cancer has classified formaldehyde as a carcinogen since it is extremely dangerous to humans even in low quantities. When consumed, it can cause vomiting, nausea, stomach pain, low body temperature, shallow breathing, a weak irregular pulse, memory loss, insomnia, mood swings, and even blindness due to optic nerve damage (Singh & Gandhi, 2015).

Potassium Dichromate

Despite being a widely used milk preservative, potassium dichromate is also known to irritate the skin, induce rhinitis, and trigger allergic contact dermatitis. It is also a potent biocide that may cause sewage problems and elevated levels of chromium in drinking water (Kroger, 1985). Because it is also a corrosive toxin and a carcinogen in humans, it is rapidly being replaced by alternative preservatives (Singh & Gandhi, 2015).

Antibiotics

Antibiotics are primarily used to treat several ailments, and antibiotics are utilised widely to treat mastitis disease. Milk contains antibiotics in the form of antimicrobial residues. Tetracycline, aromatic amines, gentamicin residue after mastitis therapy, neomycin residues, sulfamethazine residues, chloramphenicol residues, and other milk adulterants are also of great concern (Chauhan *et al.*, 2019). A common cause of milk contamination is intramammary antibiotic infusion for mastitis treatment. The presence of these medications' residues in milk causes major health risks, including allergic responses, an increase in the number of antibiotic-resistant bacteria, interference with gut flora, and some of them may have carcinogenic potential. It may also result in tissue damage. It interferes with the bacterial fermentation process, resulting in significant losses in fermented products. Urticaria may be caused by a very small quantity of penicillin residual (Maheshwara Reddy *et al.*, 2017).

Detergents

Because detergents are very inexpensive and widely available household goods, reports of adulteration of milk and infant formula by detergent powders are prevalent (Chauhan *et al.*, 2019). Milk adulteration through the use of soaps and detergents lowers the nutritional value of the milk and poses a serious risk to human health in the form of hypotension, respiratory irritation, seizures in children, nausea, and digestive problems like stomach pain and vomiting (Singh & Gandhi, 2015).

Water

Water is the most common adulterant used to increase the amount of milk, which reduces its nutritional content. However, if polluted water is mixed into milk, it poses a major health risk to the milk-consuming population (Maheshwara Reddy *et al.*, 2017). Because pond water has greater specific gravity than tap water and is a strong source of nitrates, some unscrupulous individuals may use it to adulterate milk. In humans, drinking milk tainted with polluted water can induce stomach problems in the elderly and major health problems in newborns and children whose primary diet is milk.

Vegetable Fats or Oil

Milk fat has been a target of fraudulent methods due to greater prices and increased demand, such as substitution with cheaper or more easily accessible nonmilk fats of plant and animal origin. Vegetable oils are frequently employed for this purpose. Vegetable oil is also a major source of fat in synthetic milk. When polyunsaturated fatty acids in vegetable oil are exposed to oxygen and/or UV radiations during processing, they auto-oxidize, leading to the formation of inflammatory peroxides and hydroperoxides that induce cell inflammation, digestive disorders, and, finally, cancer (Singh & Gandhi, 2015).

Extraction Methods of Dairy-based Foods for Adulterant Detection

The biggest challenges in detecting adulterants in food, such as antibiotics and others, are due to the strong non-covalent connections created with matrix proteins and the amount of fatty material present (Gentili *et al.*, 2012). Therefore, the success of the adulteration detection methods depends greatly on the efficiency of the sample extraction methods. Some of the sample extraction protocols are discussed below.

Non-steroidal anti-inflammatory drugs (NSAIDs) are used to treat lactating cow mastitis, for long-term analgesia, and to decrease inflammation. To counteract the strong interaction between NSAIDs and dietary proteins during their detection using the LC-MS technique, Gentili *et al.* (2012) developed a method for extracting NSAID from bovine milk samples using a deproteination/extraction process. In this method, an aliquot (5 mL) of milk was transferred into a centrifuge tube, and 10 mL of acetonitrile was added. The mixture was vortexed for 2 minutes before being treated in the ultrasonic bath for 10 minutes and then centrifuged at 7,000 rpm for 10 minutes at 0 °C. The supernatant fraction was concentrated in a water bath at 30 °C under a moderate nitrogen stream to a final volume of 5 mL to eliminate the organic solvent. The extract was centrifuged at 7,000 rpm for 10 minutes at 0 °C to obtain the final extracted sample.

Similarly, to detect melamine in milk and milk products, 1% Trichloro acetic acid solution, 2.2% lead acetate solution, and the sample were mixed in a ratio of 10:1:2 to eliminate protein and to extract the sample. 30 mL of the mixture was thoroughly mixed and then centrifuged at 10,000 rpm for 10 minutes. A supernatant volume of 10 mL was put into a preconditioned PCX-SPE cartridge. Melamine was eluted using 6 mL of a 25% ammonia solution–methanol (1:20, v/v) solution. The eluate was evaporated to dryness under a nitrogen stream at 50 °C, and the residue was redissolved in 0.5 mL of methanol-water (1:4, v/v). The solution was then filtered over a 0.45 μm filter for HPLC analysis (Sun *et al.*, 2010). These sample preparation processes are time-consuming, labor-intensive, and have low selectivity.

Molecular imprinting is a popular process for creating selective recognition sites in stable polymers, and it's also a good way to make selective sorbents. The extraction of melamine from milk using molecularly imprinted polymers (MIPs) that employed melamine as a template molecule has been described and they were all more selective in their recognition (He *et al.*, 2014). Details on this extraction methods are available in He *et al.* (2014). Rapid extraction method for the melamine from powdered milk sample for direct electrospray ionization tandem mass spectrometry analysis is available (Domingo *et al.*, 2015).

Nascimento *et al.* (2015) developed a dispersive liquid-liquid microextraction (DLLME) process in a flow-batch system for determining formaldehyde in milk rapidly. The Hantzsch process was utilised to derivatize formaldehyde, and the product was extracted using the ionic liquid (IL) trihexyltetradecylphosphonium chloride as a dispersant. The flow-batch chamber was built of stainless steel and had resistive heating to accelerate the derivatization reaction. Spectrophotometric measurements in the organic phase were performed directly with an optical fibre spectrophotometer. The detection limit and coefficient of variation were reported to be 100 g/L and 3.1%, respectively.

Real-time PCR is now widely utilized as an effective quantitative method for detecting adulterants in milk and milk products. However, the effectiveness of the PCR experiment is heavily dependent on the quality of the isolated DNA. Liao *et al.* (2017) created a DNA extraction technique for milk powder that can identify cow milk in goat milk powder. The milk powder was rinsed with purified water and PBS first, followed by centrifugation. The sediment was then combined with extraction buffer (pH 8, comprising Tris-chloride, NaCl, and EDTA), SDS, and proteinase K before being incubated at 56 °C for 4 hours. Following incubation, an equivalent amount of Tris-phenol was added and centrifuged. To get the supernatant, an equal volume of phenol: chloroform: isoamyl alcohol (25:24:1) and chloroform: isoamyl alcohol (24:1) was added and centrifuged. Finally, DNA was precipitated with ice-cold 100% ethanol and washed with ethanol: water solution (7:3). To dissolve the DNA, 25 microliters of Tris-EDTA was added. Likewise, the selectivity of tetraborate-EDTA buffer for extracting plant proteins from milk was exploited to construct a quick turbidimetric detection technique to identify insoluble plant protein in skimmed milk powder adulterated with soy, pea, brown rice, and hydrolyzed wheat protein (Scholl *et al.*, 2014).

DART ionisation in combination with high-resolution mass spectrometry (DART-HRMS) enables quick discrimination of milk samples from various species (cow, goat, sheep), classification of cow's milk produced in organic and conventional farming, and detection of foreign fat in dairy products (soft cheese) (Hrbek *et al.*, 2014). For DART-HRMS technique the milk sample preparation comprises dilution of milk with an equivalent amount of methanol, hand-shaking (2 minutes), and centrifugation (5 min, 11,000 rpm). In another process for isolating non-polar fraction milk, milk samples were hand-shaken for 2 minutes with an equivalent quantity of toluene before being centrifuged for 5 minutes (11,000 rpm). The soft cheese samples (1 g) were extracted with 6 mL of toluene using an Ultraturrax laboratory homogenizer for 1 minute and then centrifuged (5 min, 11,000 rpm).

The use of DNA analysis to identify animal species has been more efficient in detecting adulterations in commercial dairy products. DNA from somatic milk cells, mostly leucocytes, survives in cheese and can be used to differentiate between species (Feligini *et al.*, 2005). Adulteration detection in Italian Mozzarella cheese DNA extraction was carried out by combining the material with lysis buffer and proteinase K solution and then incubating at 42 °C overnight. Phenol was added to the digested suspensions and centrifuged for 30 minutes at 14000 rpm and 4 °C. The supernatant was then extracted in a 1:1 volume ratio using phenol and chloroform. Sodium acetate and ethanol were used to pellet the DNA, which was then centrifuged for 30 minutes at 14000 rpm at 4 °C. Finally, the pellets were suspended in 200 µL of double-distilled water.

Matrix-assisted Laser Desorption/Ionization Time-Of-Flight Mass Spectrometry (MALDI-TOF-MS) allows for the rapid detection of cow milk adulteration in feta cheese and sheep milk yoghurt (Kritikou *et al.*, 2022). The samples were diluted with 0.1 % trifluoroacetic acid extraction solvent (TFA). Following a 10-minute homogenization with an overhead shaker, the samples were treated to a 15-minute ultrasonic-assisted extraction at room temperature (Cozzolino *et al.*, 2001; Kritikou *et al.*, 2022).

Methods for Detection of Adulteration

Milk quality control tests are critical for ensuring adulterant-free milk for consumption. Milk adulteration lowers the quality of milk and can possibly make it harmful. These are often categorized as qualitative or quantitative detection techniques. Color-based chemical reactions (biochemical) are used in qualitative detection procedures, but quantitative detection methods are sophisticated and diverse. Biotechnological and electrical approaches are examples of this (Azad & Ahmed, 2016).

Detection of Adulteration in Liquid Milk & Its Chemistry/Principle

Milk is a fluid that can have any additive added to it without changing its physical appearance. These compounds are introduced in such a way that they defy the lactometer test, which is used to detect any differences in the specific gravity or solids-not-fat content of milk. Milk is often adulterated with additives such as common salt, starch, sugar, wheat flour, baking soda, urea, and washing soda. However, these are not permitted by law, and their discovery is critical. Many adulterants are added to milk to increase its volume, while others are added to maintain and improve its shelf life.

Rapid Color-based Methods

Sucrose or Cane Sugar

Principle: Seliwanoff's reagent detects added sucrose in milk (0.5 % resorcinol solution. 0.5 g resorcinol in 40 ml of distilled water, add 35 ml of concentrated HCl and make the volume up to 100 ml using distilled water). Diluted hydrochloric acid hydrolyzes sucrose into glucose and fructose upon reagent addition. Boiling causes the interaction between fructose with resorcinol, resulting in a red tint. If no sucrose is present, the sample will remain white (Gandhi *et al.*, 2020).

Procedure: Add 1 mL of 0.5 % resorcinol solution to equal quantity of milk in a test tube and stir. Place the test tube in a boiling water bath for 5 minutes. The red colour confirms the presence of cane sugar (Kamthania *et al.*, 2014).

Starch or Other Cereal Flours

Principle: Blue colour developement in the presence of iodine is used to identify starch in milk. Iodine solution in potassium iodide generates triiodide, which reacts with starch to give a violet-blue colour complex (Gandhi *et al.*, 2020).

Procedure: In a test tube, 3 ml of a well-mixed milk sample is taken. It is brought to a boil over a flame and then cooled to room temperature. A drop of 1% iodine solution is added and mixed. The presence of blue colour shows the presence of starch, which disappears when boiled and reappears when cooled (Kamthania *et al.*, 2014).

Urea

A. Detection of Urea Using dimethyl Aminobenzaldehyde (DMAB) Reagent

Principle: In low acidic solution at room temperature, urea forms a yellow complex with DMAB (Gandhi *et al.*, 2020). The intensity of the yellow colour may be measured at 425 nm, allowing for a quantitative procedure (Sharma *et al.*, 2012).

Procedure: In a test tube, place 5 mL of milk and 5 mL of DMAB reagent (1.6 % in ethyl alcohol with 10% HCI). The presence of additional urea is indicated by the development of a characteristic yellow colour. Because of the presence of natural urea in milk, the pure milk sample develops a faint yellow colour. When performed on protein-free filtrate produced with 24% trichloroacetic acid (TCA), the test is more sensitive (Kamthania *et al.*, 2014).

B. Detection of Urea Using Urease Enzyme

Principle: When the urease enzyme works on urea, it releases ammonia, which causes the solution to become blue in the presence of a bromothymol blue indicator (Gandhi *et al.*, 2020).

Procedure: In a test tube, place 5 mL of milk and 0.2 mL of urease (20 mg/ml). Shake vigorously at room temperature before adding 0.1 mL Bromothymol Blue (BTB) solution (0.5 %). The presence of urea in milk is indicated by the appearance of a blue colour after 10-15 minutes (Azad & Ahmed, 2016).

4.1.1.4 Glucose

Principle: To identify glucose adulteration in milk, Barfoed's test is used. Glucose transforms cupric to cuprous in acidic conditions. Phosphomolybdic acid (colourless) is transformed into phosphomolybdous acid (blue color) in the presence of these cuprous ions. Heating time should be shorter than 3 minutes following the addition of cupric acetate, since extended heating time may convert disaccharides to monosaccharides (Gandhi *et al.*, 2020).

Procedure: Fill a test tube with 1 mL of milk and add 1 mL of Barfoed's reagent. In a boiling water bath, heat the mixture for exactly 3 minutes. Cool quickly under running water. To the turbid solution, add 1 mL of phosphomolybdic acid reagent. The presence of glucose is indicated by the immediate emergence of a deep blue tint. In the case of pure milk, just a faint bluish colour appears as a result of diluted Barford's reagent (Azad & Ahmed, 2016).

Maltodextrin

Principle: Maltodextrin is a polysaccharide composed of 3–20 glucose units connected by α(1-4) glycosidic bonds. It is typically used as a food additive and is produced through partial starch hydrolysis. Iodine when reacts with no or little amylase containing waxy starch gives red-brown color (Gandhi *et al.*, 2020).

Procedure: Fill a test tube with 5 mL of milk sample. Add 2 ml of iodine solution (0.05 N) and mix well. The presence of dextrin/maltodextrin is indicated by the development of a chocolate red-brown colour (Gandhi *et al.*, 2020).

Pond Water – Diphenyl Amine Test

Principle: This technique detects the presence of nitrates/nitrites in pond water. Nitrates enter the pond water from fertilisers applied to the crops. Nitrates react with diphenylamine to produce a blue colour in an acidic environment (Azad & Ahmed, 2016).

Procedure: Fill a test tube with 2 mL of milk. Drain the milk from the test tube after rinsing it with it. Two to three drops of the reagent should be placed down the edge of the test tube. Take note of the evolved colour. In the presence of nitrate in the milk sample, a deep blue colour will appear. There will be no colour development in a pure milk sample (Kamthania *et al.*, 2014).

Vegetable Fat

Procedure: Dissolve 1–2 g of molten heat clarified ghee in 2–3 ml of hexane. Add 1.5–2.0 mL colour development reagent (water, nitric acid, and sulfuric acid in the ratio of 20:14:6). Shake vigorously and leave undisturbed until two layers separate. The presence of vegetable oils is indicated by the presence of characteristic orange colour in the top layer (Gandhi *et al.*, 2020).

Mastitic/ Abnormal Milk – Bromothymol Blue Test

Procedure: In a test tube, place 5 mL of fresh milk. 1 mL of bromothymol blue (0.04% aqueous solution) Mix thoroughly and look at the colour. The development of a greenish-blue or blue tint indicates the presence of mastitic milk (Gandhi *et al.*, 2020).

Skim Milk Powder in Natural Milk

Principle: The use of skimmed milk powder (SMP) to adjust the solid-to-fat ratio in the sale of cow/buffalo or mixed milk is not permitted by law. The reaction of phosphomolybdic acid with milk protein is used to detect SMP addition in whole milk. Because of the presence of reducing groups in the proteins of skim milk powder, coagulation of reconstituted skim milk powder with acetic acid, followed by heating and addition of phosphomolybdic acid, produces a strong blue color (Gandhi *et al.*, 2020).

Procedure: In a centrifuge tube, place 50 mL of milk. For 15 minutes, centrifuge at 5000 rpm. Remove the cream layer and gently pour 5 mL of skim milk into a test tube. For coagulation, add 2 mL of 4% acetic acid and wash the curd with purified water. Add 2 mL of 1 % phosphomolybdic acid to the curd. Thoroughly combine the ingredients and heat in a water bath at boiling temperature for 5 minutes before cooling. Take note of the colour of the fluid in the tube. The curd made from pure milk should be greenish in colour, but the curd derived from skimmed milk powder should be bluish in colour. The amount of skim milk powder in the sample determines the intensity of the bluish colour (Sharma *et al.*, 2012).

Detergents in Milk

Principle: Detergents are not present in milk, but they enter when synthetic milk is added to it or when milk contact surfaces are not adequately rinsed or cleansed to remove detergent. Methylene blue indicator is used to identify detergents in milk. The detecting mechanism is that methylene blue is soluble in water but forms a blue-coloured complex with anionic detergents. The complex is blue in colour and soluble in chloroform (Gandhi *et al.*, 2020).

Procedure: Take 1 mL of milk, add 1 mL of methylene blue solution (In 100 ml of pure water, add 12.5 mg of methylene blue dye), and then 2 mL of chloroform. After a 15-second vortex, the sample was centrifuged at 1100 rpm for 3 minutes. The presence of detergent is indicated by the development of a more intense blue colour in the lower layer, whereas the lack of detergent is shown by the development of more intense blue colour in the upper layer (Sharma *et al.*, 2012).

Common Salt

Principle: Adulteration of milk with salt is detected based on the reaction between silver nitrate and sodium chloride in the presence of potassium chromate indicator (Gandhi *et al.*, 2020).

$AgNO_3 + NaCl \rightarrow AgCl + NaNO_3$

$2AgNO_3 + K_2 CrO_4 \rightarrow Ag_2 CrO_4 + 2KNO_3$

Procedure: Mix 5 mL of milk and 2 mL of 0.1 N silver nitrate solution in a test tube. Add 0.5 mL of 10% potassium chromate solution to the contents and mix well. Pure milk has a chocolate brown or brick red tint, indicating the absence of dissolved solids whereas yellow colour indicates the existence of dissolved chlorides (Azad & Ahmed, 2016).

Ammonia Compounds

Nessler's reagent detects ammonia compounds in milk by creating a brown colour.

Procedure: In a test tube, place 5 mL of milk and 1 mL of Nessler's reagent (Take 8 g mercuric chloride in 150 mL distilled water, 60 g NaOH in 150 mL distilled water, 16 g KI in 150 mL distilled water. Mix the chemicals and make up the volume to 500 mL using distilled water). Examine the colour after mixing the contents. Ammonium compound-containing milk becomes dark, whereas pure milk turns yellow (Sharma *et al.*, 2012).

Soya Powder

Principle: The urea is converted into carbon dioxide and ammonia by the urease enzyme found in soybean powder. The ammonia makes the medium alkaline, which causes the solution to become pink.

Procedure: In a test tube, place 5 mL of milk and 1 mL of soybean powder detection solution (In distilled water, make a 0.1 % phenol red solution. Allow the solution to stand overnight before filtering it. In 100 ml of phenol red solution, dissolve 1 g of urea. Keep the solution in a cold, dry location). Mix the materials and set them aside for 10–15 minutes. The presence of a pink tint in milk suggests the presence of soybean powder (Gandhi *et al.*, 2020).

Sulfates

Principle: The sulphates are identified by the interaction of sulphates with barium chloride, which produces barium sulphate, which settles as a white precipitate.

Procedure: Pour 20 mL of milk into a beaker and add 1 mL of a 5% lactic acid solution. Bring the contents to a boil and then let them cool. Filter the solution and transfer the filtrate to a test tube. Mix in 3–4 drops of 1% barium chloride solution. The presence of white precipitates indicates the existence of sulphates, whereas no precipitates indicate the absence of sulphates (Gandhi *et al.*, 2020).

Formalin

Principle: In the Hehner and Leach test, added formaldehyde produces a distinctive violet colour when combined with ferric salts and other oxidising agents.

A. Hehner Test

Procedure: Fill a test tube with 10 mL of milk. Gently pour in 0.5 mL of 10% ferric chloride solution. Pour 5 mL of pure sulfuric acid through the test tube's sides so that without combining with milk, it creates a distinct layer at the bottom. The formation of violet to purple coloured ring at the interface of the two liquids shows the presence of formalin (Gandhi *et al.*, 2020).

B. Leach Test

Procedure: In a test tube, pour roughly 5.0 mL of milk and add equal quantities of strong HCl containing 1 ml of 10% ferric chloride solution (500 ml HCl + 1 ml $FeCl_3$). Heat for 5 minutes over a flame. Rotate the tube to break the curd and observe the colour. The violet tint suggests the presence of formaldehyde (Sharma *et al.*, 2012).

C. Chromotropic Acid Test

Procedure: Fill a test tube with 1 mL of milk. 1 mL of chromotropic acid reagent (saturated solution of 1, 8-dihydroxy naphthalene-3, 6-disulfonic acid in approximately 72 % sulfuric acid (about 500 mg/100 ml) resulted in a light straw-colored solution) is added and thoroughly mixed. The presence of formalin in milk is indicated by the emergence of a grey tint, whilst the control sample stays colourless (Azad & Ahmed, 2016).

Hydrogen Peroxide

Principle: The principle of oxidation of para-phenylenediamine by hydrogen peroxide is used to determine it. Para-phenylenediamine, also known as 1,4-diaminobenzene, is a yellow compound that, when oxidised with hydrogen peroxide, transforms into quinone diamine, which is blue in colour (Gandhi *et al.*, 2020).

Procedure: In a test tube, place 2 mL of milk and an equal proportion of alcohol. Shake thoroughly before adding 5 drops of 2% para-phenylenediamine solution. The development of blue colour indicates that the sample contains hydrogen peroxide. Use a milk sample that is devoid of hydrogen peroxide to perform a blank test. There will be no blue colour in this situation (Azad & Ahmed, 2016).

Neutralizers

A. Rosalic Acid Test

Principle: Rosalic acid is an indicator that becomes red in alkaline conditions and orange-brown in acidic conditions. Since this test is dependent on the change in pH of milk after neutralisation, it may not function if the neutralisation of milk is not done properly or if the milk develops acidity (Gandhi *et al.*, 2020).

Procedure: In a test tube, place 10 mL of milk and an equal proportion of ethyl alcohol (95 %). Add a few drops of rosalic acid solution (0.1 % —100 g rosalic acid in 30 mL ethyl alcohol and distilled water to produce the amount of 100 mL). Combine the ingredients. The appearance of rose-red colour shows the existence of a neutralizer, whereas the brown colour indicates the absence of a neutralizer. Rose-red colour will show on the addition of NaOH, KOH, and $Ca(OH)_2$ whereas pink colour will arise with the $NaHCO_3$, KCO_3, and $CaCO_3$ addition (Kamthania *et al.*, 2014).

B. Alkalinity of Ash Test

Principle: The alkalinity of the ash content is measured after the added neutralizer is neutralised by the produced acidity. Because of the presence of a neutralizer in milk, the ash level of milk will be higher than in control milk (Gandhi *et al.*, 2020).

Procedure: In a silica crucible, place 20 mL of milk. First, evaporate the water to dryness, then burn the content to ash in a muffle furnace at 550°C. Add 10 mL of distilled water to the ash and titrate against 0.1 N HCl using phenolphthalein as an indicator. More than 1.2 mL of 0.1 N HCl will be consumed by milk containing neutralizer (Kamthania *et al.*, 2014).

Table 1: Limit of detection (LOD) for different adulterants in milk using rapid color-based methods

Adulterant	Method	Limit of detection	Reference
Sucrose	Resorcinol test	0.2%	(Azad & Ahmed, 2016; Kamthania *et al.*, 2014)
Starch	Iodine solution test	0.02%	(Azad & Ahmed, 2016; Gandhi *et al.*, 2020)
Urea	Dimethyl amino benzaldehyde test	0.2%	(Azad & Ahmed, 2016; Sharma *et al.*, 2012)
Glucose	Barfoed's reagent test	0.1%	(Azad & Ahmed, 2016; Gandhi *et al.*, 2020)
Maltodextrin	Iodine reagent test	0.3%	(Gandhi *et al.*, 2020; Sharma *et al.*, 2012)
	Citric and iodine reagent test	0.1%	
Pond water	Diphenylamine test	0.02%	(Kamthania *et al.*, 2014; Sharma *et al.*, 2012)
Detergents	Methylene blue test	0.0125%	(Gandhi *et al.*, 2020)
Common salt	Silver nitrate test	0.02%	(Azad & Ahmed, 2016)
Ammonium compounds	Nessler's reagent test	0.15%	(Gandhi *et al.*, 2020)
Sulfates	Barium chloride test	0.05%	(Gandhi *et al.*, 2020)
Formalin	Leach test	0.1%	(Gandhi *et al.*, 2020; Sharma *et al.*, 2012)
	Chromotropic acid test	0.05%	
Hydrogen peroxide	Para-phenylene diamine test	0.025%	(Sharma *et al.*, 2012)
	Potassium iodide and starch reagent test	0.004%	
Neutralizers	Rosalic acid test	0.1% (NaOH, Na_2CO_3)	(Sharma *et al.*, 2012)
		0.2% ($NaHCO_3$)	

Instrument-based Methods

It is an essential requirement to provide an appropriate level of food quality and safety in order to adequately protect customers and allow business. A broad network of effective quality assurance programs must be in place to check the quality and safety of these foods before they reach consumers. This is primarily attainable with the use of newly created instrument-based methods. Modern methods should provide fast, sensitive, and accurate methods for detecting and analysing adulterants (Das *et al.*, 2016).

Cryoscopic Method for Detection of Added Water in Milk

The addition of water to milk alters its density, the refractive index of the lactoserum, and, most crucially, its freezing point (FP). The thermistor cryoscopic measurement of raw milk FP is perhaps the most extensively used technology for assessing milk purity.

Commercial devices for such determinations can handle a high number of samples, are entirely automatic, and deliver cost-effective and quick results.

A test sample of milk is super-cooled to an adequate temperature and crystallisation is induced using a technique developed by the International Dairy Federation. When the temperature of the frozen milk does not vary by more than 0.5 milli degree Celsius (m°C) in the previous 20 seconds, the FP is regarded to be obtained. This FP should not be greater than -520 m°C for heat-treated bovine milk for human consumption, or it should be near to the average FP of raw milk reported in the area where the drinking milk was obtained (Poonia *et al.*, 2017).

Although it is a simple procedure, it does have drawbacks. Because the molar concentration of solutes in the milk of different species varies, detecting additional water in mixed milk using this approach alone will be challenging. Furthermore, the inclusion of preservatives and other solutes such as sugar and salts reduces the FP of diluted milk and so escapes detection as adulteration. Furthermore, the increase in acidity might jeopardise the FP findings.

Infrared Spectroscopy

A. Mid-infrared Spectroscopy

The development of chemometric techniques and infrared spectroscopic (IR) technology has made this methodology a powerful tool for assessing the quality and authenticity of food. The key benefits of IR include its ability to analyze materials with little to no sample preparation, ease of use, rapid data gathering, high sensitivity and specificity, and potential to serve as a "fingerprint" technology (Santos *et al.*, 2013b). He *et al.* (2010) tested MIR spectroscopy paired with two-dimensional (2D) correlation for detecting adulteration of tetracycline hydrochloride, melamine, urea, glucose in milk. The spectra revealed differences in peak locations and shapes between the control and adulterated milk, indicating the method's ability to identify milk adulteration. In another study attenuated total reflectance mid-infrared microspectroscopy (ATR MIR-microspectroscopy) and soft independent modelling of class analogy (SIMCA) analysis was used as a quick method for detecting and quantifying synthetic milk as well as milk adulterated with synthetic urine, whey, urea and hydrogen peroxide. Samples were placed on a 192-well microarray slide, air-dried, and MIR-microspectroscopy was used to acquire spectra as depicted in figure 1. The main spectral variations between the control and adulterated samples were seen in the protein (amide I and II) vibration bands at 3500–3300 and 1640–1500 cm^{-1}. Samples loaded with whey, urea, synthetic milk, and synthetic urine exhibited notable alterations due to the impact of the adulteration process on the milk protein spectral profile (Santos *et al.*, 2013b).

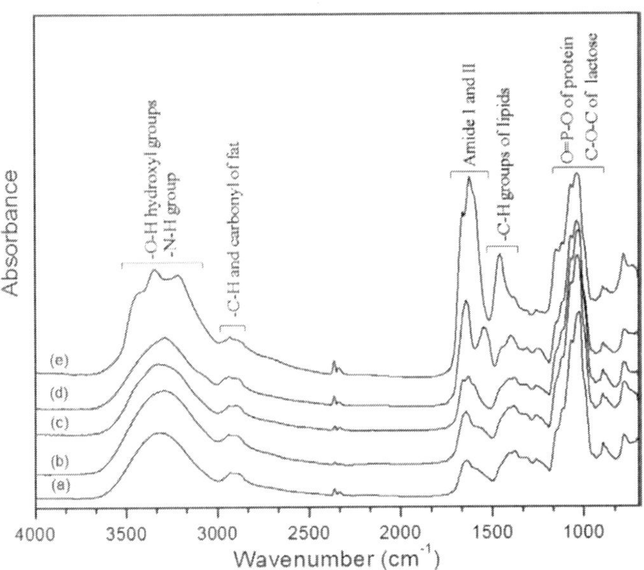

Fig. 1: MIR-microspectroscopy spectra of dried films for, **(a)** control milk, **(b)** milk adulterated with hydrogen peroxide (0.15 g/L), **(c)** synthetic milk (0.8 g/L of urea), **(d)** milk adulterated with whey (30 g/L) and **(e)** milk adulterated with urea (12.5 g/L) (Source: Santos *et al.*, 2013b)

Botelho *et al.* (2015) effectively detected water, starch, sodium citrate, formaldehyde, and sucrose in milk at concentrations ranging from 5000 to 10,000 ppm using the ATR and MIR methods. In this case, multivariate supervised classification (partial least squares-discriminate analysis, PLS-DA) was employed to evaluate outcomes, and the report included selectivity and specificity data. Melo Vieira *et al.* (2016) proposed a viable, sample preparation-free, and quick verified approach for detecting and quantifying cheese whey as an adulterant in raw milk utilising mid-infrared spectroscopy with Fourier transform (FTMIR) and the chemometric technique of partial least squares (PLS). The PLS model was developed using FT-MIR spectroscopy of a typical set of raw milk samples comprising 6.0–28.0 % cheese whey. Similarly, MIR/ATR was employed to detect broad group of substances such as formaldehyde, hydrogen peroxide, bicarbonate, carbonate, chloride, citrate, hydroxide, hypochlorite, starch, sucrose, and water in milk. The data was examined using SIMCA, which was found to be an effective initial screening step before employing a one-class model (Gondim *et al.*, 2017). By measuring GMP protein with MIR, a quick approach for detecting and quantifying milk powder adulteration by whey was evaluated. The study revealed the MIR spectroscopy as a quick and reliable tool for identifying milk powder fraud using cheese whey (de Carvalho *et al.*, 2015).

B. Near-infrared Spectroscopy

Near-infrared (NIR) spectroscopy with chemometrics is now widely employed in food analysis, particularly milk evaluation as it is quick, non-destructive, and low-cost technology with simple operation. Since it may be used to determine many milk constituents simultaneously, the NIR method in particular has been studied for the quantitative measurement of numerous milk constituents (Kasemsumran *et al.*, 2007). Jha & Matsuoka (2004) employed short-wavelength NIR in the area of 700–1124.8 nm to identify synthetic milk. Kasemsumran *et al.* (2007) evaluated the possibility of employing a near-infrared (NIR) approach to identify and quantify water and whey as adulterants in natural milk. The NIR spectra of a combination of samples and natural milk samples were obtained in the range of 1100 – 2500 nm. Two wide bands were depicted between 1450 and 1900 nm. The 1450 nm band is primarily caused by the interaction of OH symmetric and antisymmetric stretching modes of water. The other band is allocated to the OH group stretching and deformation vibrations in combination mode. The intensity of the water bands increases as the amount of water added to the milk samples increases while the intensity of the band in the 2300 – 2400 nm wavelength range varies with the amount of whey in the samples as shown in figure 2.

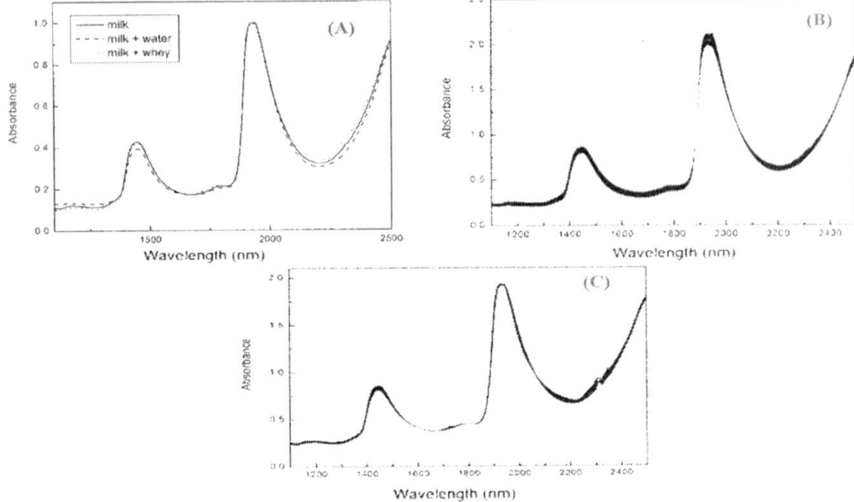

Fig. 2: (A) NIR spectra in the region of 1100 – 2500 nm of natural milk (solid line), milk adulterated by water (dash line), and milk adulterated by whey (dotted line) (B) NIR spectra in the region of 1100 – 2500 nm of milk adulterated by water (C) NIR spectra in the region of 1100 – 2500 nm of milk adulterated by whey (Source: Kasemsumran *et al.*, 2007).

Similarly, Mazivila *et al.* (2020) propose a potential analytical strategy for milk authenticity based on first-order NIR spectroscopic data paired with data-driven soft independent modeling of class analogy (DD-SIMCA). This one-class classifier accurately classified all genuine milk powder samples as members of the target class from milk powder adulterated with melamine and sucrose at concentrations of 0.8–2 % and 1–3 %, respectively. Non-linear pattern recognition approaches based on near-infrared spectroscopy can be also used to quickly identify melamine, urea, and ammonium nitrate adulteration in cow milk (Zhang *et al.*, 2014). Likewise, the least-squares support vector machine (LS-SVM) is suggested as a multivariate calibration approach for the simultaneous quantification of starch, whey, and sucrose identified in powdered milk samples, employing near-infrared spectroscopy with direct diffuse reflectance measurements (Borin *et al.*, 2006). One-class partial least squares and NIR spectroscopy were combined to produce a quick and effective analytical method for melamine screening in liquid milk. This approach is easy, ecologically sound, and eliminates the need for time-consuming and tedious sample preparation, making it an alternate methodology to commonly used procedures (Chen *et al.*, 2017).

C. Fourier Transform Infrared Spectroscopy (FTIR)

Fourier transform infrared spectroscopy (FTIR) is a quick biochemical fingerprinting technology that might substantially shorten the sample processing time. It is an effective method for researching molecular interactions and conformational changes in complex structures. FTIR differs from normal IR detection technology in that it uses interferometric modulation of radiation to measure many frequencies at the same time. Using complex algorithms, the resultant interferogram is transformed back to the original form of spectrum. The main benefits of FTIR over dispersive-based instruments are better sensitivity due to a higher signal-to-noise ratio, faster analysis, higher energy throughput and superior accuracy. Furthermore, the FTIR technique can deliver accurate and sensitive results in a short time after minimal sample preparation (Jha *et al.*, 2015).

In order to address the demand for quick, efficient, and affordable analytical methodologies for identifying contaminated milk, Coitinho *et al.* (2017) made an effort to calibrate and verify a compact FTIR instrument (MilkoScan FT1, Foss Analytical) for monitoring raw milk adulteration. With wavelengths spanning from 2.0 to 10.8 m (5012-926 cm^{-1}), the equipment can scan the whole middle infrared region, and the spectra were exported and preserved in electronic files. It can detect adulteration because of the calibration's high sensitivity to substances including cornstarch (98%), sodium bicarbonate (100%), sodium citrate (99%), and formaldehyde (84%). Similarly, Analytical Pvt. Ltd invented the MilkoScreen™ equipment (http://indifoss.com/products/milko-screen), which is based on FTIR spectroscopy and can identify extraneous water as well as other adulterants such as urea, sugar, starch, and melamine in milk; it is already in use in many developed nations. Similar manner, the MilkoScan™ FT 120 infrared instrument developed by FOSS, Denmark (http://www.foss.dk/search/searchnps/milkoscan.html), which is also based on an FTIR interferometer (FTIR spectroscopy), scans the full spectrum in the MIR range (2500–25 000 nm), allowing distinctions among normal and abnormal milk to be identified. In more complex liquids, such as flavoured and fermented foods, a principal component analysis (PCA) calibration can be used to determine the major ingredients as well as particular sugars, lactic acid, urea, free fatty acids, and so on (Poonia *et al.*, 2017).

Cassoli *et al.* (2011) created calibrations to determine the concentrations of sodium bicarbonate, sodium citrate, and whey using automated FTIR methods with an accuracy of 0.005%, 0.009%, and 2.26%, respectively, for sodium bicarbonate, sodium citrate, and whey. It is a suitable choice for monitoring these adulterants when combined with automated equipment because of its cheap operational cost and strong analytical performance. Jawaid *et al.* (2013) intended to employ the Fourier transform infrared spectroscopy (FTIR) method of single bounce attenuated total reflectance (SB-ATR) as an effective quick instrument for the detection and quantification of melamine in milk (liquid and powder). Figure 3 compares the spectra of raw milk and melamine + milk (1:1 w/w) in the 860–660 cm^{-1} range. The absorption at 806 cm^{-1} is typical of out-of-plane bending modes of melamine's triazine ring; this band was missing in raw milk. To get the optimal regression equation, the findings were validated using statistical parameters and quantified using a well-developed PLS calibration model.

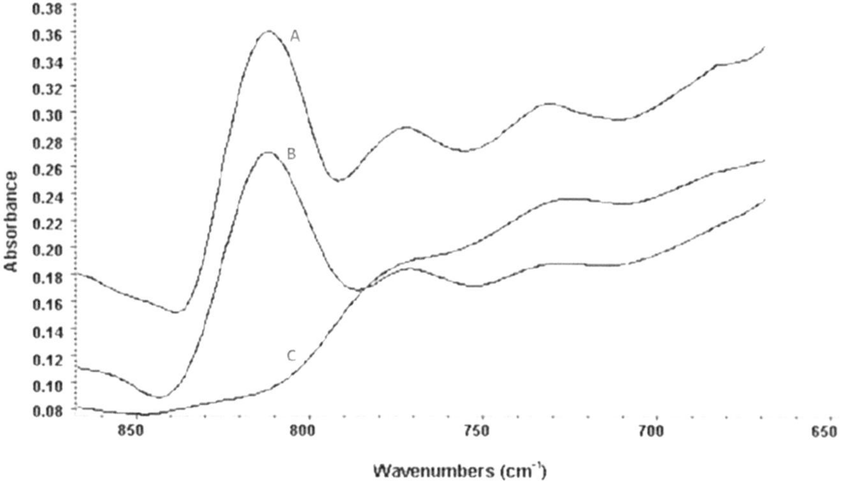

Fig. 3: ATR-FTIR spectra of (A) melamine standard, (B) milk adulterated with melamine (1:1) and (C) milk (Source: Jawaid *et al.*, 2013)

Jha *et al.* (2015) also employed ATR-FTIR to quantify added urea in milk. The amount of urea predicted well in the spectral ranges of 1649–1621 and 1611–1580 cm^{-1} using PLS regression with coefficients of determination of 0.906 and 0.879 for calibration and validation, respectively. The results of this study suggest that ATR-FTIR has the potential to develop into a rapid method for identifying and measuring additional urea in milk at concentrations as low as 100 ppm for commercial use. Nicolaou *et al.* (2010) used FTIR in conjunction with multivariate analysis to detect milk adulteration with milk from different species.

Differential Scanning Colorimetry (DSC)

DSC is a technique for measuring the temperatures and heat fluxes associated with material transitions as function of time and temperature in controlled environment. The fluctuation of glass transition temperature (Tg) is an effective tool for estimating the influence of adulteration on the physicochemical and structural characteristics of food samples (Poonia *et al.*, 2017). In a study, Tg values for pure milk and diluted milk were studied using DSC. Tg of pure milk (89.2 kg water and 10.8 kg total solids per 100 kg milk) ranged from -114 to -118 °C. Using this technology, extra water added to milk may be identified down to 0.5% (Goswami & Gupta, 2008).

Many different chemometric data categorization methods were utilized in recent years to check the quality and legitimacy of food items. Soft independent modelling by class analogy (SIMCA), partially least squares discriminant analysis (PLS-DA) and k-nearest neighbours (k-NN) are the most often used chemometric approaches for data categorization (Farah *et al.*, 2021). Although these approaches can deal with high dimensionality, high collinearity, and small size difficulties, they do not perform well when dealing with high nonlinearity. To address this limitation, machine-learning techniques have become an essential component in the field of chemometrics (Song *et al.*, 2020). As they can effectively capture nonlinear relationships between observations and classes, nonlinear classifiers like artificial neural networks (ANN), random forests (RF), and gradient boosting machines (GBM) excel at classifying complex data, leading to higher classification accuracy and greater robustness (Farah *et al.*, 2021).

To identify adulteration of raw bovine milk containing whey, urea, formaldehyde, and starch, DSC was combined with GBM, RF and multilayer perceptron (MLP). The boiling and crystallisation enthalpy of the starch-adulterated samples significantly varied from the control, demonstrating that DSC may detect starch adulteration in milk at 1 g/L. The peak temperature and onset temperature of crystallisation were the critical criteria for detecting formaldehyde in cow milk using DSC. All DSC values, except onset crystallisation temperature and crystallisation enthalpy, differed for milk contaminated with whey. Significant changes in initial boiling temperature, initial melting temperature, boiling enthalpy and peak boiling temperature were identified for raw milk samples contaminated with urea. MLP and GBM were able to categorize 100% of contaminated samples, however, RF performed optimally, with prediction and recognition capabilities of 88.5 % and 100%, respectively. Generally speaking, the most significant predictor for the MLP model was the peak temperature of boiling while the most significant discriminating predictor for the GBM and RF models was the peak temperature of crystallization (Farah *et al.*, 2021).

Chromatography

Earlier detection of foreign proteins in milk was limited to methods based on protein immunological properties, such as enzyme-linked immunosorbent assay (ELISA), in which different fractions of milk or whey proteins could be detected by monoclonal or polyclonal antibodies, or by analysing the differences in the electrophoretic properties of milk fraction proteins and adulterant proteins. These methods, however, were inadequately sensitive and accurate at low concentrations (Poonia *et al.*, 2017). Nowadays detection of nitrogen-rich adulterants and non-dairy proteins in raw milk can be easily identified using chromatographic techniques.

A. High-Performance Liquid Chromatography (HPLC)

High-performance liquid chromatography, earlier known as high-pressure liquid chromatography, is an analytical chemistry technique for separating, identifying, and quantifying each component of a mixture. Pumps are used to move a pressured liquid solvent containing the sample mixture through a column containing a solid adsorbent material. Each component in the sample interacts somewhat differently with the adsorbent material, resulting in varied flow rates and separation of the components as they flow out of the column. The primary distinction between traditional LC and HPLC is that the mobile phase of liquid chromatography travels due to gravity, but the mobile phase of HPLC flows due to the high pressure provided by the pump, which reduces analysis time (Hamilton & Sewell, 1982). Using HPLC, detection of melamine with the other nitrogen compounds in milk can be conducted. Wei *et al.* (2009) employed an HPLC method to determine cyromazine (an insecticide) and melamine residues in milk. The technique employs an NH_2 column and a 97% acetonitrile eluate at a 1 mL/minute flow rate. After being treated with NaOH, the samples were extracted with acetonitrile and 20% NH_4OH. The retention times for cyromazine and melamine were 8 and 12 minutes, respectively. Both substances had detection limits of 0.2 ng and quantitation limits of 0.02 mg/kg.

Another technique for quantifying cow milk adulteration in goat milk was developed, which was based on solvent separation of whey proteins followed by HPLC/ESI-MS. The cow milk adulteration was confirmed using the molecular marker β-lactoglobulin. The LC separation was carried out on a Jupiter C4 column at a 0.2 mL/min flow rate. Mobile phase A was formic acid (0.2 %) in water, whereas mobile phase B was formic acid (0.2 %) in acetonitrile. The gradient program was created in 28 minutes and ranged from 30% B to 44% B. Retention time and the molecular mass created by multiple-charged molecular ions were used to identify the adulterants. Two peaks from goat milk, corresponding to α-lactalbumin and β-lactoglobulin, elute at retention times of 14.8 and 20.1 min, respectively, and three peaks from cow milk, corresponding to α-lactalbumin, β-lactoglobulin variant B, and β-lactoglobulin variant A, elute at 19.0, 20.8, and 22.2 min, respectively. It is demonstrated that the suggested approach can detect cow milk as low as 5% level in goat milk (Chen *et al.*, 2004). Similarly, Jablonski *et al.* (2014) devised a simple, nontargeted UHPLC-UV (Ultra high-performance liquid chromatography- UV) technique for detecting adulteration of SMP by foreign-protein-containing compounds with 215 nm detection.

RP-HPLC ion-pairing chromatography with UV detection can be utilized for the detection of melamine in milk. Melamine was discovered at very low amounts, namely 0.01 μg/mL. It was feasible to detect and quantify melamine at food-safe concentrations in 25 minutes while remaining simple, sensitive, and robust, and allowing for the analysis of a large number of samples without loss in column performance. Very little noise was detected, highlighting the efficiency of the clean-up technique for a complex matrix (Salman *et al.*, 2012). In another approach, a sensitive and reliable SPE–HPLC (Solid-phase extraction- HPLC) technique for melamine residue measurement in liquid milk was developed. The suggested approach is sensitive, reliable, and accurate, and it can detect melamine residues in various liquid milk at levels as low as 60 µg/kg. The approach may be used to routinely determine melamine residue in various liquid milk samples (Sun *et al.*, 2010).

Ferreira & Oliveira (2003) devised an RP-HPLC technique for detecting glycomacropeptide (GMP) to identify the addition of rennet whey to milk intended for powder manufacture. A polystyrene-divinylbenzene column and gradient elution with two solvents were employed in the optimised method. Solvent A included 0.1% TFA in water, while solvent B had 95% acetonitrile, 5% water, and 0.1% TFA. The flow rate was set at 1 mL/min. A UV detector at 214 nm was used to separate two peaks corresponding to the aglyco components of the two main genetic variations (A and B), which eluted after the less well resolved GMP components. The measurements were taken in the 15–200 µg/mL linear range. The detection limit was found to be 2 µg/mL. Ferreira & Caçote (2003) used RP-HPLC to identify and quantify β-lactoglobulins in blends of bovine, caprine, and ovine milk, and quantification of milk species was possible in the 5–95 % concentration range. Gradient elution was performed at a flow rate of 0.5 ml/min and a temperature of 45 °C using a mixture of two solvents: solvent A (0.1 % TFA in water) and solvent B (0.09 % TFA in 80% aqueous acetonitrile, v/v). At 215 nm, the effluent was observed. Different chromatographic patterns were found for bovine, ovine, and caprine whey proteins under the circumstances utilised. The retention durations for β-lactoglobulin peaks varied according to milk type.

Rapid identification of adulterants in fresh milk has also been accomplished using liquid chromatography-tandem mass spectrometry. Abernethy & Higgs (2013) developed a technique for detecting the economically motivated adulteration of fresh milk using a variety of small, nitrogen-containing molecules, such as melamine, ammeline, ammelide, etc. As an internal standard, $^{15}N_2$–urea was employed. An ultra-high performance liquid chromatograph was used for the analysis, which was conducted at a flow rate of 1 mL/min with mobile phases of 0.1 % ammonium formate in water (A) and 0.1 % formic acid in acetonitrile (B) with a rapid gradient of 1.2 min. A mass spectrometer in multiple reaction monitoring (MRM) mode was used for detection, using electrospray ionisation (ESI) at 600 °C. The adulteration of milk with urea-based fertiliser can be identified by employing the contaminant biuret and triuret as markers which compose up to 0.4- 1% of synthetic urea.

Sensitive, reliable, and specific techniques for determining aglyco-CMP_A (κ-Casein macropeptide genetic varient A), aglyco-CMP_B (κ-Casein macropeptide genetic varient B), and total CMP_{AB} were established using RP-HPLC connected to a mass spectrometer fitted with an electrospray ionisation source (ESI-MS). Aglyco-CMP_A and aglyco-CMP_B were quantified from 1359^{5+}, 1698^{4+}, 2264^{3+} and 1352^{5+}, 1690^{4+}, 2253^{3+} m/z specific ions, respectively, using multiple ion monitoring (MIM) mode. Total CMP_{AB} was determined by UV, single ion monitoring (SIM), and multiple reaction monitoring (MRM) detection on peptide κ -CN (162–169) liberated from pepsin hydrolysis. MRM detection yielded a high degree of selectivity; the limit of quantification was 10 pmol within the standard curve range of 10–1000 pmol (Mollé & Léonil, 2005).

B. Gas chromatography

Miao *et al.* (2009) used gas chromatography-tandem mass spectrometry (GC-MS/MS) to create an analytical technique for simultaneous qualitative and quantitative detection of melamine and triazine-related byproducts in milk and milk products. Melamine and triazine-related by-products, such as ammelide, ammeline, and cyanuric acid, were extracted from the samples using a solvent combination of diethylamine, water, and acetonitrile (10:40:50). Following centrifugation, an aliquot of the supernatant was evaporated to dryness under a moderate stream of nitrogen gas, and melamine and triazine-related by-products were derivatized with *N,O*-Bis(trimethylsilyl) trifluoroacetamide (BSTFA) containing 1% trimethylchlorosilane (TMCS). Melamine derivatives and analogues were determined using gas chromatography/ tandem mass spectrometry with multiple reactional monitoring (MRM) and 2,6-Diamino-4-chloropyrimidine (DACP) as an internal reference. Results showed that melamine, ammelide, ammeline, and cyanuric acid had linear detectable ranges ranging from 0.004 mg/kg to 1.6 mg/kg. Melamine, ammelide, ammeline, and cyanuric acid detection limits in milk powder were 0.002 mg/kg, while the limit of quantification was 0.005 mg/kg.

Headspace gas chromatography (HS-GC) has been widely used in the research of volatile chemicals in a nonvolatile sample matrix. Using HS-GC detection, Xie *et al.* (2019) proposed a robust and automated approach for quantifying urea in milk with a column flow rate of 2.7 mL/min and nitrogen as the carrier gas. The HS-GC technique analyses CO_2 generated by the enzymatic hydrolysis process of urea in milk samples after being treated with the enzyme urease at 35 °C for 40 min. The CO_2 produced by the enzymatic hydrolysis of urea in the presence of the enzyme urease was quantified by GC. In another investigation, a high metrological order approach for analyzing urea in milk and milk powder was created employing isotope dilution (ID) linked to GC/MS. The combination of GC/MS and ID is a globally recognized reference approach that produces excellent specificity and trueness. The method was used to measure urea concentrations in milk and milk powder after samples were pre-treated with acetonitrile to eliminate proteins. For chromatographic separation, a fused-silica capillary column coated with OV-5 was utilized. Helium was used as the carrier gas, with a flow rate of 1.0 mL/min and a split ratio of 1:50. The ionization energy for MS analysis was 70 eV, while the ion source temperature was 250 °C (Dai *et al.*, 2010).

Immunoassays

Immunoassays use an immunological response to measure a specific molecule in very low quantities. Because of their specificity, simplicity, and sensitivity, among other advantages, immunoassays have been the most extensively used test for regulatory purposes in determining food authenticity. Some typical examples are radioimmunoassays, enzyme immunoassays, immunoprecipetin, and fluorescence immunoassays. Recent years have seen the rapid development of numerous commercial immuno-assay kits for use in the food and feed industries as a result of advancements in ELISA technology. The same immuno-assay principles as ELISA tests are used by lateral flow tests. Both types of tests are excellent for food screening testing of a large number of samples due to their simplicity and the quick time required for analysis (Bonwick & Smith, 2004). Hurley *et al.* (2004) created an assay capable of identifying adulteration of high-quality milk with milk from less expensive sources. An indirect, competitive ELISA was developed to quickly identify cow's milk in goat, sheep, and buffalo milk. The test employs a monoclonal antibody developed against bovine IgG. This antibody detects a species-specific epitope on the heavy chain of bovine IgG1 and IgG2 antibodies. When assaying various combinations of milk contaminated with cows' milk, a peroxidase-conjugated anti-mouse IgG antibody was utilized to identify binding monoclonal antibody, and subsequent enzymatic conversion of substrate resulted in obvious changes in absorbance. The test had detection limits of 1.0 g/mL of bovine IgG or 0.1 % (vol/vol) adulteration with cows' milk. The same study team created a similar anti-bovine IgG sandwich ELISA that is more sensitive than the competitive assay for detecting cows' milk in goat, sheep, and buffalo milk cheese. The detection threshold of this method was reported to be 0.001% cows' milk adulteration of sheep or buffalo milk and 0.1% cows' milk adulteration of goat milk (Hurley *et al.*, 2006).

In another experiment to detect cow milk adulteration in goat milk, Song *et al.* (2011) modified the polyclonal antibodies against bovine β-casein by combining them with lyophilized goat β-casein before being tested by indirect ELISA. The results demonstrated that the blocked antibodies had little cross-reactivity with bovine α, κ-casein, and milk whey protein. The detection threshold of this method was 2% for several types of milk, including raw milk, heat-treated milk, and goat caseins and the adulteration may be measured in a range of 2% to 50%. Similarly, an indirect ELISA for the detection of cow's milk (1-25%) in goat's milk has been developed. Polyclonal antibodies produced in rabbits against bovine whey proteins were used in the test (Castro *et al.*, 1992). Likewise, a sandwich ELISA for detecting specific quantities of cows' milk in ewes' milk and cheese used polyclonal antibodies which was raised in goats against bovine caseins (Rodriguez *et al.*, 1993). Using a polyclonal rabbit anti-glycomacropeptide antibody, sandwich ELISA may also be used to identify and quantify whey in raw milk. The detection limit of the technique was reported to be 0.047 % and the quantification limit to be 0.14 % (Chávez *et al.*, 2012). To detect melamine in raw milk and milk powder, Yin *et al.* (2010) developed an indirect competitive ELISA approach based on the production of monoclonal antibodies. Melamine detection limits in milk and milk powder were 0.1 and 0.2 mg/L, respectively. The ELISA technique has a lot of potential for food product verification, but so far there hasn't been much progress in expanding its authentication capabilities. The drawbacks of the ELISA method include the difficulty in generating an antibody specific to a given protein at first. However, when the selectivity of the approach is considered, this is a relatively modest hurdle to overcome. ELISA techniques, like chromatography and mass spectrometry, are impractical for on-site detection because of extensive incubation times, unnecessary washing processes, and apparatus complexity (Jha *et al.*, 2016).

Fluorescence polarisation immunoassay (FPIA) is a competitive immunoassay approach that relies on the polarisation of the fluorescence of a tiny fluorescein-labeled hapten (tracer) when it is bound by a particular antibody. For melamine determination, FPIA based on a polyclonal antibody was developed. Three fluorescein-labeled melamine tracers with varying structures and spacer bridges were created. FPIA's LOD (10 % inhibition) was 9.3 ng/mL, while its IC_{50} (50 % inhibition) was 164.7 ng/mL. The FPIA antibody demonstrated 21.2 % cross-reactivity to the fly-killing pesticide cyromazine but no cross-reactivity to other naturally structurally similar chemicals (Wang *et al.*, 2011). Similarly, Luminex-based immunoassays can be a valuable method in food control by simultaneously detecting proteins from multiple plant sources in milk powder at the fraud (%) level using polyclonal antisera generated against soy, pea, and soluble wheat proteins. The Luminex 100 flow analyzer and the MultiAnalyte Profiling (xMAP) technology are used in multiplex-based immunotechnology that involves the covalent coupling of the antigen or antibody on carboxylated polystyrene microspheres (5.6 m beads) internally dyed with a red and orange fluorophore. Up to 100 distinct color-coded bead sets may be identified by altering the ratio of the two fluorophores, and each bead set can be connected to a different biological probe. A reporter molecule [a fluorophore (e.g., Alexa) attached to the antibody or antigen] is used to detect and quantify the immunocomplex (Haasnoot & du Pré, 2007).

Western blot is a frequently used technique to separate and identify proteins. Through gel electrophoresis, a mixture of proteins is sorted based on molecular weight, and hence on type, in this approach. These findings are then transported to a membrane, where each protein forms a band. The membrane is then treated with antibodies that recognize the protein of interest. The unattached antibody is washed away, leaving just the antibody that is bound to the protein of interest. After that, the attached antibodies are recognized by processing the film. Exclusively one band should be observed since the antibodies only bind to the protein of interest. Because the thickness of the band correlates to the quantity of protein present, performing a standard can reveal the amount of protein present (Mahmood & Yang, 2012). Western blot analysis can be used for GMP detection, using anti-GMP polyclonal antibodies. This test is simple, quick, sensitive, and specific, detecting 0.001% (w/v) GMP, 0.50% (v/v) liquid cheese whey, and 0.001% (w/v) dehydrated cheese whey (Chávez *et al.*, 2008). Oancea (2009) presented a rapid antibody-based analytical method for identifying unauthorized sweet whey addition in fresh, pasteurized, UHT, and condensed milk. Using this method, GMP was extracted from k-casein and other non-GMP compounds by deproteinization with 20% TCA, and 1/1000 diluted samples were colorimetrically developed on immunosticks with monoclonal antibodies highly specific to GMP/κ-casein. This method may detect a whey addition of 1–2 % in milk.

Polymerase chain reaction

Molecular approaches have been effectively employed to improve the sensitivity and specificity of authenticity testing. Polymerase chain reaction (PCR) is an *in vitro* technology that amplifies a certain DNA sequence in numerous folds and may successfully identify the milk of various species in a mix (Jha *et al.*, 2016; Poonia *et al.*, 2017). Real-time PCR is another type of PCR that allows for detection and quantification at the same time. It has been successfully used to identify and quantify goat's milk adulteration with cow's milk in the range of 0.5–100% (Dabrowska *et al.*, 2010). Darwish *et al.* (2009) used primers based on cow mitochondrial 12S rRNA genes to detect bovine milk in buffalo milk at levels as low as 0.5 %. Furthermore, in a more sophisticated PCR format, duplex PCR using two sets of primers targeting the mitochondrial D-loop region has been effectively employed to detect up to 1% bovine milk in caprine milk (Kotowicz *et al.*, 2007). Using primers targeting the mitochondrial 12S ribosomal RNA gene, the polymerase chain reaction was used to identify goats' milk in sheep's milk. Using goat-specific primers, a 122-bp fragment was recovered from goat milk DNA, but no amplification signal was found from sheep, cow, or water buffalo milk DNA. With a sensitivity threshold of 0.1 %, polymerase chain reaction analysis of raw and heat-treated sheep/goat milk binary mixtures permitted the specific detection of goats' milk (López-Calleja *et al.*, 2005). For species-specific identification of cow and buffalo milk, a polymerase chain reaction amplifying a segment of the mitochondrial DNA D loop region was designed. In both simplex and multiplex polymerase chain reaction, a common forward primer was employed with two separate species-specific reverse primers, resulting in amplification of 126 bp and 226 bp products for cow and buffalo, respectively. The detection limit of the assay was reported to be 0.1 % adulteration of cow and buffalo milk (De *et al.*, 2011).

Electromigration Methods

Based on variations in molecular weights in milk from various species, polyacrylamide gel electrophoresis (PAGE) can identify the presence of adulterant proteins. Tamime *et al.* (1999) used the PAGE method to demonstrate the presence of bovine milk (25 % or more) in the caprine milk product *Kishk*. SDS-PAGE was utilised to separate the κ-,β-, and α-caseins from bovine and caprine milk. Native-PAGE (polyacrylamide gel electrophoresis) was employed in another investigation for the simultaneous qualitative and quantitative examination of bovine milk adulteration in caprine and ovine milk utilising whole milk samples as well as their whey protein fraction. The band strength of bovine β-lactoglobulins in all milk combinations and bovine α lactalbumin in caprine/bovine milk blends was measured for quantification (Pesic *et al.*, 2011). This approach is rather quick and detects bovine whey proteins in nearly every bovine/caprine or bovine/ovine milk mixture. It is, nevertheless, time-consuming and labor-intensive as a typical examination method. In addition, tiny molecules are not well visible (Poonia *et al.*, 2017).

Capillary electrophoresis is another type of electrophoresis in which a complex array of large and tiny molecules is separated in a narrow-bore fused silica capillary. Adulterants in bovine, ovine, and caprine milk blends have been effectively detected using capillary zone electrophoresis of caseins and whey proteins (Poonia *et al.*, 2017). Cartoni *et al.* (1999) performed capillary electrophoresis of certain whey protein fractions to determine the presence of cows' milk in goats' milk and cheese. The minimal amount of cows' milk identifiable was 2% in milk blends and 4% in cheeses.

Isoelectric focusing (IEF), is a method for separating distinct molecules based on isoelectric point differences. Špoljarić *et al.* (2013) developed an isoelectric focusing technique for detecting bovine milk in ovine and caprine cheeses. The approach comprises casein separation from cheese, isoelectric focusing of γ2- and γ3-casein derived from the hydrolysis of β-casein by plasmin, and lastly, densitometry detection and quantification of c-casein in bovine, ovine, and caprine cheeses. The approach may detect raw and heat-treated bovine milk in fresh and matured cheeses prepared from ovine or caprine milk, or a combination of these kinds of milk.

Table 2: Instrument-based methods coupled with chemometrics for detection of adulterants in milk

Technique of detection	Target adulterants	Reference
Mid-infrared (MIR) spectroscopy paired with 2D correlation	Melamine, urea, glucose, tetracycline hydrochloride	(He *et al.*, 2010)
Attenuated total reflectance mid-infrared (ATR-MIR) microspectroscopy coupled with soft independent modelling of class analogy (SIMCA)	Synthetic milk, whey, hydrogen peroxide, urea	(Santos *et al.*, 2013a)
ATR-MIR coupled with partial least squares (PLS) discriminate analysis	Water, starch, sodium citrate, formaldehyde, sucrose	(Botelho *et al.*, 2015)
Fourier transform mid-infrared (FT-MIR) spectroscopy coupled with PLS	Cheese whey in raw milk	(Melo Vieira *et al.*, 2016)
ATR-MIR coupled with SIMCA	Formaldehyde, hydrogen peroxide, neutralizers, salts, starch, sucrose, water	(Gondim *et al.*, 2017)
Near-infrared spectroscopy (NIR)	Synthetic milk	(Jha & Matsuoka, 2004)
	Water and whey	(Kasemsumran *et al.*, 2007)
NIR coupled with data-driven- SIMCA	Melamine, sucrose	(Mazivila *et al.*, 2020)
NIR coupled with least squares support vector machine	Starch, whey, sucrose	(Borin *et al.*, 2006)
NIR coupled with PLS	Melamine	(Chen *et al.*, 2017)
Milkoscan FT1, Foss Analytical (based on FTIR)	Corn starch, sodium bicarbonate, sodium citrate, formaldehyde	(Coitinho *et al.*, 2017)
Milkoscreen TM (based on FTIR)	Water, urea, sugar, starch, melamine	(Poonia *et al.*, 2017)
FTIR coupled with single bounce ATR and PLS	Melamine	(Jawaid *et al.*, 2013)
ATR-FTIR with PLS	Urea	(Jha *et al.*, 2015)

Technique of detection	Target adulterants	Reference
Differential scanning colorimetry (DSC)	Water	(Goswami & Gupta, 2008)
DSC coupled with random forest, gradient boosting machine and multilayer perceptron	Formaldehyde, whey, urea, starch	(Farah *et al.*, 2021)
HPLC	Cryomazine and melamine	(Wei *et al.*, 2009)
HPLC with electrospray ionization (ESI) mass spectrometry	Presence of cow milk in goat milk	(Chen *et al.*, 2004)
Reversed phase- HPLC	Soya protein in bovine milk	(Espeja *et al.*, 2001; Kruså *et al.*, 2000)
	Glycomacropeptides	(Ferreira & Oliveira, 2003)
RP-HPLC with UV detection	Melamine	(Salman *et al.*, 2012)
Solid phase extraction- HPLC	Melamine	(Sun *et al.*, 2010a)
Ultra-high performance liquid chromatography with ESI	Melamine, ammeline, ammelide, urea	(Abernethy & Higgs, 2013)
Gas chromatography tandem mass spectrometry (GC-MS/MS)	Melamine, triazine related products	(Miao *et al.*, 2009)
Head space gas chromatography (HS-GC)	Urea	(Xie *et al.*, 2019)
GC-MS with isotope dilution	Urea	(Dai *et al.*, 2010b)
Indirect competitive ELISA	Cow's milk in goat, sheep and buffalo milk	(Hurley *et al.*, 2004)
	Melamine	(Yin *et al.*, 2010)
	Cow's milk in goat milk	(Castro *et al.*, 1992; Song *et al.*, 2011)
Sandwich ELISA	Cow's milk in ewe milk	(Rodriguez *et al.*, 1993)
	Whey in raw milk	(Chávez *et al.*, 2012)
	Melamine	(Hurley *et al.*, 2006)
Fluorescence polarization immunoassay (FPIA)	Melamine	(Wang *et al.*, 2011)
Western blot assay	Glycomacropeptides	(Chávez *et al.*, 2008; Oancea, 2009)
Real-time - polymerase chain reaction (RT-PCR)	Cow's milk in goat milk	(Dabrowska *et al.*, 2010)
Capillary electrophoresis	Cow's milk in goat milk	(Cartoni *et al.*, 1999)
Isoelectric focusing	Cow's milk in ovine and caprine milk	(Špoljarić *et al.*, 2013)

Paper Strip-based Methods

Test strips are a simple, effective, and low-cost analytical instrument used in the field or on-site for the quick detection of an analyte. A test strip that has been coated with a chemical that interacts with the analyte is either dipped in the sample solution or put onto the test strip. The use of test strips for in-process detection is useful because of their ease of use and high specificity. Test strip findings are based on either eye inspection or quantitative analysis utilising digital readers.

Methods Based on Dry Chemistry

Milk adulteration detection procedures based on wet chemistry, which include fundamental experimental techniques such as measuring, mixing, weighing, and boiling chemicals, need competent staff to prepare the reagents and a variety of glassware. Dry reagent-based strip tests are an appealing alternative to traditional wet chemical procedures since they have longer stability and are less prone to squander chemicals. It also has reduced expenses, is easier to use, has a shorter analysis time, and can make rapid choices without the usage of a traditional laboratory. Paper strips are impregnated with dry chemicals and the milk sample is added in the dry chemistry strip-based procedures. Dry reagent test strips can be used as an alternative for emergency testing and do not require expert workers because the findings can be understood with the naked eye based on the colour shift (Gautam, 2016).

Kumar *et al.* (2000) developed a dry reagent strip to detect urea in tainted milk. Strips were made by dissolving urease enzyme in 25mM tris-acetate buffer and combining it with chromogen solution. The enzyme's stability was increased by β-mercaptoethanol, and the cross-linking reagent was 0.01 % gluteraldehyde. Following 5-minute centrifugation at 10000g, the supernatant was immobilized on Whatman no. 1 paper and dried at 30°C to obtain a pale-yellow strip. The strip changes colour from yellow to magenta based on the quantity of urea in milk and can test urea concentrations as low as 0.1 g/L. The strip functions on the concept that after dipping into milk samples, immobilized urease turns urea into ammonia and carbon dioxide. The ammonia that has been released then interacts with dye, changing the colour of the strip. The blue colour on the strip had to be observed following immersion in milk and dye solution for the test. The anionic detergent reacts with the cationic dye, resulting in a distinct colour change. The test is sensitive enough to identify the presence of 0.1 % detergent in milk. As anionic detergents are employed in the manufacturing of synthetic milk, the test can be used to distinguish synthetic milk from pure milk as well as pure milk that has been tainted with detergent (Rajput *et al.*, 2015).

Mal *et al.* (1988) invented the 'diastix strip' for detecting glucose in milk. The enzyme glucose oxidase catalysed the synthesis of gluconic acid and hydrogen peroxide from the oxidation of glucose, while the second enzyme peroxidase catalysed the oxidation of potassium iodide contained in the strip by hydrogen peroxide to generate free iodine. It was reported that the technology would detect a minimum of 0.05 % glucose. Pal *et al.* (1989) improved the above-mentioned technique for detecting added sucrose in milk. Sucrose was hydrolyzed to glucose and fructose by the 'invertase enzyme,' and the resulting glucose was identified using the diastix strip. Narang & Khokhar (2004) developed a urea spot test based on dry reagent chemistry. The testing procedure includes placing a drop of indicator solution containing 1.6-1.8% *p*-dimethyl aminobenzaldehyde, 0.6-0.7% methyl red, and 0.3-0.4% bromocresol green, as well as an equivalent amount of milk to be tested, on filter paper. Panchal (2013) created an enzymatic strip employing horseradish peroxidase to detect hydrogen peroxide in milk. The lowest detection limit was stated to be 0.005 %.

Gautam (2016) developed dry reagent strips for the detection of various adulterants added to milk such as neutralizers, urea, hydrogen peroxide, and sucrose. The neutralizer detection strip was created on the assumption that adding neutralizers to milk would alter the natural pH of the milk. As a result, for the construction of a neutralizer detection strip, a suitable pH-sensitive chromogen was adsorbed on filter paper. The colour of the strip changes from yellow to green in the presence of neutralizer. The urea detection strip was created by immobilizing urease and chromogen together. Urease destroys the urea in milk when it comes into touch with it, and the resultant chemicals raise the pH of the milk. The rise in milk pH was detected by chromogen, which caused it to change colour. The colour of the strip changed from yellow to red in the presence of urea. The sucrose detection test strip had two important components: chromogen and enzymes such as invertase, glucose oxidase, and peroxidase. The basic idea was that sucrose is hydrolyzed into glucose and fructose in the presence of invertase, and glucose in the presence of glucose oxidase interacts with oxygen and water to form hydrogen peroxide. The resulting hydrogen peroxide combines with 2,4,6-tribromo-3-hydroxy benzoic acid to make 2,4-dibromo-3,6-dioxocyclohexa-1,4-dienecarboxylic acid, which then reacts with 4-amino antipyrine to give a red colour molecule. The hydrogen peroxide detection strip worked by releasing iodine from potassium iodide in the presence of hydrogen peroxide, which was catalysed by the enzyme peroxidase. The strip's colour changed to brown in the presence of hydrogen peroxide. The limit of detection for NaOH, urea, sucrose, and H_2O_2 have been reported to be 0.03 %, 0.08 %, 0.1 %, and 0.001 %, respectively.

Lateral Flow Assay

Singer and Plotz invented the lateral flow test in 1956. It is also known as an immunochromatographic assay since it is designed to identify the presence or absence of a target analyte in a sample. Color-producing particles are attached to the assay's particular biological components (antigen/antibody/aptamer). The findings are evaluated on the reaction matrix or read by eye or with the aid of a reader. Lateral flow assays (LFAs), which consist of prefabricated strips of a carrier material holding dry reagents that are activated by adding a fluid sample, are used for diagnostic reasons such as detecting organ failure as well as detecting pollutants and adulterants in food and feed. The main concept behind LFA is the passage of a liquid sample containing the target analyte along a strip of polymeric material via capillary flow over numerous reactive zones where molecules having more or less specific interactions with the analytes have been attached. The surface layer of the LFA strip transports the sample from the sample pad to the conjugate pad down the strip, encountering the detection zone and ending at the absorbent pad. By collecting excess liquid, absorbent pads help to maintain appropriate liquid flow. The labelled analyte (colloidal gold, dyed polystyrene, phosphor, dye-encapsulating liposomes, quantum dots, colorful or fluorescent or paramagnetic monodisperse latex particles or liposomes or colloidal carbon) is dried on the conjugate pad and interacts with the fluid flow once the sample is added. On the nitrocellulose membrane strip, two lines are sprayed: a test line and a control line. The identification of the sample analyte and the reporter at the test line will result in the desired response. The accumulation of labelled conjugate on the test and control lines by the immunological recognition system provides observable signals, which may be seen with the naked eye or with a lateral flow strip reader. The control line ensures that the liquid flows properly through the strip (Gautam, 2016). Assays are offered in a variety of formats. These are selected based on the type of analyte to be recognised as well as the ligand employed for recognition. When only antibodies are employed as recognition elements, the test is known as a 'lateral flow immunoassay,' and it occurs in two formats: competitive (indirect) assays and sandwich (direct) assays (Sharma *et al.*, 2021).

A simple, quick, specific, and reliable method for quantifying cow milk in goat milk was developed by Liu *et al.* (2019). A competitive lateral flow immunoassay strip was created by utilising a particular monoclonal antibody (mAb) tagged with colloidal gold nanoparticles (GNPs) for binding to cow milk casein. The detection limit was 0.07 % cow milk in goat milk, offering identical specificity and greater sensitivity than ELISA. The working principle was explained as follows; when there was no cow casein in the goat milk sample, the detection reagent was caught by the capture reagent, forming a readily visible test line. On the contrary, if cow casein is present in the sample, it will bind to the immobilised capture reagent with the restricted quantity of detection reagent in a competitive manner. If there is enough cow casein in the sample, it will fully block the reaction with the capture reagent, resulting in no visible test line on the nitrocellulose membrane. The control line was always visible, indicating that the test strip and technique were effective. Similarly, in another study Galan-Malo *et al.* (2018) validated lateral flow strips designed for the identification of cow milk in water buffalo, sheep, or goat milk. The main antibodies for strip synthesis were created in response to cow milk IgG. None of the buffalo, sheep, or goat samples tested positive in the selectivity investigation. All of the samples spiked with 0.5 % bovine milk tested positive, demonstrating the specificity of the created strips. This study's detection limit was determined to be 0.5 %.

The presence of soymilk in whole bovine milk has been determined using a lateral flow detection approach. The technology employs commercially available rabbit anti-soy protein antibodies coupled to gold nanoparticles (AuNPs), with soymilk protein in contaminated milk competing for limited antibodies. Anti-rabbit immunoglobulin was immobilised in the control line to ensure the flow characteristics of antibody—conjugated AuNPs. Soymilk (antigen) was coated over the test line in the design. The assay was formulated on the basis of a competitive reaction idea, in which free analytes in the sample and soymilk coated at the test line compete for a limited quantity of colloidal gold-labeled anti-soy antibodies. If the analyte is lacking or below the detection limit in the sample, the soymilk coated at the test line will bind gold labelled anti-soy antibodies, causing an accumulation of gold to form a red line. If the concentration of soymilk in the sample exceeds the detection limit, the full colloidal gold-labeled-anti-soy antibody will attach to the sample soymilk and hence will not be accessible for soymilk at the test line, resulting in no line formation. The detection limit for soymilk in whole bovine milk was 1.75 % (v/v), and results were obtained in 5 minutes (Gautam *et al*., 2017) The concept is illustrated in the figure 4.

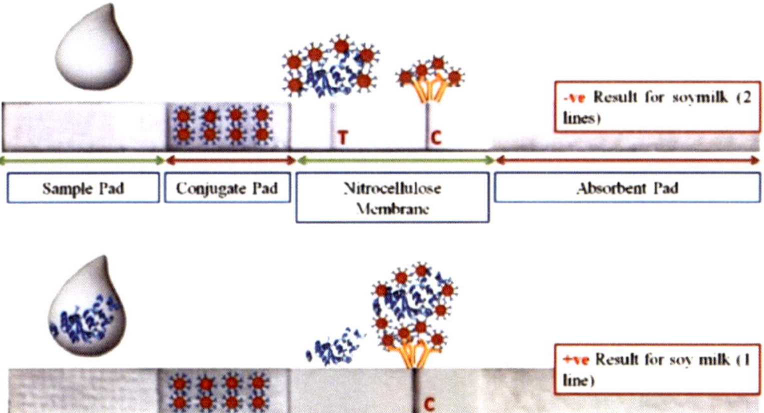

Fig. 4: Competitive LFA format for detection of soymilk in milk (*Source*: Gautam, 2016)

A competitive type lateral flow strip for field detection of buffalo milk in cow milk was developed. Antibodies against buffalo milk were produced and purified using a protein A column. After the Hansa test, Double Immunodiffusion assay, and ELISA were used to ensure the purity of the antibodies, they were labelled with carbon nanoparticles. The control line was printed with a secondary antibody produced in goats against the primary antibodies, while the test line was coated with buffalo milk antigen. For the positive samples, both the half strip (containing only nitrocellulose membrane and absorbent pad) and the whole strip (including sample-cum-conjugate pad instead of separate sample and conjugate pad) exhibited a single line. The test was reported to have a detection time of 10 minutes and a threshold of 5% level (Sharma *et al*., 2021).

For the rapid detection of bovine rennet whey in liquid milk and milk powder, an immunochromatographic lateral-flow test dipstick test was developed by Martín-Hernández *et al*. (2009) based on the binding of casein glycomacropeptide (cGMP) by two specific anti-bovine κ-casein monoclonal antibodies. Antibovine κ-casein monoclonal antibody 6A10 mounted on red colloidal particles interacted with cGMP found in milk sample extracts. This combination (colloidal particles/antibody/cGMP) moved to the reaction region through chromatography and interacted with additional specific anti-bovine κ-casein monoclonal antibodies (4G10) immobilised on the strip. A red test line was formed as a result of this reaction. Simultaneously, a similar reaction occurred with blue colloidal particles, resulting in the creation of a blue control line, demonstrating that the test was correctly performed. The test had a visual detection limit of roughly 15 ng/ mL for cGMP and 1 % (v/v) for rennet whey in milk.

Detection of Adulteration in Ghee

Milk fat, which is one of the most expensive edible fats, is more typically manipulated during adulteration. It can be blended with or substituted with less expensive vegetable or animal oils and fats. The fatty acid composition, triacylglycerol profile, and distinct fractions of other minor lipid components, mostly from the unsaponifiable fraction, are the most efficient techniques to identify the foreign fats in milk (Poonia *et al.*, 2017). To identify the addition of vegetable oils to milk fat, a quick gas chromatographic (GC) approach was designed for identifying the sterol fraction of vegetable oils, and milk fat, or combinations (Alonso *et al.*, 1997). Other supplementary approaches for identifying milk fat include pyrogram fingerprints generated from pyrolysis, GC using a flame ionisation detector, and headspace GC for determining chemicals contributing to milk fat aroma. Since the unique distribution of the fatty acid moieties on the glycerol backbone is genetically regulated and hence retained, the examination of intact triglycerides for the identification of adulteration is regarded as favourable. Because the triacylglycerol composition of milk fat varies according to season and place, GC classification of milk fat into groups of equal numbers of acyl-carbon atoms (C24–C54) has been shown to be a more effective criterion for distinguishing their origin than fatty acid content. An international standard (IDF, 2002) provides a reference technique for detecting foreign lipids in dairy products, including both animal and vegetable fats such as tallow and lard, employing GC analysis of triglycerides. Vegetable and animal fats are recognized qualitatively and quantitatively in pure milk fat using established triglyceride formulas, regardless of feeding or lactation circumstances (Poonia *et al.*, 2017). Vegetable fat in the form of oils such as palm oil, sunflower oil, soybean oil, and others are the most typical and notable adulterants in ghee. There are numerous approaches, including physicochemical constant-based methods, chromatographic methods, spectroscopic methods, and trace component-based methods for identifying the adulteration of ghee by vegetable oils.

Rapid Methods

Reichert-Meissl Value and Polenske Value

RM value is primarily a measure of butyric (4:0) and caproic (6:0) fatty acids while PV is a measure of caprylic (C8:0) and capric (C10:0) acid in ghee (Gandhi *et al.*, 2014). The RM value of milk fat falls between 17 to 35, which is typically greater than the value of all other fats and oils (generally 1), except for palm kernel and coconut oils, which range from 4 to 8 (Naik, 2021). Several studies have reported the use of RM value for the adulteration of vegetable and animal body fats in ghee (Gandhi *et al.*, 2014; Upadhyey, 2014). Ahmed *et al.* (2020) revealed that utilising RM value assessment, hydrogenated vegetable oils, tallow, coconut oil, and soyabean oil adulterated at 5% or above could be discovered in ghee samples. According to Kumar (2008) the adulteration of buffalo ghee with palm and soybean oils may be identified at a 15% level using PV determination.

Iodine Value

The iodine value is a measure of the amount of unsaturated bonds in a lipid. Kumar (2008) revealed that this constant might be a helpful criterion in identifying more than 10% adulteration of buffalo and cow ghee with various vegetable oils (excluding palm oil). Similarly, Upadhyey (2014) revealed that adulteration of groundnut oil at 5% level could be identified when applied singly in cow ghee, however, detection threshold in buffalo ghee was 10% and higher utilizing iodine value. Also, at a 25% level, adulteration of pure ghee (either buffalo or cow) with coconut oil was identified using iodine value (Salem *et al.*, 2019).

Saponification Value

The saponification value indicates the average molecular weight of the fatty acids present. Kumar (2008) found from his research that, based on saponification value, adulteration of pure cow ghee with either animal body fats or vegetable oils could be recognised at 10% levels of adulteration, except for buffalo body fat, which could be detected at 5% levels of adulteration. In the case of buffalo ghee, only 15% of adulteration with animal body fats (excluding pig body fat) or vegetable oils (except palm oil) could be discovered. Similarly, a minimum of 25% adulteration of pure ghee, either buffalo or cow ghee, with coconut oil was detected (Salem *et al.*, 2019).

Refractive Index and Butyrorefractometer Reading

The degree of bending of light waves flowing through a liquid or transparent substance is determined by this attribute. Milk fat has a refractive index ranging from 1.4538 to 1.4578 and B.R. readings ranged from 40-45. Salem *et al.* (2019) reported that on the basis of RI estimation, adulteration of pure ghee, either buffalo or cow, with coconut oil at a 25% level or more could be detected. Based on BR value measurement palm oil adulteration in buffalo and cow ghee could be identified at least 10% levels (Chaudhari *et al.*, 2021).

Apparent Solidification Time (AST) Test

The apparent solidification time is the time it takes for melted fat to become apparently solidified at a particular temperature. The test can be performed as follows: In a test tube, place 3.0 g of totally melted fat and keep it at 60°C for 5 minutes. Place the test tube in a chilled water bath set at 18 ± 0.2°C and start the stopwatch at the same time. Keep an eye on the test tube until the fat sample seems to solidify, as evidenced by no movement of the fat sample when the test tube is tilted. At this point, pause the timer and note the time it takes for the fat to seem to solidify (Kumar *et al.*, 2009). While the AST values for pig body fat, goat body fat, and vanaspati were 1 minute 30 seconds, 40 seconds, and 1 minute 50 seconds, respectively, the average AST times for buffalo and cow milk fat were 2 minutes 40 seconds and 3 minutes 10 seconds, respectively. On the other hand, all of the examined vegetable oils continued to be liquid forever. The technique could detect the addition of individual vegetable oils at all levels in the case of cow milk fat but not in the case of buffalo milk fat, whereas the adulteration of buffalo milk fat with vanaspati (hydrogenated vegetable fat) was detectable at levels greater than or equal to 10% but not in cow milk fat. This was discovered after taking into account the overall range of AST values at 18°C for both buffalo and cow pure milk fat. Goat body fat was found at amounts above or equal to 10% in both cow and buffalo milk fat, however pig body fat was only found at those same levels in buffalo milk fat (Kumar *et al.*, 2009). Gandhi *et al.* (2018) used this test to a solid fraction of milk fat produced at 15°C (S15) following solvent fractionation and found that a combination of palm olein and sheep body fat (1:14) could be identified at a level of 20% in milk fat.

Complete Liquefaction Time (CLT) Test

The CLT is defined as the time it takes for the solid fat to totally melt at 44°C. To perform the test, place 3.0 g of melted fat in a test tube and keep it at 60°C for 5 minutes. Refrigerate the test tube holding the fat sample (6-8°C) for 45 minutes to allow the melting fat sample to solidify. Place the test tube in a water bath maintained at 44±0.1°C and start the stopwatch at the same time. Keep an eye on the test tube until the fat sample has totally liquefied. Stop the timer at this point and note the time it takes for the fat to completely liquefy. CLT values for pure cow ghee samples varied from 2 minutes 12 seconds to 3 minutes 15 seconds, whereas CLT values for pure buffalo ghee samples ranged from 2 minutes 35 seconds to 3 minutes 15 seconds (Kumar, 2008). At 44°C, the addition of vegetable oils (palm, rice bran, and soybean) resulted in a drop in CLT values in both pure cow and buffalo ghee. The quantity of adulterants used determined whether CLT values decreased or increased. The more the amount of adulterant added, the higher the CLT values (Naik, 2021).

CLT has been utilized to identify adulteration of groundnut oil and goat body fat in ghee. Adulterants added singly may be identified at greater quantities (15 %, w/w, groundnut oil in cow ghee; and 10%, w/w, goat body fat in buffalo ghee), however fractionation decreased the detection limit to the lowest level (10 %, w/w) (Upadhyay, Kumar, *et al.*, 2016). According to Gandhi *et al.* (2018) adulteration of cow ghee with palm olein could be identified at a level of 15% employing CLT. On the other hand, sheep body fat could be easily recognized in cow and buffalo ghee samples independently tainted with sheep body fat at all levels investigated except the 5% level. The blend of palm olein and sheep body fat was discovered at levels of 6+14 (20%) and 9+21 (30%), but not at 3+7 (10) %. The lower level of 3+7 (10) % was also traceable after fractionation.

Crystallization Time Test (CTT)

In a stoppered test tube, place 0.8 mL of melted fat and add 2.5 mL of an acetone-benzene solvent combination (3.5:1.0). Slowly combine the materials. To equilibrate the temperature, place the test tube in a water bath maintained at 20°C for 3 minutes. The tube should then be placed in another water bath that is maintained at 17 + 0.2°C until crystallization begins. Take note of the time it takes for crystallization to occur. It is a good way to find adulteration of ghee with animal body fat (Panda, 1996). Kumar (2008) performed this test and discovered that when ghee samples were adulterated with vegetable oils, the crystallization time increased. This rise was proportional to the degree of adulteration with vegetable oil. The crystallization time of pure buffalo ghee, cotton tract ghee, buffalo ghee adulterated at 10% with pig body fat, cow body fat, buffalo body fat and vegetable oils is 18-20 minutes, 10.5-12.5 minutes, 11.5-12.5 minutes, 4.5-5.5 minutes, 3.0-4.0 minutes, and 26-36 minutes, respectively.

Differential Scanning Colorimetry

Ghee adulteration with foreign fats is a common occurrence in the literature. Vegetable oils and animal fats are thought to be significant adulterants in milk fat. There are few analytical procedures available when animal body fat is used as an adulterant in ghee. As a result, the use of DSC becomes critical since it is a sensitive, quick, and reliable approach for detecting oil and fat foods (Farah *et al.*, 2018). Differential scanning calorimetry was utilized to identify adulteration of pure ghee with caprine body fat when it was added alone (at 5, 10, and 15%) and in combination with groundnut oil (at 5, 10, and 15 %). The transition behaviour of the samples was examined using crystallizing and melting curves. Melting and crystallization are the two most prevalent physical phenomena that characterize the thermal behaviour of lipids in thermal analysis curves. Fats and oils have concave-shaped melting curves due to their polymorphism, which is characterized by their peak maxima (mid-point). In the cooling zone, pure cow ghee, pure buffalo ghee, and caprine body fat all displayed two crystallization peaks; however, groundnut oil did not show any notable crystallisation peak. Thus, adulterated ghee samples revealed a change in the midrange temperature of thermal curves when compared to pure ghee, indicating the inclusion of foreign fats. The findings demonstrated that, regardless of the type of adulterants, identification of adulteration was achievable at the lowest level of 5 % (Upadhyay *et al.*, 2017). Similar results have been reported by Lambelet & Ganguli (1983). When modest quantities of pig or buffalo body fat are added to cow or buffalo ghee, an additional peak appears at high temperature in the melting and crystallization curves, as determined by the differential scanning calorimetry (DSC) method. The crystallization diagrams clearly show ghee adulterations with various animal fats at amounts as low as 5%.

Colorimetric Methods

According to Bector & Sharma, (2002), palm oil added to ghee may be identified at a level of 5% using potassium ferricyanide & ferric chloride reagent based on tannins, which are thought to be naturally present as contaminants in palm oil. To perform this test, 2 mL of melted contaminated ghee was combined with 5 mL of acetonitrile in a test tube. The contents are thoroughly shaken and centrifuged. Acetonitrile is appropriately mixed with I mL of 0.008M ferric chloride and 0.3 mL of 0.003 M potassium ferricyanide reagent solution in the lower layer. The presence of palm oil in milk fat is indicated by the prussian blue colour. Sharma *et al.* (2007) showed that utilizing a quick color-based test, a minimum 7.5 % amount of vegetable oil in ghee may be easily recognized. In a combination of molten fat and hexane (1:1.5), they utilized 1.0 ml of a colour development reagent, namely (H_2O: H_2SO_4: HNO_3; 20:6:14). The presence of vegetable oil was revealed by the formation of a pronounced orange tint in the top layer.

Ramani *et al.* (2018) created a chromogenic technique to identify ghee adulteration using palm oil. They utilised 1.0 mL of pure ghee or adulterated ghee samples (5, 10, 15, and 20% adulteration) and allowed them to react with 2.0 mL of DPPH solution 50 mg/100 mL ethanol. They stated that the colour of pure ghee stayed violet while that of contaminated ghee shifted from violet to yellow. The developed chromogenic test has a detection level of up to 5%. A simple and quick method for detecting palm oil in ghee was established. In this approach, 2.0 mL of pure ghee or contaminated ghee (5, 10, 15, and 20% level of adulteration) samples were combined with 1.0 mL of chromogenic solution, which consisted of 1.0 mL of Ferric Chloride solution (0.008 M) followed by the addition of 0.3 ml Potassium Ferricyanide solution (0.03 M). After mixing the contents for 30 seconds, the colour change was noted. The colour of genuine ghee remained green, whereas the colour of tainted ghee was altered from green to blue. The aforementioned technique also found a 5% level of palm oil adulteration in ghee (Ramani *et al.*, 2019).

Advanced Instrument-based Methods

Infrared Spectroscopy

The infrared (IR) region, which ranges from 0.8 to 1,000 μm (14,000 to 20 cm^{-1} in wave number), is classified into three categories: near-infrared (0.8-2.5 μm), mid-infrared (2.5 to 15.38 μm), and far-infrared (15-1,000 μm). Infrared absorption has been widely employed in the examination of lipids, particularly for the determination of cis- and trans-isomers. Unsaturated fatty acids from natural vegetable oils and fats are isolated (non-conjugated) and in cis- configuration. Natural trans-isomers may be found in modest concentrations in animal and marine tats. In comparison to hydrogenated vegetable oils, which can include up to 50% trans fatty acids due to non-stereospecific hydrogenation, bovine milk fat has a low amount (5%) of trans fatty acids (Akashamrut, 2011). The vibrations of molecules in a specific sample are measured using infrared spectroscopy. Each bond in a molecule has a distinct vibrational frequency, which may be used to identify the functional groups present in a sample. When the effects of the various functional groups are combined, the result is a unique molecular "fingerprint" that may be used to establish the identification of a sample. The mid-infrared (MIR; 4000–400 cm^{-1}) range may be used to analyze fundamental vibrations and related rotational–vibrational structure, whereas the near-infrared (NIR; 14 000–4000 cm^{-1}) region can stimulate overtone or harmonic vibrations, yielding more complicated structural information (Antony *et al.*, 2018). Near-infrared is often utilized for molecular studies, and FT-NIR spectroscopy offers significant promise for the development of quick and easy procedures for determining the authenticity and inherent quality criteria of edible fats.

Konevets *et al.* (1987) used IR spectroscopy to study the cis-trans configurations of individual fats (milk fat, animal body fat, vegetable fat, and hydrogenated fat) and their combinations and found that additions of up to 10% animal, vegetable, and hydrogenated fats to milk fat could be identified. Sato *et al.* (1990) employed a near-

IR spectroscopic approach to identify as little as 3% foreign fat in milk fat. Kumar (2003) stated that using IR spectroscopy, as little as 5% of vanaspati added to ghee may be identified due to elevated levels of trans isomers in ghee. Antony *et al.* (2018) measured the FT-MIR spectra of ghee and vegetable oils using horizontal attenuated total reflectance (HATR) in the 4000-650 cm^{-1} range. They compared the absorbance by carbon-hydrogen (C-H) stretch in the fatty acid chain at 3.48 μm and the absorbance by carbonyl (C-O) stretch in the ester bond at 5.7 μm in ghee and vegetable oils. The fingerprint area revealed strong variations in the spectra of ghee and those of vegetable oils, which may be very successfully used to develop FT-MIR spectroscopy as a potential method for the detection of vegetable oil adulteration in ghee.

Another investigation was conducted to create FT-NIR spectroscopy in conjunction with chemometrics for the detection of adulteration in ghee. A combination FT-IR spectrometer that covers the whole NIR spectral range and for chemometrics analysis, the software programme AssureID was utilized, which employs two approaches for matching spectra: COMPARE and SIMCA. To acquire NIR spectra, a sample accessory was filled with 10 ml of ghee, oils, fats, or contaminated ghee and scanned in transmittance mode with 32 scans and a resolution of 8 cm^{-1} in the region of 10000–4000 cm^{-1}. The NIR spectra of cow ghee and buffalo ghee measured between 10000 and 4500 cm^{-1} revealed three wide bands in the ranges of 9000 to 8000, 7500 to 6150, and 6000 to 5700 cm^{-1}. The developed approach predicted the presence of foreign oil or fat in ghee at a rate of 2.0%. However, when blended at a concentration of 20% or more, all adulterants may be identified and quantified. The success rate below 20% varied based on the foreign oils/fats used, the rate of mixing, and/or the type of ghee used. The approach may be used to identify and quantify foreign oils/fats in ghee efficiently and easily (Aparnathi *et al.*, 2019).

Upadhyay *et al.* (2018) used ATR-FTIR spectroscopy with chemometrics to identify the presence of pig body fat in pure ghee. Pure mixed ghee was spiked with pig body fat at 3, 4, 5, 10, and 15% levels. The spectra of pure and spiked materials were recorded in MIR from 4000 to 500 cm^{-1}. Using principal component analysis at a 5% level of significance on the given wavenumber range, separate clusters of data were generated. The finding revealed that spike of pig body fat in pure ghee may be detected as low as 3%. Similarly, ATR-FTIR spectroscopy in conjunction with chemometrics was used to determine the presence of goat body fat in ghee (at 1, 3, 5, 10, 15, and 20% levels). The spectra of clean and spiked samples were collected in the 4000–500 cm^{-1} wavenumber region. Using principal component analysis at a 5% level of significance, separate clusters of pure ghee and spiked samples were generated in the chosen wavenumber ranges (1786–1680, 1490–919, and 1260–1040 cm^{-1}). SIMCA was used to classify the samples, and pure ghee had a classification performance of 100 %. The study found that spiked samples of goat body fat could be identified in ghee at a 1% level (Upadhyay *et al.*, 2016).

Chromatographic Methods

The International Organization for Standardization (ISO) has defined a reference method (ISO 17678:2010) for detecting cow milk fat purity that is based on the profile of triacylglycerols with even carbon values ranging from C_{24} to C_{54} using gas-liquid chromatography. However, utilising this method of triglyceride analysis, adulteration of vegetable oils and animal depot fats could not be identified below a 2% level (Pathania *et al.*, 2020). Triacylglycerol profiles in milk and nonmilk fat have been determined using GC. Because milk fat triacylglycerol composition varies according to season and place, GC classification of milk fat into groups of equal numbers of acyl-carbon atoms (C_{24}–C_{54}) has been shown to be a more effective criterion for detecting their origin than fatty acid content (Sukhija & Palmquist, 1988). A gas chromatography method to identify RM- adjuster in cow and buffalo ghee using nitrogen as the carrier gas and CP-Ultimetal SimDist CP7532 as the column with a column flow rate of 3.00 mL/min of nitrogen carrier gas was established. The detector used was the flame ionization detector. They found the formation of significant-sized signature peaks in the retention time range of 4.5-6.5 minutes for RM-adjuster and adulterant fat, but not in pure cow and pure buffalo ghee samples. The inclusion of RM-adjusted highly adulterant fat to a minimum level of 0.5 % in both cow and buffalo ghee could be identified using the established approach (Pathania *et al.*, 2020).

High levels of plant sterols (phytosterols), notably β-sitosterol, indicate the presence of vegetable fat in milk. LC-GC has been utilised successfully to identify cotton and rapeseed oil in milk fat using β-sitosterol as a marker. Trans-esterification of the fat, pre-separation of the sterol fraction from other lipid components, and online transfer to the capillary GC system are all part of the process. The online method eliminates the need for time-consuming sample preparation processes prior to GC analysis. It is also feasible to measure cholesterol in milk fat at the same time (Kamm *et al.*, 2002). Povolo *et al.* (1999) used GC to identify the presence of beef tallow in milk fat by determining a mixture of more than one fatty component (diglycerides, 3,5-cholestadiene, and triacylglycerols). When the findings were examined independently, the approaches did not appear to be capable of lowering the LOD below 5%. However, by using multivariate statistical approaches, the power of these studies was enhanced, and the statistical model generated was able to drop the detection limit of beef tallow to 2% (w/v).

Upadhyay (2014) found that utilising Reversed-Phase Thin Layer Chromatography (RP-TLC), ghee adulteration with vegetable oils may be easily identified as low as 1% on the RP-18 and RP-8 plates, whereas the RP-2 plate did not aid in the detection of adulteration. Similarly, to determine the purity of milk fat, RP-TLC methodology was developed. The study used the sterol separation method using C18 stationary phase. Ghee adulteration with coconut oil up to 7.5%, soybean oil, sunflower oil, groundnut oil up to 1%, and designer oil up to 2% may be identified using the standardised RP-TLC methodology (Rani *et al.*, 2015). Milk adulteration with up to 5% coconut oil and up to 1% other vegetable oils such as rice bran oil, sunflower oil, soybean oil, and groundnut oil may be identified utilising a standardised RP-TLC methodology on RP-18 and RP-8 plates (Sukthija, 2018). Paradkar *et al.* (2001) developed an HPTLC technique for detecting groundnut oil and palm oil, which are the primary components of synthetic milk. The Rose–Gottlieb technique was used to remove milk fat, and HPTLC was performed on aluminium plates precoated with silica gel $^{60}F_{254}$. The oils' minimum detectable amount was reported to be 80 mg/100 mL.

Electronic Nose

Electronic noses (e-noses) are devices that simulate the sense of smell. These devices generally employ an array of sensors to precisely identify and discriminate scents in complicated samples at a cheap cost. Research employing flash gas chromatography electronic nose combined with multivariate chemometrics looked at the feasibility of detecting soybean oil adulteration (from 10% to 100%) in pure cow ghee. A flash gas chromatography electronic nose (FGCEN) apparatus was used to acquire volatile chemical fingerprint information. The pattern recognition chemometric model's principal component analysis (PCA), soft independent modelling of class analogy (SIMCA), and discriminant function analysis (DFA) were used to investigate the classification outcomes. Acetaldehyde (*Z*)-4-heptenal, 2-propanol, ethyl propanoate, and pentan-2-one had the strongest fingerprint odours of all the samples detected by FGCEN analysis. The odour analysis inquiry was completed in 90 seconds on average. The PCA and DFA chemometric approaches achieved a clear distinction of all samples with an outstanding classification accuracy of more than 99 %. The SIMCA model, on the other hand, revealed that SIMCA could only be utilised to identify ghee adulteration at greater concentration levels (30 % to 100 %) (Roy *et al.*, 2022).

For the detection of varying quantities of sunflower oil and cow body fat combined with pure cow ghee, an electronic nose system was utilised (10%, 20%, 30%, 40%, and 50%). To accomplish this, the principal components analysis (PCA) and artificial neural networks (ANNs) methodologies were applied. The accuracy of the principal components analysis of sunflower oil and cow body fat was 96 % and 97 % of the data variance, respectively, based on the results. According to the findings, artificial neural networks correctly recognised the adulteration of sunflower oil and cow body fat with 91.3 % and 82.5 % accuracy, respectively (Ayari *et al.*, 2018).

Table 3: Instrument-based methods coupled with chemometrics for detection of adulterants in ghee

Technique of detection	Target adulterants in ghee	Reference
Infrared spectroscopy	Vanaspati	(Kumar, 2003)
FT-MIR spectroscopy coupled with ATR	Vegetable oil	(Antony *et al.*, 2018)
FT- Near-infrared spectroscopy with SIMCA	Animal body fat, vegetable oil	(Aparnathi *et al.*, 2019)
ATR-FTIR coupled with Principal Component Analysis (PCA) and SIMCA	Goat and pig body fat	(Upadhyay *et al.*, 2018; Upadhyay, *et al.*, 2016)
Gas liquid chromatography	Groundnut oil, sunflower oil, soybean oil	(Kumar, 2003)
Gas-liquid chromatography	Goat body fat, groundnut oil	(Upadhyay, 2014)

Technique of detection	Target adulterants in ghee	Reference
Gas chromatography	RM-adjusted highly adulterant fat	(Pathania *et al*., 2020)
Gas chromatography	Beef tallow	(Povolo *et al*., 1999)
Reversed-phase- thin layer chromatography (RP-TLC)	Coconut oil, sunflower oil, rice bran oil, groundnut oil	(Rani *et al*., 2015; Sukthija, 2018)
High pressure- thin layer chromatography	Groundnut oil, palm oil	(Paradkar *et al*., 2001)
Flash gas chromatography electronic nose coupled with PCA and discriminant function analysis	Soybean oil	(Roy *et al*., 2022)
Electronic nose with PCA and artificial neural network	Cow body fat and sunflower oil	(Ayari *et al*., 2018)

Future Prospects

Future advances in analytical procedures are likely to place a greater focus not only on the refining of existing analytical methods but also on the previous phases of sample preparation, which are typically difficult to automate. The future of milk authentication approaches encompasses the domain of mixture analyses employing automated methods, perhaps in conjunction with separation techniques, eliminating laborious sample preparation processes. Similarly, when determining distinct features of a commodity, a blend of diverse analytical approaches may be more effective than depending on a single method. Due to the complexities of some issues, univariate statistics must be replaced by newer approaches such as SIMCA, ANN, PCA, PLS, and so on to help with pattern recognition and classification of actual and counterfeit items. Finally, further efforts are still needed to validate new analytical approaches. This validation of new processes, as well as the creation and evaluation of new reference materials through joint trials, will remain a critical concern. It is also envisaged that the evolution of official milk testing procedures would be based on advanced instrumental techniques in the future. Such technologies include the electronic nose and mobile phone-based software, which are generally inexpensive, rapid, and simple to use. The future appears to be associated with the further development of analytical systems that combine strong analytical instruments with data processing software.

Conclusion

Though the economic benefit is seen as one of the primary causes for milk and milk products adulteration, insufficient supply for the world's growing population has also paved the way for this. Because of a lack of competent monitoring and law enforcement, this problem is especially serious in emerging and underdeveloped nations. Existing standard detection procedures are not always practical or available in these nations, making it difficult to handle the many methods of fraudulent adulteration in milk and its products. This necessitates a collaborative effort from scientific communities and regulatory bodies in the development, implementation, and distribution of improved techniques for detecting adulteration. Furthermore, understanding and access to information can play a critical part in overcoming this issue in these locations. Because unethical producers or suppliers have established a thorough grasp of the fundamental concepts in the detection of adulteration, they have been able to modify their blend in many situations to come near to matching the natural product specifications. As a result, the goal is to create simple and cost-effective procedures for detecting adulteration in milk and milk products that can be employed repeatedly.

References

Abernethy, G., & Higgs, K. (2013). Rapid detection of economic adulterants in fresh milk by liquid chromatography-tandem mass spectrometry. Journal of Chromatography A, 1288, 10–20.

Ahmed, S., Hamid, M. A., & Rahman, M. M. (2020). Assessment of ghee adulterated with oils and fats in Bangladesh. Journal of Advanced Veterinary and Animal Research, 7(4), 678.

Akashamrut, P. M. (2011). Validation of methods for detection of ghee adulteration with animal body fat. M.Tech thesis submitted to ICAR-National Dairy Research Institute (Deemed University), Karnal, India.

Alonso, L., Fontecha, J., Lozada, L., & Juárez, M. (1997). Determination of mixtures in vegetable oils and milk fat by analysis of sterol fraction by gas chromatography. JAOCS, Journal of the American Oil Chemists' Society, 74(2), 131–135.

Antony, B., Mehta, B. M., Sharma, S., Ratnam, K., & Aparnathi, K. D. (2018). Comparative appraisal of ghee and common vegetable oils for spectral characteristics in FT-MIR reflectance spectroscopy. Journal of Food Science and Technology, 55(9), 3632–3639.

Antony, B., Sharma, S., Mehta, B. M., Ratnam, K., & Aparnathi, K. D. (2018). Study of Fourier transform near infrared (FT-NIR) spectra of ghee (anhydrous milk fat). International Journal of Dairy Technology, 71(2), 484–490.

Aparnathi, K. D., Sharma, S., Antony, B., & Mehta, B. M. (2019). Development of method for detection and quantification of foreign oils and fats in ghee (heat clarified milk fat) using FT-NIR spectroscopy coupled with chemometric. Indian Journal of Dairy Science, 72(1), 12–22.

Ayari, F., Mirzaee-Ghaleh, E., Rabbani, H., & Heidarbeigi, K. (2018). Detection of the adulteration in pure cow ghee by electronic nose method (case study: sunflower oil and cow body fat). International Journal of Food Properties, 21(1), 1670-1679.

Azad, T., & Ahmed, S. (2016). Common milk adulteration and their detection techniques. International Journal of Food Contamination, 3(1), 1–9.

Azeez, O. H., Alkass, S. Y., & Persike, D. S. (2019). Long-term saccharin consumption and increased risk of obesity, diabetes, hepatic dysfunction, and renal impairment in rats. Medicina, 55(10), 681.

Barham, G. S., Khaskheli, M., Soomro, A. H., & Nizamani, Z. A. (2014). Extent of extraneous water and detection of various adulterants in market milk at Mirpurkhas, Pakistan. IOSR Journal of Agriculture and Veterinary Science, 7(3), 83-89.

Bector, B. S., & Sharma, V. (2002). Physico chemical characteristics of ghee adulterated with palm oil and its detection. NDRI Annual Report, 44–45.

Bonwick, G. A., & Smith, C. J. (2004). Immunoassays: their history, development and current place in food science and technology. International Journal of Food Science & Technology, 39(8), 817–827.

Borin, A., Ferrão, M. F., Mello, C., Maretto, D. A., & Poppi, R. J. (2006). Least-squares support vector machines and near infrared spectroscopy for quantification of common adulterants in powdered milk. Analytica Chimica Acta, 579(1), 25–32.

Botelho, B. G., Reis, N., Oliveira, L. S., & Sena, M. M. (2015). Development and analytical validation of a screening method for simultaneous detection of five adulterants in raw milk using mid-infrared spectroscopy and PLS-DA. Food chemistry, 181, 31-37.

Cartoni, G., Coccioli, F., Jasionowska, R., & Masci, M. (1999). Determination of cows' milk in goats' milk and cheese by capillary electrophoresis of the whey protein fractions. Journal of Chromatography A, 846(1–2), 135–141.

Cassoli, L. D., Sartori, B., & Machado, P. F. (2011). The use of the Fourier Transform Infrared spectroscopy to determine adulterants in raw milk. Revista Brasileira de Zootecnia, 40, 2591-2596.

Castro, C., Martín, R., García, T., Rodrguez, E., Gonzalez, I., Sanz, B., & Hernandez, P. E. (1992). Indirect enzyme-linked immunosorbent assay for detection of cow's milk in goat's milk. Food and Agricultural Immunology, 4(1), 11–18.

Chaudhari, A., Sindhav, R., Sudheendra, C. H. V. K., Solanki, A., Hazra, T., & Ramani, V. (2021). Efficacy of selected fat constant analysis to ascertain the purity of ghee during storage. International Journal of Current Microbiology and Applied Sciences, 10(03), 1014–1019.

Chauhan, S. L., Priyanka, K. D. M., Paul, B. R., & Maji, C. (2019). Adulteration of milk: A Review. Int. J. Chem. Stud, 7, 2055-2057.

Chávez, N. A., Jauregui, J., Palomares, L. A., Macías, K. E., Jiménez, M., & Salinas, E. (2012). A highly sensitive sandwich ELISA for the determination of glycomacropeptide to detect liquid whey in raw milk. Dairy Science and Technology, 92(2), 121–132.

Chávez, N. A., Salinas, E., Jauregui, J., Palomares, L. A., & Macías, K. (2008). Detection of bovine milk adulterated with cheese whey by Western blot immunoassay. Food and Agricultural Immunology, 19(4), 265–272.

Chen, H., Tan, C., Lin, Z., & Wu, T. (2017). Detection of melamine adulteration in milk by near-infrared spectroscopy and one-class partial least squares. Spectrochimica Acta - Part A: Molecular and Biomolecular Spectroscopy, 173, 832–836.

Chen, R. K., Chang, L. W., Chung, Y. Y., Lee, M. H., & Ling, Y. C. (2004). Quantification of cow milk adulteration in goat milk using high-performance liquid chromatography with electrospray ionization mass spectrometry. Rapid Communications in Mass Spectrometry, 18(10), 1167–1171.

Coitinho, T. B., Cassoli, L. D., Cerqueira, P. H. R., da Silva, H. K., Coitinho, J. B., & Machado, P. F. (2017). Adulteration identification in raw milk using Fourier transform infrared spectroscopy. Journal of Food Science and Technology, 54(8), 2394–2402.

Cozzolino, R., Passalacqua, S., Salemi, S., Malvagna, P., Spina, E., & Garozzo, D. (2001). Identification of adulteration in milk by matrix-assisted laser desorption/ionization time-of-flight mass spectrometry. Journal of Mass Spectrometry, 36(9), 1031–1037.

Dabrowska, A., Wałecka, E., Bania, J., Zelazko, M., Szołtysik, M., & Chrzanowska, J. (2010). Quality of UHT goat's milk in Poland evaluated by real-time PCR. Small Ruminant Research, 94(1–3), 32–37.

Dai, X., Fang, X., Su, F., Yang, M., Li, H., Zhou, J., & Xu, R. (2010). Accurate analysis of urea in milk and milk powder by isotope dilution gas chromatography-mass spectrometry. Journal of Chromatography B: Analytical Technologies in the Biomedical and Life Sciences, 878(19), 1634–1638.

Darwish, S., Allam, H., Amin, A., & 2009, undefined. (2009). Evaluation of PCR assay for detection of cow's milk in water buffalo's milk. World Applied Sciences Journal, 7(4), 461–467.

Das, S., Goswami, B., & Biswas, K. (2016). Milk Adulteration and Detection: A Review. Sensor Letters, 14(1), 4–18.

de Carvalho, B. M. A., de Carvalho, L. M., dos Reis Coimbra, J. S., Minim, L. A., de Souza Barcellos, E., da Silva Júnior, W. F., Detmann, E., & de Carvalho, G. G. P. (2015). Rapid detection of whey in milk powder samples by spectrophotometric and multivariate calibration. Food Chemistry, 174, 1–7.

De, S., Brahma, B., Polley, S., Mukherjee, A., Banerjee, D., Gohaina, M., Singh, K. P., Singh, R., Datta, T. K., & Goswami, S. L. (2011). Simplex and duplex PCR assays for species specific identification of cattle and buffalo milk and cheese. Food Control, 22(5), 690–696.

Domingo, E. D. C., Tireli, A. A., Nunes, C. A., Batista, A. V., Guerreiro, M. C., & Pinto, S. M. (2015). Rapid extraction of melamine in powdered milk for direct electrospray ionization tandem mass spectrometry analysis. Talanta, 132, 535–540.

Ellis, D. I., Brewster, V. L., Dunn, W. B., Allwood, J. W., Golovanov, A. P., & Goodacre, R. (2012). Fingerprinting food: current technologies for the detection of food adulteration and contamination. Chemical Society Reviews, 41(17), 5706–5727.

Espeja, E., García, M. C., & Marina, M. L. (2001). Fast detection of added soybean proteins in cow's, goat's, and ewe's milk by perfusion reversed-phase high-performance liquid chromatography. Journal of Separation Science, 24(10–11), 856–864.

Farah, J. S., Cavalcanti, R. N., Guimarães, J. T., Balthazar, C. F., Coimbra, P. T., Pimentel, T. C., Esmerino, E. A., Duarte, M. C. K. H., Freitas, M. Q., Granato, D., Neto, R. P. C., Tavares, M. I. B., Calado, V., Silva, M. C., & Cruz, A. G. (2021). Differential scanning calorimetry coupled with machine learning technique: An effective approach to determine the milk authenticity. Food Control, 121, 107585.

Farah, J. S., Silva, M. C., Cruz, A. G., & Calado, V. (2018). Differential calorimetry scanning: current background and application in authenticity of dairy products. Current Opinion in Food Science, 22, 88–94.

Feligini, M., Bonizzi, I., Cubric Curik, V., Parma, P., Greppi, G. F., & Enne, G. (2005). Detection of Adulteration in Italian Mozzarella Cheese Using Mitochondrial DNA Templates as Biomarkers. Food Technol. Biotechnol, 43(1), 91–95.

Ferreira, I. M. P. L. V., & Caçote, H. (2003). Detection and quantification of bovine, ovine and caprine milk percentages in protected denomination of origin cheeses by reversed-phase high-performance liquid chromatography of beta-lactoglobulins. Journal of Chromatography A, 1015(1–2), 111–118.

Ferreira, I. M. P. L. V. O., & Oliveira, M. B. P. P. (2003). Determination of caseinomacropeptide by an RP-HPLC method and monitoring of the addition of rennet whey to powdered milk. Journal of Liquid Chromatography and Related Technologies, 26(1), 99–107.

Francis, A., Dhiman, T., & Sumana Mounya, K. (2020). Adulteration of Milk: A Review. Journal of Science and Technology, 5(6), 37–41.

Galan-Malo, P., Mendiara, I., Razquin, P., & Mata, L. (2018). Validation of a rapid lateral flow method for the detection of cows' milk in water buffalo, sheep or goat milk. Food Additives and Contaminants - Part A Chemistry, Analysis, Control, Exposure and Risk Assessment, 35(4), 599–604.

Gandhi, K., Kumar, A., & Lal, D. (2018). Solvent fractionation technique paired with apparent solidification time (AST) test as a method to detect palm olein and sheep body fat in ghee (clarified milk fat). Indian J Dairy Sci , 71(3), 246–251.

Gandhi, K., Sharma, R., Gautam, P. B., & Mann, B. (2020). Quality Assessment of Raw Milk. Chemical Quality Assurance of Milk and Milk Products, 33–67.

Gandhi, K., Upadhyay, N., Aghav, A. D., Sharma, V., & Lal, D. (2014). Detection of adulteration of ghee (clarified milk fat) with palmolein and sheep body fat using Reichert-Meissl (RM) value coupled with solvent fractionation technique. Indian J. Dairy Sci., 67(5), 387–393.

Gautam, P. B. (2016). Development of strip-based tests for detection of selected adulterants and indigenous enzymes in milk and milk products. M.Tech thesis submitted to ICAR-National Dairy Research Institute (Deemed University), Karnal, India.

Gautam, P. B., Sharma, R., Lata, K., Rajput, Y. S., & Mann, B. (2017). Construction of a lateral flow strip for detection of soymilk in milk. Journal of Food Science and Technology, 54(13), 4213–4219.

Gentili, A., Caretti, F., Bellante, S., Mainero Rocca, L., Curini, R., & Venditti, A. (2012). Development and validation of two multiresidue liquid chromatography tandem mass spectrometry methods based on a versatile extraction procedure for isolating non-steroidal anti-inflammatory drugs from bovine milk and muscle tissue. Analytical and Bioanalytical Chemistry, 404(5), 1375–1388.

Gondim, C. de S., Junqueira, R. G., Souza, S. V. C. de, Ruisánchez, I., & Callao, M. P. (2017). Detection of several common adulterants in raw milk by MID-infrared spectroscopy and one-class and multi-class multivariate strategies. Food Chemistry, 230, 68–75.

Goswami, T. K., & Gupta, S. K. (2008). Detection of dilution of milk with the help of glass transition temperature by differential scanning calorimetry (DSC). African Journal oF Food Science, 2(1), 7–10.

Haasnoot, W., & du Pré, J. G. (2007). Luminex-based triplex immunoassay for the simultaneous detection of soy, pea, and soluble wheat proteins in milk powder. Journal of Agricultural and Food Chemistry, 55(10), 3771–3777.

Hamilton, R. J., & Sewell, P. A. (1982). Introduction to high performance liquid chromatography. Introduction to High Performance Liquid Chromatography, 1–12.

He, B., Liu, R., Yang, R., & Xu, K. (2010). Adulteration detection in milk using infrared spectroscopy combined with two-dimensional correlation analysis. Proc. SPIE 7572, Optical Diagnostics and Sensing X: Toward Point-of-Care Diagnostics, 75720P (26 February 2010).

He, D., Zhang, X., Gao, B., Wang, L., Zhao, Q., Chen, H., Wang, H., & Zhao, C. (2014). Preparation of magnetic molecularly imprinted polymer for the extraction of melamine from milk followed by liquid chromatography-tandem mass spectrometry. Food Control, 36(1), 36–41.

Hrbek, V., Vaclavik, L., Elich, O., & Hajslova, J. (2014). Authentication of milk and milk-based foods by direct analysis in real time ionization–high resolution mass spectrometry (DART–HRMS) technique: A critical assessment. Food Control, 36(1), 138–145.

Hurley, I. P., Coleman, R. C., Ireland, H. E., & Williams, J. H. H. (2004). Measurement of Bovine IgG by Indirect Competitive ELISA as a Means of Detecting Milk Adulteration. Journal of Dairy Science, 87(3), 543–549.

Hurley, I. P., Coleman, R. C., Ireland, H. E., & Williams, J. H. H. (2006). Use of sandwich IgG ELISA for the detection and quantification of adulteration of milk and soft cheese. International Dairy Journal, 16(7), 805–812.

IDF (International Dairy Federation). (2002). Milk: Determination of Freezing Point. Thermistor Cryoscope Method (Reference Method). FIL-IDF Standard no. 108:2002/ISO 5764. Brussels: IDF.

Jablonski, J. E., Moore, J. C., & Harnly, J. M. (2014). Nontargeted detection of adulteration of skim milk powder with foreign proteins using UHPLC-UV. Journal of Agricultural and Food Chemistry, 62(22), 5198–5206.

Jawaid, S., Talpur, F. N., Sherazi, S. T. H., Nizamani, S. M., & Khaskheli, A. A. (2013). Rapid detection of melamine adulteration in dairy milk by SB-ATR-Fourier transform infrared spectroscopy. Food Chemistry, 141(3), 3066–3071.

Jha, S. N., Jaiswal, P., Borah, A., Gautam, A. K., & Srivastava, N. (2015). Detection and quantification of urea in milk using attenuated total reflectance-fourier transform infrared spectroscopy. Food and Bioprocess Technology, 8(4), 926–933.

Jha, S. N., Jaiswal, P., Grewal, M. K., Gupta, M., & Bhardwaj, R. (2016). Detection of Adulterants and Contaminants in Liquid Foods—A Review. Critical Reviews in Food Science and Nutrition, 56(10), 1662–1684.

Jha, S. N., & Matsuoka, T. (2004). Detection of adulterants in milk using near infrared spectroscopy. Article in Journal of Food Science and Technology-Mysore, 41(3), 313–316.

Kamm, W., Dionisi, F., Hischenhuber, C., Schmarr, H. G., & Engel, K. H. (2002). Rapid detection of vegetable oils in milk fat by on-line LC-GC analysis of β-sitosterol as marker. European Journal of Lipid Science and Technology, 104(11), 756–761.

Kamthania, M., Saxena, J., axena, K., & Sharma, D. K. (2014). Milk Adultration: methods of detection & remedial measures. International Journal of Engineering and Technical Research, 1, 15–20.

Karima, M. F. A., Ragaa, S. H. M. S. D., & Abeer, A. A. (2015). Detection of some chemical hazards in milk and some dairy products. African Journal of Food Science, 9(4), 187–193.

Kasemsumran, S., Thanapase, W., & Kiatsoonthon, A. (2007). Feasibility of Near-Infrared Spectroscopy to Detect and to Quantify Adulterants in Cow Milk. Analytical Sciences, 23, 907–910.

Konevets, V. I., Roganoua, Z. A., & Smolyanski, A. L. (1987). Use of infra-red spectroscopy for analysing fats of different origin. Izvestiya Vyrshikh O Chebnykh Zavadenii Pischevaya Tekhnologiya, 1, 64–67.

Kotowicz, M., Adamczyk, E., & Bania, J. (2007). Application of a duplex-PCR for detection of cows' milk in goats' milk. Annals of Agricultural and Environmental Medicine, 14(2), 215–218.

Kritikou, A. S., Aalizadeh, R., Damalas, D. E., Barla, I. v., Baessmann, C., & Thomaidis, N. S. (2022). MALDI-TOF-MS integrated workflow for food authenticity investigations: An untargeted protein-based approach for rapid detection of PDO feta cheese adulteration. Food Chemistry, 370, 131057.

Kroger, M. (1985). Milk Sample Preservation. Journal of Dairy Science, 68(3), 783–787.

Kruså, M., Torre, M., & Marina, L. M. (2000). A reversed-phase high-performance liquid chromatographic method for the determination of soya bean proteins in bovine milks. Analytical Chemistry, 72(8), 1814–1818.

Kumar, A. (2003). Comparative studies on physico-chemical properties of various oils and fats with a view to detect adulteration in ghee. M.Tech thesis submitted to ICAR-National Dairy Research Institute (Deemed University), Karnal, India.

Kumar, A. (2008). Detection of adulterants in ghee. M.Tech thesis submitted to ICAR-National Dairy Research Institute (Deemed University), Karnal, India.

Kumar, A., Ghai, D. L., Seth, R., & Sharma, V. (2009). Apparent solidification time test for detection of foreign oils and fats adulterated in clarified milk fat, as affected by season and storage. International Journal of Dairy Technology, 62(1), 33–38.

Kumar, A., Goyal, S. K., Pradhan, R. C., & Goyal, R. K. (2015). A study on status of milk adulterants using in Milk of District Varanasi. South Asian J. Food Technol. Environ, 1(2), 140–143.

Kumar, H., Kumar, A., Kumari, P., Jyotirmai, S., & Tulsani, N. B. (2000). A rapid estimation of urea in adulterated milk using dry reagent strip. Indian Journal of Chemical Technology, 7, 146–147.

Lambelet, P., & Ganguli, N. C. (1983). Detection of pig and buffalo body fat in cow and buffalo ghees by differential scanning calorimetry. Journal of the American Oil Chemists' Society, 60(5), 1005–1008.

Liao, J., Liu, Y. F., Ku, T., Liu, M. H., & Huang, Y. (2017). Qualitative and quantitative identification of adulteration of milk powder using DNA extracted with a novel method. Journal of Dairy Science, 100(3), 1657–1663.

Liu, B., Si, J., Zhao, F., Wang, Q., Wang, Y., Li, J., Li, C., & Li, T. (2019). Rapid detection of cow milk adulteration/contamination in goat milk by a lateral flow colloidal gold immunoassay strip. Journal of Dairy Research, 86(1), 94–97.

López-Calleja, I., González, I., Fajardo, V., Martín, I., Hernández, P. E., García, T., & Martín, R. (2005). Application of Polymerase Chain Reaction to Detect Adulteration of Sheep's Milk with Goats' Milk. Journal of Dairy Science, 88(9), 3115–3120.

Maheshwara Reddy, D., Venkatesh, K., & Venkata Sesha Reddy, C. (2017). Adulteration of milk and its detection: A review. International Journal of Chemical Studies, 5(4), 613–617.

Mahmood, T., & Yang, P. C. (2012). Western Blot: Technique, Theory, and Trouble Shooting. North American Journal of Medical Sciences, 4(9), 429.

Mal, A. K., Chakraborty, S. M., Das, M. K., & Bandyopadhyay, T. S. (1988). Rapid detection of urea and glucose in milk. Indian Journal of Dairy Science, 41(3), 373–374.

Marchal, L. M., Beeftink, H. H., & Tramper, J. (1999). Towards a rational design of commercial maltodextrins. Trends in Food Science & Technology, 10(11), 345–355.

Martín-Hernández, C., Muñoz, M., Daury, C., Weymuth, H., Kemmers-Voncken, A. E. M., Corbatón, V., Toribio, T., & Bremer, M. G. E. G. (2009). Immunochromatographic lateral-flow test strip for the rapid detection of added bovine rennet whey in milk and milk powder. International Dairy Journal, 19(4), 205–208.

Maurya, S., Kumar, K., Ahmad, S., Kumar, A., Khan, S., Gupta, P., & Kumar, P. (2017). Investigation of adulterants in milk and its products from Lucknow City. Journal of Biological Sciences and Medicine, 3(2), 14-18.

Mazivila, S. J., Páscoa, R. N. M. J., Castro, R. C., Ribeiro, D. S. M., & Santos, J. L. M. (2020). Detection of melamine and sucrose as adulterants in milk powder using near-infrared spectroscopy with DD-SIMCA as one-class classifier and MCR-ALS as a means to provide pure profiles of milk and of both adulterants with forensic evidence: A short communication. Talanta, 216, 120937.

Melo Vieira, S., Maria de Souza, L., Silva França, A., Soares Oliveira, L., & Borges Neto, W. (2016). FTMIR-PLS as a promising method for rapid detection of adulteration by waste whey in raw milk. Springer, 96, 123–131.

Miao, H., Fan, S., Wu, Y. N., Zhang, L., Zhou, P. P., Li, J. G., Chen, H. J., & Zhao, Y. F. (2009). Simultaneous determination of melamine, ammelide, ammeline, and cyanuric acid in milk and milk products by gas chromatography-tandem mass spectrometry. Biomedical and Environmental Sciences, 22(2), 87–94.

Mollé, D., & Léonil, J. (2005). Quantitative determination of bovine κ-casein macropeptide in dairy products by Liquid chromatography/Electrospray coupled to mass spectrometry (LC-ESI/MS) and Liquid chromatography/Electrospray coupled to tamdem mass spectrometry (LC-ESI/MS/MS). International Dairy Journal, 15(5), 419–428.

Moore, J. C., Spink, J., & Lipp, M. (2012). Development and Application of a Database of Food Ingredient Fraud and Economically Motivated Adulteration from 1980 to 2010. Journal of Food Science, 77(4), R118–R126.

Moosavy, M. H., Kordasht, H. K., Khatibi, S. A., & Sohrabi, H. (2019). Assessment of the chemical adulteration and hygienic quality of raw cow milk in the northwest of Iran. Quality Assurance and Safety of Crops and Foods, 11(5), 491–498.

Naik, Y. K. (2021). Development of color sensor based rapid method for detection of vegetable oils adulteration in ghee. M.Tech thesis submitted to ICAR-National Dairy Research Institute (Deemed University), Karnal, India.

Narang, G., & Khokhar, R. S. (2004). A process of testing urea in milk and an indicator solution thereof (Patent No. 822/DEL/2004). Indian patent office.

Nascimento, C. F., Brasil, M. A., Costa, S. P., Pinto, P. C., Saraiva, M. L. M., & Rocha, F. R. (2015). Exploitation of pulsed flows for on-line dispersive liquid–liquid microextraction: spectrophotometric determination of formaldehyde in milk. Talanta, 144, 1189-1194.

Nicolaou, N., Xu, Y., & Goodacre, R. (2010). Fourier transform infrared spectroscopy and multivariate analysis for the detection and quantification of different milk species. Journal of Dairy Science, 93(12), 5651–5660.

Oancea, S. (2009). Identification of glycomacropeptide as indicator of milk and dairy drinks adulteration with whey by immunochromatographic assay. Romanian Biotechnological Letters, 14(1), 4146–4151.

Pal, P. K., Bhattacharya, A. M., Das, M. K., & Bandyopadhyay, T. S. (1989). Rapid detection of added sugars in milk. Indian Journal of Dairy Science, 42(2), 405.

Panchal, B. R. (2013). Development of strip-based methods for detection of common adulterants in milk. M.Tech thesis submitted to ICAR-National Dairy Research Institute (Deemed University), Karnal, India.

Panda, D. K. (1996). Detection of adulteration of foreign fats in milk fat. M.Tech thesis submitted to ICAR-National Dairy Research Institute (Deemed University), Karnal, India.

Paradkar, M. M., Singhal, R. S., & Kulkarni, P. R. (2001). An approach to the detection of synthetic milk in dairy milk: 3. Detection of vegetable oil and sodium bicarbonate. International Journal of Dairy Technology, 54(1), 34–35.

Pathania, P., Sharma, V., Arora, S., & Rao, P. S. (2020). A novel approach to detect highly manipulated fat adulterant as Reichert–Meissl value-adjuster in ghee (clarified butter) through signature peaks by gas chromatography of triglycerides. Journal of Food Science and Technology, 57(1), 191–199.

Pesic, M., Barac, M., Vrvic, M., Ristic, N., Macej, O., & Stanojevic, S. (2011). Qualitative and quantitative analysis of bovine milk adulteration in caprine and ovine milks using native-PAGE. Food Chemistry, 125(4), 1443–1449.

Poonia, A., Jha, A., Sharma, R., Singh, H. B., Rai, A. K., & Sharma, N. (2017). Detection of adulteration in milk: A review. International Journal of Dairy Technology, 70(1), 23–42.

Povolo, M., Bonfitto, E., Contarini, G., Toppino, P. M., & Daghetta, A. (1999). Study on the Performance of Three Different Capillary Gas Chromatographic Analyses in the Evaluation of Milk Fat Purity. Journal of High Resolution Chromatography, 22(2), 97–102.

Rajput, Y. S., Singh, G., & Sharma, R. (2015). A strip based test for detection of detergent in milk (Patent No. 750/DEL/2015). Indian Patent Office.

Ramani, A., Ch, S. V. K., Hazra, T., Sudheendra, C. V. K., Hariyani, A. S., Prasad, S., & Ramani, V. M. (2018). Comparative appraisal of ghee and palm oil adulterated ghee on the basis of chromogenic test. Researchgate.Net, 7(12), 623–627.

Ramani, A., Hazra, T., Parmar, M. P., Sindhav, R. G., & Ramanai, V. M. (2019). A simple rapid technique for detection of palm oil in ghee. Indian Journal of Dairy Science, 72(4), 441–444.

Rani, A., Sharma, V., Arora, S., Lal, D., & Kumar, A. (2015). A rapid reversed-phase thin layer chromatographic protocol for detection of adulteration in ghee (clarified milk fat) with vegetable oils. Journal of Food Science and Technology, 52(4), 2434–2439.

Rodríguez, E., Martín, R., García, T., González, I., Morales, P., Sanz, B., & Hernández, P. E. (1993). Detection of cows' milk in ewes' milk and cheese by a sandwich enzyme-linked immunosorbent assay (ELISA). Journal of the Science of Food and Agriculture, 61(2), 175-180.

Roy, M., Doddappa, M., Yadav, B. K., Jaganmohan, R., Sinija, V. R., Manickam, L., & Sarvanan, S. (2022). Detection of soybean oil adulteration in cow ghee (clarified milk fat): An ultrafast study using flash gas chromatography electronic nose coupled with multivariate chemometrics. Journal of the Science of Food and Agriculture, 102(10), 4097-4108.

Salem, E. R., Awad, R. A., & el Batawy, O. I. (2019). Detection of milk fat adulteration with coconut oil depending on some physical and chemical properties. International Journal of Dairy Science, 14(1), 36–44.

Salman, M., Hamed, E. S. S., Al-Amoudi, M. S., Salman, L., Alghamdi, M. T., & Bazaid, S. A. (2012). Identification and determination of melamine in milk by high performance liquid chromatography–UV-Detector. Der Pharma Chemica, 4(2), 737-748.

Santos, P. M., Pereira-Filho, E. R., & Rodriguez-Saona, L. E. (2013a). Application of Hand-Held and Portable Infrared Spectrometers in Bovine Milk Analysis. Journal of Agricultural and Food Chemistry, 61(6), 1205–1211.

Santos, P. M., Pereira-Filho, E. R., & Rodriguez-Saona, L. E. (2013b). Rapid detection and quantification of milk adulteration using infrared microspectroscopy and chemometrics analysis. Food Chemistry, 138(1), 19–24.

Sato, T., Kawano, S., & Iwamoto, M. (1990). Detection of foreign fat adulteration of milk fat by near infrared spectroscopic method. Journal of Dairy Science, 73, 3408–3413.

Scholl, P. F., Farris, S. M., & Mossoba, M. M. (2014). Rapid turbidimetric detection of milk powder adulteration with plant proteins. Journal of Agricultural and Food Chemistry, 62(7), 1498–1505.

Sharma, R., Rajput, Y. S., Barui, A. K., & Naik, Lakshmana. N. (2012). Detection of Adulterants in Milk-A Laboratory Manual (Revised Edition). www.ndri.res.in

Sharma, R., Verma, A., Shinde, N., Mann, B., Gandhi, K., Wichers, J. H., & van Amerongen, A. (2021). Adulteration of cow's milk with buffalo's milk detected by an on-site carbon nanoparticles-based lateral flow immunoassay. Food Chemistry, 351, 129311.

Sharma, V., Lal, D., & Sharma, R. (2007). Colour based platform test for the detection of vegetable oils/fats in ghee. http://krishi.icar.gov.in/jspui/handle/123456789/60716

Singh, P., & Gandhi, N. (2015). Milk preservatives and adulterants: processing, regulatory and safety issues. Food Reviews International, 31(3), 236-261.

Singuluri, H., & Sukumaran, M. (2015). Milk adulteration in Hyderabad, India-a comparative study on the levels of different adulterants present in milk. Indian Journal of Dairy Science, 68(2), 190–192.

Song, H., Xue, H., & Han, Y. (2011). Detection of cow's milk in Shaanxi goat's milk with an ELISA assay. Food Control, 22(6), 883–887.

Song, W., Jiang, N., Wang, H., & Guo, G. (2020). Evaluation of machine learning methods for organic apple authentication based on diffraction grating and image processing. Journal of Food Composition and Analysis, 88, 103437.

Spink, J., & Moyer, D. C. (2011). Defining the Public Health Threat of Food Fraud. Journal of Food Science, 76(9), 157–163.

Špoljarić, J., Mikulec, N., Plavljanić, D., Radeljević, B., Havranek, J., & Antunac, N. (2013). Proving the adulteration of ewe and goat cheeses with cow milk using the reference method of isoelectric focusing of γ-casein. Mljekarstvo, 63, 115–121.

Sukhija, P. S., & Palmquist, D. L. (1988). Rapid method for determination of total fatty acid content and composition of feedstuffs and feces. Journal of Agricultural and Food Chemistry, 36(6), 1202–1206.

Sukthija, M. P. (2018). To study the efficiency of reversed-phase thin layer chromatographic method to detect vegetable oil/fat in milk. M.Tech thesis submitted to ICAR-National Dairy Research Institute (Deemed University), Karnal, India.

Sun, H., Wang, L., Ai, L., Liang, S., & Wu, H. (2010). A sensitive and validated method for determination of melamine residue in liquid milk by reversed phase high-performance liquid chromatography with solid-phase extraction. Food Control, 21(5), 686–691.

Tamime, A. Y., Barclay, M. N., Law, A. J., Leaver, J., Anifantakis, E. M., & O'connord, T. P. (1999). Kishk-a dried fermented milk/cereal mixture. 2 Assessment of a variety of protein analytical techniques for determining adulteration and proteolysis. Le Lait, 79(3), 331–339.

Upadhyay, N. (2014). Detection of vegetable oil and animal body fat adulteration in ghee using solvent fractionation technique. M.Tech thesis submitted to ICAR-National Dairy Research Institute (Deemed University), Karnal, India.

Upadhyay, N., Goyal, A., Kumar, A., & Lal, D. (2017). Detection of adulteration by caprine body fat and mixtures of caprine body fat and groundnut oil in bovine and buffalo ghee using differential scanning calorimetry. International Journal of Dairy Technology, 70(2), 297–303.

Upadhyay, N., Jaiswal, P., & Jha, S. N. (2016). Detection of goat body fat adulteration in pure ghee using ATR-FTIR spectroscopy coupled with chemometric strategy. Journal of Food Science and Technology, 53(10), 3752–3760.

Upadhyay, N., Jaiswal, P., & Jha, S. N. (2018). Application of attenuated total reflectance Fourier Transform Infrared spectroscopy (ATR–FTIR) in MIR range coupled with chemometrics for detection of pig body fat in pure ghee (heat clarified milk fat). Journal of Molecular Structure, 1153, 275–281.

Upadhyay, N., Kumar, A., Goyal, A., & Lal, D. (2017). Complete liquification time test coupled with solvent fractionation technique to detect adulteration of foreign fats in ghee. International Journal of Dairy Technology, 70(1), 110-118.

Wang, Q., Haughey, S. A., Sun, Y. M., Fremin, S. A., Li, Z. F., Liu, H., Xu, Z. L., Shen, Y. D., & Lei, H. T. (2011). Development of a fluorescence polarization immunoassay for the detection of melamine in milk and milk powder. Analytical and Bioanalytical Chemistry, 399(6), 2275–2284.

Watt, B. E., Proudfoot, A. T., & Vale, J. A. (2012). Hydrogen peroxide poisoning. Toxicological Reviews 2004 23:1, 23(1), 51–57.

Wei, R., Wang, R., Zeng, Q., Chen, M., & Liu, T. (2009). High-performance liquid chromatographic method for the determination of cyromazine and melamine residues in milk and pork. Journal of Chromatographic Science, 47, 581–584.

Xie, W. Q., Yu, K. X., & Gong, Y. X. (2019). Rapid and quantitative determination of urea in milk by reaction headspace gas chromatography. Microchemical Journal, 147, 838–841.

Yin, W., Liu, J., Zhang, T., Li, W., Liu, W., Meng, M., He, F., Wan, Y., Feng, C., Wang, S., Lu, X., & Xi, R. (2010). Preparation of monoclonal antibody for melamine and development of an indirect competitive ELISA for melamine detection in raw milk, milk powder, and animal feeds. Journal of Agricultural and Food Chemistry, 58(14), 8152–8157.

Zhang, L. G., Zhang, X., Ni, L. J., Xue, Z. bin, Gu, X., & Huang, S. X. (2014). Rapid identification of adulterated cow milk by non-linear pattern recognition methods based on near-infrared spectroscopy. Food Chemistry, 145, 342–348.

16

Adulteration Detection in Milk Products: An Overview

Richa Singh, Rajan Sharma, Heena Sharma

ICAR-National Dairy Research Institute, Karnal, Haryana, India

1. Introduction

Food is an essential requirement for survival. We work hard to earn money so that we can feed ourselves. However, at the end of the day, many of us are unsure about the quality of food we consume. The allure of money, along with a general disregard for humanity, has resulted in the addition of adulterants to food ranging from simple stones in rice to the more harmful chemicals in food of daily use. Surprisingly, food fraud is not a modern issue. There are several references to fraudulent food behaviour as early as the Romans, but it is usually considered to be the Victorian who places food fraud on the map. Food adulteration is defined as the process of adding ingredients that are harmful to one's health, or removing substances that are beneficial to one's health. Economic adulteration requires at least 3 % (by weight) substitution to offer a decent financial return for the adulterators. Lesser substitution than this may not fetch a profitable price to them. Larger percentage of adulteration not only represents a higher rate of return on the adulterator's investment, but it also means a greater danger of detection. Generally, a common pattern is used for the deception which begin with a low proportion of substitution and then, as word of the profit opportunity spreads, other adulterators will adulterate at an increasingly higher proportion until greed raises the amount of adulteration to the point where the deception becomes obvious to the consumer through a different appearance. Adulteration of food not only cheats the consumer but also pose a serious risk to their health. Humans are highly sensitive to any type of adulterants and it may result in acute side effects including vomiting, diarrhea, dysentery etc. Food poisoning, abdominal pain, and other food diseases, generally accompanied by fever and chills, can be caused by improperly pasteurized/ heat-treated milk. Besides these, consumption of adulterated milk and milk products can have long term serious health effects. Adulterants such as urea or excessive amount of preservative have been correlated with colon, peptic ulcers, cirrhosis liver, heart diseases, various blood disorders, bone marrow abnormality and damage to kidneys.

2. Status of Adulteration of Milk Products

Milk and milk products are valuable source of proteins, fatty acids (FA), minerals (e.g., calcium, phosphorus, magnesium), vitamins (B2, B12, A, B6, and D), and bioactive peptides (Haug *et al.*, 2007) and nutritionally required for all age groups including children, pregnant women and elderly people who are considered to be very vital due to high nutritional demand. The authenticity and the detection of adulterations in dairy products are currently a primary priority to ensure a system for traceability of dairy products from dairy plant to consumer table. Presence of one or more unwanted ingredients in the milk and milk products, which is also known as adulteration, is categorized under two broad categories; unintentional adulteration and intentional adulteration. Unintentional adulteration refers to the presence of inferior substances due to carelessness or lack of adequate facilities and hygiene. For instance, entry of dirt, foreign substance in the packaging material. While, intentional adulteration of milk and milk products is a serious offence in which inferior substances (physical or biological in nature) having almost similar properties to the ingredients, are added. For instance, addition of water to the milk or removal of milk solids. As per one survey, about 68% of milk sold in the market do not meet the minimum FSSAI standards (FSSAI, 2012). In more recent survey done by FSSAI, results do indicate that out of 6432 liquid milk samples analysed throughout the country, 7.08% samples were non-compliant having food safety issues while 41.1% samples failed due to quality related non-compliant issues (FSSAI, 2019). This survey indicated the presence of adulterants such as hydrogen peroxide, detergent, urea, neutralizers, maltodextrin, sugar etc in liquid milk. Although, systematic survey on the presence of adulterants in milk products has not been done, however, media reports as well as research publication (Amrutkala, 2012; Bhandekar *et al.*, 2019) do indicate the presence of adulterants in dairy products also. The major dairy products like ghee, butter, cheese, paneer, yoghurt/dahi, milk powder etc., are majorly encountered with the following categories of adulteration

1. ***Adulteration of proteins by replacing milk protein with cheaper alternatives:*** Milk protein is superior to practically all other food proteins for human nutrition based on its protein digestibility corrected amino acid score (Rutherfurd *et al.*, 2015). As milk protein is more valuable than vegetable protein, milk products are possible target for adulteration with cheaper vegetable proteins as soy or pea protein (ISO/IDF, 2006).

2. ***Adulteration of fat by replacing milk fat with vegetable or animal body fat:*** Milk fat is one of the costliest components of milk and majorly adulterated by cheaper vegetable oils such as vanaspati (hydrogenated plant oils), animal body fats (tallow, lard), interesterified fats, Reichert-Meissl value-adjusted specific adulterant oil, Mineral oils – solid paraffins etc. The adulteration of milk fat with above mentioned sources is point of concern from customer point of view as it affects its nutritional quality. The use of animal body fat as an adulterant is a key topic of concern from a religious standpoint also.

3. ***Adulteration of starch in ghee, butter, khoa, paneer/channa, dahi and ice-cream:*** Starch is considered to be common adulterant in many dairy products. Its adulteration is driven by cheap availability of bulk amount of starch from different sources.
4. ***Adulteration of starch and maltodextrin in milk powder:*** Starch and maltodextrin are common adulterants found in milk where they are added to raise S.N.F. content to mimic the addition of water. When adulterated milk is converted into powder, these adulterants are concentrated.
5. ***Adulteration of thickeners like starch and gelatin in cream:*** As per FSSAI, the minimum fat content in cream is 10%. More the fat in cream, thicker will be its consistency. Therefore, some scrupulous producers add starch and gelatin to enhance rich consistency of adulterated cream.
6. ***Adulteration of saccharin in ice-cream:*** Non-nutritive sweeteners, aspartame and sucralose, are permitted to be added in ice-cream. Saccharin, cheaply available non-nutritive sweetener, is not allowed by FSSAI to be used in ice-cream.
7. ***Adulteration of admixing of milk used for dairy product preparation:*** Nowadays consumers are demanding milk prepared from specific milk (like goat, sheep and camel milk) due to perceived health benefits. Some products like cheese also require use of specific species milk due to its characteristic flavour and other physico-chemical properties. However, non-availability or premium price of milk of a particular species motivates producers to substitute higher value milk with undeclared milk to fulfil the demand of milk products. Though mixing milk from different animal species is the simplest way to adulterate milk and expected not to cause any harmful health effect but for consumer who is paying higher price for specific product (made exclusively from goat, cow or buffalo milk), it is considered as an economic adulteration.

3. Methods for Detection of Adulteration in Milk Products

Different traditional and instrument-based methods have been developed for detection of adulteration in dairy products. Besides, traditional methods, nowadays infrared spectroscopy coupled with multivariate analysis is gaining popularity as one of the important analytical techniques in detection of adulteration of milk products. Infrared spectroscopy is based on the study of the interaction of electromagnetic waves with matter, and the electromagnetic radiation is divided into distinct wavelength zones of near-infrared (NIR; 800 to 2,500 nm) and mid-infrared (MIR; 2,500 to 25,000 nm) spectrum (O'Donnell & Cullen, 2014). When quantification or detection of a specific component is not possible, infrared technologies can still be used to discriminate samples by identifying peculiar spectral differences using supervised methods like artificial neural networks and discriminant analysis, as well as unsupervised methods like hierarchical clustering, demonstrating that infrared spectra can discriminate between adulterated and non-adulterated samples. The application of untargeted analysis of adulteration detection will be requirement of the hour. Adulterations would be out of control if only targeted analysis methods are used, and the analysis would be locked in a loop of "adulteration, targeted analysis, and fresh adulterations," and so on. Therefore, untargeted detection approaches are necessary to allow for the identification of a variety of known and unknown adulterants in dairy products (Brereton, 2007). The Infrared spectroscopy has emerged as widely used technique for this non-targeted detection of adulteration in milk products.

3.1 Ghee

In Indian sub-continent, milk fat is mainly consumed in the form of ghee and main applications are its use as a cooking or frying medium, table top item for garnishing purpose and also for religious purposes. The main features of ghee for its consumers include its unique pleasant flavour, texture (semihard to granular) and colour (golden yellow to light brown). Because of its popularity and high demand, ghee often remains in short supply and demands high price. Ghee is also a unique natural fat which has short chain fatty acids such C4, C6, C8 and C10 which are usually absent in other oils and fats. Compared to regular vegetable oils, the ghee is priced at least 3 times high and thus for economic reasons it is adulterated. Media reports do indicate the problem of adulteration of ghee with cheaper vegetable oils which is perhaps the main adulterant of ghee. To overcome the adulteration of ghee, food regulatory body of India i.e., Food Safety Standards Authority of India (FSSAI) has set standards of ghee which are mainly aimed for curbing the adulteration rather than on its other quality parameters. Since physico-chemical parameters of ghee often change with geographical region, feed, animal species, season etc, often those engaged in the unethical business of adulteration ghee take benefit of the same. Noting this point, earlier FSSAI has prescribed Butyro-refractometer reading and Reichert meissl value different for different regions of India (Table 1 and 2). But in the present FSSAI regulations, all India one BR and RM value has been prescribed. (Table 3)

Table 1: Old FSSAI standards for ghee, butter oil and anhydrous milk fat

Type of Physico-chemical parameter	Ghee	Butter oil	Anhydrous milk fat/anhydrous butter oil
Baudouin test	Negative	Negative	Negative
Butyro-refractometer reading at 40°C	Values as per Table 2	40-44	40-44
Reichert-Meissl value, minimum	Values as per Table 2	28	28
Polenske value	-	1.0-2.0	1.0-2.0
Moisture content (%)	Not more than 0.5	Not more than 0.4	Not more than 0.1
Milk fat (%), minimum	99.5	99.6	99.8
Free fatty acids (% oleic acid), maximum	3.0	0.4	0.3
Peroxide value (milliequivalent of oxygen/ Kg fat), maximum	-	0.6	0.3

Table 2: Value of Butyro-refractometer reading at 40°C and Reichert-Meissl value of ghee for various geographical areas of India

Sr.No.	Name of the state & U.T. as per old FSSAI standards	Butyro-refractometer reading at 40°C	Reichert-Meissl value, minimum
1.	Bihar, Chandigarh, Delhi, Jharkhand, Punjab, Haryana (Areas other than cotton tract areas), West Bengal (Areas other than cotton tract), Sikkim	40 to 43	28
2.	Arunachal Pradesh, Assam, Goa, Haryana (Cotton Tract Areas), Himachal Pradesh, J & K, Kerala, Lakshadweep, Maharashtra (Areas other than Cotton Tract Areas), Manipur, Meghalaya, Mizoram, Nagaland, Orissa, Rajasthan (areas other than Jodhpur Distt.), Tripura, U.P., Uttarakhand	40 to 43	26
3.	Chhattisgarh, Karnataka (Belgaum Distt.), Madhya Pradesh (Areas other than cotton tract areas), Pondicherry	40 to 44	26
4.	Andhra Pradesh, Daman & Diu, Dadar & Nagar Haveli, Karnataka (Areas other than Belgaum Distt.)	40 to 43	24
5.	Andaman & Nicobar Island, Tamil Nadu	41 to 44	24
6.	Gujarat (Areas other than cotton tract)	40 to 43.5	24
7.	Gujarat (Cotton tract areas), Madhya Pradesh (Cotton tract areas), Maharashtra (Cotton tract areas), Rajasthan (Jodhpur sub division), West Bengal (Bishnupur sub division)	41.5 to 45	21

Apart from the parameters listed in Table 1 and 2, the number of other physico-chemical parameters have been reported in the literature for detection of adulteration of ghee. Readers are encouraged to take note of one of the recent reviews in this direction (Sharma *et al.*, 2020).

3.1.1 Butyro-Refractometer (BR) Reading of Ghee

Refractive index (RI) of ghee is usually measured in the form of BR reading which can be derived mathematically from RI. BR reading of ghee is estimated in a specialized equipment at 40°C and does not have any units. The general values for BR reading of milk fat (40 to 43) and vegetable oils and fats (above 50) are so wide apart that this property can be employed as an indicator for milk fat adulteration with vegetable oils and fats, except coconut oil (38 to 39) and palm oil (39 to 40). The BR reading of animal body fats are in the range of 44 to 51. Availability of the modern Peltier controlled BR machines have made the estimation of BR reading of fat quite simple and this can be considered as a rapid method which can be employed as a preliminary test to assess the quality of ghee (BIS, 1966). However, BR test should be used with caution as careful adulteration of ghee can lead to BR reading of ghee within its natural range. Earlier as per FSSAI, ghee is required to have BR reading in the range of 40-43 but this range is wide (40-45) in those areas of the country wherein dairy animals are fed extensively with cottonseed oil/cake (Table 2). Now FSSAI as given single range of 40-44 throughout country.

3.1.2 Reichert-Meissl (RM) and Polenske Value

Estimation of RM value and Polenske value of ghee are some of the classical methods employed for checking the adulteration of ghee. Both of these values measure short chain fatty acids in a given sample of oil/fat and thus are very specific to ghee as such fatty acids are present only in ghee. RM value is the number of ml of 0.1 N aqueous alkali solution required to neutralize the water-soluble, steam volatile fatty acids distilled from 5 g of ghee under the precise conditions specified in the method. Polenske value is the number of ml of 0.1 N aqueous alkali solution required to neutralize the water-insoluble, steam volatile fatty acids distilled from 5 g of ghee under the precise conditions specified in the method. RM value is a measure of the lower fatty acids of ghee like butyric (C4:0), and caproic (C6:0). Butyric acid contributes about ¾ and caproic acid ¼ to the RM value, while the Polenske value is made up of ¼ caprylic (C8:0) and 3/4 capric acid (C10:0). Since the presence of lower fatty acids is peculiar to milk fat, the RM and PV are important characteristics of ghee. In general, RM value and Polenske value for cow and buffalo ghee is about 28 and 1.5, respectively. Of the common vegetable oils, only coconut and palm kernel contain steam volatile acids, and both exhibit RM of 7 and Polenske value of 13. Earlier, as per FSSAI ghee is required to have RM value of not less than 28 but in case of cotton tract areas, this permissible limit has been lowered to accommodate variation due to feeding of animal with cottonseed oil/cake (Table 2). Now FSSAI as given single value of 26 throughout country. Although, RM and Polenske value test can be done in a basic chemical laboratory, however, tests require a specific type of glass apparatus (BIS, 1966) and expertise to produce reproducible results. In recent year, reports do indicate the manipulation of RM value of ghee by addition of RM-adjustor also known as cutting oil, the addition of which fraudulently raise the RM value of oil/fat and thus limiting the use of RM value. To counter this, dairy industry is using an innovative method wherein the RM value of alcohol washed ghee is estimated and as per informal industry standards, the difference between normal RM value of ghee and alcohol washed RM value of ghee should not be more than 2.0 units.

3.1.3 Baudouin Test

Baudouin test is a test used for detection of presence of hydrogenated vegetable oil (HVO) popularly known as vanaspati. Since texture of HVO and ghee are quite similar, the addition of HVO to ghee is very much practiced by unscrupulous traders. To overcome the adulteration of ghee with HVO, Vegetable Oil Products (Regulation) Order, 1998, Government of India make it obligatory to add sesame oil in HVO. This order states that "vanaspati shall contain raw or refined sesame oil (til oil) not less than 5 per cent by weight but sufficient so that when the vegetable oil product is mixed with refined groundnut oil in the proportion of 20:80, the red colour produced by the Baudouin test shall not be lighter than 2:0 Red units in a 1 cm cell on the Lovibond scale. Baudouin test is qualitative test and is quite simple to do and results are available in 10 min (BIS, 1966).

3.1.4 Instrument Based Methods

3.1.4.1 Fatty Acid Profile of Ghee

As indicated earlier, ghee has unique fatty acid profile and this can be used for the ascertaining the authenticity of ghee. Usually values of various fatty acids in ghee remain in a particular range and any deviation in this range can be used for evaluation of quality of ghee. Gas liquid chromatography (GLC) has become a useful tool in the quantitative analysis of the fatty acids composition of natural lipids. The fatty acids are usually analyzed after conversion to the corresponding methyl esters. In this method, absolute fatty acids contents are not estimated rather, mass fraction of individual fatty acids as free acid are expressed as g/100 g of total fatty acids. The method (IDF, 2002 a, b) requires GLC equipped with a suitable injection system, a capillary column and a flame ionization detector (FID). Fatty acid profile of different milk fat and different vegetable oils and animal body fats has been presented by Sharma *et al* (2020).

3.1.4.2 Triglycerides (TGs) profile of Ghee

In this method, milk fat is directly analysed by GLC to determine its TGs based on the total carbon number. The method (IDF, 2019) requires GLC equipped with a suitable injection system, a packed or short capillary column and an FID. The method estimates the percentage of TGs of different carbon numbers of C24 to C54 (even carbon numbers only) and these obtained values are put into various TGs equations to obtain S-values which is sum of weighted TGs mass fractions. If a given ghee sample exceeds the range set for a S-value, the sample is considered adulterated with fat of non-milk origin.

3.1.4.3 Detection of β-sitosterol in Ghee

In this method, unsaponifiable matter obtained from ghee are subjected to reverse phase – high pressure liquid chromatography (RP-HPLC) to obtain the sterol profile of ghee. The presence of β-sitosterol peak in ghee indicate adulteration of ghee with vegetable oil. Method required HPLC equipped with C18 column and UV detector (FSSAI Order, 2019).

For evaluating the quality of ghee BR reading, RM value and Polenske value are simple parameters and can be done in a basic laboratory. Further these parameters are not much affected during the storage of ghee (Gosewade *et al.*, 2017). After the preliminary screening of the ghee by above mentioned three parameters, ghee must be analysed by instrument-based methods for its fatty acid profile, presence of β-sitosterol peak and triglyceride profiling to conclusively rule out the adulteration of ghee.

Table 3: New FSSAI standards for ghee, butter oil and anhydrous milk fat

Type of Physico-chemical parameter	Ghee	Butter oil	Anhydrous milk fat/ anhydrous butter oil
Moisture content (%)	Not more than 0.5	Not more than 0.4	Not more than 0.1
Milk fat (%), minimum	99.5	99.6	99.8
Butyro-refractometer reading at 40°C	40-44	40-44	40-44
Reichert-Meissl value, minimum	26	26	26
Polenske value	1.0-2.0	1.0-2.0	1.0-2.0
Free fatty acids (% oleic acid), maximum	1.5	0.4	0.3
Peroxide value (milliequivalent of oxygen/ Kg fat), maximum	-	0.6	0.3
Baudouin test	Negative	Negative	Negative
Iodine value	25-38	-	-
Saponification Value	205-235	-	-
Presence of β-sitosterol	Absent	Absent	Absent
Fatty acid profile (expressed as percentage of total fatty acids)			
Saturated Fatty acid			
C4:0, Butyric acid	1.0-4.0	-	-
C6:0, Hexanoic acid (Caproic acid)	0.5-2.0	-	-
C8:0, Octanoic acid (Caprylic acid)	0.4-1.0	-	-
C10:0, Decanoic acid (Capric acid)	0.8-3.5	-	-
C12:0, Dodecanoic acid (Lauric acid)	1.5-4.0	-	-
C14:0, Tetradecanoic acid (Myristic acid)	6.0-10.0	-	-
C15:0, Pentadecanoic acid	0.5-1.2	-	-
C16:0, Hexadecanoic acid (Palmitic acid)	22-32	-	-
C17:0, Heptadecanoic acid	0.2-0.6	-	-
C18:0, Octadecanoic acid (Stearic acid)	10-16	-	-
Mono-unsaturated fatty acid			
C16:1 (Cis 9), (-Hexadecanoic acid (Palmitoleic acid)	0.9-1.8	-	-
C18:1 (cis 9) 9-Octadecenoic acid (Oleic acid)	19-32	-	-
C18:1 (cis 6) 6-Octadecenoic acid	0.3-0.9	-	-
C18:1 (cis 11) 11-Octadecenoic acid	0.5-2.5	-	-
Poly-unsaturated fatty acid			
C18:2 (cis 9,12), 9,12-Octadecadienoic acid (Linoleic acid)	0.5-3.5	-	-
C18:2 (cis 9, trans 11), 9-cis,11-trans-octadecadienoate	0-3.0	-	-
C18:3 (cis 9,12,15) 9,12,15-Octadecatrienoic acid	0.3-1.0	-	-

3.2 Butter

Similar to ghee, butter is also a fat rich milk product and the most common adulteration encountered in butter is also replacement of milk fat with vegetable oil. As per FSSAI, the extracted fat from butter shall meet the standards for RM value and BR reading as prescribed for ghee; therefore, all methods applied for detection of vegetable oil adulteration in ghee like determination of melting profile of fat, determination of physico-chemical constants related to fat like RM value, PV, BR reading, fatty acid profile, triglyceride profile and sterol estimation can be applied for detection of adulteration in butter. Fourier-transform MIR spectroscopy with attenuated total reflectance has also been explored for determination of chicken fat adulteration in butter due to difference in peak intensity between butter and chicken fat at frequency 1,200 to 1,000 cm^{-1} (Nurrulhidayah *et al.*, 2015).

Sometimes, starch (mashed potato) is also found as an adulterant in fat rich dairy products like ghee and butter. It is added to provide rich appearance to the adulterated products. Addition of starch in ghee and butter can be easily detected by adding few drops of iodine, if the brown hue of the ghee changes to pink, it confirms presence of starch (FSSAI, 2016).

3.3 Paneer

Paneer is a heat-acid coagulated product. Paneer's popularity in Indian cuisine is growing as a result of its numerous health benefits and it is now considered a national delicacy. One of the major adulterants in paneer is replacement of milk fat in paneer with vegetable oils/fat. Such adulteration can be detected by measuring the physic-chemical constants of extracted fat as indicated for ghee or butter. To meet the growing customer demand, paneer is also purposefully adulterated with adulterants like as starch, coal tar dyes and urea to improve texture and appearance (Bhandekar *et al.*, 2019). Starch adulteration in paneer is detected by using iodine. For detection, approximately one gram of paneer is placed in test tube and 5 to 10 ml of distilled water is added. The content is boiled by holding the test tube over a flame. After cooling to room temperature, one to two drops of 1% iodine solution are added. Appearance of a blue colour indicates presence of starch. For detection of coal tar dyes adulteration in approximately 5 gram of paneer, 5 ml of dilute sulphuric acid is added. Appearance of pink colour indicates presence of coal tar dyes (FSSAI, 2012).

3.4 Milk Powder

Milk powder has been one of the world's most extensively consumed and commercially important dairy product. Since during trading of milk powders, milk protein is one of the important quality parameters, this parameter is mostly abused. Unscrupulous people deliberately add nitrogen powders, like dicyandiamide and melamine or starch and maltodextrin, in milk powder irrespective of their toxicity (Balabin and Smirnov, 2011). Melamine (2,4,6-triamino-1,3,5-triazine), is a nitrogenous chemical compound found in fertilisers and plastics. It's high nitrogen content (66 percent by mass) led to its illegal use to boost the perceived protein content of animal feed and human food ingredients (Gossener *et al.*, 2009). Maltodextrin is a hydrolyzed starch derivative and commonly employed as an adulterant due to its high solubility and low cost. Different attempts have been made to detect adulteration in milk powder.

Maltodextrin and starch in powder can be detected after reconstituting milk powder with water and adding iodine solution into it. If blue colour develops, it indicates presence of starch and if red-brown colour develops it indicates adulteration of maltodextrin (FSSAI, 2016).

The addition of nitrogenous powder can be detected using kjeldahl methods after separating non-protein nitrogen from true protein. However, detection accuracy is not very great because the composition and amount of non-protein nitrogen are affected by the kind and concentration of protein precipitation methods used. As a result, high-precision advanced analytical techniques are required for detection of adulteration in milk powder. Raman chemical imaging in the wavenumber region of 2538 to 102 cm^{-1} has been used for the detection of adulteration in milk powder. This technique recognizes characteristics bands for urea (1009 cm^{-1}), ammonium sulfate (973 cm^{-1}), melamine (673 cm^{-1}), and dicyandiamide (212 cm^{-1}) that can be used for detection of adulteration of above chemicals in milk powder (Qin *et al.*, 2017). Ultra-High-Performance Liquid Chromatography (UHPLC) has also been applied for detection of adulteration of soy, pea, brown rice, and hydrolyzed wheat protein in skimmed milk powder (Jablonski *et al.* 2014). A rapid turbidimetric method has been developed for detection of plant proteins in milk powder. The method is based on the differential solubility of milk proteins and plant proteins in tetraborate-EDTA buffer (Scholl *et al.*, 2014). Near-infrared spectroscopy (NIRS) have successfully applied for detecting multi-adulterants in skim milk powder with acid whey (1–25%), starch (2 and 5%), and maltodextrin (2 and 5%). They have applied both targeted (regression models) and non-targeted (class modelling) approach for the same. The untargeted class modelling approach offers the advantage of detecting several adulterants with a single model, including unknown adulterants. However, the sensitivity was higher in targeted regression method. The wavelengths 1,950, 1,980, 2,040, 1,650 to 1,720, and 2,250 to 2,315 nm were found to be essential for predicting the proportion of added acid whey in milk powder. The significant peaks for starch adulteration were 1,410, 1,910, 1,950, 2,180, 2,260, and 2,280 nm; for maltodextrin, they were 1,440, 1,915, 2,040, and 2,420 nm (Capuano *et al.*, 2015).

Another important adulterant in milk powder is melamine. A melamine-contaminated infant formula scandal in China in 2008 was the most visible example of melamine contamination in food. Surprisingly, such a marketed chemical for industrial use was unlawfully combined with dairy products to illude a rise in the protein content, producing the baby formula melamine controversy (Chen, 2009). Some dairy farmers and manufacturers illegally mixed their milk with water to increase the milk volume, and melamine was purposefully added to the diluted milk to increase the perceived protein content to meet the specified protein content required for milk (3%), which would be measured by relatively inexpensive nitrogen-content-based methods such as the Kjeldahl or other methods. According to the Chinese Ministry of Health's official data, melamine-contaminated milk caused 294,000 paediatric patients to develop urinary illness, such as kidney stones, and six new-born deaths (Gossner *et al.*, 2009). After the melamine disaster, many nations have implemented control criteria for establishing the maximum melamine limits for baby formula and other foodstuffs. In October 2008 the Chinese government issued a melamine restriction for infant foods with 1 mg/kg and for other milk products with a 2.5 mg/kg limit, which became official for all goods of the Chinese food industry in April 2011. As detection of melamine has become important, the FDA has recommended LC–MS/MS method for determining melamine residues at concentrations of 0.25 mg/kg in infant formula and other dairy-containing food items or ingredients. However, lower concentrations of melamine require use of standard additional net analyte signalling method (SANAS) (Tang *et al.*, 2016) hyperspectral imaging in spectral range of 990–1700 nm (Lim *et al.*, 2016).

Another risky adulterant in milk powder is adulteration with non-food-grade proteins, with one recent example of concern being hydrolyzed leather protein (Liu and Zhou, 2017). Because leather scraps may contain chromium from the tanning process, hydrolyzed leather protein poses a potential food safety risk if used as an adulterant. Methods for detecting this type of adulteration are mostly based on analysis of the amino acid L-hydroxyproline, which is relatively abundant in collagen, skin, tendons, and animal hides (Wang *et al.*, 2009). The natural level of L-hydroxyroline in milk protein is very low (Li *et al.*, 2011), and therefore, can be used for detection of hydrolyzed leather protein.

Whey powder adulteration to milk powder is also practiced. Whey is a low-cost by-product of the cheese production process. Although adding whey powder to milk powder, which is utilised in many product formulations, does not pose a health risk, it does have economic, nutritional, and legal ramifications. Whey powder adulteration in milk powder can be detected by measuring casein and whey protein ratio using kjeldahl method and by electrophoretic analysis (Souza *et al.*, 2000). The other method used for its detection is based on detection of glycomacropeptide (GMP), a specific ingredient of the cheese whey that can be used as an excellent chemical marker for detecting whey adulteration in milk powder either as such (Ferreira & Oliveira, 2003, Martín-Hernández *et al.*, 2009; Sharma *et al.*, 2021) or sialic acid associated with it (Neelima *et al.*, 2012; Rao *et al.*, 2012).

3.5 Fermented Product (Yoghurt/Dahi)

Yoghurt/dahi is a lactic acid coagulated milk product and considered to be a rich product in terms of fat, protein, calcium, phosphorus, B vitamins etc. Most of the adulteration in yoghurt/dahi have been directed towards milk fat, where various vegetable oils are used to replace milk fat. To detect adulteration of vegetable oil in milk fat liquid and gas chromatography can be used, where extracted fat is subjected to fatty acid analysis and phytosterol analysis as mentioned in detection of adulteration in ghee. In fermented products, the adulteration of protein is also observed. To detect protein adulteration of gelatin and soy protein in fermented products, application of near-infrared (NIR) spectroscopy with chemometrics was used that can detect adulterations of yoghurt containing 1 % edible gelatin, 2 % industrial gelatin, and 2 % soy protein powder (Xu *et al.*, 2013). Sometimes, starch is also added to enhance appearance of dahi/yoghurt. This can be easily detected by adding iodine solution. If blue colour appears it confirms presence of starch in dahi (FSSAI, 2012).

3.6 Khoa

Khoa is a heat desiccated product, used as base material for making Indian delicacies. The increase consumption of khoa during festive seasons has led to its adulteration. As per FSSAI, starch is a potential adulterant used in khoa and a method a has also been recommended for its detection in khoa. In this method, 11 grams of grated khoa sample is made into slurry with 20 mL hot water (80-90°C) and further volume is made to 50 mL. From this prepared sample, 5 mL of sample is added with 0.2 ml of iodine solution and development of blue colour indicates presence of starch (FSSAI, 2016). One survey was conducted in 2012 was conducted to analyse lipid composition of Khoa samples in relation to identify possible adulteration using GC. It was reported that only 57% of khoa samples exhibited milk fat characteristics, while 43% of khoa samples exhibited fatty acid composition that did not conform to milk fats. The presence of nonmilk lipids in Khoa samples was confirmed by measuring cholesterol and triglycerides (Amrutha Kala, 2012). Many of the test used for detection of adulteration in milk are also applied successfully in detection of adulteration in khoa (Parekh *et al.*, 2020). Media reports indicate that khoa may also be adulterated with blotting paper/cellulose. The method for detection of blotting paper/cellulose is not available for khoa but mentioned for rabdi by FSSAI. The tea spoonful of rabdi is diluted with water and mixed with 3 mL concentrated hydrochloric acid. The contents are stirred with glass rod and observed for the presence of fine particles on glass rods to confirm presence of blotting paper/cellulose in rabdi (FSSAI, 2012).

3.7 Cream

In cream, adulteration of thickeners is observed. Thickeners in cream and similar kind of products such as reconstituted cream, recombined cream, prepared cream (pre-packaged liquid cream, whipping cream, cream packed under pressure, whipped cream, fermented/cultured/sour cream, acidified cream) and malai are added to enhance its consistency without increasing fat content. Generally, starch and gelatin are added as thickeners in cream type products. The starch can be detected by addition of iodine solution that gives the appearance blue colour indicating the presence of starch, which disappears when sample is boiled and reappears on cooling. The gelatin is detected by Stokes' test, in which the sample is mixed with water and Stokes reagent (mercury solution in dilute nitric acid) and filtered. A saturated solution of picric acid is then added to the filtrate and the colour development is observed. Yellow precipitate is produced in the presence of gelatine, and cloudiness is observed when gelatine is present in smaller amounts (FSSAI, 2012).

3.8 Ice-cream

Different media reports indicate presence of metanil yellow as a major non-permitted food colour widely used in India. In dairy products, sweets and ice-creams are potentially adulterated with metanil yellow. Long-term ingestion of metanil yellow produces neurotoxicity in the brain. However, its detection is easy as, after extraction, development of red colour is observed after addition of few drops of concentrated hydrochloric acid (FSSAI, 2012).

Artificial sweetener, Saccharin is also a potential adulterant added in ice-cream to replace sugar. Saccharin is a non-caloric sweetener like aspartame and sucralose but FSSAI has allowed only aspartame and sucralose as permitted non-caloric sweetener that can be added in ice-cream. However, some producers deliberately add saccharin that belongs to a class of compounds known as sulphonamides and can cause allergic reactions. On consumption, saccharin can leave a bitter taste at the end. Saccharin is extracted in ether after adding few drops of concentrated hydrochloric acid. After evaporating ether, residual extract is mixed with water and sweetness is measured (FSSAI, 2012).

3.9 Cheese

Cheese is a heat-acid coagulated product majorly popular in western countries but also gaining popularity among Indian consumers. The cheese is a protein concentrated product and mainly liked due to typical flavour of the product. The adulteration in cheese is mainly related to admixing of different types of milk used for its preparation. Every cheese variety is liked for its typical flavour as it is produced from specific type of milk. Manchego, Roquefort, Pecorino, and Feta are the distinct cheese varieties made from sheep milk exclusively. Goat cheese, often known as chèvre in French, is cheese prepared solely from goat's milk. Because goat milk has less casein, it yields small, soft curds that crumble easily. Traditional goat milk cheeses include chévre, drunken goat cheese, Crottin de Chavignol, Valençay, and Geitost. Goat's milk can be used to manufacture any variety of cheeses, including Gouda, Cheddar that will have softer texture and tangier flavour than cow cheese. Geographical area authenticity is also important in cheese, where it is mentioned on a label. For example, "Emmentaler Switzerland" which has a status as a "protected appellation of origin" that assures the manufacture and processing of the product in a designated geographical area. "Emmental" cheese is also produced in France, Germany and Finland. As "Emmentaler Switzerland" cheese is expensive and attracts a considerable premium price over Emmental cheese made elsewhere and therefore required analytical strategies for such differentiation among two. Infrared spectroscopy (mid-infrared) has been proven to be one of the most helpful techniques for identifying between regions of origin from Emmental cheeses (Karoui *et al.*, 2004; Pillonel *et al.*, 2005).

3.10 Detection of Mixing of Interspecies Milk for Preparation of Dairy Products with Emphasis on Goat Milk Products

Goat milk is the world's third most widely produced milk after cow and buffalo milk (FAOstat, 2018). The increased interest in goat milk and its derivatives is due to the nutritional benefits provided by these products, which are attributed to changes in amino acid content, secondary structure of milk proteins, and smaller fat globules (Jacquot *et al.*, 2015). When compared to cow milk, these alter the digestion and nutritional value of milk (Clark and Garca, 2017; Hodgkinson *et al.*, 2018). Fermented items (yoghurt, cheese, buttermilk), ice cream, butter, condensed/dry products, and desserts can all be made with goat milk (Fazilah *et al.*, 2018). Cheese derived from goat milk is popular commercially all around the world. France leads the world in goat cheese production due to its traditional industries spread across its region (Picon *et al.*, 2019). When compared to cow cheese, the great digestibility and low caloric supply of goat cheese drive demand (Haenlein and Anke, 2011). The nutritional benefits of goat milk, as well as the low amount produced, are the key reasons for the higher prices of these produced. As a result, goat dairy products could be vulnerable to commercial fraud. The Near Infrared (FT-NIR) paired with chemometric methods has been assesses to detect goat's yoghurt and cheese contaminated with cow milk. Principal Component Analysis (PCA), Q-control chart and Partial Least Squares-Discriminant Analysis (PLS-DA) were able to detect admixing of cow milk at 10, 15 and 20% in goat milk and its further processing into cheese and yoghurt (Teixeira *et al.*, 2021).

Conclusion

The adulteration of milk and milk products is a menace to the society. Adulteration of dairy products not only affects its nutritional value but may lead to health hazard. Moreover, whenever adulteration is practiced, the hygiene is often compromised. The perception of wide spread use of adulterants in dairy products may affect the consumption of such products by general public and thus may be detrimental to whole dairy industry. To curb the menace of adulteration, more systematic surveys should be done to know the status of use and type of adulterants in dairy products. Regulatory bodies may play an important role in bringing the confidence in general public by adopting newer techniques and standards for dairy products. In this direction it is worth mentioning that in recent years, FSSAI has given new standards of ghee wherein some more parameters have been added (Table 1) with a view to control the adulteration of ghee. Analytical methods are of immense importance in the hands of regulators to mitigate adulteration. Robust analytical methods may become deterrent for those engaged in the unscrupulous business of adulteration. In this direction, rapid and instrument methods may play an important role as such methods may help in taking faster and objective decision. Research institutes should focus their research in developing such analytical method to strengthen the hands of regulatory bodies.

References

Amrutha Kala, A. L. (2012). A survey of lipid composition of Khoa samples in relation to possible adulteration. International Journal of Dairy Technology, 65(3), 444-450.

Balabin, R. M., & Smirnov, S. V. (2011). Melamine detection by mid-and near-infrared (MIR/NIR) spectroscopy: a quick and sensitive method for dairy products analysis including liquid milk, infant formula, and milk powder. Talanta, 85(1), 562-568.

Bhandekar, K; Atkare V.G. and Ramteke, V.M. (2019). Incidence of adulteration in the paneer samples collected from Nagpur city. International Journal of Chemical Studies, 7(3): 1202-1204

BIS (1966). Indian standards. Methods for sampling and test for ghee (butterfat). IS 3508: 1966, Reaffirmed 2018. New Delhi, India: Bureau of Indian Standards.

Brereton RG (2007) Applied chemometrics for scientists. John Wiley & Sons, New Jersey, USA.

Capuano, E., Boerrigter-Eenling, R., Koot, A., & van Ruth, S. M. (2015). Targeted and untargeted detection of skim milk powder adulteration by near-infrared spectroscopy. Food Analytical Methods, 8(8), 2125-2134.

Cevoli, C., Fabbri, A., Gori, A., Caboni, M. F., & Guarnieri, A. (2013). Screening of grated cheese authenticity by nir spectroscopy. Journal of Agricultural Engineering, 44(s2).

Chen, J. S. (2009). A worldwide food safety concern in 2008—melamine-contaminated infant formula in China caused urinary tract stone in 290 000 children in China.

Clark, S., & García, M. B. M. (2017). A 100-year review: Advances in goat milk research. Journal of Dairy Science, 100(12), 10026-10044.

FAOstat, 2018. Food and Agriculture Organization of the United Nations Database. From <http://www.fao.org/faostat/en/#home>. Retrieved September 24, 2021.

Fazilah, N. F., Ariff, A. B., Khayat, M. E., Rios-Solis, L., & Halim, M. (2018). Influence of probiotics, prebiotics, synbiotics and bioactive phytochemicals on the formulation of functional yogurt. Journal of Functional Foods, 48, 387-399.

Ferreira, I. M., & Oliveira, M. B. P. P. (2003). Determination of caseinomacropeptide by an RP-HPLC method and monitoring of the addition of rennet whey to powdered milk. Journal of Liquid Chromatography & Related Technologies, 26(1), 99-107.

FSSAI Order. (2019). Method for detection of adulteration in ghee (clarified milk fat) with vegetable oils. File No. 1-90/FSSAI/SP (MS&A)/2009 dated 25th March, 2018. Food Safety Standards Authority of India, New Delhi.

FSSAI. (2012). Quick Test for Some Adulterants in Food Instruction Manual- Part II. Food Safety and Standards Authority of India. New Delhi: Ministry of Health and Family Welfare, Government of India.

FSSAI. (2016). Manual of methods of analysis of foods. Milk and milk products. Lab manual 1. Food Safety and Standards Authority of India. New Delhi: Ministry of Health and Family Welfare, Government of India.

Gossner, C. M. E., Schlundt, J., Ben Embarek, P., Hird, S., Lo-Fo-Wong, D., Beltran, J. J. O., Theo, K.N. & Tritscher, A. (2009). The melamine incident: implications for international food and feed safety. Environmental Health Perspectives, 117(12), 1803-1808.

Haenlein, G. F. W., & Anke, M. (2011). Mineral and trace element research in goats: A review. Small Ruminant Research, 95(1), 2-19.

Haug, A., Høstmark, A. T., & Harstad, O. M. (2007). Bovine milk in human nutrition–a review. Lipids in Health and Disease, 6(1), 1-16.

Hodgkinson, A. J., Wallace, O. A., Boggs, I., Broadhurst, M., & Prosser, C. G. (2018). Gastric digestion of cow and goat milk: Impact of infant and young child in vitro digestion conditions. Food Chemistry, 245, 275-281.

IDF (2002a) IDF Standard 182, Milk fat. Preparation of fatty acid methyl esters. International Dairy Federation, Brussels;

IDF (2002b) IDF Standard 184, Milkfat. Determination of the fatty acid composition by gas-liquid chromatography. International Dairy Federation, Brussels.

IDF (2019) IDF Standard 202, Milk and milk products — Determination of milk fat purity by gas chromatographic analysis of triglycerides. International Dairy Federation, Brussels.

Jacquot, S., Karoui, R., Abbas, K., Lebecque, A., Bord, C., & Aït-Kaddour, A. (2015). Potential of multispectral Imager to characterize anisotropic French PDO cheeses: A feasibility study. International Journal of Food Properties, 18(1), 213-230.

Karoui, R., Dufour, É., Pillonel, L., Picque, D., Cattenoz, T., & Bosset, J. O. (2004). Determining the geographic origin of Emmental cheeses produced during winter and summer using a technique based on the concatenation of MIR and fluorescence spectroscopic data. European Food Research and Technology, 219(2), 184-189.

Li, X., Rezaei, R., Li, P., & Wu, G. (2011). Composition of amino acids in feed ingredients for animal diets. Amino Acids, 40(4), 1159-1168.

26. Lim, J., Kim, G., Mo, C., Kim, M. S., Chao, K., Qin, J., ... & Cho, B. K. (2016). Detection of melamine in milk powders using near-infrared hyperspectral imaging combined with regression coefficient of partial least square regression model. Talanta, 151, 183-191.

Martín-Hernández, C., Muñoz, M., Daury, C., Weymuth, H., Kemmers-Voncken, A.E., Corbatón, V., Toribio, T. and Bremer, M.G., 2009. Immunochromatographic lateral-flow test strip for the rapid detection of added bovine rennet whey in milk and milk powder. International Dairy Journal, 19(4), pp.205-208.

Neelima, Rao, P.S.; Sharma R. and Rajput, Y.S. (2012) Direct estimation of sialic acid in milk and milk products by fluorimetry and its application in detection of sweet whey adulteration in milk. Journal of Dairy Research.79 (4): 495-501.

Nurrulhidayah, A. F., Arieff, S. R., Rohman, A., Amin, I., Shuhaimi, M., & Khatib, A. (2015). Detection of butter adulteration with lard using differential scanning calorimetry. International Food Research Journal, 22(2).

O'Donnell, C. P., & Cullen, P. J. (2014). Emerging PAT Technologies. Process Analytical Technology for the Food Industry, 247-267.

Parekh, S., Shaikh, A. I., & Aparnathi, K. D. Detection of common adulterants in khoa using qualitative tests reported for milk. Indian Journal of Dairy Science, 73(6).

Picon, A., López-Pérez, O., Torres, E., Garde, S., & Nuñez, M. (2019). Contribution of autochthonous lactic acid bacteria to the typical flavour of raw goat milk cheeses. International Journal of Food Microbiology, 299, 8-22.

Pillonel, L., Badertscher, R., Casey, M., Meyer, J., Rossmann, A., Schlichtherle-Cerny, H., Tabacchi, R. & Bosset, J. O. (2005). Geographic origin of European Emmental cheese: Characterisation and descriptive statistics. International Dairy Journal, 15(6-9), 547-556.

Qin, J., Kim, M. S., Chao, K., Dhakal, S., Lee, H., Cho, B. K., & Mo, C. (2017). Detection and quantification of adulterants in milk powder using a high-throughput Raman chemical imaging technique. Food Additives & Contaminants: Part A, 34(2), 152-161.

Rao, P.S.; Kundu, M.; Sharma, N.; Sharma, R. and Rajput, Y.S. (2012). A spectrophotometric method for the detection of cheese whey in milk. Indian Journal of Dairy Science. 65(4): 300-304.

Rohenkohl, J. E., Corrêa, G. F., de Azambuja, D. F., & Ferreira, F. R. (2011). O agronegócio de leite de ovinos e caprinos. Indicadores Econômicos FEE, 39(2).

Scholl, P. F., Farris, S. M., & Mossoba, M. M. (2014). Rapid turbidimetric detection of milk powder adulteration with plant proteins. Journal of Agricultural and Food Chemistry, 62(7), 1498-1505.

Sharma, N.; Sharma, R.; Rajput, Y.S.; Mann, B. and Gandhi, K. (2021) Distinction between glycomacropeptide and β-lactoglobulin with 'stains all' dye on Tricine SDS-PAGE gels. Food Chemistry. 340: 127923

Sharma, Rajan, Kamal Gandhi, Surendra Nath Battula, and Bimlesh Mann (2020). "Detection of Milk Fat Adulteration." In Dairy Fat Products and Functionality, pp. 109-131. Springer, Cham, 2020.

Souza, E. M. T. D., Arruda, S. F., Brandão, P. O., & Siqueira, E. M. D. A. (2000). Electrophoretic analisys to detect and quantify additional whey in milk and dairy beverages. Food Science and Technology, 20, 314-317.

Tang, B. C., Cai, C. B., Shi, W., & Xu, L. (2016). Rapid quantification of melamine in different brands/types of milk powders using standard addition net analyte signal and near-infrared spectroscopy. Journal of Analytical Methods in Chemistry, 2016.

Teixeira, J. L. D. P., Caramês, E. T. D. S., Baptista, D. P., Gigante, M. L., & Pallone, J. A. L. (2021). Rapid adulteration detection of yogurt and cheese made from goat milk by vibrational spectroscopy and chemometric tools. Journal of Food Composition and Analysis, 96, 103712.

Turnipseed, S., Casey, C., Nochetto, C., & Heller, D. N. (2008). Determination of melamine and cyanuric acid residues in infant formula using LC-MS/MS. US FDA laboratory information bulletin, 24, 4421.

Wang, Y. H., Li, P. B., Deng, B. N., Nan, W. K., & Jin, H. (2009). Detection of hydrolyzed animal proteins in milk by hitachi L-8800 high speed amino acid analyzer [J]. Chinese Journal of Health Laboratory Technology, 11.

Xu, L., Yan, S. M., Cai, C. B., Wang, Z. J., & Yu, X. P. (2013). The feasibility of using near-infrared spectroscopy and chemometrics for untargeted detection of protein adulteration in yogurt: removing unwanted variations in pure yogurt. Journal of Analytical Methods in Chemistry, 2013.

17

Livestock Meat and Fish Presence of Potential Adulterants, Contaminants, and Their Detection

Sanjay Yadav[1], Bibha Kumari[2], Neha Thakur[1], Suresh Kumar[3] Vijay K. Bharti[4]

[1]*Department of Livestock Products Technology, Lala Lajpat Rai University of Veterinary and Animal Sciences (LUVAS), Hisar, Haryana, India*
[1]*Department of Zoology, Magadh Mahila College, Patna University, Patna, India*
[3]*National Institute of Biologicals (NIB), Ministry of Health & Family Welfare Govt. of India, Sector-62, Noida, Uttar Pradesh, India*
[4]*DRDO-Defence Institute of High Altitude Research (DIHAR) Leh, UT Ladakh, India*

1. Introduction

Livestock meat and fish are rich sources of high-quality protein and other valuable nutrients such as polyunsaturated fatty acids, vitamins, and minerals and are an essential part of our healthy diet (Vladau *et al.*, 2008; Mahmud *et al.*, 2018; Mohanti *et al.*, 2019; Mei *et al.*, 2019). Food fraud and adulteration is an age-old distortive practice. The rising incidence of food contaminants and adulterants-induced health issues has challenged consumers to identify and consume safe livestock meat and fish products. Its significance has increased in recent years due to global trade, awareness among consumers, and sociopolitical issues. Therefore, various government policies and guidelines are being developed to provide consumers with safe and wholesome meat and fish. The significant fraudulent practices experienced in the meat sector are substituting inferior quality meat to fine quality meat, substituting spoiled meat to fresh meat, and adding non-meat ingredients.

Similarly, the indiscreet use of chemical substances to increase the self-life of fish has created various toxicity and fish-borne diseases in the consumers. Livestock meat and fish are perishable organisms because they are high in protein and fat, which cause decay soon after death, so chemical adulteration is very common in the livestock meat and fish industry. So, the presence of these chemical contaminants, toxicants, and adulterants in meat poses a severe threat to human life.

Therefore, rapid, efficient, and reliable analytical methods for screening and detecting adulterants and contaminants in meat and fish products have considerable potential to address the issue of animal origin food safety (Lo and Shaw, 2018; Kumar *et al.*, 2021; Khatun *et al.*, 2021). Recently, molecular technologies have provided a breakthrough strategy for species detection in food authentication. Applying DNA-based adulteration detection techniques where universal polymerase chain reaction (PCR) is commonly used would give consumer safety and regulatory bodies more capacity to identify fraud in meat, fish, etc. Therefore, the correct application of these methods to validate the genuineness of meat foods can help protect consumer interest and public health. So, in the above context, this chapter supplies relevant information regarding adulterants found in meat and their detection methods.

2. Status of Adulteration and Contaminants in Livestock Meat and Fish

India has the world's largest livestock population. It produces around 5.3 million metric tons of meat annually, among the largest producer of buffalo meat and the second-largest producer of goat meat. The current processing level in poultry is 6%, while it stands at 21% for meat. The meat industry of India is mainly controlled by the unorganized sector, which helps the practice of adulteration or substitution of meat in meat products. A similar situation is with the marketing of fish products. It has been estimated that about 25-30% of meat sold in India is adulterated. Besides biological adulteration, soil and groundwater contamination with heavy metals is a significant source of chemical contamination in livestock meat and fish products. These contaminants are present in levels greater than the permissible level in the water and ultimately find their way into the food chain, where they surpass the legal standards as per FSSAI (Table1). Other sources of chemical contamination of meat and meat products are unsupervised and excessive antibiotic usage, veterinary drug and growth supplement residues, preservatives, etc., and are discussed later in the chapter.

Fish and other seafood deteriorate rapidly due to enzymatic reactions and rapid microbial breakdown (Boziaris, 2014; Getu *et al.*, 2015; Mahmud *et al.*, 2018). This deterioration is associated with such spoilage is grouped into microbial metabolic activities, endogenous enzymatic activities, and chemical oxidation of lipids, which shorten the shelf life of these food items. Hence, if not stored at the correct temperature of 5 degrees Celsius or preserved properly, it will be damaged. Therefore, both chemical and artificial preservatives have been developed to extend the shelf life of fish and their products to provide the consumer with the best quality fish. However, undesirably, sellers now use various chemicals such as formalin and ammonia to prevent spoilage and prolong shelf life. These contaminants and excessive preservatives become potential adulterants and are difficult to detect. Recently, there have been reports of fish marketing with added harmful chemicals such as formaldehyde, ammonia, etc. (Hoque *et al.*, 2018). According to India's Food Safety and Standards (FSSAI), regulation 2011, formalin is not permitted in foods. Formaldehyde is a toxic aldehyde listed as a carcinogen by the International Agency for Research on Cancer. While ammonia is not a carcinogen, continued ingestion of ammonia can lead to many health problems, including damage to the lining of the mouth, throat, esophagus, and stomach (Priya *et al.*, 2018). Hence, fish marketing adulterated with these chemicals and preservatives can cause serious health problems for fish eaters.

3. Biological Adulterations for Economic Benefits and Quality Concealing

3.1. Livestock Meat

Adulteration in meat and meat products is of two types, i.e., firstly, economic/ intentional, and secondly, unintentional adulteration. As the name suggests, economic or intentional adulteration is the deliberate tampering of food for the financial gains of the seller. Biological adulterations could be due to mixing two or more livestock meat/ fish species for economic benefits/quality concealing. It also might be done by illegally adding inferior or fake ingredients, e.g., grain proteins in meats, or by duplicitously substituting real ingredients with counterfeit ones, e.g., adding pork in halal meats. In a densely populated nation like India, there is a considerable gap between foods' production and consumption points, especially meat and meat products. This gives enough scope for dishonest sellers to introduce deceitful items in genuine food products, compromising food safety.

The second type of adulteration results from the irresponsible, inadequate, and unhygienic production line where unwanted chemical contaminants are introduced into the meat. Cross-contamination with microorganisms of zoonotic importance like E.coli, Salmonella, Campylobacter, Listeria spp. also falls under this category. Most global food recalls are often a result of the contamination of undesirable pathogenic microorganisms in meat. This unintentional adulteration also includes the spoilage by vermin, entry of physical contaminants like dust/dirt, and finally, residues from packaging material; these all are hazardous to consumer health. The practice of injecting water in carcass by the butchers for economic gains has also been observed by regulatory authorities like FSSAI, which may cause more microbial load and limit shelf-life.

It is well known that different socio-cultural and religious values guide consumer preference for specific kinds of meat consumption. Therefore, specific livestock species are most preferred for meat in particular groups and society without much attention to their nutritional value, e.g., in India, chevon is the highest-priced meat and so often adulterated with cheaper meats like beef, carabeef. Hence, this kind of biological adulteration, i.e., mixing two or more meat species into superior market value meat and their products, is prevalent in many parts of the world. Some examples of closely related animals whose meat is difficult to distinguish by the naked eye due to similar pigmentation, e.g., beef and horse meat, beef, mutton, poultry, and pork. The chances of this kind of adulteration are higher in processed meats because it is easier to mask the cheaper ingredient inexpensive ingredients when meat is broken down into small pieces while preparing meat products.

3.2. Fish

Replacing expensive fish with cheaper alternatives and falsely labeling frozen fish fillets as fresh are two significant concern scams in the seafood industry (Calagari *et al.*, 2020). Furthermore, consumers generally prefer captured fish over farmed fish. In aquaculture, organic farmed fish have generally considered healthier and better quality in animal welfare and the environment than traditional farmed fish. Replacing high-quality fish species with cheaper alternatives and mislabeling their geographic origin is part of the most common fraud in the seafood sector (Hassoun *et al.*, 2020). This is why labeling farmed fish as wild or traditionally farmed fish as organic is considered fraud. This kind of biological adulteration in fish is a long-known problem worldwide. Seafood is often mislabeled because it is one of the most internationally traded foods through complex and opaque supply chains. Due to the globalization of the fish industry, consumers are encountering more and more multi-fish species adulteration. Hence, labeling information is essential for processed fish products because recognizable external morphological features are usually removed, making consumers rely on product labeling for details about their contents (Vindigni *et al.*, 2014).

Therefore, detecting this biological adulteration is difficult by commonly used chemical methods. Furthermore, the lack of rapid and cheap point of care (POC) or field detection tools makes it difficult to ensure food safety and quality. Recently, duplex PCR using species-specific mitochondrial *Cyt b* gene primers has been employed for multispecies meat adulteration (Khatun *et al.*, 2021). It is an issue of concern for food technologists across the globe, as with the increase in demand for processed and ready to eat (RTE) meat products like nuggets, sausages, patties, etc., the chances of adulteration are also increasing.

4. New Generation Adulterants and Contaminants: Radioisotopes/ Chemical Hazardous Preservatives/Food Colors/Undetectable Chemicals/ etc.:

Meat, fish, and their products are highly perishable due to high water activity and rich nutrient profile, making them highly suitable for microbial growth. Hence, various food chemicals and additives are added to enhance the shelf-life of these products. The chemicals and adulterants in meat and fish products pose serious health threats even at trace levels.

4.1. Chemical Contaminants and Toxicants Based Adulterants in Meat and Fish

The chemical contaminants in meat can be categorized into micropollutants and process-induced toxicants. The former results from human activity, veterinary drug residues, or effluents of industrial plants. These toxicants then gain entry into the food chain and are amplified at each trophic level. In contrast, the process-induced contaminants result from meat processing techniques like canning, smoking, curing, or cooking (MaiaMeurillon *et al.*, 2018). Table 1 is summarized FSSAI standards for various common contaminants in poultry, meat, and fish products, which is helpful for the quality check to reduce chemical contaminations into the meat.

4.1.1. Micropollutants

Sources and types of commonly occurring micropollutants in meat, meat products, and fish are:

1. **Mycotoxins:** These are a heterogeneous group of secondary metabolites produced by molds. They have a detrimental effect on human health due to their potent pharmacological effects. Livestock and fish feeds are the most common vehicle for mycotoxin toxicity in foods of animal origin (Marroquín-Cardona *et al*., 2014). Poorly managed fish ponds and aquaculture processing points are other sources of potential mycotoxins and adulteration in fish.
2. **Toxic heavy metals:** These metals enter into livestock meat and fish through contaminated feed, water, industrial processing, and preservatives used for extending self-life (Kan and Meijer, 2007). These metals have profound health implications when ingested and accumulated in excess, or some of them even in trace quantities, e.g., arsenic, mercury, lead, and cadmium. Therefore, to reduce these contaminations in meat and fish, there is a need to reduce their presence in sources, e.g., feedstuffs should be tested and ensured that they contain permissible levels of these as per FSSAI (Table 1).
3. **Dioxins and dioxin-like products:** These chemicals are another class of undesirable by-products of manufacturing processes and are listed as persistent organic pollutants (POPs). They are further classified into three groups: PCDDs (polychlorinated dibenzo-p-dioxins), PCDFs (polychlorinated dibenzofurans), and PCBs (polychlorinated biphenyls). Exposure to these compounds directly or indirectly has been reported to cause hazardous effects on health. One of the main contamination routes in the diet comes from consuming meat products and fish from animals raised in the open in polluted regions (ANSES, 2011).
4. **Brominated flame retardants (BFRs):** These chemicals are a class of compounds that help in reducing the flammability of products. Examples include polybrominated diphenyl ethers (PBDEs), hexabromocyclododecanes (HBCD), tetrabromobisphenol A (TBBPA), and other phenols, polybrominated biphenyls (PBBs), and other brominated flame retardants. Meat and meat products can be contaminated when food animals are reared in establishments when such toxicants are used disproportionately (ANSES, 2012).
5. Veterinary drugs such as antibiotics, anthelmintics, insecticides, hormones, fish pond disinfectants, etc.
6. Pesticide residues are common in meat products widely used in agricultural fields (ANSES, 2011).
7. **Formalin:** formalin in fish has become the preferred preservative in the fish food industry. Once the ingredient is mixed with fish or other seafood, seafood stays intact for several days, promising better market value of the old stock. The

effects of ingested fish or fish products containing this preservative have a slow poison effect in the consumer body (Pandey *et al.*, 2000). Hence, it should be immediately put consumers off. However, most of the formalin used to preserve shellfish is so diluted that the odor is not easily noticeable.

8. **Ammonia and sodium benzoate:** There are several toxic compounds in fish; ammonia and sodium benzoate are at the forefront (Bandyopadhyay *et al.*, 2020). Ammonia is a colorless gaseous compound with a distinctive strong odor. It is highly soluble in water and toxic to aquatic organisms. According to Indian and international regulations, fresh fish and seafood are mainly preserved in ice. But in fact, using some substance other than ice to extend shelf life is a fraud. Ammonia is used inexpensively in the ice-making process to reduce the cost of ice. This slows down the melting process and makes fish look dramatically longer and fresher. However, according to the FDA, it is considered "generally regarded as safe" and can be used in foods at concentrations above 0.1% (Lennerz *et al.*, 2015).

4.1.2. Process-induced Toxicants

These toxicants are produced during meat and fish processing. Polycyclic aromatic hydrocarbons (PAHs) are a class of compounds produced during meat processing like smoking. They can induce mutations or cause cancerous effects as they form DNA adducts. In smoked meats, a classic example of PAH is benzo[a]pyrene (MaiaMeurillon *et al.*, 2018).

Table 1: FSSAI standards for common contaminants in poultry, meat, and fish products (FSSAI, 2011)

Contaminant	**Food Product**	**Permissible level**
Metal Contaminants		
Lead	Canned fish, canned meats	5.0 ppm
	Corned beef, luncheon meat, Cooked Ham, Chopped meat, Canned chicken, Canned mutton and Goat meat, and other related meat products	2.5 ppm
Copper	Meat and meat products	30.0 ppm
Arsenic	Meat and meat products	1.1 ppm

Contaminant	**Food Product**	**Permissible level**
Tin	Corned beef, Luncheon meat, Cooked Ham, chopped meat, Canned chicken, Canned mutton, and Goat meat	250.0 ppm
Zinc	Meat and meat products	50.0 ppm
Cadmium	Meat and meat products	0.5 ppm
Methyl Mercury	Meat and meat products	0.25 ppm
Crop contaminants and naturally occurring toxic substances		
Aflatoxin	Meat and meat products	30.0 μg/kg
Aldrin	Meat	0.2 ppm
	Eggs	0.1 ppm (shell-free basis)
Carbaryl	Fish	0.2 ppm
D.D.T. (The limits apply to D.D.T., D.D.D., and D.D.E. singly or in any combination)	Meat, poultry, and fish	7.0 ppm (whole product basis)
Endosulfan	Fish	0.20 ppm
Fenitrothion	Meat	0.03 ppm
Hexachlorocycle hexane and its isomers (α, β, γ and δ)	Fish	0.25 ppm
Chlorienvinphos	Meat and Poultry	0.2 ppm (carcass fat)
Chlorpyrifos	Meat and Poultry	0.1 ppm (carcass fat)
2,4D	Meat and Poultry	0.05 ppm
	Eggs	0.05 ppm (shell-free basis)
Ethion	Meat and Poultry	0.2 ppm (carcass fat)
	Eggs	0.2 ppm (shell-free basis)
Monocrotophos	Meat and Poultry	0.02 ppm (carcass fat)
	Eggs	0.02 ppm (shell-free basis)
Trichlorfon	Meat and Poultry	0.10 ppm
Carbendazim	Eggs	0.10 ppm (shell-free basis)
	Meat and Poultry	0.10 ppm (carcass fat)
Benomyl	Eggs	0.10 ppm (shell-free basis)
	Meat and Poultry	0.10 ppm (carcass fat)
Carbofuran	Meat and Poultry	0.10 ppm (carcass fat)
Cypermethrin	Meat and Poultry	0.20 ppm (carcass fat)
Edifenphos	Meat and Poultry	0.02 ppm (carcass fat)
	Eggs	0.01 ppm (shell-free basis)
Fenthion	Meat and Poultry	2.0 ppm (carcass fat)
Fenvalerate	Meat and Poultry	1.0 ppm (carcass fat)
Phenthoate	Meat and Poultry	0.05 ppm (carcass fat)
Phorate	Meat and Poultry	0.05 ppm (carcass fat)

Contaminant	Food Product	Permissible level
	Eggs	0.05 ppm (shell-free basis)
Pirimiphos-methyl	Meat and Poultry	0.05 ppm (carcass fat)
	Eggs	0.05 ppm (shell-free basis)
Antibiotic and other pharmacologically active substances		
Tetracycline	Kinds of seafood including shrimps, prawns, or any other variety of fish and fishery products	0.1 ppm
Oxytetracycline	-do-	0.1 ppm
Trimethoprim	-do-	0.05 ppm
Oxolinic acid	-do-	0.3 ppm
All Nitrofurans including Furaltadone, Furazolidone, Furylfuramide, Nifuratel, Nifuroxime, Nifurprazine, Nitrofurantoin, Nitrofurazone, Chloramphenicol, Neomycin, Nalidixic acid, Sulphamethoxazole, Aristolochiaspp and preparations thereof, Chloroform, Chlorpromazine, Colchicine, Dapsone, Dimetridazole, Metronidazole, Ronidazole, Ipronidazole, Other nitroimidazoles, Clenbuterol, Diethylstilbestrol, Sulfanoamide drugs (except approved Sulfadimethoxine, Sulfabromomethazine and Sulfaethoxypyridazine), Fluoroquinolones and Glycopeptides.	Should not be detected in any seafood	

In fish, commonly used chemical preservatives to preserve fish are formaldehyde and sodium nitrite. Formaldehyde, commonly labeled as formalin, is widely used to store corpses in morgues and increase the shelf life of fresh food and fish (Nowshad *et al.*, 2018). Because fish is located far from the point of sale from the place of catch, formalin is used as a preservative while ammonia mixed with water is used to maintain the freshness of the fish (Daw and Chioma 2019). Another preservative of fish is sulfites; it functions as antimicrobial and antioxidant agents, while sulfur dioxide is used in iced water. Further, to prevent microbial spoilage of fish, nitrites, sulfides, erythorbate, ascorbate, and organic acids are added to fish or fish products (Samanta and Choudhary 2019).

Some fish market traders apply a red dye to the fish gills to make it look fresh. Attracted by their "freshness," some customers pay more for these fish only to realize, once cooked, that they are not fresh at all. They showed that the red substance did not quickly fade or peel off when they touched the gills with their finger to check the freshness of the fish, so they did not doubt that it was an artificial substance. Hence, fish contaminated with chemicals in excess has a deleterious effect on consumer health, which needs to be stopped and controlled through regulations.

4.2. Radioisotopes Adulterants Based on Adulterants in Meat and Fish

The adulterated dairy and meat products are a new important food source for radiation contamination during intentional and unintentional events like radiological emergencies, grazing and consumption on contaminated grass and water (Adeniji *et al.*, 2013). It is also called "economically motivated adulteration" and anti-state motivated adulteration." The radioactive strontium, uranium-238, radium-226, thorium-232, potassium-40 are important radionuclides agents contaminated to livestock food through the soil-plant-animal food chain. These radioisotopes accumulate into the liver, bone, lungs, kidney, brain, muscles, etc., and when consumed, it is causing a serious threat to consumers' health. However, this kind of adulteration has no immediate and near impact on consumers and their health but has a deleterious effect on a long-term basis and a specific consumer. This can be done intentionally and targeted, so very difficult to detect. Hence, immediate attention should be paid to awareness among the consumers, food professionals, and other stakeholders on this new kind of meat and fish adulteration. Recently, International Atomic Energy Agency (IAEA), Viena, has developed some detection methods for helping various state agencies estimate various radionuclides presence in food items. Plans for an emergency response to an accidental release of such radioactive materials should specifically include the use of the FAO International Radionuclide Action Levels for Foods or the guideline levels determined by the Codex Alimentarius Commission (IAEA. 1986; FAO. 1987; Codex Alimentarius Commission. 1992).

5. Effect of Chemical Contaminants and Adulterants on Human Health

The acute, accidental, and malicious consumption of contaminated and adulterated meat and fish is hazardous and may result in death. The chronic ingestion of these foods also elevates the risk of life-threatening conditions like cancer, immune disorders, endocrine, nervous dysfunction, etc. Other new kinds of adulterants like heavy metals present in the food when consumed may cause constipation, colic, anemia, cardiovascular diseases, hypertension, abnormal fetal development, chronic renal dysfunction, etc. These metals are a potent human carcinogen, and their bioaccumulation into the human body has genotoxic and carcinogenic effects (Audebert *et al.*, 2012). These chemicals, after ingestion, undergo activation in the liver, intestine, and extrahepatic tissues and interfere with target organ functioning leading to fatal effects. It damages DNA and induces mutations by forming DNA adducts, and later increases the incidence of cancers, endocrine disruption, etc. (Ramesh *et al.*, 2004). Mycotoxins also have carcinogenic, mutagenic, teratogenic, and hepatotoxic effects in both animals and human beings.

6. Important Methods of Sampling and Extraction for Detection of Various Kinds of Adulterants in Meat and Fish

The meat and fish adulterants (chemicals and biological) are introduced at various levels, from production to consumption. So, to detect these adulterants in the finished products, the protocols can be broken down into three significant steps, i.e., sampling, purification, and identification.

6.1. Sampling

The use of solid-liquid extraction (SLE) and liquid-liquid extraction (SLE) techniques are age-old methods used to sample pesticide and other residues in food products. The latter is suitable for milk and milk products, whereas the former is preferred for meat and meat product analysis. Sample preparation usually involves triturating the meat/fish sample and dissolving it in suitable organic solvents at high rpm. The recovery of the chemicals in the sample depends upon multiple factors like the nature of solvent and sample, combinations used, evaporation losses, time, temperature. SLE is a laborious and time-consuming technique, much like the traditional Soxhlet extraction method. Some alternatives to these techniques with good results for meat and fish products are: supercritical fluid extraction (SFE), accelerated solvent extraction (ASE), microwave-assisted extraction (MAE), matrix solid-phase dispersion (MSPD), and graphitized non-porous carbon (GNPC) extraction (LeDoux, 2011). Though these techniques are time-saving than traditional ones, their high cost and skill set requirement outweigh this advantage. There is a continuous improvement in sampling methods which the users/laboratory should consult.

6.2. Purification

This step is necessary to eliminate the background noise and artifacts in the results due to co-extraction and eventual co-elution of the impurities or irrelevant constituents. It improves the sensitivity and repeatability of the detection procedure and minimizes matrix issues like lipid incorporation in meat products. Freezing centrifugation, solid-phase extraction (SPE), liquid-liquid partitioning, solid-phase micro-extraction (SPME), and gel permeation chromatography (GPC) are some analytical techniques that can be utilized to obtain desired results.

6.3. Separation and Detection

Gas chromatography (GC) and liquid chromatography (LC) are two important techniques to identify and quantify various analytes in meat, fish and other animal origin foods. GC is based on the principle of separation of volatile compounds in food on their different partition coefficient, i.e., the ratio of solubility of a substance in the gaseous mobile phase and stationary liquid phase. The LC mobile phase is liquid, and the stationary phase is solid. The variants of these separation techniques include High-performance liquid chromatography (HPLC), Thin-layer Chromatography (TLC), and Paper Chromatography. GC and LCs are further equipped with detection systems based upon the principle of mass spectrophotometry. Mass spectrometric detectors (MSDs) like single quadrupole, ion trap, tandem-MS, and time-of-flight-MS (TOF-MS) help quantify an analyte.

7. Methods for Detection of Adulteration/Cold Cut and their Chemistry/ Principle

Various detection methods, based on morphological/anatomical characteristics, sensory markers (smell, color, texture), and chemical tests have been developed to authenticate marketed food products and verify adulterants. The adulterants can be detected in several ways depending on the type of adulterant being chosen. These methods include analytical, physical, chemical, and advanced protein and DNA-based molecular methods.

In fish, the visual or organoleptic method is widespread for differentiating spoiled/ deteriorated frozen fish from fresh or acceptable frozen fish. The list of visual quality parameters of fresh and old/iced fish is presented in Table 2. Some of the most common methods for detecting formaldehyde/formalin are based on spectrophotometry. The presence of organic and inorganic chemicals in fish, such as sulfur dioxide, aldehydes, and other amines, can interfere with this detection method. Among these, the most sensitive of these methods is flow injection, which has a detection limit of 9 ppt (0.011 g/m3) and is high-performance liquid chromatography, which offers a detection limit of 0.0017 ppm (0.002 mg/m3).

Table 2: Characteristics of fresh fish and old/iced fish

S.N.	Quality Parameter	Characteristics of Body Parts	Specification	
			Fresh Fish	**Old/ Iced Fish**
1	General appearance	Scale/Skin	Bright, shining	Shed, Dull, Dry, Flaky
		Blood spots on Gills	Up to 10-30%	More than 30%
		Stiffness	Stiff, Elastic	Soft, Firm
		Belly	Firm, Soft	Belly burst
2	Eyes	Clarity	Clear, shining	Hazy/ cloudy
		Shape	Normal	Sunken
3	Gills	Color	Fresh red	Dark red, Maroon, Maroonish black
4	Fins	Condition	Healthy look	Ragged, decay
5	Smell	Status	No smell, muddy	Foul, Metallic, Pungent, Stale meat/rancid

Further, the methods for meat adulteration can be broadly classified into two categories: destructive methods and non-destructive methods (Fig. 1) (Li *et al.*, 2020). Thorough sample preparation is the prerequisite of the destructive methods, and they can further be grouped into three broad categories, i.e., DNA-based, protein-based, and metabolite-based. On the other hand, the non-destructive methods, as the name suggested, are non-invasive and do not require sample preparation.

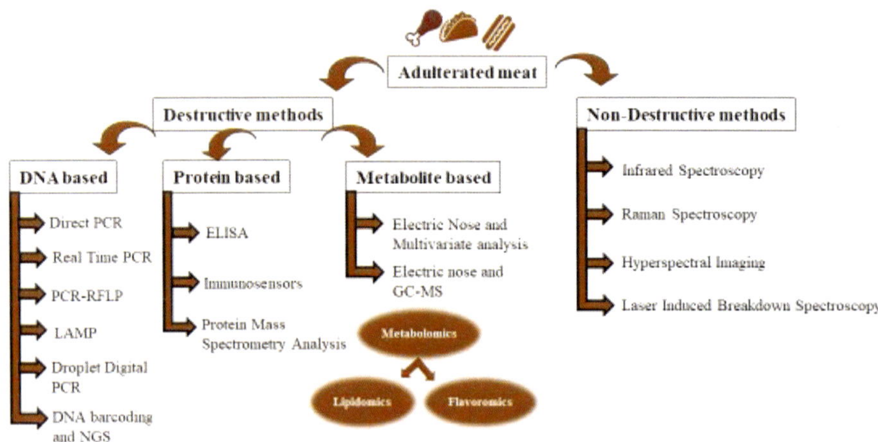

Fig. 1: Classification of methods of detection of meat adulteration (Source: Li *et al.*, 2020)

The detection methods using high-performance liquid chromatography, gas chromatography, infrared detection, colorimetry, polarography, fluorimetry, etc., are commonly used with high sensitivity.

7.1. Destructive Methods

7.1.1. DNA Based Methods

DNA is the highly conserved storehouse of genetic information, making it a reliable and powerful tool for detecting fraudulent biological practices in the meat and fish processing industry (Xiang *et al.*, 2017; Kumari *et al.*, 2019). Most of the DNA techniques developed in the past include amplification of DNA, and mitochondrial DNA is usually preferred for species identification. Most DNA-based techniques are PCR-dependent due to their simplicity, sensitivity, and high accuracy. The limitation of these methods includes professional operator requirements, high detection costs, time consumption, and unsuitability for field conditions.

Direct PCR (dPCR) is an age-old DNA amplification method and can be used directly from the sample without DNA isolation and purification. The primary difference between conventional PCR and dPCR is the use of tailored buffers that enable robust amplification defying the PCR inhibitors present in the raw sample. It has high specificity but is vulnerable to carry-over contamination. Single species dPCR is of low commercial value as often more than one species are mixed in meats. The multiplex PCR assay is an alternative to dPCR, which allows simultaneous detection of more than one species in a single reaction (Bohme *et al.*, 2019). These methods can further be improved using biochip or microchip electrophoresis instead of traditional gel electrophoresis with improved detection limits of up to a few nanograms (Li *et al.*, 2020).

Real-time PCR quantifies the DNA in the biological sample after each cycle as the reaction progresses. It uses DNA-binding dyes and sequence-specific labeled probes based on the principle of fluorescence. The linear relationship between the amount of DNA and the C_t value is used to quantify real-time-PCR, known as qPCR. There is negligible PCR contamination in this technique. So, highly specific and sensitive results for cross-species contamination in meat have been obtained using this technique, for example, chevon in pork (Wang *et al.*, 2019) and pork adulteration in meatballs (Rajharo *et al.*, 2019).

Polymerase chain reaction-restriction fragment length polymorphism (PCR-RFLP) is another DNA-based detection method that analyses variations based upon amplifying specific sequences of conserved portions of DNA using molecular scissors (restriction endonucleases). It can be effectively used to detect closely related species in meat. Examples of applications include detection of canine meat in burger formulation (Rahman *et al.*, 2015), macaque meat in the food chain (Rashid *et al.*, 2015), rabbit, rat, and squirrel meat in frankfurter products (Ali *et al.*, 2018), sea snake meat adulteration in meat products (Suntrarachun *et al.*, 2018), etc.

Loop-mediated isothermal amplification (LAMP) is a comparatively newer, more sensitive, and faster cross-species meat adulteration detection method. LAMP assay amplifies DNA in isothermal conditions, not requiring complex equipment. Additionally, the naked eye can visualize the results by using DNA binding dyes, eliminating the need for gel electrophoresis. Species-specific LAMP assay using species-specific primer set based on mt. D loop DNA sequence is quite effective in identifying meat tissues (Kumari *et al.*, 2019). Examples include onsite detection of buffalo meat (Kumari *et al.*, 2020), onsite detection of pork in processed meat products (Li *et al.*, 2016), and chicken in processed meat products (Sul *et al.*, 2019). Another interesting and novel application is paper-based LAMP for identifying tissues of cattle origin (Jawla *et al.*, 2021).

The droplet digital PCR method works on the principle of splitting the biological sample into multiple droplets (up to 20,000), and each droplet undergoes independent PCR amplification. It is a digital PCR method using a water-oil emulsion droplet system where massive sample partitioning is followed to get several discrete measurements. Alaska pollock in seafood products (Noh *et al.*, 2019), chicken in sheep or goat meat products (Ren *et al.*, 2017), and beef and pork in meat products (Cai *et al.*, 2017) detection are some examples where ddPCR has been successfully used. Often, DNA-based methods cannot identify the meats/tissues of unknown origin in the given samples. DNA barcoding and next-generation sequencing overcome this limitation by enabling multispecies detection in severely adulterated samples. This could be the future standard tool for meat quality labeling.

7.1.2. Protein-based Methods

The DNA-based methods are superior, accurate, and highly sensitive, making them the method of choice for regulatory authorities like FSSAI, FAO to ensure food safety and quality. However, factors like DNA degradation while product processing, poor trace analysis, and sampling pollution negatively affect PCR-based methods' outcomes (Naveena *et al.*, 2017). Furthermore, their complex nature and high investment and sample processing cost render them little or no use at retail or consumption stages. To beat these limitations, protein-based methods are of enormous importance. Proteins are the primary component of meat, and their specific composition, intricate structure, conservative nature, and species specificity make them a suitable substrate for meat adulteration detection. The precise nature of proteins can also help identify the nature and origin of additive/adulterants like connective tissue, plasma, milk, or other secretions and excretions in meat and meat products (Jiang *et al.*, 2018). The protein-based methods can be converted to POC tools enabling onsite detection and monitoring. Since protein-based devices are quicker, reliable, and are easy to operate. They can be broadly categorized into ELISA-based methods on antigen-antibody reaction, protein band characterization methods based upon electrophoretic analysis, and methods based upon the principle of mass spectrophotometry. Conventional electrophoretic methods are infrequently used as they are not as sensitive as newer tools like ELISA, MS-based techniques and hence tend to be insufficient to detect adulterants. Consequently, meat adulteration detection via protein-based tools mainly consists of immunoassays and mass spectrometric analysis techniques. However, ELISA kits and small potable device-dependent spectroscopy techniques are more suited to meet the needs of onsite monitoring for quick and reliable results.

Enzyme-linked immunosorbent assay (ELISA) is a biochemical technique used mainly in immunology to detect the presence of an antibody or an antigen in the given sample. Direct ELISA uses a primary antibody and is comparatively less sensitive than indirect ELISA. It has been successfully used to detect pork adulteration in beef (Seddaoui and Amine, 2020). Indirect ELISA uses a secondary antibody to amplify the signals for more precise results. Studies on the use of Indirect ELISA in the detection of pork fat protein in other animal meats (Kim *et al.*, 2017), fat adulteration in cooked and noncooked pork, beef, and chicken (Park *et al.*, 2015). Another variant is sandwich ELISA where two layers of antibodies (capture and detection antibody) bind to the antigen in the sample. This technique eliminates the sample purification step and is 2-5 times more sensitive than the previous two ELISAs. Some examples of applications of sandwich ELISA in recent years include detection of cooked wild rat meat in pork, beef, and chicken (Chen *et al.*, 2020), heated mammalian meat adulterated in poultry meats (Jiang *et al.*, 2020), and cooked chicken/turkey in pork, horse, goat, or sheep meat (Thienes *et al.*, 2019).

One major limitation of ELISA-based techniques is the occurrence of false positives due to species cross-reactivity or breakdown of proteins due to thermal processing. Use of immunosensors that uses a biosensor to convert the immune response signal into electrical, optical or any other detectable signal and then transmit and amplify the same. Although the use of immunosensors in detecting residues and adulterants in common foods like milk is established, its use in meat detection is still in its early stages. Immunosensors have been reported to detect horse meat adulteration in meat products (Masiri *et al.*, 2017), pork adulteration in raw meat (Lim and Ahmed, 2016), bovine adipose tissue in meat products (Hsieh and Gajewski, 2015).

An advanced protein-based detection method is protein mass spectrometry analysis which resolves problematic issues like false positivity and cross-reactivity among closely related species. Protein and peptide-based analysis tend to be more suitable for the meat processing industry due to the higher stability of proteins and peptides in thermally processed meat products. It is based upon the principle of identification and quantification of molecules in simple and composite samples based upon the mass to charge (m/z) ratio of ions. Gas chromatography (GS) and liquid chromatography (LC) are used for sample preparation and separation before analysis by mass spectrophotometry. This technique helps detect non-meat contaminants in meat as well. Examples include detection of duck, goose, and chicken in processed meat products (Fornal and Montowska, 2019), Buffalo, sheep, and goat meat in minced meat and meat products (Naveena *et al*., 2018), and grain proteins adulteration into meat products (Jira and Munch, 2019). The only limitation of this method is the requirement of highly trained technicians with solid mathematical backgrounds for accurate interpretation of results.

7.1.3. Metabolite Based Methods

Other metabolites, excluding proteins unique to meats, can be used for detection and identification purposes. Lipids (lipidomics) and flavor (flavoromics) compounds in meat are major metabolite profiles that can be exploited for this purpose. Fatty acid profile and volatile compounds in meat are often unique to a particular meat type and give highly sensitive results that can be generalized for identification purposes. The application involves an electric nose and GC-MS to detect adulteration in meat (Haddi *et al*., 2015). These methods are not yet widespread due to their non-quantitative nature and high multifactorial variation in these compounds.

7.2. Non-destructive Methods

Meat adulteration identification techniques based upon DNA, protein, and metabolite involve a specific sample pretreatment protocol, such as tissue disruption, extraction of the target analyte, and isolation and separation. These pretreatment protocols are invasive, cumbersome, and time-consuming. This led the researchers to look for newer, more straightforward, and non-invasive methods. Non-destructive analytical methods can be defined as techniques that involve zero alteration of the physical and chemical properties of the sample during detection and quantification of the analyte of interest. The samples subjected to such analysis suffer no modification or losses, yet they are rarely introduced to the bulk product post-analysis. They are becoming increasingly critical analytical tools in food quality testing and ensuring food safety. Spectroscopic techniques are the most common type of non-destructive technique used in meat adulteration identification. The basic principle of spectroscopic analysis of meat and meat products is that the different meat components like proteins, lipids, carbohydrates, or inorganic elements yield different spectra when subjected to different wavelengths (Rady & Adedeji, 2018). Besides their non-destructive nature, these techniques offer advantages like time-saving, minimal sample preparation, and no provision for sample pretreatment. The following techniques are explained in brief below: infrared spectroscopy (IRS), Raman spectroscopy (RS), hyperspectral imaging (HSI), and laser-induced breakdown spectroscopy (LIBS).

Infrared spectroscopy (IRS) is an optical technique that involves interaction between light (infrared) and mass (sample/analyte). The infrared light at varying frequencies is shone on the sample, leading to absorption of specific frequencies by it, and this selective absorption of light helps identify the substance. Organic compounds in meat differ in their molecular structure, chemical bonds, and these fine differences are picked up by this technique (Fu and Ying, 2016). The differences in absorption spectra of various meats can be exploited to develop and standardize meat adulteration determination.

Another non-destructive optical technique complementary to IRS that can be used is Raman Spectroscopy (RS). It is based upon the principle of the Raman effect, which involves the scattering of light at different wavelengths due to molecular vibrations of the sample. The Raman spectra obtained after lighting a sample using a high-intensity laser beam can be analyzed in a spectrometer to arrive at accurate conclusions regarding the nature of additives in a meat sample. Partial least squares-discriminant analysis (PLS-DA) model and RS based on functional groups of essential and specific constituents of various proteins, lipids, peptides of meat as well as non-meat ingredients can help in the detection of chemical as well as biological ingredients in meat products (Nunes *et al*., 2019). Insensitivity to water, no sample preparation, non-invasiveness, and small sample volume requirement are a few characteristics of RS analysis. Most importantly, it is suitable for opaque samples (Lee *et al*., 2018).

IRS and RS methods are less practical if the adulterants are unevenly distributed in a sample as these techniques measure the sample locally or a specific portion of it. In such cases, spectroscopic tools need to be combined with imaging techniques to build a concrete image highlighting the actual characteristics of the sample of interest. The use of multispectral imaging (MSI) and hyperspectral imaging (HSI) can help overcome the shortcomings of IRS and RS. HIS and MSI use the findings of spectral measurement like amount and wavelength emitted or reflected and combine them with digital imaging. The results are extremely sensitive to the chemistry of the material, physical characteristics, and biological attributes of the analytes. HIS and MSI in multiple linear regression (MLR) models have been used to detect pork adulteration in beef (Kamruzzaman *et al*., 2015).

Lastly, Laser-induced breakdown spectroscopy (LIBS) is a rapid, advanced, and non-invasive chemical tool that uses a short laser pulse to create a microplasma (via laser ablation) on the surface of the given sample and analyte. While cooling down, this microplasma emits light with discrete spectral peaks that are detected, analyzed, and ultimately help in sample characterization. This technology is high-speed (only a few seconds needed), accurate, versatile (a vast range of covered elements), and a thin sample analysis without substrate interference. As meat and meat products have different compositions and inherently different elemental characteristics, spectral fingerprinting techniques like LIBS can also help detect minute tamperings. Other advantages are similar to the other optical tools mentioned earlier in the text. Table 3 summarizes various detection methods for different adulterants of meat and can be adopted for fish with slight modification.

Table 3: Various detection methods based on Calorimetric, Visual/Organoleptic, Physical/Manual, PCR-DNA, HPLC, GC-MS, etc. techniques used for different adulterants

Technique	Principle	Adulterant	Species/ Product	Reference
Calorimetric	Differential scanning calorimetry (DSC)	Wild boar meat	Beef meatballs	Guntarti *et al.*, 2017
	Gold nanoparticle-based biosensor	Pork	Halal meat products	Kuswandi *et al.*, 2017
DNA based methods (Lab-based)	Multiplex PCR	Fox, dog, mink, and rabbit	Any other meat	Wu *et al.*, 2018
	TaqMan real-time PCR	Cattle and horse	Any other meat	Wang *et al.*, 2019
	SYBR Green real-time PCR	Wild boar meat	Meatball formulation	Arini *et al.*, 2018
	Eva Green real-time PCR	Pork	Food samples	Lubis *et al.*, 2018
	PCR-RFLP	Beef, buffalo, and pork	Food samples	Hossain *et al.*, 2016
DNA based point of care methods (POCs) / Visual method	RPA-LFA	Extended-spectrum β-lactamase (ESBL)-producing Escherichia coli	Pork	Kanokudom *et al.*, 2021
	LAMP-LFA	Horse and donkey meat	Any other meat	Zhang *et al.*, 2019
	SEA LFA	Pork and duck	Any other meat	Wang *et al.*, 2020
	RPA-LFA	Sheep	Any other meat	Li *et al.*, 2019
	CPA-LFA	Cow, sheep, arctic fox, and pig meat	Any other meat	Zheng *et al.*, 2020
	PCR-LFA	Duck	Beef	Qin *et al.*, 2019
	Paper-based LAMP-LFA	Pork	Any other meat	Jawla *et al.*, 2021(a)
	Paper-based LAMP-LFA	Beef	Any other meat	Jawla b *et al.*, 2021 (b)
	PCR-LFA	Turkey	Any other meat	Wang *et al.*, 2020

Protein-based	Nano-LC-QTOFMS/MS	Chicken, duck, and goose meat	Processed meat products	Montowska and Fornal, 2017
	HPLC-MS/MS	Grain proteins	Processed meat products	Jira and Munch, 2019
	Electrochemical immunosensor	Pork	Halal meat	Mandli *et al.*, 2018
	Sandwich ELISA	Cooked wild rat meat	Pork, beef, and chicken	Chen *et al.*, 2020
Metabolite based	Electric nose	Pork	Halal meat	Sarna *et al.*, 2020
Spectroscopic analysis	FT-IRS	Pork	Beef and mutton	Yang *et al.*, 2018
	NIRS	Pork, chicken, dog, and horse meat	Minced lamb and beef	Lopez Maestresalas *et al.*, 2019

7.3. Other Methods of Detection are Available for Fish

Traditional techniques for analyzing harmful chemicals are time-consuming, so recently, ICAR's Central Institute of Fisheries Technology (ICAR-CIFT, 2017) has developed paper strip methods for rapidly detecting formaldehyde and ammonia. These detection methods are simple, fast, and based on visual observation of color development in 2 minutes. The kit contains 25 test strips, a reagent solution, and a standard card for color comparison. The strip should be rubbed on the surface of the fish, and then 1-2 drops of the supplied reagent solution should be added to achieve a color change.

7.3.1. Schiff's Method for Detection of Formalin

Formalin detection becomes difficult since it is injected deep into the flesh; the smell may not be noticeable until cut. Therefore, choosing alternative tests to popular types will reduce the chances of finding formalin-treated fish. Schiff's reagent is very useful for formalin detection, as Schiff's reagent reacts with formaldehyde or formalin and changes the solution to pink or violet.

7.3.2. Nessler's Method for Detection of Ammonia

The Nessler's reagent (K2HgI4) reacts with the ammonia in the sample (under highly alkaline conditions) to produce a yellow-colored species. The intensity of the color is directly proportional to the concentration of ammonia in the fish sample.

8. Food Recall Protocol and Regulation by FSSAI

Fresh, wholesome, unadulterated, hygienic, uncontaminated, safe, and correctly labeled meat product is the duty of the produce and right of the consumer. So, any violation of these provisions needs to be brought under the scrutiny of regulatory authorities (Iammarino *et al.*, 2017). FSSAI is vested with this responsibility in India and continuously develops various standards and guidelines to monitor food quality.

Hence, any food business operator (FBO) under the FSSAI Act, 2006 can be subjected to regulatory action. A food recall is effective against adulterated/defective/risky products. It involves removing the concerned product that risks consumer safety from sale, distribution, and consumption. The product can be removed from the market at any stage of the food chain. Figure 2 attempts to provide a representative overview and step-by-step breakdown of a food recall procedure and its implementation (FSSAI, 2017).

Fig. 2: A Step-by-step breakdown of food recall protocol (FSSAI, 2017)

9. Commercially Available Kits for Detection of Adulterants in Meat

Several DNA and protein-based kits are already available in the market for various meat species (Table 4).

Table 4: Commercially available kits for detection of adulterants in meat

Manufactured By	**Trade name**	**Principle**
Thermofisher Scientific	RapidFinder™ Quant Multi-Meat Set for use with following kits: ➢ RapidFinder™ Beef ID Kit ➢ RapidFinder™ Pork ID Kit ➢ RapidFinder™ Halal ID Kit ➢ RapidFinder™ Poultry ID Kit ➢ RapidFinder™ Chicken ID Kit ➢ RapidFinder™ Turkey ID Kit ➢ RapidFinder™ Equine ID Kit ➢ RapidFinder™ Sheep ID Kit ➢ RapidFinder™ Goat ID Kit ➢ RapidFinder™ Fallow Deer ID Kit	Real-time PCR detection of DNA
ELISA Technologies	➢ ELISA-TEK™ Raw Meat Species Test Kits ➢ ELISA-TEK™ Cooked Meat Species Test ➢ Kits MELISA-TEK™ Test Kit (Pork or Ruminant) ➢ MELISA-TEK™ Ruminant High Sensitivity Extraction Kit	ELISA
Bio-Check	Raw Meat FlowThrough™	Visually assessed immunoassay
Perkin Elmer	AuroFlow™ Fluoroquinolone Strip Test Kit for Meat	Lateral flow assay
Neogen	BioKits for Speciation Identification	Sandwich-based ELISA
Neogen	Reveal® for Sheep	Immunochromatographic lateral flow assay
Nelson Jameson	Premi®Test Antibiotic Test Kits (Antibiotic test for meat, fish, eggs and feed)	Microbiological-based screening tube-test
ICAR-CIFT kits for formaldehyde and ammonia in fish	Kit contains 25 test strips	Visually assessed color change

10. Conclusion

Although surplus technological tools to detect adulterants in raw and cooked livestock meat and fish have been developed, their commercial adaptation is low to negligible due to their high cost and complicated nature. Firstly, the upcoming research should bring down the cost and complexity of existing detection techniques. Secondly, the development of improved tools with minimal sample pre-processing steps should be considered instead of bulky equipment, miniature user-friendly devices that integrate sample pre-processing and concentration steps and easily visualize results interpreted at retail and consumer levels. Thirdly, integrating advanced biotechnological tools like next-generation sequencing with innovative information technologies is the future of meat quality and safety. From a legislative viewpoint, redefinition and stringent regulation of existing standards for additives and preservatives in meat and fish products by competent global regulatory authorities is the need of the hour. It is noteworthy that the research in meat and fish products authentication is highly dynamic, constantly amalgamating analytical and molecular techniques to get quicker, accurate, reliable, and highly sensitive results with excellent repeatability. Lastly, an important facet is better surveillance, monitoring, and, if needed, recall protocols to ensure high-quality meat and fish products to the end consumers.

References

Adeniji, A.E., Alatise, O.O., Nwanya, A.C. 2013. Radionuclide concentration in some fruit juices produced and consumed in Lagos, Nigeria. American Journal of Environmental Protection, 2 (2), 37-41.

Ali, M. E., Ahamad, M.N.U., Asing, Hossain, M.A.M., & Sultana, S. (2018). Multiplex polymerase chain reaction-restriction fragment length polymorphism assay discriminates of rabbit, rat and squirrel meat in frankfurter products. Food Control, 84, 148–158.

ANSES (2011). Etude de l'alimentationtotalefrançaise 2 (EAT 2). Avis de l'ANSES, rapport d'expertise.

ANSES (2012). Avis de l'Agencenationale de sécurité sanitaire de l'alimentation, de l'environnement et du travail relatifauX analyses de retardateurs de flammebromés (RFB) à mettreenœuvre dans le cadre des prochains plans de surveillance. Avis de l'Anses (Saisine n° 2010-SA-0225).

Arini, R. L., Ramadhani, D., Pebriyanti, N. W., Sismindari, S., & Rohman, A. (2018). The use of species-specific primer targeting on D-loop mitochondrial for identification of wild boar meat in meatball formulation. Journal of Advanced Veterinary and Animal Research, 5, 361.

Audebert, M., Zeman, F., Beaudoin, R., Pery, A., &Cravedi, J. P. (2012). Comparative potency approach based on H2AX assay for estimating the genotoXicity of polycyclic aromatic hydrocarbons. Toxicology and Applied Pharmacology, 260 (1), 58–64.

Böhme, K., Calo-Mata, P., Barros-Velázquez, J., &Ortea, I. (2019). Review of recent DNA-based methods for main food-authentication topics. Journal of Agricultural and Food Chemistry, 67(14), 3854-3864.

Cai, Y., He, Y., Lv, R., Chen, H., Wang, Q., & Pan, L. (2017). Detection and quantification of beef and pork materials in meat products by duplex droplet digital PCR. PLoS One, 12, e0181949.

Chen, X., Ran, D., Zeng, L., & Xin, M. (2020). Immunoassay of cooked wild rat meat by ELISA with a highly specific antibody targeting rat heat resistant proteins. Food and Agricultural Immunology, 31, 533–544.

Codex Alimentarius Commission. 1992. Section 6.1: Guideline levels for radionuclides in foods following accidental nuclear contamination for use in international trade, Codex Alimentarius, 1. General requirements, Rome, FAO/WHO, (2nd ed.)

FAO. 1987. Report of the Expert Consultation on Recommended Limits for Radionuclide Contamination of Food, Rome, 1-5 December 1986. ESN/MISC/87/1, Rome, FAO.

Food safety and standards (contaminants, toxins and residues) regulations, 2011.

FSS (Recall Procedure) Regulations, 2017

Fu, X., & Ying, Y. (2016). Food safety evaluation based on near infrared spectroscopy and imaging: A review. Critical Reviews in Food Science and Nutrition, 56, 1913–1924.

Guntarti, A., Rohman, A., Martono, S., & Yuswanto, A. (2017). Authentication of wild boar meat in meatball formulation using differential scanning calorimetry and chemometrics. Journal of Food and Pharmaceutical Sciences, 5(1), 8-12.

Haddi, Z., El Barbri, N., Tahri, K., Bougrini, M., El Bari, N., Llobet, E., Bouchikhi, B. (2015). Instrumental assessment of red meat origins and their storage time using electronic sensing systems. Analytical Methods, 7, 5193–5203.

Hossain, M. A., Ali, M. E., Abd Hamid, S. B., Asing, Mustafa, S., Mohd Desa, M. N., & Zaidul, I. S. (2016). Double gene targeting multiplex polymerase chain reaction-restriction fragment length polymorphism assay discriminates beef, buffalo, and pork substitution in frankfurter products. Journal of Agricultural and Food Chemistry, 64, 6343–6354.

Hsieh, Y. P., & Gajewski, K. (2015). Rapid detection of bovine adipose tissue using lateral flow strip assay. Food Science & Nutrition, 4, 588–594.

IAEA. 1986, Derived intervention levels for application in controlling radiation doses to the public in the event of a nuclear accident or radiological emergency. Safety Ser. No. 81. Vienna, IAEA.

Iammarino, M., Marino, R., &Albenzio, M. (2017). How meaty? Detection and quantification of adulterants, foreign proteins and food additives in meat products. International Journal of Food Science & Technology, 52(4), 851-863.

Jawla, J., Kumar, R. R., Mendiratta, S. K., Agarwal, R. K., Kumari, S., Saxena, V., & Rana, P. (2021b). based loop-mediated isothermal amplification and lateral flow (LAMP-LF) assay for identification of tissues of cattle origin. Analytica Chimica Acta, 1150, 338220.

Jawla, J., Kumar, R. R., Mendiratta, S. K., Agarwal, R. K., Singh, P., Saxena, V., & Rana, P. (2021a). On-site paper-based Loop-Mediated Isothermal Amplification coupled Lateral Flow Assay for pig tissue identification targeting mitochondrial CO I gene. Journal of Food Composition and Analysis, 104036.

Jiang, X., Fuller, D., Hsieh, Y. P., & Rao, Q. (2018). Monoclonal antibody-based ELISA for the quantification of porcine hemoglobin in meat products. Food Chemistry, 250, 170–179.

Jiang, X., Rao, Q., Mittl, K., & Hsieh, Y. P. (2020). Monoclonal antibody-based sandwich ELISA for the detection of mammalian meats. Food Control, 110, 107045.

Jira, W., & Munch, S. (2019). A sensitive HPLC-MS/MS screening method for the simultaneous detection of barley, maize, oats, rice, rye and wheat proteins in meat products. Food Chemistry, 275, 214– 223.

Kamruzzaman, M., Makino, Y., &Oshita, S. (2015). Hyperspectral imaging in tandem with multivariate analysis and image processing for non-invasive detection and visualization of pork adulteration in minced beef. Analytical Methods, 7, 7496–7502.

Kan, C. A., & Meijer, G. A. L. (2007). The risk of contamination of food with toXic substances present in animal feed. Animal Feed Science and Technology, 133, 84–108.

Kanokudom, S., Assawakongkarat, T., Akeda, Y., Ratthawongjirakul, P., Chuanchuen, R., & Chaichanawongsaroj, N. (2021). Rapid detection of extended spectrum β-lactamase producing Escherichia coli isolated from fresh pork meat and pig cecum samples using multiplex recombinase polymerase amplification and lateral flow strip analysis. Plos one, 16(3), e0248536.

Khatun, MA., Hossain, A., Hossain, M.S., Munshi, Huque R. 2021. Detection of species adulteration in meat products and Mozzarella-type cheeses using duplex PCR of mitochondrial *cyt b* gene: A food safety concern in Bangladesh. Food Chemistry: Molecular Sciences, 2, 100017.

Kim, J., Kwon, I. J., Kim, M., Chang, J. Y., & Shim, W. (2017). Production and preliminary characterization of monoclonal antibodies highly specific to pork fat protein. Food Control, 79, 80–86.

Kumar, Y., &Narsaiah, K. (2021). Rapid point-of-care testing methods/devices for meat species identification: A review. Comprehensive Reviews in Food Science and Food Safety, 20(1), 900-923.

Kumari, S., Kumar, R. R., Mendiratta, S. K., Kumar, D., Rana, P., Kumar, D., &Jawla, J. (2019). Species-specific loop-mediated isothermal amplification (LAMP) assay for identification of tissue of cattle origin by targeting mitochondrial gene sequences. 3 Biotech, 9(3), 69.

Kuswandi, B., Gani, A. A., Kristiningrum, N., & Ahmad, M. (2017). Simple colorimetric DNA biosensor based on gold nanoparticles for pork adulteration detection in processed meats. Sensors & Transducers, 208(1), 7.

LeDoux, M. (2011). Analytical methods applied to the determination of pesticide residues in foods of animal origin. A review of the past two decades. Journal of chromatography A, 1218(8), 1021-1036.

Lee, J. Y., Park, J. H., Mun, H., Shim, W. B., Lim, S. H., & Kim, M. G. (2018). Quantitative analysis of lard in animal fat mixture using visible Raman spectroscopy. Food Chemistry, 254, 109–114.

Lee, S.-Y., Kim, M.-J., Hong, Y., & Kim, H.-Y. (2016). Development of a rapid on-site detection method for pork in processed meat products using real-time loop-mediated isothermal amplification. Food Control, 66, 53–61.

Li, Y. C., Liu, S. Y., Meng, F. B., Liu, D. Y., Zhang, Y., Wang, W., & Zhang, J. M. (2020). Comparative review and the recent progress in detection technologies of meat product adulteration. Comprehensive Reviews in Food Science and Food Safety, 19(4), 2256-2296.

Lim, S. A., & Ahmed, M. U. (2016). A label free electrochemical immunosensor for sensitive detection of porcine serum albumin as a marker for pork adulteration in raw meat. Food Chemistry, 206, 197 203.

Lo, Y.T., Shaw, P.C. 2018. DNA-based techniques for authentication of processed food and food supplements. Food Chemistry, 240, 767-774.

Lopez-Oceja, A., Nuñez, C., Baeta, M., Gamarra, D., & de Pancorbo, M. M. (2017). Species identification in meat products: A new screening method based on high resolution melting analysis of cyt b gene. Food Chemistry, 237, 701–706.

Lubis, H., Salihah, N. T., Norizan, N. A., Hossain, M. M., & Ahmed, M. U. (2018). Fast and sensitive real-time PCR-based detection of porcine DNA in food samples by using EvaGreen dye. Food Science and Technology Research, 24, 803–810.

MaïaMeurillon, Jérémy Ratel, & Erwan, Engel (2018) How to secure the meat chain against toxicants? Innovative Food Science and Emerging Technologies 46, 74–82.

Mandli, J., El Fatimi, I., Seddaoui, N., & Amine, A. (2018). Enzyme immunoassay (ELISA/ immunosensor) for a sensitive detection of pork adulteration in meat. Food Chemistry, 255, 380–389.

Marroquín-Cardona, A. G., Johnson, N. M., Phillips, T. D., & Hayes, A. W. (2014). MycotoXins in a changing global environment - A review. Food and Chemical Toxicology, 69, 220–230.

Masiri, J., Benoit, L., Thienes, C., Kainrath, C., Barrios-Lopez, B., Agapov, A., Samadpour, M. (2017). A rapid, semi-quantitative test for detection of raw and cooked horse meat residues. Food Control, 76, 102–107.

Montowska, M., & Fornal, E. (2017). Label-free quantification of meat proteins for evaluation of species composition of processed meat products. Food Chemistry, 237, 1092–1100.

Naveena, B. M., Jagadeesh, D. S., Babu, A. J., Rao, T. M., Kamuni, V., Vaithiyanathan, S., Rapole, S. (2017). OFFGEL electrophoresis and tandem mass spectrometry approach compared with DNAbased PCR method for authentication of meat species from raw and cooked ground meat mixtures containing cattle meat, water buffalo meat and sheep meat. Food Chemistry, 233, 311–320.

Noh, E. S., Park, Y. J., Kim, E. M., Park, J. Y., Shim, K. B., Choi, T. J., Kang, J. H. (2019). Quantitative analysis of Alaska pollock in seafood products by droplet digital PCR. Food Chemistry, 275, 638– 643.

Nunes, K. M., Andrade, M. V. O., Almeida, M. R., Fantini, C., & Sena, M. M. (2019). Raman spectroscopy and discriminant analysis applied to the detection of frauds in bovine meat by the addition of salts and carrageenan. Microchemical Journal, 147, 582–589.

Park, B. S., Oh, Y. K., Kim, M. J., & Shim, W. B. (2015). Skeletal muscle troponin I (TnI) in animal fat tissues to be used as biomarker for the identification of fat adulteration. Korean Journal for Food Science of Animal Resources, 34, 822–828.

Qin, P., Qiao, D., Xu, J., Song, Q., Yao, L., Lu, J., & Chen, W. (2019). Rapid visual sensing and quantitative identification of duck meat in adulterated beef with a lateral flow strip platform. Food Chemistry, 294, 224–230.

Rady, A., & Adedeji, A. (2018). Assessing different processed meats for adulterants using visible-near-infrared spectroscopy. Meat Science, 136, 59–67.

Raharjo, T. J., Chudori, Y. N. C., & Agustina, F. W. (2019). TaqMan probe real-time polymerase chain reaction targeting the ATPase 6 gene for the detection of pork adulteration in meat and meatballs. Journal of Food Safety, 39(6), e12715.

Raharjo, T. J., Nuryanti, I., Patria, F. P., & Swasono, R. T. (2018). Mitochondrial ND-1 gene-specific primer polymerase chain reaction to determine mice contamination in meatball. International Food Research Journal, 25, 638–642

Rahman, M. M., Ali, M. E., Hamid, S. B. A., Bhassu, S., Mustafa, S., Al Amin, M., & Razzak, M. A. (2015). Lab-ona-chip PCR-RFLP assay for the detection of canine DNA in burger formulations. Food Analytical Methods, 8, 1598–1606.

Ramesh, A., Walker, S. A., Hood, D. B., Guillen, M. D., Schneider, K., &Weyand, E. H. (2004). Bioavailability and risk assessment of orally ingested polycyclic aromatic hydrocarbons. International Journal of Toxicology, 23(5), 301−333.

Rashid, N. R., Ali, M. E., Hamid, S. B., Rahman, M. M., Razzak, M. A., Asing, & Amin, M. A. (2015). A suitable method for the detection of a potential fraud of bringing macaque monkey meat into the food chain. Food Additives and Contaminants Part A, 32, 1013–1022.

Ren, J., Deng, T., Huang, W., Chen, Y., & Ge, Y. (2017). A digital PCR method for identifying and quantifying adulteration of meat species in raw and processed food. PLoS One, 12, e0173567.

Sarno, R., Triyana, K., Sabilla, S. I., Wijaya, D. R., Sunaryono, D., & Fatichah, C. (2020). Detecting Pork Adulteration in Beef for Halal Authentication Using an Optimized Electronic Nose System. IEEE Access, 8, 221700-221711.

Seddaoui, N., & Amine, A. (2020). A sensitive colorimetric immunoassay based on poly(dopamine) modified magnetic nanoparticles for meat authentication. LWT-Food Science and Technology, 122, 109045.

Sul, S., Kim, M. J., & Kim, H. Y. (2019). Development of a direct loop-mediated isothermal amplification (LAMP) assay for rapid and simple on-site detection of chicken in processed meat products. Food Control, 98, 194–199.

Suntrarachun, S., Chanhome, L., & Sumontha, M. (2018). Identification of sea snake meat adulteration in meat products using PCR-RFLP of mitochondrial DNA. Food Science and Human Wellness, 7, 170–174.

Thienes, C. P., Masiri, J., Benoit, L. A., Barrios-Lopez, B., Samuel, S. A., Meshgi, M. A., Samadpour, M. (2019). Quantitative detection of chicken and turkey contamination in cooked meat products by ELISA. Journal of AOAC International, 102, 557–563.

Wang, W. J., Liu, J. J., Zhang, Q. D., Zhou, X., & Liu, B. (2019). Multiplex PCR assay for identification and quantification of bovine and equine in minced meats using novel specific nuclear DNA sequences. Food Control, 105, 29–37.

Wang, X., Wang, X., Shi, C., Ma, C., & Chen, L. (2020). Highly sensitive visual detection of nucleic acid based on a universal strand exchange amplification coupled with lateral flow assay strip. Talanta, 216, 120978.

Wang, Z. C., Wang, Z. Y., Li, T. T., Qiao, L., Liu, R., Zhao, Y., Chen, A. L. (2019). Real-time PCR based on single-copy housekeeping genes for quantitative detection of goat meat adulteration with pork. International Journal of Food Science and Technology.

Wang, Z., Wang, Z., Li, T., Qiao, L., Liu, R., Zhao, Y., & Chen, A. (2020). Real-time PCR based on single-copy housekeeping genes for quantitative detection of goat meat adulteration with pork. International Journal of Food Science & Technology, 55(2), 553-558.

Wu, Q. Q., Xiang, S. N., Wang, W. J., Zhao, J. Y., Xia, J. H., Zhen, Y. R., & Liu, B. (2018). Species identification of fox-, mink-, dog- , and rabbit-derived ingredients by multiplex PCR and real-time PCR assay. Applied Biochemistry and Biotechnology, 185, 1–12.

Yang, L., Wu, T., Liu, Y., Zou, J., Huang, Y. M., Babu, V. S., & Lin, L. (2018). Rapid identification of pork adulterated in the beef and mutton by infrared spectroscopy. Journal of Spectroscopy, 6, 1–10.

Zhang, C., Zhang, X., Liao, G., Shang, Y., Ge, C., Chen, R., Xu, W. (2019). Species-specific TM-LAMP and trident-like lateral flow biosensor for on-site authenticity detection of horse and donkey meat. Sensors and Actuators B: Chemical, 301, 127039.

Zheng, F., Li, S., Wang, S., Feng, T., Jiang, Z., & Pan, J. (2020). Crosspriming isothermal amplification combined with nucleic acid test strips for detection of meat species. Analytical Biochemistry, 597, 113672.

18

Adulterations and Contaminants in Dietary Supplements/ Nutraceuticals An Overview

K R Anilakumar

Defence Food Research
Defence Research and Development Organization, Mysore, Karnataka, India

In recent years dietary supplements have gained attractiveness within fitness and wellness circles. The other commonly used terms areNutraceuticals, dietary supplements or wellness-focussed foods. Being highly nutritious they are associated with a number of potential therapeutic effects. Nutrient-rich ingredients like fruits, vegetables, nuts, seeds, and grains are often considered dietary supplements as well. Oats fibre called, beta glucan, which has been shown to reduce inflammation, enhance immune function, and perk up heart health.

1. Conventional and Non-conventional Dietary Supplements

Nutrient-rich foods like fruits, vegetables, and legumes are often considered dietary supplements. Meanwhile, modified foods have been fortified with micro and macro nutrients to boost health benefits of foods.

Table 1: Conventional dietary supplements

S.No.	Conventional dietary supplements	Examples
1.	Fruits	Berries, citrus and non-citrus, etc.
2.	Vegetables	Broccoli, Cauliflower, Cabbage, Kale, Okra, Spinach, Beet roots
3.	Nuts	Almonds, Cashews, Pistachios, Brazil nuts
4.	Seeds	Chia seeds, Flax seeds, hemp seeds, Pumpkin seeeds, Watermelon seeds
5.	Legumes	Various kinds of beans and Lentils
6.	Whole grains	Oats, Barley, Buckwheat, Brown rice, Black rice, Bamboo rice
7.	Seafood	Salmon, Sardines, Mackerel, Anchovies, Cod
8.	Fermented foods	Tempeh, Kombucha, Kimchi, Kefir, Sauerkraut

S.No.	Conventional dietary supplements	Examples
9.	Spices	Turmeric, Cinnamon, Ginger, Cayenne pepper, Saffron, Fenugreek, Coriander
10.	Beverages	Coffee, Green tea, Black tea, Olong tea

The fortified foods along with commonly used fortificants are given in Table 2.

Table 2: Fortified foods and fortificants

S.No.	Fortified Foods	Fortificants
1.	Fortified juices	Micro nutrients (Vitamins,Minerals,etc) and Electrolytes
2.	Dairy products that are fortified	Vitamin A
3.	Fortified Milk alternatives	Vitamin A
4.	Fortified grains	Vitamins and minerals
5.	Forified cereal and granola	Vitamins and minerals
6.	Fortified eggs	Fatty acids, and Carotenoids

2. Fortified Foods

It gives opportunities to enjoy the nutrient benefits filling up the gaps of nutrient deficiencies.

a. Fortified cereal

Fortified foods contain added micro-nutrients like vitamins and trace elements that are not naturally present in them. Fortification is meant to get better people's levels of particular nutrients and is general for foods that adults and children typically eat, such as grains, milk, and juice. Cereal is one of the most commonly fortified foods. For example, 1 40 grams of fortified total cereal boasts 40 mg of iron which is 100% of the 'daily value' .As the same size serving of an unfortified wheat cereal meets only 10% of the 'daily value'. Here, much of iron content in breakfast cereals' may be due to fortification. It's important to checkour nutrient intake, as many people in the developed countries also don't consume enough iron, calcium, or vitamins A, C, D, and E. As a result deficiencies may lead to negative health effects. Breakfast cereals are commonly fortified with the nutrients: retinol, thiamine, riboflavin, niacin, pyridoxines, cyanocobalamine, vitamins D_2 and $D_{3,}$ folic acid, zinc, iron, calcium, etc.

Food companies frequently fortify ready-to-eat, pre-packaged cereals—and sometimes hot cereals like oatmeal. Conversely, fortified cereals are not inherently healthy. While some are made with whole grains and high in fibre and protein, others contain more or less no nutrients.For example, fortified cereals from some affluent manufacturers offer nofibre and only 1 gram of protein per 29 grams. One can tell if a cereal is fortified or not because the added nutrients will be labelled on the packaging. Often, below the ingredient list, there's a list of vitamins and minerals used to fortify the product. It is important to note that fortification varies from region to region. It's more regular to find fortified cereals in Western countries. What's more, certain countries, including the United States, Canada, Costa Rica, Chile, and South Africa, mandate the fortification of *atta* with vitamin B_9 so it's more common to find folic-acid-enriched cereals in these places. Conspicuously, cereals that are less heavily processed are less likely to be fortified as the method employed will help in retaining the nutrients to a certain extent. For example, muesli usually consists of unfortified whole oats, nuts, seeds, and dried fruit.

Health benefits of fortified cereals

a. Improved nutrient intake

Many people in the United States do not meet the dietary recommendations for certain vitamins and minerals. As such, eating fortified foods may aid in getting these micro nutrients. A recent study noted that eating fortified foods boosted the intake of folic acid and vitamins A and C. Some people, such as young children, vegetarians, and pregnant or breastfeeding women, may benefit particularly from fortified cereals due to their increased nutrient needs. That said, fortified foods may increase your risk of exceeding certain nutrient recommendations.

b. Reduced birth defects risk

Fortifying cereal grains with folate, a synthetic form of folic acid — has successfully reduced the incidence of neural tube defects, which are one of the most common birth defects in North America. Folate is a B vitamin necessary for proper growth and development. In reality, all women of childbearing age are suggested to consume 400 mcg of B_9 daily and eat folate-rich foods. Therefore, fortified cereal may gain women who are or may become pregnant.

Downsides of fortified cereal

While fortification can enhance nutrient content, cereal is still not essentially healthy.

a. Loaded with sugar and refined carbs

Many fortified cereals are high in added sugar and refined carbohydrates. Added to that, most people eat more than the recommended serving size. In fact, a study in 72 adults determined that people ate 200% of the labelled serving size, on average. For instance, 53 grams of Raisin Bran Crunch packs 13 grams of added sugar. Doubling that portion size would provide a whopping 26 grams of added sugar.

According to the American Heart Association (AHA), women and men should limit their daily intake of added sugar to 25 and 37.5 grams, respectively. This means that a bowl or two of fortified cereal could easily put you close to — or even above — your daily sugar limit.

Not only do Americans tend to already go beyond guidelines for sugar intake, but diets high in added sugars are also correlated with an increased risk of persistent conditions like obesity, heart disease, and diabetes.

b. Misleading health claims

Many manufacturers label their cereals with misleading health claims, such as "low-fat" or "whole-grain". This is deceptive because the primary ingredients are usually refined grains and sugar. For example, Honey Nut Cheerios are marketed as helping lower cholesterol. Yet, a 37.0 gram serving contains 12 grams of sugar. Research suggests that diets high in added sugar raise your risk of heart disease. Such misleading claims may direct people to overindulge in foods that aren't healthy. What's more, many fortified cereals are marketed to young adults and children. Studies reveal that advertisements have an effect on children's taste preferences and may contribute to the risk of obesity. As such, one should read labels carefully to avoid any deceptive claims. It's best to choose cereals that are low in sugar and high in fibre. It is advised to look for types with fewer than 6 grams of sugar and at least 3 grams of fibre per serving. Fibre can help boost fullness and reduce cholesterol levels, among other benefits. Since many cereals lack protein, include a protein source to create a more satisfying, balanced meal. Consider adding Greek yogurt, nuts, or peanut butter. On the other hand the best option for a nutrient-rich breakfast is whole, unprocessed foods, such as oatmeal, yogurt, fruit, or eggs.

The rise and evolution of dietary supplements

The global nutraceutical market is projected to reach USD 826,426.3 million by 2028, expanding at a CAGR of 8.9% over the forecast period. Rising awareness regarding calorie lessening and weight loss in countries, including U.S., China, and India, is expected to support the application of nutraceuticals, which, in turn, will have a significantimpact on the industry growth. Demand for nutraceuticals is increasing at the global level as it improves the digestive and immune systems and enhances the cognitive behaviour of consumers. Growing drift among consumers to vary dietary habits is likely to boost up the demand for nutraceuticals. The consumer believes that improper diet results in an increase in costs on pharmaceuticals are anticipated to boost the demand for nutraceuticals. This would also support the government as it would result in lesser expenditure on healthcare and low social security costs. The mount in disposable income, increasing consumer awareness concerning health issues, and rapid urbanization are likely to augment the nutraceutical market growth over the predicted years. Positive outlook towards medical nutrition owing to increasing weight management programs, along with cardiovascular diseases, is anticipated to propel the demand for nutraceuticals. The rise and evolution of wellness-focused diets, viz. keto and paleo, are driving food producers to organize their products in this direction. Functional food products, such as probiotics and omega-3, are highly used in yogurt and fish oils in order to reduce the risk of cardiovascular diseases and develop the quality of intestinal microflora, which is further projected to stimulate the market growth of the nutraceutical products over the coming years. Based on dietary supplement product, prebiotics and probiotics are projected to expand at a CAGR of 10.1% over the forecast period in consequence of increasing usage of probiotics as a most important ingredient among the majority of food manufacturing companies to supply better nourishment and lessen health problems caused by harmful bacteria. Micro- nutrients such as vitamins and trace elements jointly accounted for over 40.71% share in 2019 owing to increasing consumer interest in the association between diet and health. This, in turn, has increased demand for these products. In terms of functional food product, vitamins emerged as the largest segment in 2019 and are expected to get bigger at a CAGR of 5.8% in the coming years. In terms of product, functional food accounted for the largest share in 2019 and generated revenue of USD 187.51 billion. Rising healthcare costs, in concert with growing geriatric population across the world, are anticipated to drive the dietary supplement segment over the forecast period.Asia Pacific is expected to spectate the fastest growth over the forecast period. Expansion of the trade chain in countries, such as India, Japan, and China, has increased the consumption of functional foods and nutraceuticals in this region. The dietary supplement industry represents a highly reasonable landscape. Key players dominate the nutraceutical market space and have been focusing on various strategic initiatives, including mergers and acquisitions, product modernization, and portfolio extension.

3. Adulterants and Contaminants

a. Heavy metals and other contaminants (Deleterious Substances, filth and foreign matter) in herbal preparations and nutraceuticals

Herbal medicines are being used as dietary supplements worldwide. *Ayurveda* is a traditional medicine system, is native to India, and is also used in many parts of world as an alternative to standard healing regimens. In the recent past the use of these supplements in the form of decoctions, pills and powders are used for alleviating generative diseases. They are utilized as traditional folk remedies for thousands of years (Maiga *et al.*, 2005). Medicinal plants attained an increasingly significant role in the pharmaceutical, dietary supplements, and natural cosmetic industries (Kim *et al.*, 2016). It was reported that a total of 60,107 COVID-19 cases (85.20% of the total cases) in China were treated by Chinese herbal medicines with positive results in all infectivity stages. This included ofsymptoms, lessened rates of deterioration and mortality, quick recovery as well as disease prevention on February 17, 2020. However, concerns raisedonthe safety of herbal medicines after studies indicated that extraordinary levels of heavy metals were present in some botanical medicines. This was a delinquent more recurrently encountered in traditional medical therapy, considered to be a cause of several health disorders (Maiga *et al.*, 2005; Brookes *et al.*, 2019). Samples taken from both developed and developing countries have shown enormous levels of hypothetically toxic heavy metals in products available to the public (Awodele *et al.*, 2013). It is known that numerous traditional medicines might give rise to adverse renal pathology. The underlying mechanism is yet to be acertained but it has been associated with toxicity by heavy metals (Awodele *et al.*, 2013). It is known that the uptake of heavy metals by plants and subsequent accumulation along the food chain is a potential hazard to living symptoms (Jiwan and Kalamdhad, 2011). Heavy metals are known to cause low renal excretion rates, resulting in adverse effects in humans even at too low concentrations (Jamroz *et al.*, 2015). They produce toxic effects due to their interfering in many known normal biochemical and metabolic pathways (Kurban *et al.*, 2015). Several health disorders were linked to unwarranted uptake of dietary heavy metals, which includes decreased immunity,heart related problems, fetal malformation, impaired psychosocial and neurological behavior (Harris *et al.*, 2011). Lead and cadmium are not essential elements that are required neither in the human body nor in plants, and which effect various bimolecular opposing functional effects at low level doses (Maiga *et al.*, 2005). Though an essential component of many enzymes, excessive intake of copper can cause dermatitis, irritation of the UR tract, abdominal discomfort, and hepatic diseases (Harris *et al.*, 2011). Arsenic and mercury can harm pulmonary, nervous, renal and respiratory systems, as well as causing skin pathology (Jarup, 2003; Mahurpawar, 2015). It may also induce disorders in the central nervous system, liver, lungs, heart, kidney and brain. Leading to high blood pressure, abdominal discomfort, dermatitis, ulcer and is related with various types of cancers (Paulsen *et al.*, 2012). To look at and guarantee the safety of herbal medicines, multiple studies regarding heavy metal contamination in herbal medicines have been carried out in China (Ye *et al.*, 2013; Li *et al.*, 2018), India (Singh *et al.*, 2016), Iran (Behnam *et al.*, 2017), Egypt (Abou and Abou Donia, 2000), South Africa (Mulaudzi *et al.*, 2017), United States (Harris *et al.*, 2011), Brazil (Leal *et al.*, 2013), and Australia (Kim *et al.*, 2016), etc. Given harsh consequence it may bring to wellbeing and surroundings, revelation risk assessment (Kirkham *et al.*, 2010; Okatch *et al.*, 2012), vulnerability quotient (Li *et al.*, 2018), and environmentalmenaceevaluation (Awodele *et al.*, 2013), have been conducted by workers. This showed that heavy metal defect in herbal medicines is an area requiring urgent attention, with impendingjeopardy to human health (Sandilyan and Kathiresan, 2014). Studies show that experience of heavy metals through general dietary consumption giveharmfully to human health (Wei and Cen, 2020). In addition, specific recognition of metals is required for precisejudgmentbecause of considerable overlap between various clinical syndromes connected with heavy metal toxicity (Kurban *et al.*, 2015).

Heavy metal contamination in herbal medicines and herb incorporated decoctions and nutraceuticals is a global risk to human beings in particular at levels above known threshold concentrations. The concentrations of five heavy metals cadmium, lead, arsenic, mercury and copper were investigated using Inductively Coupled Plasma Optical Mass Spectrometry with 1773 samples around the world (Luo *et al.*, 2021). According to Chinese Pharmacopoeia, 30.51% samples were detected with at least one over-limit metal. Totally 25 kinds of herbs present at an unacceptable risk as assessed with the Hazard Quotient or Hazard Index. It is important to note that, the carcinogenic risks were all under acceptable limits. Especially, as posed the highest danger in all indicators counting estimated daily intake, hazard index, and carcinogenic risks (Luo *et al.*, 2021).Here, we report the case of a 58-year-old woman who presented with abdominal pain, anemia, liver function abnormalities, and an elevated blood lead level. The patient was found to have been taking the Ayurvedic medicine Jambrulin prior to presentation. Chemical analysis of the medication showed high levels of lead. Following treatment with an oral chelating agent, the patient's symptoms resolved and laboratory abnormalities normalized. This case highlights the need for increased awareness that some Ayurvedic medicines may contain potentially harmful levels of heavy metals and people who use them are at risk of developing associated toxicities (Gunturu *et al.*, 2011).Ayurvedic medicine is a long-established system native to India (Chopra and Doiphode, 2002). This system stresses the use of natural plant-based medicines, and minerals including sulfur, arsenic, lead, copper and gold are often added to formulations in the doses prescribed in ancient books viz. *Ashtangahridaya, Charakasamhitha,* etc. with the belief that these metals are essential components of vital molecules within the human body. In India, more than 100 colleges offer degrees in traditional Ayurvedic medicine and in western countries, Ayurvedic medicine is gaining attractiveness as complementary management to modern medicine. Ayurvedic medicines are used to indulge a wide spectrum of diseases from headaches to cancer. Currently, the United States does not specify a documentation requirement for Ayurvedic practitioners, though many training programs are being offered through state-approved institutions. Some Ayurvedic preparations have been found to contain containe lead and/or mercury at 100 to 10,000 times greater than acceptable limits (Saper *et al.*,2008). Although not common in western societies, lead exposure through dietary sources is a well-recognized phenomenon and in past years, calcium supplements have been a source of lead poisoning (Kim *et al.*, 2003). In addition to Ayurvedic medicine, other traditional medicines originating from Asian, Middle Eastern and Hispanic cultures have been found to contain lead and other heavy metals [Gogtay *et al.*, 2002)]. Though many health supplements are now subject to limited government regulation in the U.S. these medicines are readily obtainable as herbal remedies in health food centres. Their safety and efficacy are not regulated by government agencies such as the U.S.FDA(Marcus, Grollman. 2002).]. Thus, without sufficient public awareness, the risk of heavy metal exposure in individuals taking these supplements is quite high. A case of lead poisoning secondary to ingestion of Indian Ayurvedic medicine, Jambrulin, was demonstrated.

The properties of herbs and spices used in nutraceutical formulations or Ayurvedic preparations are given in Table 3.

Table 3: Herbs and spices used in nutraceutical formulations or Ayurvedic preparations

Herbs/ Spices		**Properties**
1.	Ajwain	• Possess antibacterial and antifungal properties • Hypocholesterolemia, hypotriglyceridemia, hypotensive, anti-ulcerative and detoxicant (Anilakumar *et al.*, 2009) • Preventing coughing and improve airflow
2.	Ashwagandha	• Enhances memory and cognition • Maintains thyroid levels • Cardio-respiratory endurance enhancer • Immune boosters • Promotes muscle strength and support the comfortable joint movement. • Promotes sexual function in both men and women

Herbs/ Spices		Properties
3.	Brahmi	• Neuroprotective • Decreases cortisol levels in the body • Reduces stress and anxiety • Antioxidant • Calms the *Vata* in mind, reduces excess *Kapha* in the body.
4.	Cardamom	• Works as a natural tranquilizer • Treats indigestion, food poisoning asthma, dysuria, and bad breath • Antihypertensives and hypoglycemic • Antifungal • Protects hepatomegaly and reduces risk of fatty liver diseases
5.	Cumin	• Improves digestion • Maintains BMI • Increases levels of high-density lipoprotein • Antioxidant, anti-inflammatory, and Antimicrobial (Anilakumar *et al.* 2007). • Reduces addictive behaviour.
6.	Licorice	• Provides relief in stomach ulcers, food poisoning, and heartburn. • Antibiotic properties against *H. Pylori.* • Enhances body's production of healthy mucus to support respiratory system. • Cures hoarse throat. • Potent anti-inflammatory, anti-ageing, antioxidant, anti-bacterial, anti-arthritic, analgesic and antacid. • Useful in treatment of prostatitis in old age and menopause issues.
7.	*Manjistha*	• Detoxifies the body by purifying the lymph and blood. • Effective medicine to clean the liver, regulate kidney functions and clear acne. • Eases the symptoms of dermatitis, blood sugar issues, ear and eye pain, inflammatory problems, urinary tract, and gynaecological disorders.
8.	Neem	• Antiseptic, antibacterial, and antifungal. • Potent blood purifier and detoxifier • Cures acne, eczema, skin diseases, skin damage, and wounds. • Alleviates Pitta disorders • Reduces fever, particularly malaria fever • Maintain proper dental hygiene and care • Strengthens hair roots and promotes healthy hair growth • Prevents vomiting, nausea, and fever • Carminative effect
9.	*Shatavari Asparagus resemosus)*	• Diuretic, laxative, antitumor, aphrodisiac, anti-depressant (Phani kumr *et al.*, 2015) adaptogenic, rejuvenative, antacid, antispasmodic, antioxidant, demulcent, galactagogue, anti-microbial, anti-inflammatory and immunomodulator. • Alleviates cough, Cures stomach, intestine or oesophagus ulcers • Maintains blood sugar levels • Prevents collagen breakdown and maintains skin's elasticity.

Herbs/ Spices		**Properties**
10.	Turmeric	• Purifies the blood and channels of the body. • Enhances the complexion of the skin and circulating blood and nutrients to the skin. • Anti-inflammatory • Balances *Pitta* • Maintains healthy blood sugar levels. • Strengthens the immune system
11.	*Haritaki*	• Enhances the vital functions of the body, such as strong digestive fire, adequate nutrient absorption, healthy waste elimination, etc. • Improves complexion • Improves digestion and eases bowel movement • Aid to memory and intelligence • Fat scraper • Reduce cholesterol levels in the body • Prevents lung infections in individuals with cystic fibrosis • Beneficial in neuro-muscular weakness and diminished vision • Cures sore throat, obesity, burns, leucoderma, bleeding gums, caries, cough, tuberculosis, prolapsed uterus, and stomach problems such as dysentery. • Useful in folk treatments for piles and haemorrhoids • Anti-inflammatory • Bolsters heart health and protects against blood disorders. • Traditional natural hair dye
12.	Guduchi	• Adaptogen • Anti-stress properties • Balances Dosha or Dhatu up to normal levels in the body. • immunomodulator, • Anti-Inflammatory • Hypoglycemic • Antioxidant, • Antimutagenic, • Detoxifier • Hematogenic. • Treat digestive ailments, such as hyperacidity, colitis, abdominal pain, appetite loss and liver ailments like hepatitis. • Cure chronic fever and enhances appetite and energy. • Reduces blood urea levels and get rid of renal calculi. • Treat gout and rheumatic ailments when used along with ginger
13.	Amlaki (*Emblica officinalis*)	• Antioxidant and promotes detoxification (Anilakumar *et al.*, 2004). • It neutralizes excess Pitta and regulates gastrointestinal tract. • Neuroprotective • Promote healthy hair healthy eyes, nails, and skin. • Immune booster • Promotes healthy blood sugar levels

Herbs/ Spices		Properties
14.	*Aloe vera*	• Anti-inflammatory and antibacterial (Anilakumar *et al.*2010). • It supports digestion. • Improves heart health. • healthy blood sugar levels, • Normalizes menstrual cycle • Blood purification • Cure hair, skin and eye issues. • Have mild laxative effect.
15.	*Triphala*	• Digestive tonic that improves the digestion process, cleanses the digestive tract (Anilakumar *et al.*, 2020) • Improves bowl movement. • A natural laxative • Anti-inflammation • Anti-cancer • Improves metabolism, helps in weight loss and management. • Anti-Fatigue • Antioxidant • Treat dermatological problems like acne and eczema. • Preserve the health and beauty of hair and skin.
16	Coriander seeds	• Helps lower blood sugar • Rich in immune-boosting antioxidants. • Benefits heart health, brain health. • Promotes digestion and gut health. • May fight infections and detoxification (Anilakumar *et al.*, 2001)
17	Cinnamon	• Hypothermic (Chaithanya *et al.*, 2018) • Antioxidant (Anilakumar *et al.*, 2016), diuretic, anti-inflammatory, anti-ulcerative, Hypoglycemic

b. Detection of contaminants and methods

The American Herbal Products Association describe that there are as many as 3000 plant species in trade. The FDA describe about 85,000 dietary supplement products in the open market. The innovation of manufactured goods far exceeds that of growth and substantiation of analytical methods, with novel ingredients, matrices, and their blends. This apparently results in an analytical group that has been incompetent to keep up. Only after meticulous validation procedures to guarantee that methods are robust being indented should they be used in a usual setting to confirm product authenticity and excellence.

As the use of herbal and other botanical supplements are on the rise, command for safe, high-quality products, and regulatory requirements, manufacturers must have confidence in the use of analytical methods helping to verify the eminence and genuineness of their ingredients and products. Knowing as much as possible about the raw materials and finished products they will be using either to manufacture a product or that will be studied in a clinical trial beforehand. This would allow manufacturers and researchers to generate rational decisions about analytical needs and approaches that they can use to devise products or studies (Betz and Hardy, 2014). Chemical recognition methods are being designed to support ward off adulteration of herbal supplements. There is a diversity of techniques that can be used and when coupled with discriminate analysis permit for the detection of known adulterants and contaminants.

4. FSSA Initiative: An Indian Scenario

Nutraceuticals are considered as foods by the FSS Act, 2006, Rules and Regulations, 2011. The Food Safety and Standards Authority of India (FSSAI) has concerned regulations with regard to licensing and registration of food business, productionisation, packing and nutritional labelling, food product standard and quickly. In FSSA in India (FSSAI) has classified regulatory guidelines for approval of nutraceuticals in the Indian market. In the Nutraceuticals Regulations, 2016, under the 'General Requirements,' the FSSAI has mentioned that mere combination of vitamins and minerals formulated in tablets, capsules, syrup formats shall not be covered in any of the categories of these regulations except when vitamins and minerals are added. Food Safety Standards Regulations, 2011, issued guidance for licensing and registration of food businesses; packaging and labelling; food product standards and food additives; prohibition and restriction on sales; contaminants, toxins and residues; laboratory and sampling analysis.

FSSAI has directed Nutraceuticals Regulations to stop using 14 ingredients lacking scientific data for safe usage. The country's apex food regulator ordered FBOs to discontinue the use of raspberry ketones which is used for anti-obesity, silica, *Angelica sinensis*, *Paulliniacupana, Saw palmetto, Notoginsing*, Chlorella growth factor, pine bark extracted to *Pinus radiate*, pine bark extracted from *Pinuspinaster*, Vitamin D_3-veg, chaga extract, oxalobacterformigenes, phytavail iron and tea tree oil. FBOs were also asked to discontinue antichoke, kale powder, salvia hispanica, cashewfruit, passion fruit, kiwi fruit extracts, broccoli and enzymes, including pectinase and xylanase, as health supplements. However, their use in products was not prohibited, but food business operators cannot claim that the products are supplement or Nutraceuticals. No further manufacturing of products using these ingredients will be allowed until these ingredients are assessed and approved by FSSAI. Further, food business operators are directed to provide information and data in respect of these ingredients within one month. The label of Nutraceuticals products should state "Nutraceuticals" Recommended Usage Warning for the risk of overindulgence consumption. Prohibitions on labelling claims must mention for Nutraceuticals Products Cure of disease claims, e.g. "Prevents bone fragility in post-menopausal women," implied cures for disease claims through pictures, vignettes or symbols. It needs regulatory authorities on product quality and safety in order to minimise the adverse events, toxicity, adulteration, misuse, overdose during human consumption. As food products are reaching from one country to another, maintaining safety and quality standards as per various regulatory guidelines set by the respective governments becomes important. To preserve quality and safety concerns of Nutraceuticals all the new companies and existing companies should follow the regulatory guidance of FSSAI for better use.

5. USFDA Regulations

The regulatory authorities worldwide are focusing on the Quality of Products and Safety with utmost care. As food products are getting from one country to another, maintaining safety and quality standards as per various regulatory guidelines set by the respective governments becomes imperative; which can be a genuine driver for the industryexpansion. Foods and food habits in today's lifestyle have showed the way to the disturbances in an ideally nutritionally balanced body. Consequently in such state if "food be our medicine" then it would be immense to achieve a healthy body and mind.

6. Impacts of Adulteration on Health and the Producers

The problems of adulteration create the food items not fit for consumption due to poor handling (Asrat *et al.*, 2012). In the past few decades, adulteration of food has become one of the grim problems and eating of adulterated food causes serious diseases like cancer, diarrhoea, asthma, ulcers, etc. In general, adulteration of food items has a very severe impact on producers/farmers, food processors or manufacturers and enterprises, consumers and ultimately government. Enterprises are impacted by a loss of confidence in customer in their products, evokes and destruction of contaminated products, grievance expenses and increases of insurance premium remittance and costs related to apparatus replacement or cleaning. A supplier's mistake is inevitably reported in the mass media, it casts doubt on the reputation of company (Pandpal *et al.*, 2012). This influences not only the sales of that fastidious product, but also the sales of many other products supplied by the same company's warehouse or retailers. It might also result in the products that can be banned or discarded mechanically. The effects of such outlaws on the food production industry are manifold, intense, and across-the-board. A producer that depends on a banned imported groceries and other foods not only go through economic loss to the crashed product but also faces lost sales caused by failure of public confidence. The resulting brand damage can be demoralizing, and recuperation can require significant time and expense when consumers have moved on to other suppliers' products (Ibens, 2014). Impacts on farmers/producers adulteration not only haveacause on big enterprises but also farmers or producers (like dairy, honey, coffee, wheat, etc) can be affected by the weakest relation in the industry chain. Many farmers suffer enormous losses, cost increases due to feed costs, milk cow shortage caused by mass sales or slaughter during the disaster, for example in the case of China dairy Scandal and lack of acceptance of the products (Qian *et al.*,2011 and Nie, 2008).

7. Micro-Nutrient Excess: A Boon or Bane for Human Health?

In biological systems metal ions endorse responses that range from deficiency to toxicity. Some, such as iron and zinc, have a known finest intake range for normal, healthy individuals. Metal ions contained within well-designed molecules already comprise a great advantage for the medicinal pharmacopoeia. However, whether essential or not, the threshold for toxicity can be very short. One of the challenges of designing metal-based drugs is to equilibrium the potential toxicity of an active formulation with the substantial positive collision of these increasingly common therapeutic and diagnostic aids.

The safety and efficacy for vitamin and mineral supplements are crucial but it can be separately dealt with. Misinterpretation of "safe and adequate" to mean "safety limit" would inflict restrictions on micro nutrient intakes that are not needed ascertain safety. Limitation of intakes to the RDAs would exclude reductions in disease risk from these nutrients. The edge of safety between the usual dietary intake and the intake that would produce unpleasant effects varies greatly among the different nutrients. Very high intakes of vitamins A and D, niacin, pyridoxine, and selenium have produced adverse effects. Many widely discussed presumed undesirable effects of vitamin C, vitamin E, and trivalent chromium have little factual basis. There is no evidence of adverse effects from beta-carotene supplements except in current heavy smokers as on 1977 (Hathcock , 1997).

There are many physiological effects of caffeine that is found in sports and energy drinks. Caffeine deeds as a boon and curse/bane as it owns valuable and undesirable effects (Kumar *et al.*,2018).

The evolution of biomolecules to supply the zinc ions was vitally significant in reducing the toxicity in cells which facilitates the numerous cellular functions for this element (Jerome, 2007). Arsenic and selenium are unusual metalloids as they both induce and cure cancer. Both of these minerals cause carcinogenesis, pathology, cytotoxicity, and nontoxicity in humans, with reactive oxygen species playing a significant role. While As induces adverse effects by decreasing DNA methylation and affecting protein 53 expression, Se induces adverse effects by modifying thioredoxinreductase. However, they can react with glutathione and *S*-adenosylmethionine by forming an As–Se complex, which can be secreted extracellular. It is hypothesized that there are two types of interactions between As and Se. At low concentration, Se can lessen As toxicity via excretion of As–Se compound [(GS3)2AsSe]−, but at high concentration, excessive Se can enhance As toxicity by reacting with *S*–adenosylmethionine and glutathione, and modifying the structure and activity of arsenite methyltransferase. This was summarized with respect to their toxicity mechanisms and the interaction between As and Se toxicity, and to provide suggestions for future investigations (Hong-Jie Sun, 2014).

Future Prospects

Food adulteration has harmful impacts on public health. The causes for food adulteration are profit margin and dishonesty of producers, retailers and food processors. This eventually results that the consumer is either cheated or frequently turn into sufferer of diseases. The following recommendations hold good: (a) In developing countries, adequate laws, funding and staffing should be appeared to control above discussed challenges (b)Diverse stakeholders should participate in a role to develop appropriate sampling programs based on statistical validity and sound sampling methodologies (c) Further study should be conducted on quantification of adulterants and contaminants in dietary supplements and (d) Adaptation of effective analytical methods to detect fraudsters.

Although evidence-based phytonutrients are pharmacologically successful, Nutraceuticals and dietary supplements of plant / herbal origin appear to be an accepted aspect of disease-preventive food components in pure form as single components as in Nutraceuticals. In dietary supplements it's in the way of extracts or enriched fractions in dietary supplements. These products are in general effective in helping to maintain healthy status of our bodies and, hence, leading to disease prevention. On the other hand, the risks and benefits of Nutraceuticals/dietary supplements/herbals/ medicinal products is not as well-known as for conventional drugs. There is an absence of reported adverse effects and effects of drug interactions do not mean that the mentioned products are devoid of these properties. There are innumerable formulations in *Ayurveda/ Siddha,* etc that is capable of reducing diseases. The formulations are based on conventional methods and the chances of adulteration or contamination are very rare. However it is proposed to test its purity, toxicity, etc. by chemical and biological means. On the other hand, all we should be aware of is the fact that the pharmaceutical quality of herbal preparations seems to be inadequate since their regulatory and licensing mechanisms are weak in many countries due to the great amount of adulteration or substitution. Our literature review indicated that many products claiming to be "herbal or natural" have been in reality identified with compelling proof of adulteration of "synthetic adulterants" or "undeclared synthetic drugs". Or else, contamination has become a hazardous phenomenon in various parts of the world. Therefore, uncontrolled use of herbal preparations along with pure compounds obviously creates a public health risk and all countries are suggested to implement more strict regulations and licensing processes. This might develop regulatory standards for all kinds of analysis creating adequate pre- and post-marketing inspections to safe guard their community.

References

Abou, A. A. K., and AbouDonia, M. A. (2000).Heavy metals in Egyptian spices and medicinal plants and the effect of processing on their levels. J. Agric. Food Chem. 48, 2300–2304. doi:10.1021/jf990508p\

Andrae-Marobela, L. M. D., Das, S., and Silva, E. B. D. (2018).Metal concentrations in traditional and herbal teas and their potential risks to human health. Sci. Total Environ. 633, 649–657. Doi: 10.1201/9780429275876-1.

Anilakumar,K. R., Sudarshana Krishna, K. R., Chandramohan, G., Ilaiyaraja, N., Farhath Khanum, Bawa, A. S. (2010). Effect of Aloe vera extract on antioxidant enzymes and azoxymethane-induced oxidative stress in rats. Ind. J. Expt. Biol. 48, 837-842.

Anilakumar , K. R. and Naveen, S. (2016). Cinnamon-a promising spice with nutraceutical properties. World Journal of Pharmaceutical Research. 5: 453-466.

Anilakumar, K. R. and Semwal, A. D. (2020). Immuno-Modulatory Properties of Triphala - A Combination of Three Dried Fruits . Ind. Food Ind. Mag . 2, No (2), Mar-Apr 2020.

Anilakumar, K. R., Nagaraj,N. S., Santhanam, K. (2001). Effect of coriander seeds on hexachlorocyclohexane-induced lipid peroxidation in rat. Nutr. Res., 21, 1455-62.

Anilakumar, K. R., Nagaraj,N. S., Santhanam, K. (2004). Protective effects of amla on oxidative stress and cytotoxicity in rats challenged with dimethyl hydrazine. Nutr. Res . 24, 313-319.

Anilakumar, K. R., Saritha, V., Farhath Khanum, Bawa, A. S. (2007). Effect of cooking of total phenols, flavonoids and antioxidant activity of some selected spices of Indian culinary. J. Food Sci.Technol. 44, 357-359.

Anilakumar, K. R., Saritha, V., Farhath Khanum, Bawa, A. S. (2009). Ameliorative effect aof ajwain extract on hexachlorocyclohexane-induced lipid peroxidation in rat liver. Food Chem .Toxicol.47,279-282.

Asrat, A., Zelalem Y., and Ajebu N. (2012). Quality of fresh whole milk produced in and around Boditti, Wolaita, South Ethiopia. Afri. J. Anim. Biomed. Sci., 7(2): 95-99.

Awodele, O.,Popoola,T. D., and Amadi, K. C. (2013).Traditional medicinal plants in Nigeria--remedies or risks. J. Ethnopharmacol. 150, 614–618. doi:10.1016/j.jep.2013.09.015

Behnam, A., Ghorbanpour, M., and Nikabadi, S. (2017). Heavy metals in contaminated environment: destiny of secondary metabolite biosynthesis, oxidative status and phytoextraction in medicinal plants. Ecotoxicol. Environ. Saf. 145, 377–390. doi:10.1016/j.ecoenv.2017.07.035

Betz J.M., Hard, M. (2014). Evaluating the Botanical Dietary Supplement Literature. Herbal Gram; 101:58–67.

Brookes, L., Dong, L., and Huang, Q. (2019). Detection and risk assessments of multi-pesticides in 1771 cultivated herbal medicines by LC/MA–MS and GC/MS-MS. Chemosphere 12, 74–77. doi:10.5772/57257.

Chaitanya Pandit, Sai Latha, S., Usha Rani, T., Anilakumar, K. R. (2018). Pepper and cinnamon improve cold induced cognitive impairment via increasing non-shivering thermogenesis; a study, International Journal of Hyperthermia DOI: 10.1080/02656736.2018.1511835.

Chopra, A., Doiphode VV. (2002). Ayurvedic medicine. Core concept, therapeutic principles, and current relevance. Med Clin North Am. 86:75–89. doi: 10.1016/S0025-7125(03)00073-7.

Gogtay, N.J., Bhatt H.A., Dalvi S.S., and Kshirsagar N.A. (2002). The use and safety of non-allopathic Indian medicines. Drug Saf. 25:1005–1019. doi: 10.2165/00002018-200225140-00003

Gunturu, K., S., Nagarajan P., McPhendran P., Goodman T.R., Hodson M. E.,and Strout M P. (2011). Ayurvedic herbal medicine and lead poisoning.JHematolOncol.4:51.

Harris, E. S., Cao, S., and Littlefield, B. A. (2011). Heavy metal and pesticide content in commonly prescribed individual raw Chinese Herbal Medicines. Sci. Total Environ. 409, 4297–4305. doi:10.1016/j.scitotenv.2011.07.032

Hathcock, J.N.,(1997). Vitamins and minerals: efficacy and safety. The American Journal of Clinical Nutrition, Volume 66, Issue 2, Pages 427–437, https://doi.org/10.1093/ajcn/66.2.427.

Ibens D. (2014). The great melamine scare. The scandal draws attention to standards and leads to development of new tests. Food Quality & Safety, Current Issue

Jamroz, D., Safa, A. K., and Husna, R. (2015). Determination of heavy metals concentration in traditional herbs commonly consumed in the United Arab Emirates. J. Environ. Health 15, 1–6. doi:10.1016/j.fct.2018.01.048

Järup, L. (2003). Hazards of heavy metal contamination. Br. Med. Bull. 68, 167. doi:10.1093/bmb/ldg032

Jerome, N. (2007). Zinc Toxicity in Humans, School of Public Health, University of Michigan. P 1-7. Elsevier B.V.

Kim, M., Kim C., and Song I. (2003). Analysis of lead in 55 brands of dietary calcium supplements by graphite furnace atomic absorption spectrometry after microwave digestion. Food AdditContam. 20:149–153. doi: 10.1080/0265203021000053588.

Kim, H. S., Seo, B. H., and Bae, J. S. (2016).An integrated approach to safer plant production on metal contaminated soils using species selection and chemical immobilization. Ecotoxicol. Environ. Saf. 131, 89–95. doi:10.1016/j.ecoenv.2016.05.008

Kirkham, C. A., Cave, M. R., and Wragg, J. (2010) Determination of the bioaccessibility of chromium in Glasgow soil and the implications for human health risk assessment. Sci. Total Environ. 409, 267–277. doi:10.1016/j.scitotenv.2010.09.007

Kumar, V., Kaur, J., Panghal, A., Kaur, S. and Handa, V. (2018). "Caffeine: a boon or bane", Nutrition & Food Science, Vol. 48 No. 1, pp. 61- 75. https://doi.org/10.1108/NFS-05-2017-0100.

Kurban, H. N. S., Abdou, K.A., and Hassan, E.H.Y. (2015).Impact of toxic heavy metals and pesticide residues in herbal products. BeniSuef Univ. J. Bas. App. Sci. 11, 27–39. doi:10.1016/j.bjbas.2015.10.001

Leal, A. S., Prado, G., and Gomes, T. C. B. (2013).Determination of metals in medicinal plants highly consumed in Brazil. Braz. J. Pharm. Sci. 499, 599–607. doi:10.1016/j.mex.2018.01.003.

Li, X., Chi, W., and Tian, H. (2018).Probabilistic ecological risk assessment of heavy metals in western Laizhou Bay, Shandong Province, China. PLoS One 14, e0213011. doi:10.1371/journal.pone.0213011.

Li, X., Kong, D., and Wang, R. (2019). Safety evaluation of heavy metals contaminated Xiaochaihu Tang using health risk. China J. Chin.Mater. Med. 93, 121–129. doi:10.19540/j.cnki.cjcmm.20191015.201

Luo, L., Wang B., Jiang J., Fitzgerald M., Huang Q., Yu Z., Li H., Zhang J., Wei J., Yang C., Zhang H., Dong L., and Chen S. (2021). Heavy Metal Contaminations in Herbal Medicines: Determination, Comprehensive Risk Assessments, and Solutions. Front. Pharmacol. 11:595335. doi: 10.3389/fphar.2020.595335

Mahurpawar, M. (2015).Effects of heavy metals on human health. Int. J. Regul. Govern.530, 2394–3629. doi:10.1016/j.fct.2018.01.048

Maiga, A., Diallo, D., and Bye, R. (2005).Determination of some toxic and essential metal ions in medicinal and edible plants from Mali. J. Agric. Food Chem. 53, 2316–2321. doi:10.1021/jf040436o.

Marcus, D. M, Grollman A P. (2002) Botanical medicines--the need for new regulations. N Engl J Med347:2073–2076. doi: 10.1056/NEJMsb022858.

Mulaudzi, R. B., Tshikalange, T. E., and Olowoyo, J. O. (2017). Antimicrobial activity, cytotoxicity evaluation and heavy metal content of five commonly used South African herbal mixtures. South Afr. J. Bot. 112, 314–318. doi:10.1080/03067319.2019.1702652.

Nagaraj, N. S., Anilakumar, K. R., O.V. Singh (2010). Dially disulfide causes caspase-dependent apoptosis in human cancer cells through a Bax-triggered mitochondrial pathway. J Nitr. Biochem. 21(5), 405-412.

National Pharmacopoeia Commission.(2020). Pharmacopoeia of the People's Republic of China, Part IV. Beijing, China: China Medical Science Press.

Nie, Y.L. (2008). Collections of speeches in the Summit Forum of dairy industry development in China. China Dairy. 12:10–19.

Pandpal, S.D., SirvastavaA.k., and Negi K.S. (2012). Estimation of quality of raw milk (opened and branded) by milk adulteration testing kit. Ind. J. Commu. Health, 24(3), pp188.

Paulsen, M., Gatebe, E., and Gitu, L. (2012).Profile of heavy metals in selected medicinal plants used for the treatment of diabetes, malaria and pneumonia in Kisii region, Southwest Kenya. Global J. Pharmacol. 6, 245–251. doi:10.1080/15569543.2016.1225768.

Phani Kumar, G. , Akhil Balakrishnan, Mahadeva Naika, Anilakumar, K .R. (2015). Effect of Asparagus racemosus on amelioration of weight loaded forced swim stress on experimental animal model, J. Pharmacy Res. 9(4), 306-312.

Qian, G., Guo X., Guo J., and Wu J. (2011). China's dairy crisis: impacts, causes and policy implications for a sustainable dairy industry. IJSDWE.Vol. 18(5), No. 5, pp434–441.

Yigrem, S, Beyene F, Tegegne A and Gebremedhin B (2008).

Sandilyan, S., and Kathiresan, K. (2014).Decline of mangroves—a threat of heavy metal poisoning in Asia. Ocean Coast Manag. 102, 161–168. doi:10.1016/j.ocecoaman.2014.09.025

Saper, R.B, Phillips RS, Sehgal A, Khouri N, Davis RB, Paquin J, Thuppil V, Kales SN. (2008). Lead, mercury, and arsenic in US- and Indian-manufactured Ayurvedic medicines sold via the Internet. JAMA. 300:915–923. doi: 10.1001/jama.300.8.915.

Singh, N. K., Raghubanshi, A. S., and Upadhyay, A. K. (2016). Arsenic and other heavy metal accumulation in plants and algae growing naturally in contaminated area of West Bengal, India. Ecotox. Environ. Saf. 130, 224–233. doi:10.1016/j.ecoenv.2016.04.024

Wang, G., Guo, Y., and Qiao, R. (2014). Study on the transfer rates of heavy metals of medicinal plants before and after decocting. Anhui Med. Pharm. J. 18 (11), 2072–2074. doi:10.1201/9781420003963-32

Wang, J., Chen, X., and Shi, Y. (2009). Evaluation of heavy metal accumulation in greenhouse soils in Shenyang, Northeast China. Proc. SPIE Int. Soc. Opt. Eng. 7491, 26.doi:10.1007/978-94-011-6763-5_26

Wei, J., and Cen, K. (2020). Content and dietary exposure of cadmium among residents in Northeast China: a case study based on the 5th China Total Diet Study. Environ. Sci. Pollut. Res. 11, 24–29. doi:10.1007/s11356-019-07016-7not

19

Global Trends in Food Authenticity, Traceability, Branding and Consumer Preference

***Sheetal Thakur*[1] *and Priyanka Minhas*[2]**

[1]*University Centre for Research and Development, Department of Biotechnology Chandigarh University, Gharuan-Mohali, Punjab, India*
[2]*College of Veterinary Science, Guru Angad Dev Veterinary and Animal Sciences University, Ludhiana, Punjab, India*

Introduction

It is an eminent actuality that the diets of end users are being advanced globally; however, the patterns of food procurement and food safety protocols diverge across countries. The fundamental rationale of the trend is that the consumer awareness with regard to food safety, quality and variety has been proliferated greatly. Looking back to the history of food industry, it is unquestionable that the past has a full frame of misfortune, indignity and many disastrous incidents. Moreover, the entire muddle was not always related with microorganisms; numerous of them were associated with environmental problems, technological advancements or certain alterations in production protocols and management. There are various incidents that have been recorded in the past. For an instance, in 1979, mercury poisoning was detected in oranges; In UK, baby food products were found contaminated with pin, glass and caustic soda in 1989 due to which the product was evoked from the market. There are several such episodes related with food blunders that are still in the remembrance of common people (Gregory, 2000). In the recent period as well, the food contamination incidents have been observed more often. As published by UK Food Laws News (2002), nitrofurans were found in chicken samples traded in Northern Ireland from Brazil and Thailand. Corn infected with Aflatoxin in Kenya affected 317 people with the severity of hepatic failure (Azziz-Baumgartner *et al.*, 2005). In 2011, the illegal alcohol was found contaminated that resulted into more than 100 deaths in West Bengal, India. It was claimed that the alcohol may have the traces of methanol and/or ammonium nitrate. Australian strawberries contamination with needles in 2018; dioxin contamination of eggs in Indonesia reported in 2019 are the most recent disasters recorded in food contamination incidents so far.

The detailed knowledge about such food fraud incidents has been made available to public by the extreme social network spread over the past few years, due to which, the consumer awareness on the subject of food safety, quality and variety has been propagated greatly. In order to meet these gradually more diverse necessities by considering the choice of consumer, the authentic food with perceptible clarity in labeling and branding has become an inevitable prerequisite in global food markets.

This is the fundamental reason that the food authenticity and traceability have become a global matter for which the numerous countries throughout the world have been addressing this concern at a multilateral level, such as Food Modernization and Safety Act (FSMA, 2011) in USA; tracking and tracing software Grapenet for selling of table grapes from India to EU, Rapid Alert system for Food and Feed (RSFF) introduced by EU (Bujji *et al.* 2012). The enduring alterations are motivating food supply chains to espouse closer synchronization among manufacturer and trader, in order to smooth the progress of customizing the food products to achieve the demands of end user.

In order to maintain the food safety and to overcome the cases of food contamination for the human wellbeing, it is of utmost importance to define the actual meaning of "authenticity" and "traceability" so that the same can act as a restriction or a prospect for the organization, specialists or for the end users. Globalization, a rapid drift which is progressing for decades, has changed the insight for different sectors, including the initial step of procurement of raw material from agricultural sector to the consumption of the final product and lastly, the recycling of the waste and this entire chain covers the intermediate practices of production, processing and dispersal. With an upsurge in globalization of food industry, the food marketplaces are progressively acting in response to the preferences of consumers at each level of food production, processing, packaging and supply chain.

Traceability works as a system that provides information about the product history and helps in keeping the lucidity in food production chain by which the food can be traced and monitored. The apparent example of such practice is product recall; a means of revoking the items non-conforming with food quality and safety standards (Spence *et al.*, 2018). A traceability procedure is indispensable to be followed from farm (from where the raw ingredients are obtained) to fork (means the consumption by end user) and even beyond that as there can be the possibility of fault in the goods consumed. Moreover, the authenticity of certain food items under specified situations, like packaged produce, may not be possible for the consumer to examine, understand or validate with no trouble. There is the only possibility for the consumer to evaluate such condition through certain indications given by food makers and monitors that help in signifying the authenticity attribute (Martinez & Epelbaum, 2011). Therefore, consumer rely on the suitable and impartial improvement methods, efficient implementation protocols of regulatory bodies and their confidence on food chain associates and the set of rules followed by them for product production are certain measuring factors that are indicative of food reliability (Meijboom *et al.*, 2006).

Industries always make endeavor to get hold of more customers along with the grip on existing ones by using the protocol of branding their products with novel strategies of magnetize new buyers without affecting the previous expression in the market. The overall effect of branding the food products has been observed to help food industries for increasingly expanding a great market worldwide. With the vision of providing safe food to the consumers to meet their requirements, the producers are following the rule of being realistic and evident with regard to the clear trade in global food market.

Global Drift in Food Sector as a Point of Well being

With an emergence in the technology, the food sector is flourishing in every aspect with a great enhancement in providing the authentic information through branding and labeling that helps in building the trust of consumer. Although the past time incidents on food frauds have been recorded immensely, the traceability verification and maintenance is under great consideration from past few years in order to avoid the accidents and to provide the safe and authentic food for consumer. In food sector, like all other industries, the chief influential factors are the technological advancement and alterations in the consumer requirement and wish (Harmsen *et al.*, 2000). Progress in technology has facilitated the food industry to manufacture the safe food with enhanced sensory and quality attributes, better nutritional behavior, easy processing procedures, extended shelf life and superior storage and transportation stability (Grunert *et al.*, 2019). However, this is not always the case that each food product that is launched in the market has a great success story. It is apparent that there are certain food products that were introduced by using enhanced technology, but the product establishment got failed and the product disappeared swiftly, regardless of the technology. The deep study on the reasons of this failure concluded that not only the technology, launching industry and production related strategies work for the success of overall business in food sector, but the targeted consumer prerequisite and wish is a fundamental factor that defines the overall accomplishment (Brown & Eisenhardt, 1995). Considering the facts associated with food related consumers from more than a few decades, there would be conformity from most of the people that many changes have been observed in relation to consumer requirements and wish at various nations throughout the world (Grunert, 2017). For a few decades, it has been noticed that the consumer become more health conscious and this is the reason that the palatability and other sensory related motives are considered secondary for majority of population as they concentrate more on good physical condition irrespective of the fact that whether the food is appetizing or not (Ares *et al.*, 2014). Likewise, talking about the swamped lifestyle of consumers, the drift has been shifted towards the convenience from past few years. The consumers being in the target-pressured jobs prefer to save time rather than going out for shopping as well as spending hours in the kitchen for the preparation of food.

The food manufacturers are taking into account the expectations of consumers related with their health and well being and are actively creating a market to progressively offer the novel food products that meet the consumer requirements (Gray *et al.*, 2003). Therefore the selection of food products is done on the basis of the ability of the food to prevent or reduce the risk of health related problems (Asioli *et al.*, 2017). In order to simplify the concept, a model explaining the hierarchy of consumer expectation to satisfaction where the various food product related and non product related factors plays the role (Fig. 1). Aware consumer is concerned about the raw ingredients and the place of origin of the raw material. Also, the process used for product production, health related benefits, transparency of food supply chain are certain major aspects related with the food product which are considered by the consumer before the purchase of food. Several other non-food related expectations are equally countable, like, market value, product branding, product certification, trust and credibility about which the consumer is conscious as all these factors are ultimately gives rise to consumer satisfaction. Sensory attributes such as taste, texture, smell and appearance, role of the food in health and overall well being as well as the sustainability of the food production are food product related results that leads to consumer satisfaction. However, Authenticity and Traceability are two most important parameters that covers product related as well as non-product related prospects for achieving the consumer satisfaction.

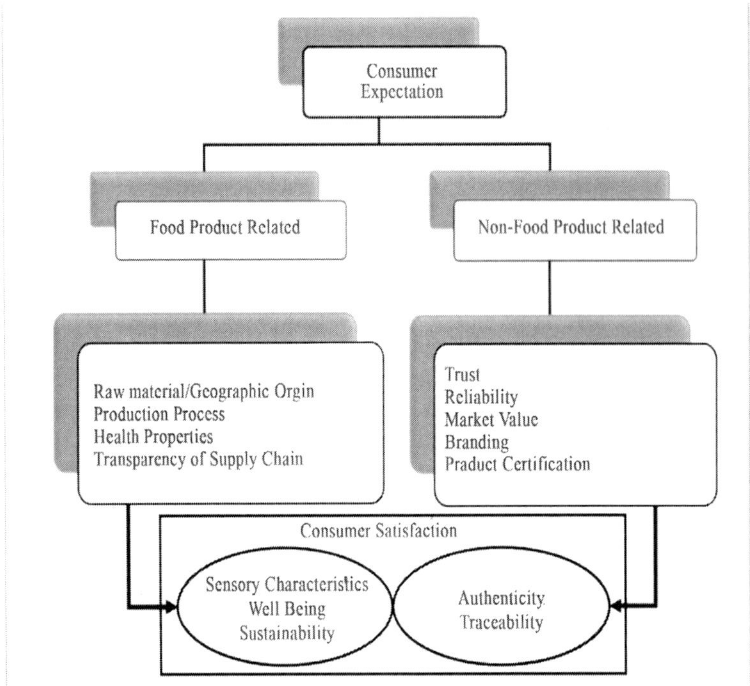

Fig. 1: A model showing the hierarchy of consumer expectation to consumer satisfaction

More recently, the interest of consumer is accelerating toward the food authenticity. In response to the globalization of food supply and an amplification of less transparent food supply chain, majority of consumers are concerned about the food product being truthful, genuine, natural, undestroyed, novel and coming from the short supply chain (Zepeda & Deal, 2009).With the concern toward authentic food, the global market is growing with the trend of providing the Local, Natural and Organic foods (Khan & Prior, 2010). Food manufacturers are now more inclined toward the production of the products being minimally processed with no artificial constituents and are of the natural appearance that give the feel of homemade products (Rozin, 2005; Lunardo & Saintives, 2013).

Additionally, it is observed that the people have started to concentrate on the sustainability of food production as the society is more alert on certain crucial issues like environmental factors, natural protein source alternatives, food waste management, ethical aspects *etc.* (Crabbe *et al.*, 2018). Under environmental issues, the climate alterations are of the great concern for many consumers, so they consider the purchase of such kind of food items that can reduce the burden of ecosystem (Motoshita *et al.*, 2015). The highlighting example in this matter is the reduction in the consumption of meat and meat substitutes as this has been observed that the consumers have shown the interest in being vegetarian rather that eating meat for the environmental well being (Siegrist *et al.*, 2015). This needs to be considered that although the change in consumer behavior towards meat consumption has benefitted to the food sustainability, human welfare is compromised. However, the global drift and recent advancements in the production of plant based protein alternatives have attracted the consumers to meet their protein based requirements (Videbaek & Grunert, 2017).

The food sector waste generation is segregated into two chief categories: waste from food and packaging waste. This is not the fact that consumers are not concerned about the food waste and they are not contributing for controlling the same but whenever the food preference, ease and healthiness comes in the way, the human priorities get changed accordingly (Aschemann-Witzel *et al.*, 2015). Similarly, the contribution of food packaging related waste in the form of single usage packages adding burden to the waste management strategies (Geueke *et al.*, 2018). Linked to the environmental concern, people have become aware and demands eco-friendly and sustainable food packaging (Singh & Pandey, 2018). In addition to this, some people set their preference for the food procurement based upon the social aspect and hence their decision of selecting those food brands that are considering the ethical values (Ladhari & Tchetgna, 2015; Crabbe *et al.*, 2018). All these factors involved in the sustainability of food products have a great impact on the overall wellness as well as a major contributor towards amendments in the product development strategies in global market.

The economical development is another major factor that has encouraged the global food sector, which in turn, is connected with urbanization that has influenced the food requirements of people. With a change in consumption pattern due to this growth, the other fundamental considerations in the food choices are shelf stability, waste reduction, and food safety concern along with the nutritional aspects, taste and expediency (Winger & Wall, 2006). This is undeniable that the consumers are much aware of the priorities with respect to food consumption and hence, they give more value to the quality life and self-accomplishment (Meulenberg & Viaene, 2005). For such kind of awareness and behavioral changes for the food preferences, there are several key factors like, earnings, age, family size, social circle and education (McCluskey, 2015). This revolution in the consumer behavior and expectations towards food preferences has challenged the global food market to develop a transparent and authentic system where the food production strategies and new product development demands a balance with food choices of consumer (Schmidt, 2005).

Nevertheless, the future food system has miscellaneous stipulations and challenges *viz.* demographic expansion, obesity, malnutrition, starvation, work atmosphere, jobs, environmental pollution, food leftover and resource exhaustion that come under the way of horizontal growth of the food market. Such hurdles are acting as the chief points that support the need of a transformative reform.

Food Authenticity- Its Vitality and Role in Safe Food Supply

Food authenticity has remained a matter of great concern from approximately past two centenaries and the issue was outlined by South Europe for the first time. However, the major action related with this subject has taken by China in previous fifteen years (Danezis *et al.*, 2016a). The authentic food is considered as the type of food that "fulfills the entire conformities described by the manufacturer or processor given on label". For the identification and/or review of the food being authentic, the following points are evaluated, as whether:

1) The components are unique or replaced with other cheaper ones in order to generate high-cost food items to cheat the customer.
2) The process of the product/ingredient and its history is reported or kept concealed.
3) The food category mentioned is correct or incorrect
4) The food description meets the rations of permissible name or not
5) The geographical origin is mentioned correctly or not.

By considering these points of detection, the assessment of food authenticity for providing the safe food with all apparent information for consumer is achievable. However, with an intensification in the industry build up as well as globalization, the food authenticity has become complicated to deal with as there is great availability of majority of the food items in market that are transformed worldwide and this is the reason of the product being vulnerable to misrepresentation (Johnson, 2014). Due to this, the area of origin of the food item has become under major consideration as a matter of food quality, consumer as well as food safety and the same is taken into account to be abiding by the regulations of resident area along with intercontinental rules and standards (Aung & Chang, 2014). The consumers' awareness towards each food item they eat and drink make them conscious to be knowledgeable about the details related with the geographical origin.

Also, with the progression in technology, the scams are designed with such innovative approaches that the conventional methods of analyzing the authenticity do not give the satisfactory results. Not only consumers, but, the genuine food manufacturers are also concerned about food authentication in order to circumvent the inequitable competition with dishonest processors who use false means of gaining financial advantage from the similar kind of food products (Arvanitoyannis, 2016). Due to such inappropriate and illicit actions, food distributors as well as the other authorities dealing with food production and processing are also concerned about authenticity and the same is documented since long time (Posudin *et al.*, 2015). Such activities have rampant effect on financial losses globally, plus the faith of consumer comes under question mark regarding the integrity of food products as given on the label.

In current era, the food authentication is a swiftly emergent subject as the general public has become increasingly attentive regarding the food quality and safety aspects. Authenticity is considered as a vital element at all stages of food production process, including the raw material procurement to end product. In the present stage, as the complexity and disintegration of food supply chains are providing the good chance to food frauds, the strict regulations regarding food authenticity have been launched and followed round the globe with the purpose of providing safe and quality food to consumer. However, it is not always easy for a consumer to observe and validate the food authenticity rightly, like, in case of packaged foods. Under these circumstances, the signaling pointers offered by the food manufacturer and the regulatory body act as an indicative path to get the attributes of food authentication (Martinez & Epelbaum, 2011). Hence, to build the trust of consumer in a food product being authentic, the effective enforcement of legislations, implementation of suitable improvement methods, reliable production practices, and consumer favored strategies by food chain supplier play an imperative part to describe the food integrity (Meijboom *et al.*, 2006). Food labeling and protocols of trading are the two domineering aspects that supply the demonstrable specifications and describe the authentication of food. Along with the significance of geographical origin of food and strategies of various food hoaxes that explains different aspects of authenticity, the consumer behavior towards safety of food coupled with traceability has also been summarized under various studies (Cicia & Colantuoni, 2010; Hansstein, 2014).

Food related characteristics of economic adulteration are also chiefly associated with the authenticity. Adulteration in foods is potentially detrimental to human health and that is why it becomes important to consider the chemical composition, also known as the fingerprints of food stuffs. The chemistry of food is considered as an excellent symbol of validating the authenticity as well as locating the adulterants in food. Any disparity in the fingerprints of foodstuff works as a clear indicative of alteration in the level of food metabolites that can happen due to various reasons, like, modification in production process, change in the factors affecting storage and shelf life of food, and most importantly, the adulteration practices (Cubero-Leon *et al.*, 2018). It has been claimed in a recent research that the food fingerprints comprise the potent means of examining the food authentication (Medina *et al.*, 2019). The amplification in economic adulteration is also motivated by the globalization of food trade, increased demand of high-value goods, cost variations and the aspects that motivate the unlawful benefits.

All these points related with the consumer health and satisfaction that is achievable by maintaining the food quality as well as safety describes the need of considering the food authenticity in global market. The implementation of an effective regulatory framework to establish the legitimate food manufacturing procedures, to describe the geographical basis of food with transparency, to execute a well-built legal agenda and to maintain the food chain integrity would facilitate a roadmap for food authentication.

Recent Advances in Food Authenticity Techniques

The complexity of food management system has augmented the need of analytical assessment of food authenticity so that an appropriate approach can be executed for the safety of consumer. In order to address the food authentication issues, there have been certain major analytical methods that can collect the required information (Cuadros-Rodríguez *et al.*, 2016):

a) Chemical approach that make use of bio-chemical markers of chemical fingerprinting.

b) Isotopic approach that set a limit with accurate information in the form of isotope ratio in food

c) Bio-molecular approach that is associated with protein/DNA

The various analytical techniques like nuclear magnetic resonance, chromatography, infrared spectroscopy, mass spectrometry and sensing devices, have been fabricated and applied throughout the supply chain with the goal of assuring food quality and safety along with ensuring the traceability process (Bianchi *et al.*, 2018).

a) Chemical methods

The food is a compound matrix and it is often difficult to find out some target constituents that are available in minute concentrations. Therefore, the sample preparation is considered as a crucial step for determining the food fingerprints in the experimental layout. Also, the selection and sensitivity of the method gets enriched with an efficient pre-treatment that further helps to shield the analytical columns, which in turn, gives enhanced end results (Mansour & Khairy, 2017). A number of frequently used extraction methods have been observed in food fingerprinting, like, solid phase extraction/micro-extraction, liquid-liquid extraction, microwave-assisted/ ultrasound-assisted extraction, stir bar sorptive extraction, extraction by packed sorbent and quick, easy, cheap, effective, rugged, and safe (QuEChERS). However, the micro-extraction methods are popular among all of the given techniques, due to low sample/solvent requisite as well as less expensive and reduced labour time making them more eco-friendly (Gómez-Ríos *et al.*, 2017).

Under analytical methodologies for determination of fingerprint of foodstuffs, spectrometry and spectroscopy-based analysis have been observed to be used in recent years (Cubero-Leon *et al.*, 2018). Different chromatographic methods, as capillary electrophoresis, gas chromatography and liquid chromatography are used to separate the various constituents of a mixture. Chromatographic technique coupled with chemometric system of analysis is commonly used in the food system with the aim of supervising food authenticity issues (Esteki *et al.*, 2018). Presently, separation techniques in combination with spectroscopic/spectrometric techniques, generally named as hyphenated techniques, are becoming popular due to their capacity of solving complicated analytical issues and are useful in certifying food authenticity (Llano *et al.* 2018).

Spectroscopic methods of analysis are based upon the interaction of electro-magnetic radiation with the substance that comprises the food matrix. The structural, physico-chemical and compositional behavior of the sample can be obtained with the detection of wavelength/frequency in the emitted/absorbed energy spectrum. The major advantages of spectroscopic methods are high-speed, automatic, inexpensive, solvent-free, non-destructive and non-invasive (Danezis *et al.*, 2016b). Several spectroscopy methods, such as vibrational spectroscopy (infrared, near-infrared and Raman); ultraviolet–visible spectroscopy; nuclear magnetic resonance; electron spin resonance; X-ray spectroscopy; electron spectroscopy; Mössbauer and thermal spectroscopy are practiced, among which, Infrared, Raman and Nuclear magnetic resonance are reported to be used more often in food authentication as well as food safety and quality analysis (Ansari *et al.*, 2018; Grassi *et al.*, 2018). Flouroscence spectroscopy is an another simple, inexpensive and non destructive analytical method having low to very low detection limits and such methods are developed to check the food adulteration, authenticity and quality of foods like, olive oil (Poulli *et al.*, 2007).

Mass spectrometry method is generally used along with gas/liquid chromatography techniques. Nonetheless, current developments in non-chromatographic mass spectrometric technique such as matrix assisted laser desorption ionization time-of-flight (MALDI-TOF-MS), direct analysis in real time (DART-MS), Desorption Electrospray Ionization (DESI-MS) and proton transfer reaction mass spectrometry (PTR-MS) have advanced the innovative techniques for detection of food adulteration and authenticity (Guo *et al.*, 2016). DART-MS and DESI-MS are collectively called as Ambient mass spectrometry techniques (AMS) that provide the situation where the complex and time consuming sample preparations remain minimal (Black *et al.*, 2016; Lu *et al.*, 2018). In addition, MALDI-MS methods also need a smallest amount of sample with no pre-treatment (Danezis *et al.*, 2016a).

Another novel technology is the use of electronic sensors that work by emulating the different senses of human, like, electronic nose (E-nose) uses a semi-selective gas sensor and replicates the olfactory sense, electronic tongue (E-tongue), a kind of liquid sensor copy the gustatory sense and electronic eye (E-eye) which is based on colorimetric methods imitate sense of vision (Patel, 2014). Electronic noses for volatiles, while, electronic tongues for non-volatiles in liquid have been created as portable array sensors for food systems (Bandyopadhyay & Sen 2011). A PEN2 portable electronic nose with ten gas sensors, for example, has been shown to identify meat adulteration and confirm the origin of wine, liquor, and edible oil (Xu *et al.*, 2017). To assess toxin levels, the array sensors were immersed in a toxin-containing liquid and produced a pattern of electrical signals, which were detected using a bespoke machine learning model (Cruz *et al.*, 2018). Hence, the multi-component mixtures can be characterized by these E-sensors that help in fulfilling both qualitative and quantitative purpose of the mixture with the advantage of low cost and time requirements as compared to other methods, however, they are limited to identify selective number of components (Gan *et al.*, 2016). Therefore, the combination of E-sensor and chemometric methods is a potentially effective approach for the detection of food adulterants and the assessment of authentication as well as traceability.

b) Isotopic methods

The term isotope is defined as the atoms of the same element that differ from each other by the mass. Stable isotopes are divided into two sets on the basis of atomic mass; light, also known as bio-elements and heavy isotopes. The ratios of different isotope groups are investigated in food authentication (Kelly *et al.*, 2005). Several methods are used for the estimation of isotopic ratios, as Isotope Ratio Mass Spectrometry (IRMS), Multi Collector – Inductively Coupled Plasma – Mass Spectrometry (MC-ICP-MS), and Thermal Ionization Mass Spectrometry (TIMS) (Danezis *et al.*, 2016b). The stable isotopic ratios tend to change with the alterations in geographical origin, climate, soil pedology etc., so, these ratios are relevant to determine food authenticity (Drivelos & Georgiou, 2012).

c) Biomolecular methods

A great advancement in the identification and detection of components in food for authenticity are Nucleotide and protein based techniques. Since, the nucleotide material *i.e.* DNA can be extracted from any tissue, this property of high stability of DNA permits the examination of abundantly processed foods to trace contaminants. DNA is alike in all somatic DNA-based techniques used for authenticity of foods, so the DNA fragments are required to amplify with high precision by Polymerase Chain Reaction (PCR), a technique fit in "genomics" as the entire genome of the sample is used (Ali *et al.*, 2014). As an example of DNA-based methods, the authenticity of olive fruits was reported to be identified by using micro-satellite markers that may act as a uniform platform pertaining cultivar specific genotypic profiling which can be made available to the researchers (Yadav *et al.*, 2021). Conversely, the term "proteomics" where proteins and peptides act as markers in determining the various properties of food products, is applicable for a systematic identification of novel proteins/peptides. The method used genomics and proteomics are highly sensitive as the quantity of the sample required for the analysis is minuscule (Sun *et al.*, 2014). Microbial fingerprinting is another budding sub-field of proteogenomic for identifying food authentication in which PCR Denaturing Gradient Gel Electrophoresis technique (PCR-DGGE) is used. The study of microbial flora fingerprints of cheese has been documented by Arcuri *et al.* (2013).

Other than these three broad categories of the analytical methods for food analysis, certain other methods are also being used. Elemental profiling is one such technique associated with the various macro-micro and rare earth minerals that are obtained by plants from the soil. Therefore, any variation in geographical origin acts as an indicative of the variation in rare earth elements and hence, the elemental fingerprints can be studied using different techniques. Inductively Coupled Plasma-Mass Spectroscopy (ICP-MS) and Atomic Emission Spectroscopy (ICP-AES) are the exclusively used techniques recognized for their ability to measure multi-elements (Gallardo *et al.*, 2013; Laursen *et al.*, 2014). Immunoassays are another analytic tool that that works with the principle of antigen-antibody interaction. Immunological techniques are popular from one decade for the verification of unique standards of different food stuffs and the ingredients used in foods. The major advantages of immunoassays include cost effective, speedy, user-friendly process with high specificity, high throughput and sensitivity. Food industry is developing antibodies against certain specific antigens like, food allergens, contaminants, toxins etc. that are then used as capture molecules to entrap target antigens (Gajewski & Hseih, 2009). One widely used and highly accepted immunological assay is Enzyme-linked immunosorbent assay, ELISA, (Asensio *et al.*, 2008) that helps to assess the food authenticity in various fields, as dairy, meat and fish products. Moreover, the detection of certain implicit treatments, like, food irradiation and the presence of genetically modified organisms (GMOs) in foods. Furthermore, the utilization of varied approaches for determining the food frauds and adulterants have been recently reported as a promising review for food authenticity as well as traceability (Amaral, 2021)

Micro and Nanobarcodes: New tool for authentication

Although it is required to protect the food quality during production, it is now mandate to detect the forged products. Such fake food items in the market leads to the shortfall of rewards for trade and also create a threat to the health of consumer since counterfeit products do not act in accordance with the legislation. Therefore, in the current times, the food authenticity can be followed by using a subsequent method: extrinsic encrypted tag, which is not visible with naked eyes by means of real-time reading and food-authorized to put into the food item as well as on packaging. In this category of extrinsic tools, micro and nano-barcodes are novel tools for traceability and authentication (Sung-Kyoung & Sang-Bok, 2009). The property of minute size of such barcodes is helpful in allowing their micro or nano size of these barcodes allows their introduction and spread in an object. There remain codes in the micro particle that is understandable by optical signals or microscopic techniques, remains unseen by eyes and allows the tracking of the food product to protect against fraud or adulteration. However, an accurate and convenient detection system is required to decrypting the particle code.

Holograms: Holography is a method where the recording of a light sensitive particle in a light field is done. The technique is somehow comparable to the photography in which an image created by a lens is documented upon an object or precisely captured with a semiconductor sensor. Hence, Hologram is a recording of interference pattern on a photographic emulsion that is illuminated by a coherent light source: a laser. Holograms are capable of replicating 3D surface impression and delusion of depth. These holograms are used for security purpose, but the reflected light patterns depend on viewing orientation. This fact adds the intricacy for examination and reproduction of patterns (Shaked *et al.* 2009). The major drawback of this technique for food industry is less security and high cost (Ghaani *et al.* 2016).

Extrinsic DNA tags: Intrinsic DNA tags are other successful technique that acquires almost all the properties needed to trace the food deception (Scarano & Rao, 2014). This technique helps to identify the biological constituents of food along with the benefit where information regarding the geographical origin of a product can be effectively tracked and the adulteration with low quality ingredients can be located. In spite of this, the fingerprinting-DNA approaches with extrinsic DNA tags have come into sight.

The simplicity of this method is that it recommences the brilliant coding capacity by accumulating four nucleotides. An encapsulated, embedded or immobilized DNA in/on a matrix is located on the product to be authenticated. However, the potential of such DNA tags is not easy to evaluate as it is complex to acquire access of the thorough technical papers, which remain classified (Scarano & Rao, 2014).

Fluorescent Barcodes: In the area of micro barcodes, fluorescence possess the benefit that particles emit light due to which their detection get enhanced, therefore, the chances to extract the location and character of particles against the background and support material. The signal produced by the fluorescence acts as an information carrier which can be encoded in several many ways like, fluorescence intensity, location and shape of the particles as well as the fluorescence spectrum (Finkel *et al.*, 2004).

***Quantum Dots*:** Also known as Semiconductor nano-crystals are semiconductor nano-particles that limit the charge carriers *i.e.* electrons in the 3D pattern. This mechanism leads to the quantum size effect in which the detention of the charge carriers (Electron gas) persuades a change in the band-gap configuration as well as in energy levels inside the materials. This phenomenon then directs the recombination mechanisms that lead to tunable photoluminescence. This permits tweaking the absorption and emission peaks by the particle size which is a unique tuning property of the photoluminescence. This mechanism is of great importance in nano-barcoding since the particles can have an exclusive characteristic which is closely related to their production process. The major application of quantum dots in the field of food authentication has emerged over the past few years for the detection of pesticides in agricultural products (Ji *et al.*, 2011, Nsibande & Forbes, 2016).

'Traceability' As a Quality Index in Food Business

The term Traceability can be described as the past of a product in terms of the attributes of that food product and/or the properties linked with the product once they undergo any specific process of value addition by means of the coupled production procedures along with the desired environmental aspects (Regattieri *et al.*, 2007). The information attained relating to the associations at origin can be used throughout the food supply chain *i.e.* during the ordering of the product in order to identify the needs of the structured product as well as at the time of the delivery process with the purpose of specifying the product characteristics. Moreover, the collected information can be utilized for reporting purpose in the food supply chain and for third party. Hence, traceability is considered as a model connecting to all the food products and all type of supply chain. In the current scenario, the market environment is dealing with the competition among the firms as a focus on achieving the factor of consumer satisfaction; traceability is an obligatory tool to get the market harmony. Undeviating advantages of this parameter are the food safety, marketing/ business benefits and supply chain optimization.

In particular, food safety is an essential aspect in food business due to which the traceability has become a relevant factor as it has been recorded that millions of people get affected by food borne illness, every year. Hence, an accurate and well organized system conveying precise, judicious and reliable information of food products throughout the supply chain can trim down the operating expense and may increase the productivity. Simultaneously, a strong and organized system enclose several food safety components, for instance, a detailed information about the origin of food ingredients, nutritional value, components added and processing history which remains transparent for the consumer that gives a feel of getting safe and healthy food. Sadly, there is lack of the universal legal obligation for the set up of traceability system in food supply chain. Therefore, in order to intend the traceability in system, a small number of established industries are creating their own schemes and methodologies, but the challenges are still dominating in context to the lack of set standards and the systems are quite differentiated and are giving varied economic results. Even though Traceability is an effective measurement for developing safe food, it is also costly; therefore the detailed evaluation of its economic impact is thoroughly required.

Latest Trends and Recent Advances in Traceability Systems

Food traceability has become increasingly important as a result of technology improvements, and traditional approaches are no longer sufficient for certain items. Establishing data links between processing and logistics units has become imperative in these terms, necessitating the acquisition of automatic identification technologies (Badia-Melis, 2015).

Radio Frequency Identification (RFID): RFID technologies integrated with block chains are an eccentric amalgamation for real-world implementation in the agri-food industry. RFID is a wireless, non-contact method of transmitting data by means of radio frequency waves. A categorized item with RFID tags tolerates users to robotically and entirely classify and track inventory and assets. Additionally, even if block-chain gives the impression to agonize from high-technology constraints and a lack of practical implementations, the possibilities afforded by block-chain systems are undoubtedly a fascinating and current topic of future research. Food products could be tracked since their creation using block-chain technology, which could track food transformations across all nodes of the supply chain (Figorilli *et al.*, 2018), and information about production, transformation, and transportation could be inserted using an info-tracing platform (Costa *et al.*, 2013). Many authors have discussed the advantages of RFID over the widely used barcode method in context of food traceability. RFID technology is an appropriate approach for the food stratification (Regattieri *et al.*, 2007), for the reason that construing segment is speedy and entirely robotic.

In addition, the two-dimensional barcode, often known as QR code, represents a comparatively low-cost tracing mechanism. The supreme luxurious technique to govern the food traceability is DNA barcoding. To quote an instance, Campanaro *et al.* (2018) stated that the barcoding of DNA regions have a sky-scraping variability pertained to intra-specific and inter-specific, which is why authentication and detection of adulteration are vivacious for food traceability. The following are some examples of how traceability technologies have been used in the food industry. In indubitable state of affairs, milk byproduct traceability has committed RFID dockets at first hand to the cheese (Barge *et al.*, 2014). Furthermore, literature provides a variety of different applicable technologies for meat product traceability, including NFC (Pigni & Conti, 2017), bar codes (straightly in collaboration effected on livestock, carcasses and poultry), NIR technology (Ning *et al.*, 2008) and DNA barcoding (Ahmed *et al.*, 2018). Some studies used nuclear magnetic resonance (Ritota *et al.*, 2013) and a combination of NFC–RFID tags to investigate the excellence and care traceability of edibles (Mainetti *et al.*, 2013). Some authors use isotope analysis to track milk, oils, and beverages (Dutra *et al.*, 2011), chemometrics analysis (Versari *et al.*, 2014), plasma mass spectrometry (Benincasa *et al.*, 2007) and DNA barcoding (Kalaitzis & El-Zein, 2016) to assess the origin.

RFID: Currently an efficient tool for traceability control: The existing improvements in RFID technique, as well as the assimilation of fundamental portions, in particular logger capabilities and integrated sensors, have prearranged an innovative dimension to the use of RFID technology in food traceability systems. As Ruiz-Garcia and Lunadei (2011) designated, the solicitation of RFID to food traceability are plentiful and speckled. RFID has arisen as a principal contender in the expansion of food stock sequence traceability arrangements during the past era, with employments intensifying at a swift stride (Costa *et al.*, 2013). RFID technology allows for effective data sharing, alteration, and handling by allowing commodities to be identified without having to come into physical touch with them (Zhang & Li, 2012). One such RFID substructure that has the rewards of programmed documentation, undeviating EPC for entirely associates, inexpensive equipment investment, and easily drawn information as compared to conventional lot numbering and interior information methods was proposed by Kelepouris *et al.* (2007).

Qian *et al.* (2012) formed a Wheat Flour Milling Traceability System (WFMTS) that included two-dimensional barcode and RFID technology to endorse the structure in a wheat flour mill in China. Small wheat flour bags were identified with QR Code labels, and wheat flour bins were identified with RFID tags that automatically recorded logistics information. Feng *et al.* (2013) trans-evaluate the bovine species traceability arrangement that united RFID technology with a PDA and a barcode printer. They were able to obtain real-time and precise information congregation and broadcast, as well as high facts tracking and tracing efficiency crosswise the bovine species supply chain. Catarinucci *et al.* (2011) applied a fusion of WSN and RFID to progress the white wine's traceability from vineyard to customer glass. WSN is a wireless expertise comparable to RFID that has been exasperated by myriad dissimilar writers to expand food quality and aquaculture traceability (Qi *et al.*, 2014).

RFID barriers: Certain Ultra High Frequency tags that could be beneficial and are now being used for cold chain nursing and traceability controller have superior meddling than other incidences. These diverse groups are solitary reachable in the United States and Europe. These broadcasts are forbidden in Japan and China (Ruiz-Garcia & Lunadei, 2011). The irrelevant mode of keyboarding evidence, the unproductive sequence of data entry and communication machinery connected with RFID reader and the high implementation cost are the key impediments to deploying the RFID system in cattle/beef traceability systems (Feng *et al.*, 2013).

Near Field Communication: NFC permits for modest and protected two-way communication between electrical campaigns. It harmonizes several accepted wireless techniques at consumer level. NFC consents plans to argument information in less than 4 cm (NFC Forum, 2014). For diverse motives such as safekeeping, anti-theft and distinct identity, these tags are fashioned at a micro level so that they can fit inside the substances. NFC is an innovative variety of RFID that is frequently adapted for making expenditures and salvaging evidence over short distances (Mainetti *et al.*, 2012). The vital benefit of NFC expertise over barcodes and QR codes, similar to RFID, is that it does not necessitate a laser beam to travel between two devices to have a solid route. Chen *et al.* (2014) envision a scenario in which a buyer can use an android gadget to recite passive material and significant principles earlier to securing and acquiring of food.

Isotope analysis: When deliberating nuclides of a precise element, the phrase stable isotope is favored; it usually refers to isotopes of the same element. Measuring the appraisal relative copiousness of such sturdy deviations (isotope analysis) harvests a variant ratio that can be utilized as an exploration instrument. Coalescing constant isotope ratio scrutiny with supplementary chemical approaches such as isotope ratio mass spectroscopy (IRMS), inductively coupled plasma mass (ICP-MS), chromatography and near infrared spectroscopy, a latent policy for agro-product legitimacy and traceability has arisen (Zhao et al, 2014).

Sturdy variant analysis has established to be an efficacious apparatus for influencing the geographical origin of agricultural harvests. Identifying and distinguishing varied agro-food products such as meat, milk, cereal crops, wine and oil has had varying degrees of accomplishment. Horacek and Min (2010) invigilated the carbon (C), nitrogen (N) and hydrogen (H) isotope proportions in the defatted dry mass of beef models from Korea, the United States Australia and New Zealand and found a clear carbon isotope trend for individual set of tasters.

Chemometrics and NIRS for food traceability: Chemometrics is the use of mathematical and statistical methods demonstrating to analytical data composed on a variety of platforms in order to extract eloquent chemical information. Chemometric representations have multifold recompenses over traditional arithmetical models, together with ease of updatability, simplicity, robustness to defective data and the capability to be easily transformed into a set of provisions that can aid in the development of verdict instructions on the legitimacy of food origin (Vandeginste, 2013). Synergistic use of influential investigative techniques and chemometrics exhibiting seems to be a probable pathway for the construction of genuineness and traceability models, as food traceability is an imperative constituent of assuring food safety and quality (Bertacchini *et al.*, 2013). In order to investigate the quality attribute of sugarcane honey syrup, the chemometric analysis was carried out that allowed the identification of 32 volatile organic compounds acting as potential markers for authenticity and traceability of syrup in international market (Silva *et al.*, 2021).

Cozzolino (2014) navigated a judgement of the rehearsal of near infrared (NIR) and mid infrared (MIR) spectroscopy in aggregation with multivariate figures analysis procedures such as linear discriminant analysis (LDA), principal component analysis (PCA) and partial least squares discriminant analysis (PLS-DA) to aid in cereal authentication and traceability. He distinguished and settled that the development of novel spectroscopy skills. To give an example, hyper spectrum imaging and micro spectroscopy, as well as new algorithms, will expand the use of spectroscopy as one of the furthermost accommodating instruments for cereal traceability and authentication. Near-infrared spectroscopy i.e. NIRS in conjunction with chemometric approach has been revealed to be able to foresee the geographic origin of wheat grain.

DNA barcoding: DNA barcoding has been publicized to be predominantly potent in the traceability of seafood (Becker *et al.*, 2011). It may also be harnessed to trace mammalian meat (Cai *et al.*, 2011) or raw milk (Arcuri *et al.*, 2013), as well as edible plant tracing (De Mattia *et al.*, 2011). Other applications include processed foods and fruit residues in chocolates, juices, purees, cookies, etc. (Sakai *et al.*, 2010). New technological advancements in food traceability can bring a significant increase in efficiency and efficacy. In present period, the procurement of indispensable evidence by the means of Radio Frequency Identification throughout the hierarchy of the edible providence and NFC as soon as the merchandise reaches the termination user. The particular wireless monitoring campaigns have been amended in recent years and they now offer *in situ* governance or in the moment while the goods are on the move; however, none are proficient of meeting all of the needs of today's heterogeneous food supply chain. Current developments (DNA barcoding, chemometrics, *etc.*) can block (roughly few of) the breaches, such as disease spread or historical control, however, such techniques are still in the challenging periods or necessitate a bulky disposition of properties that are not reasonable for all food chain agents, so they are unlikely to become an alternative in the near future. Numerous of innovative applications have been emerged in this area, as these new-fangled traceability notions argument to registering not only the source of harvests, but also new metrics merged in intellectual logistics, permitting for the acquaintance of produce excellence and consequently enlightening chain strategy and belongings movement.

Additional portable detection technologies:

Even though laboratory-based discovery can offer consistent and precise outcomes, erudite and luxurious apparatus are mandatory, which are laborious and require manpower (Hameed *et al.*, 2018). It is difficult to conduct on-site food analysis due to the shortcomings of traditional methods and this greatly hampers their application. This hindrance gives rise to the need of convenient detection methods with the properties of being less costly and easy to use. In recent times, the portable food detection technologies can be clustered, which include portable spectrometers, and smartphone-based assay.

Portable spectroscopy*:* This technique facilitates non-damaging, speedy and on-site assay of food ingredients and food safety assessment, including Raman, infrared (IR) and nuclear magnetic resonance (NMR). Infrared absorption and Raman scattering spectroscopy are two of these portable spectroscopic equipments that estimate evidence associated to chemical particles in food prototypes, however NMR spectroscopy investigates food components grounded on the magnetic body characteristics of atomic nuclei. (Hashimoto *et al.*, 2019).

Major benefit of such spectroscopic methods is their ability to examine food models through packaging without any damage to the food, lowering the detection expenditure for food traceability procedures significantly. The use of a glass vial to test the quality of Arabia coffee using a miniature IR spectroscopy has been documented by Correia *et al.* (2018). The instrument is designed with the feature of identifying the adulterants, like, corn and robusta coffee present even at lower concentrations in Arabia coffee and the instrument needs no physical contact with the sample. Although the practicality of portable spectroscopy in food traceability has been established, environmental conditions (*e.g.*, temperature and moisture) may interfere with the measurement and food compositions.

Smartphone-based analysis: The equipment has been designed with the latest technology in which the accessories like digital camera with battery, a processor, storage space and wireless data transmission are installed and the is used with the purpose of data collection. It is a good tool for food analysis because it allows for processing, interpretation and sharing. To illustrate it, a smartphone-based system was developed to trace nitrite which is found as a chemical additive in meat foodstuffs. Earlier, the colorimetric array system using a paper fluidic means was developed for the detection of nitrites and now the smartphone technology is used for the quantitative analysis of the same (Lopez-Ruiz *et al.*, 2014).

In a different invention, a smartphone-based fluorescence microscopic method was assembled for finding *Salmonella spp.* (a foodborne pathogen) rapidly where the smartphone captured and analyzed footage of fluorescence-labeled *Salmonella* travelling on a microchip with the use of an intended application interface to enumerate the bacteria quantitatively (Wang *et al.*, 2019).

Smart indicators in combination with sensors assimilation in food supply chain: To overcome the concern related with the examination of the interior of packaged food with respect to environmental conditions, smart packaging is acquired to scrutinize a merchandise's statistics crosswise the nutriment supply structure (Schaefer & Cheung, 2018). Twofold categories antiquated for the inventive packing arrangements were investigated:

(1) The sensors and indicators assist in assembling the information in relation to the alteration in environmental factors within the food packaging along with the condition history of the packaged food (Ghaani *et al.*, 2016). To exemplify it, Temperature, Gas, Freshness and Biosensor Indicator are wholly inculcated in diet wrapping.

(2) For food distribution and traceability, data carriers are used to store and transfer data. To give you an example, consider barcodes, like these are used in food authentication.

Food Product Branding; A Chief Element in the International Food Structure

Marketing is always considered an indispensable element of any trade which exhibit the characteristics displaying the ethnicity of respective organization and/or corporate values that paves the way for consumer to have faith in the scheme and strategy of the firm (McKitterick, 1957). In order to create the standards for buyer, associate and the general public, marketing is the fundamental platform that constructs, distribute, correspond and substitute the contributions of a business. Marketing of any product is described by four Ps *i.e.* product, place, price, and promotion and all these aspects have an enormous impression on consumer preferences with respect to Brand (Grönroos, 2006). Brand predilection is determined through Brand positioning which, in turn, linked with the feasible offerings based on consumer perception. Hence, Branding gives the assurance of consumer awareness (Iyer *et al.*, 2019). Positioning portray the unique competitive distinction in the market that makes the commitment irrefutable (Shafiee *et al.*, 2020). Hence, this approach of brand positioning works on the strategy of crafting the brand connections in the intellect of consumers in such a way that they can consider and evaluate a particular brand uniquely among others (Fuchs & Diamantopoulos, 2010). This concludes that the seller give rise to brand positioning through effective branding, which is an essential exercise of marketing.

The brand is described as a sign, name, word or design and/or combination of all these parameters. Brand is considered to perform two vital functions; a) It makes a distinction of items from each other b) It gives the demonstration of the origin of product (Kapferer, 2008). Brand play equally important role to traders as well as customers as it identifies the origin/maker of the product and allow the consumer to relegate responsibility to a particular producer or trader (Nalca *et al.*, 2018). The series of incidents, in context to the product and its marketing, experienced by the consumer in the past times help them to illustrate the difference between the brands that can meet their needs and the one which are not. As a result, brands act as a simplified method of indication for a buyer regarding product decisions (Cengiz & Akdemir, 2016). This makes the consumer a fundamental intent of dealer who always considers the consumer behavior as well as consumer brand perception before drafting the outline for marketing plans (Keller, 2013).

Whenever a product is introduced in the market, brand is considered a game of intelligence in which each constituent quote a narrative and all such stories are fabricated smartly and collectively in order to achieve the goal of consumer satisfaction as well as to represent a reputed picture in the mind of consumer (Farhana, 2012). Furthermore, the availability of brand in the market is considered as a supplementary aspect that influences the decisive factor of brand selection by customer (Kucuk, 2011). It is evident that the consumer preferences may diverge on the basis of food security, packaging, origin, tradition, nation, etc. (Cantallops & Salvi, 2014). Therefore, the market of developing and developed countries is extremely diverse on a global scale, where the customer value and corporate culture are primarily influenced by marketing behavior. Hence, consumer behavior, brand choices, and brand values are integral parts of efficient marketing.

Global Branding: It is obvious that global branding refers to the brands that are sold globally. In order to define global food brand, both qualitative and quantitative measures are required to be considered. The per annum sale of product, sale outside the home market and same category indicating the same brand name are major quantitative parameters. A qualitative criterion of food brands is quite challenging to maintain in global market. The perusable reason for the difficulty is the cultural differences among various regions with respect to taste and eating habits. For the global success in food branding, strategy, awareness and establishment are the qualitative measures that need to be managed carefully.

With the vision of meeting the consumer expectations and trade requirements in each nation, global food brands needs to be tailored. A business strategy is requisite by which a common pathway can be made to influence the precise trade proficiency and competitive benefits across countries. By following such approach, a centralized corporation can work flexibly across multiple nations by maintaining a balance among all the contributors including staff and management of the particular state. It must be notified that such decisive factors do not presume that formulation of the product of global brand is alike globally. Moreover there can be the variation in the product positioning as well as the brand name across the countries. This is a clear indicative of the fact that both the product and brand are interchangeable (Levitt, 1983).

The food producers always remain fascinated towards global branding. There are various reasons for this attraction, like, economic development, domestic saturation, competitive benefit, etc. As a mandate requirement, global food brands usually need a noteworthy adaptation of their product formulation as well as product positioning so that the customer expectancy and legal aspects of each nation can be achieved successfully. However, the regional food producers always get the competitive benefit as they have the availability of labor on low cost and they remain familiar with the work environment of the local area of the country. These factors may counterbalance the magnetism of global branding. As a result, the pros and cons of global food strategy must be evaluated by the food manufacturer. In order to be a food manufacturer of a global brand, a great experience of the market plays a vital role. Also, the sales level and number of flourishing brands depicts the size of food turnover and the financial performance that is achievable with the well executed organization on local as well as international level (Ramsey, 2003).

Consumer awareness with respect to the food purchase and utilization has an immense effect on global food branding. Recent trends in ready-to-serve or instant food products are in huge demand and seeking the attention of consumers round the globe. It has been reported that the novel and superior quality of the brands selling instant foods have changed the standard of living of customers (Ramasamy & Kalaivanan, 2005). The availability of local as well as global brands in market and online store present valuable offers to buyers with an aim of enhancing the way of life by protecting energy and time (Srinivasan & Nirmala, 2014). The combination of high quality along with cost efficiency offered by brands will always remain the first choice of public. Along with this, there remain various other influential parameters that can affect the nature of consumer for brand selection (Boyle *et al.*, 2018). Besides, the organization and market offerings have an enhanced success rate due to an increased identification of the significance of brand image which indicates that an outstanding brand image can enhance customer experience and satisfaction for long-lasting devotion (Ogba & Tan, 2009).

Consumer Preferences and Its Impact on Global Market

Consumer behavior is described as the analysis that create a procedure which has been followed by people or establishments in order to choose, assure, utilize and arrange the products, ideas, or experiences to study the impact of these procedures on needs of consumer and the public (Vyshnavi & Rao, 2016). A detailed investigation of consumer behavior is a fundamental point that is considered prior to the production of constituents and services which is appropriate pertaining to meet the needs of consumers as how, why and where they buy and consume a specific brand along with information regarding their preferences to other brands (Shafiee *et al.*, 2020). It has been observed that the consumer prefer the product with a brand as it is always of better quality as compared to the other non-branded products. This is the major reason that the brands provide an ease for consumer to track the product of choice and hence, speed up the decision making for purchase of goods.

Consumer perception is a module that helps public to pick, classify and identify the response of sensory receptors to various elemental stimuli like, smell, taste, color, texture, light, and sound (Carman, 1990). There are several significant factors in market mix like brand impression, allocation procedure, cost, expenses to publicize in the market that influence the consumer perception (Pappu & Quester, 2006). (Rahman, 2012) identified that consumers have apparent pictures of themselves along with the products and brands. Therefore, brand perception is an outstanding result of the understanding of consumer with respect to the brand and it is a reaction that comes through buyers, more willingly than sellers (Spieth *et al.*, 2019).

Consumer preferences on Branding: Whenever a consumer buys any item, brand preference is considered as an exclusive attribute (Kwok *et al.*, 2006). The consumer satisfaction concerning specific needs could be achieved if the overall functional value of a brand is brilliant. Conversely, the consumer preference does not remain the same for a longer period of time if the brand fails to uphold the sustainability and product excellence (Aaker, 1996; Pitta & Katsanis, 1995). Talking about the competition among the food businesses, the brand preference has become an imperative constituent as the customers always reallocate themselves to other firms that produce competition in the market (Hellier *et al.*, 2003). The attitude of customer for brand preference is the key factor that lessens the difficulty of buying decisions (Gensch, 1987). At the very beginning, customer gets opened to various other brands and follows the convoluted decision-making process (Cătălin & Andreea, 2014).

Consumer preferences regarding food quality labels: With advancement in the global food sector, the usual direct communication of buyer with retailer has been dropped and the reason is urbanization in mass food production that has developed and settled throughout the industry (Coff *et al.*, 2008). Such inclination has favored the desired changes in the decision pattern of consumers regarding food purchase as well as the search in context to information related with the product of interest (Kristensen *et al.*, 2013). The only approach available with the buyer to get this type of information is by reading the food label displaying the ingredients used and the method followed for the manufacturing along with plethora of know-how for getting the maximum advantage of the item. Food labeling works as a silent but strong indicative signal that conveys the direct aid to buyer for deciding the purchase related aspects as all the desired information is conveyed by label about different attributes. The aware consumer is now habitual to scan the entire information given on the label, sometimes, with the purpose of comparing the quality or corroborating the knowledge gained earlier. Due to vast history of scandals related with food safety, consumer is now more focused on health conscious diet and eco-friendly consumption models and takes wise decisions regarding food preferences by reciting the content given on the packaging of product, like, nutritional information, health claims, ingredients, caution and so forth (Juhl & Poulsen, 2000; Newman *et al.*, 2014). The firm uses the credible information on label that helps them to spread a quality signal in the market along with the availability of the precise attributes and this practice create a platform to get premiums (McCluskey & Loureiro, 2003).

It has been observed that the manufacturers and organizations have acted by selling eco-labeled, organic and various other quality foods by adding the labels that clearly state that the products were manufactured by keeping in view the animal welfare, rational labor practices with an environmental friendly expression. The trend of the products labeled as free from genetically modified (GM) constituents is running successfully throughout the globe. Consumers also prefer to purchase the products having labels with the information about particular safety aspect, quality characteristics along with the words telling about the geographical origin.

The preference of consumers to eco-friendly products have given rise to eco-labels where the product assessment is done on the basis of its impact on environment by comparing it with other products in the similar category. Hence, the products labeled as "environment friendly" covers eco-labeled as well as organic products and therefore this category attracts the same category of consumers. Food quality labels provides advantages to buyer as well as to seller because people prefer to purchase the products with particular value- added quality and manufacturer can promote the food products with novel quality in a better way.

Consumer preferences for health and environmental attributes: In contemporary era, a drastic change in the food consumption trends among consumers has been observed where the people are considering the health and well being above everything. The consumers' way of selecting their diet has become more selective where the preference is given to the health products. It is evident that the health phenomenon is not restricted to the diet plans or different needs associated with some pathology studies, however, many other aspects of lifestyle have been covered so far (Nielsen, 2016). Consumer is now interested in the mounting trend of nutraceuticals, functional foods, fortified products, organic foods, products made with low environmental impact and those with biodegradable packaging. The food industry is launching the products with the aim of providing quality products having health implications and climate friendly attributes for which the consumer is ready to pay a high-grade price (Markosyan & McCluskey, 2009). Presently, there are plethora of foods that fulfill these features and the certification of the products in done where the higher quality level is being attested in order to meet the consumer satisfaction.

New Directions in Global Food Market

In order to understand the competitive nature of global food industry, one needs to be aware of the consumer preferences and the efforts of food manufacturers to meet up those demands. The involvement of disparate regional, nationwide and global mediators and systems has made the task of shifting the food from farm to fork progressively more complex. Food industry is persistently progressing and the driving force is frequent changes in the consumer preferences along with the technology enhancement, connections among the people of food supply chain and competitive business environment. State and national level food businesses are now adopting the protocols of sophisticated supply chains and distribution channels. The involvement of global trading and manufacturing firms, sellers and consumers leads to the amendments and modernization in global food markets where the ongoing trends in various sections of food industry makes the entire system a riddle and hence, the entire structure has been represented in small segments. As a solution to streamline all segments, food supply chains are working as a authenticate system for signal transmission from buyer to food producers and also for transporting the products from field to the table. Four chief components of supply chain are input supply, production, manufacturing process and trading. Food suppliers select whether to add value or other quality product in the market as per the demand of consumers and work in order to meet the particular needs of consumers. Value addition in the products can be done at any of the four stages of supply chain before delivering the product to customer. In the growing global food economy, indicating the additional value of the novel product is equally important as that of making the new product. However, the high degree of cooperation and coordination among different segments of supply chain is of utmost importance and the same is achievable by developed countries and the same is serving as a model for various food markets in high-revenue countries. The recent trend in the global food market is the availability of hypermarkets and supermarkets round the globe, where multinational dealers are operating across multiple countries. Also, the online shopping system has made the availability of discrete food products among states in a country by which the consumer awareness regarding cultural and lifestyle differences has influenced the organization of retail outlets in the markets of several developed nations.

In order to maintain the command and leadership in a particular food sector, the global food market has encouraged the food producers to embrace innovative growth approaches. The competition of food manufacturers in the retail sector is based on the marketing and promotion policies of their products. The private organizations hold the formulations, processing technology, branding and trademarks linked with the manufacturing. Other legal agreements associated with marketing and licensing for trade with other national and international corporations play a vital role in deciding that under which protocol, the food product of the manufacturer is being sold in the overseas market. Another strategy of multinational businesses in global food market is to gain the support from people for maintaining the trade related activities in developing countries. The success rate of this approach has been observed to be increased in financial reporting of recent annual statements of international businesses. Moreover, as the consumer is becoming more concerned about the environmental safety along with food quality and food safety, the food manufacturers are successively determined to incorporate the plans for substantial growth of the natural resources of the world into trade strategies and responsibilities. The current scenario of the firms is to review the product- market portfolio along with their particular function in food chain.

Conclusion

With a substandard history of food frauds, an impression of distrust from consumers is very common with respect to the food quality and safety. In order to meet the consumer expectations and demands, it has become indispensable for the global food market to understand the consumer requirements and to provide the authentic food. The world food system is under the progression of several fundamental alterations with a rise in population, economic growth, technology advancement and globalization. Also, the involvement of extremely integrated supply chains for quality and value-added food products as demanded by the population of all income levels is an integral part of overall revolution in global food market. Hence, a global drift in the food industry has introduced two major aspects *i.e.* traceability and authenticity, in order to make a proficient flow of information all through the food supply chain. The innovations in the technologies for discovering the geographical origin of food, presence of adulterants in food products have paved the ways for consumers to get the safe and authentic food. The identification and quantification methods like use of chemical, isotopic, bio-moleculear markers are the recent advancements in the techniques used for food authentication. Furthermore, other quick, reliable and cost-effective methods like, fingerprinting/profiling pooled with multivariate statistical analysis, micro and nano barcode and tags are in trend as food authenticity tools.

Although, it is quite challenging to achieve the total traceability due to horde of natural as well as artificial obstructions in the food world, it will be only partial. With the numerous chief progressions recorded in past few years, it is expected that the traceability will move towards better reliability and transparency. The food traceability mechanism can be upgraded by putting into practice the genuine techniques and by managing the significant information of the food products throughout the entire food supply chain. Hence, by collaborating the traceability system with excellent planning and logistics, the definite traceability practices can be enhanced.

Along with the identification of food by following the traceability and authenticity parameters, it is a fundamental requirement of food manufacturers to achieve success in the market by gaining the faith of consumers. Therefore, various strategies have been implemented from time to time, in order to satisfy the consumer perceptions with respect to safe and quality food. Consumer preferences towards health and environmental attributes, quality labels and food branding have a great impact on global food market. Consumer awareness leads to such perceptions which in turn restrict the food producers and traders to develop the parameters and technologies giving rise to authentic and safe food products.

The food industries established in international market will likely to grow even larger in coming decade as they are expanding their business in developing countries. Therefore, an enhancement in the consumer needs and wants is opening the path for the small scale traders to compete in the same market. Although the established brand names are significant barriers for small scale businesses, the expertise in the particular processing techniques for different products and ingredients offer potential for new contestants.

References

Aaker, D. A. (1996). Building Strong Brands. The Free Press. New York, 598-614.

Ahmed, N., Sangale, D., Tiknaik, A., Prakash, B., Hange, R., Sanil, R., & Khedkar, G. (2018). Authentication of origin of meat species processed under various Indian culinary procedures using DNA barcoding. Food Control, 90, 259–265.

Ali, M. E., Razzak, M. A., & Abd Hamid, S. B. (2014). Multiplex PCR in species authentication: probability and prospects—a review. Food Analytical Methods, 7(10), 1933-1949.

Amaral, J. S. (2021). Target and Non-Target Approaches for Food Authenticity and Traceability. Foods, 172. https://www.mdpi.com/journal/foods

Ansari, M. J., Al-Ghamdi, A., Khan, K. A., Adgaba, N., El-Ahmady, S. H., Gad, H. A., ... Kolyali, S. (2018). Validation of botanical origins and geographical sources of some Saudi honeys using ultraviolet spectroscopy and chemometric analysis. Saudi Journal of Biological Sciences, 25(2), 377–382.

Arcuri, E. F., El Sheikha, A. F., Rychlik, T., Piro-Métayer, I., & Montet, D. (2013). Determination of cheese origin by using 16S rDNA fingerprinting of bacteria communities by PCR–DGGE: Preliminary application to traditional Minas cheese. Food Control, 30(1), 1-6.

Ares, G., De Saldamando, L., Giménez, A., & Deliza, R. (2014). Food and wellbeing. Towards a consumer-based approach. Appetite, 74, 61-69.

Arvanitoyannis, I. S. (2016). Authenticity of foods of animal origin. CRC Press.

Aschemann-Witzel, J., de Hooge, I., Amani, P., Bech-Larsen, T., & Oostindjer, M. (2015). Consumer related food waste: Causes and potential for action. Sustainability, 7, 6457–6477.

Asensio, L., González, I., García, T., & Martín, R. (2008). Determination of food authenticity by enzyme-linked immunosorbent assay (ELISA). Food Control, 19(1), 1-8.

Asioli, D., Aschemann-Witzel, J., Caputo, V., Vecchio, R., Annunziata, A., Næs, T., & Varela, P. (2017). Making sense of the "clean label" trends: A review of consumer food choice behavior and discussion of industry implications. Food Research International, 99, 58-71.

Aung, M. M., & Chang, Y. S. (2014). Traceability in a food supply chain: Safety and quality perspectives. Food Control, 39, 172-184.

Azziz-Baumgartner, E., Lindblade, K., Gieseker, K., Rogers, H. S., Kieszak, S., Njapau, H., ... & Aflatoxin Investigative Group*. (2005). Case–control study of an acute aflatoxicosis outbreak, Kenya, 2004. Environmental Health Perspectives, 113(12), 1779-1783.

Badia-Melis, R., Mishra, P., & Ruiz-García, L. (2015). Food traceability: New trends and recent advances. A review. Food Control, 57, 393-401.

Bandyopadhyay, D., & J. Sen. J. (2011). Internet of things: Applications and challenges in technology and standardization. Wireless Personal Communications, 58 (1), 49–69.

Barge, P., Gay, P., Merlino, V., & Tortia, C. (2014). Item-level Radio-Frequency IDentification for the traceability of food products: application on a dairy product. Journal of Food Engineering, 125, 119-130.

Becker, S., Hanner, R., & Steinke, D. (2011). Five years of FISH-BOL: brief status report. Mitochondrial DNA, 22, 3-9.

Benincasa, C., Lewis, J., Perri, E., Sindona, G., & Tagarelli, A. (2007). Determination of trace element in Italian virgin olive oils and their characterization according to geographical origin by statistical analysis. Analytica Chimica Acta, 585(2), 366-370.

Bertacchini, L., Cocchi, M., Vigni, M. L., Marchetti, A., Salvatore, E., Sighinolfi, S., ... & Durante, C. (2013). The impact of chemometrics on food traceability. In Data handling in science and technology (Vol. 28, pp. 371-410). Elsevier.

Bianchi, F., Giannetto, M., & Careri, M. (2018). Analytical systems and metrological traceability of measurement data in food control assessment. TrAC Trends in Analytical Chemistry, 107, 142-150.

Black, C., Chevallier, O. P., & Elliott, C. T. (2016). The current and potential applications of Ambient Mass Spectrometry in detecting food fraud. TrAC Trends in Analytical Chemistry, 82, 268-278.

Boyle, P. F., Kim, H. & Lathrop, E. S. (2018). The relationship between price and quality in durable product categories with private label brands. Journal of Product & Brand Management, 27(6), 647–668.

Brown, S. L., & Eisenhardt, K. M. (1995). Product development: Past research, present findings, and future directions. Academy of Management Review, 20(2), 343-378.

Bujji, M. (2012). Tracing the footsteps, FnB news (India). http://www.fnbnews.com/article/detnews.asp?articleid=26342§ionid=32

Cai, Y. S., Zhang, L. A., Shen, F. J., Zhang, W. P., Hou, R., & Yue, B. S. (2011). DNA barcoding of 18 species of Bovidae. Chinese Science Bulletin, 56, 164-168.

Campanaro, A., Tommasi, N., Guzzetti, L., Galimberti, A., Bruni, I., & Labra, M. (2018). DNA barcoding to promote social awareness and identity of neglected, underutilized plant species having valuable nutritional properties. Food Research International, 115, 1–9.

Cantallops, S., & Salvi, F. (2014). New consumer behaviour: A review of research on eWOM and hotels. International Journal of Hospitality Management, 36(2), 41–67.

Carman, J.M., (1990). Consumer perceptions of service quality: an assessment of T. Journal of Retailing, 66(1), 33-57.

Cătălin, M. C., & Andreea, P. (2014). Brands as a means of consumer self-expression and desired personal lifestyle. Procedia- Social and Behavioural Sciences, 109(18), 103-127.

Catarinucci, L., Cuinas, I., Exposito, I., Colella, R., Fernandez, J. A. G., & Tarricone, L. (2011). RFID and WSNs for traceability of agricultural goods from Farm to Fork: electromagnetic and deployment aspects on wine test-cases. In Paper presented at the software, telecommunications and computer networks (SoftCOM), 2011.

Cengiz, H. & Akdemir-Cengiz, H., (2016). Review of brand loyalty literature: 2001–2015. Journal of Research in Marketing, 6(1), 407-434.

Chen, Y. Y., Wang, Y. J., & Jan, J.K. (2014). A novel deployment of smart cold chain system using 2G-RFID-Sys. Journal of Food Engineering, 141(2014), 113-121.

Cicia, G., & Colantuoni, F. (2010). Willingness to pay for traceable meat attributes: A meta-analysis. International Journal on Food System Dynamics, 1(3), 252–263.

Coff, C., Barling, D., Korthals, M. & Nielsen, T. (2008), Ethical Traceability and Communicating Food, Springer.

Correia, R. M., Tosato, F., Domingos, E., Rodrigues, R. R. T., Aquino, L. F. M., Filgueiras, P. R., Lacerda, V., & Romao. W. (2018). Portable near infrared spectroscopy applied to quality control of Brazilian coffee. Talanta, 176, 59–68.

Costa, C., Antonucci, F., Pallottino, F., Aguzzi, J., Sarriá, D., & Menesatti. P. (2013). A review on agri-food supply chain traceability by means of RFID technology. Food and Bioprocess Technology, 6(2), 353–366.

Cozzolino, D. (June 2014). An overview of the use of infrared spectroscopy and chemometrics in authenticity and traceability of cereals. Food Research International, 60, 262-265.

Crabbe, M., Moriarty, S., & Lieberman, G. (2018). Mintel Global Consumer Trends. http://www.mintel.com/mintel-reports.

Cruz, M. G., Ferreira, N. S., Gomes, M. T. S., Botelho, M. J., Costa, S. T., Vale, C., & Rudnitskaya, A. (2018). Determination of paralytic shellfish toxins using potentiometric electronic tongue. Sensors and Actuators B: Chemical, 263, 550–6.

Cuadros-Rodríguez, L., Ruiz-Samblás, C., Valverde-Som, L., Pérez-Castaño, E., & González-Casado, A. (2016). Chromatographic fingerprinting: An innovative approach for food'identitation'and food authentication–A tutorial. Analytica Chimica Acta, 909, 9-23.

Cubero-Leon, E., De Rudder, O., & Maquet, A. (2018). Metabolomics for organic food authentication: Results from a long-term field study in carrots. Food Chemistry, 239, 760–770.

Danezis, G.P., Tsagkaris, A.S.,, Brusic, V., Georgiou, C.A., (2016a). Food authentication: State of the art and prospects. Current Opinion in Food Science, 10, 22–31

Danezis, G. P., Tsagkaris, A. S., Camin, F., Brusic, V., & Georgiou, C. A. (2016). Food authentication: Techniques, trends & emerging approaches. TrAC Trends in Analytical Chemistry, 85, 123-132.

De Mattia, F., Bruni, I., Galimberti, A., Cattaneo, F., Casiraghi, M., & Labra, M. (2011). A comparative study of different DNA barcoding markers for the identification of some members of Lamiacaea. Food Research International, 44, 693-702.

Drivelos, S. A., & Georgiou, C. A. (2012). Multi-element and multi-isotope-ratio analysis to determine the geographical origin of foods in the European Union. TrAC Trends in Analytical Chemistry, 40, 38-51.

Esteki, M., Simal-Gandara, J., Shahsavari, Z., Zandbaaf, S., Dashtaki, E., & Vander Heyden, Y. (2018). A review on the application of chromatographic methods, coupled to chemometrics, for food authentication. Food control, 93, 165-182.

Farhana, M. (2012). Brand Elements Lead to Brand Equity: Differentiate or Die. Information Management and Business Review, 4(4), 223-243.

Feng, J., Fu, Z., Wang, Z., Xu, M., & Zhang, X. (2013). Development and evaluation on a RFID-based traceability system for cattle/beef quality safety in China. Food Control, 31(2), 314-325.

Figorilli, S., Antonucci, F., Costa, C., Pallottino, F., Raso, L., Castiglione, M., Pinci, E., Del Vecchio, D., Coll,e G., Proto, A.R., Sperandio, G., & Menesatti, P. (2018). Blockchain implementation prototype for the electronic open source traceability of wood along the whole supply chain. Sensors, 18, 31–33.

Finkel, N. H., Lou, X., Wang, C. & He, L. (2004). Barcoding the microworld. Analytical Chemistry 1, 76, 19, 352A–359A.

Fuchs, C. & Diamantopoulos, A., (2010). Evaluating the effectiveness of brand-positioning strategies from a consumer perspective. European Journal of Marketing, 44(11/12), 1763-1786.

Gajewski, K. G., & Hsieh, Y. H. P. (2009). Monoclonal antibody specific to a major fish allergen: parvalbumin. Journal of Food Protection, 72(4), 818-825.

Gallardo, J. M., Ortea, I., & Carrera, M. (2013). Proteomics and its applications for food authentication and food-technology research. TrAC Trends in Analytical Chemistry, 52, 135-141.

Gan, Z., Yang, Y., Li, J., Wen, X., Zhu, M., Jiang, Y., & Ni, Y. (2016). Using sensor and spectral analysis to classify botanical origin and determine adulteration of raw honey. Journal of Food Engineering, 178, 151-158.

Gensch, D.H., (1987). A two-stage disaggregate attribute choice model. Marketing Science, 6(3), 223-239.

Geueke, B., Groh, K., & Muncke, J. (2018). Food packaging in the circular economy: Overview of chemical safety aspects for commonly used materials. Journal of Cleaner Production, 193, 491–505.

Ghaani, M., Cozzolino, C. A., Castelli, G., & Farris, S. (2016). An overview of the intelligent packaging technologies in the food sector. Trends in Food Science & Technology, 51, 1-11.

Gómez-Ríos, G. A., Vasiljevic, T., Gionfriddo, E., Yu, M., & Pawliszyn, J. (2017). Towards on-site analysis of complex matrices by solid-phase microextraction-transmission mode coupled to a portable mass spectrometer via direct analysis in real time. Analyst, 142(16), 2928-2935.

Grassi, S., Casiraghi, E., & Alamprese, C. (2018). Handheld NIR device: A non-targeted approach to assess authenticity of fish fillets and patties. Food Chemistry, 243, 382–388.

Gray, J., Armstrong, G., & Farley, H. (2003). Opportunities and constraints in the functional food market. Nutrition & Food Science. 33(5), 213–218.

Gregory, N. G. (2000). Consumer concerns about food. Outlook on Agriculture, 29(4), 251-257.

Grönroos, C., (2006). On defining marketing: finding a new roadmap for marketing. Marketing Theory, 6(4), 395-417.

Grunert, K. G. (Ed.). (2017). Consumer trends and new product opportunities in the food sector. Wageningen Academic Publishers.

Grunert, K. G., do Canto, N. R., Liu, R. & Salnikova, E. (2019). Well-Being as a Global Food Trend: Health, Sustainability and Authenticity.

Guo, T., Fang, P., Jiang, J., Zhang, F., Yong, W., Liu, J., et al. (2016). Rapid screening and quantification of residual pesticides and illegal adulterants in red wine by direct analysis in real time mass spectrometry. Journal of Chromatography A, 1471, 27–33.

Hameed, S., Xie L., & Ying, Y. (2018). Conventional and emerging detection techniques for pathogenic bacteria in food science: A review. Trends in Food Science & Technology,81, 61–73.

Hansstein, F. V. (2014). Consumer knowledge and attitudes towards food traceability: A comparison between the European Union, China and North America. Paper presented at the 2014 International conference on food security and nutrition IPCBEE.

Harmsen, H., Grunert, K. G., & Declerck, F. (2000). Why did we make that cheese? An empirically based framework for understanding what drives innovation activity. R&D Management, 30(2), 151-166.

Hashimoto, K., Badarla, V. R., Kawai, A. & Ideguchi, T. (2019). Complementary vibrational spectroscopy. Nature Communications, 10 (1), 1–6.

Hellier, P.K., Geursen, G.M., Carr, R.A., & Rickard, J.A., (2003). Customer repurchase intention. European Journal of Marketing, 37(11/12), 1762-1800.

Horacek, M., & Min, J. S. (2010). Discrimination of Korean beef from beef of other origin by stable isotope measurements. Food Chemistry, 121, 517-520.

Iyer, P., Davari, A., Zolfagharian, M., & Paswan, A., (2019). Market orientation, positioning strategy, and brand performance. Industrial Marketing Management, 81(4), 16-39.

Ji, X. H., Zhang, N. G., Cheng, W., Guo, F., Liu, W., Guo, S. S., ... & Zhao, X. Z. (2011). Integrated parallel microfluidic device for simultaneous preparation of multiplex optical-encoded microbeads with distinct quantum dot barcodes. Journal of Materials Chemistry, 21(35), 13380-13387.

Johnson, R. 2014. Food Fraud and "Economically Motivated Adulteration" of Food and Food Ingredients. Informing the legislative debate since 1914. Congressional research service. USA. 1–39.

Juhl, H.J. & Poulsen, C.S. (2000), "Antecedents and effects of consumer involvement in fish as a product group", Appetite, 34(3), 261-267.

Kalaitzis, P., & El-Zein, Z. (2016). Olive oil authentication, traceability and adulteration detection using DNA-based approaches. Lipid Technology, 28(10–11), 173–176.

Kapferer, J.N., (2008). The new strategic brand management: Creating and sustaining brand equity long term. Kogan Page Publishers.

Kelepouris, T., Pramatari, K., & Georgios Doukidis, G. (2007). RFID-enabled traceability in the food supply chain. Industrial Management & Data Systems, 107(2), 183-200.

Keller, K. L. (2013). Strategic Brand Management: Building, Measuring, and Managing Brand Equity. 4th ed., Pearson Education Limited.

Kelly, S., Heaton, K., & Hoogewerff, J. (2005). Tracing the geographical origin of food: The application of multi-element and multi-isotope analysis. Trends in Food Science & Technology, 16(12), 555-567.

Khan, F., & Prior, C. (2010). Evaluating the urban consumer with regard to sourcing local food: a heart of England study. International Journal of Consumer Studies, 34, 161-168.

Kristensen, D. B., Askegaard, S. & Jeppesen, L. H. (2013), " 'If it makes you feel good it must be right': embodiment strategies for healthy eating and risk management", Journal of Consumer Behaviour, 12 (4), 243-252.

Kucuk, S. U. (2011). Push-based brand awareness: the role of product availability and in-store merchandising. The International Review of Retail, Distribution and Consumer Research, 21(3), 201–223.

Kwok, S., Uncles, M., & Huang, Y. (2006). Brand Preferences and Brand Choices Among Urban Chinese Consumers: An Investigation of Country-Of-Origin Effects. Asia Pacific Journal of Marketing and Logistics, 18(3), 163-182.

Ladhari, R., & Tchetgna, N. M. (2015). The influence of personal values on Fair Trade consumption. Journal of Cleaner Production, 87, 469–477.

Laursen, K. H., Schjoerring, J. K., Kelly, S. D., & Husted, S. (2014). Authentication of organically grown plants–advantages and limitations of atomic spectroscopy for multi-element and stable isotope analysis. TrAC Trends in Analytical Chemistry, 59, 73-82.

Levitt, T. (1983) 'The globalisation of markets', Harvard Business Review, May/June, 92–102.

Llano, S. M., Muñoz-Jiménez, A. M., Jiménez-Cartagena, C., Londoño-Londoño, J., & Medina, S. (2018). Untargeted metabolomics reveals specific withanolides and fatty acyl glycoside as tentative metabolites to differentiate organic and conventional Physalis peruviana fruits. Food Chemistry, 244, 120-127.

Lopez-Ruiz, N., Curto, V. F., Erenas, M. M., Benito-Lopez, F., Diamond, D., Palma, A. J., & Capitan-Vallvey, L. F. (2014). Smartphone-based simultaneous pH and nitrite colorimetric determination for paper microfluidic devices. Analytical Chemistry, 86(19), 9554–9562.

Lu, H., Zhang, H., Chingin, K., Xiong, J., Fang, X., & Chen, H. (2018). Ambient mass spectrometry for food science and industry. TrAC Trends in Analytical Chemistry, 107, 99-115.

Lunardo, R., & Saintives, C. (2013). The effect of naturalness claims on perceptions of food product naturalness in the point of purchase. Journal of Retailing and Consumer Services, 20, 529-537.

Mainetti, L., Patrono, L., & Vergallo, R. (2012). IDA-Pay: a secure and efficient micropayment system based on Peer-to-Peer NFC technology for Android mobile devices. Journal of Communication Software and Systems, 8(4), 1-6.

Mainetti, L., Patrono, L., Stefanizzi, M. L., & Vergallo, R. (2013). An innovative and low-cost gapless traceability system of fresh vegetable products using RF technologies and EPC global standard. Computers and Electronics in Agriculture, 98, 146-157.

Mansour, F. R., & Khairy, M. A. (2017). Pharmaceutical and biomedical applications of dispersive liquid–liquid microextraction. Journal of Chromatography B, 1061–1062, 382–391.

Markosyan, A., McCluskey, J. J., & Wahl, T. I. (2009). Consumer response to information about a functional food product: apples enriched with antioxidants. Canadian Journal of Agricultural Economics/Revue canadienne d'agroeconomie, 57(3), 325-341.

Martinez, M. G., & Epelbaum, F. B. (2011). The role of traceability in restoring consumer trust in food chains. In Food Chain Integrity (pp. 294-302). Woodhead Publishing.

McCluskey, J. J. (2015). Changing food demand and consumer preferences. In: Responding to Future Food Demands. Paper prepared for Agricultural Symposium, Federal Reserve Bank of Kansas City, July 14 –15(p. 18).

McCluskey, J. J., & Loureiro, M. L. (2003). Consumer preferences and willingness to pay for food labeling: A discussion of empirical studies. Journal of Food Distribution Research, 34, 95-102.

McKitterick, J., (1957). What is the marketing thought and action. American Marketing Association, Chicago, IL.

Medina, S., Pereira, J. A., Silva, P., Perestrelo, R., & Câmara, J. S. (2019). Food fingerprints – a valuable tool to monitor food authenticity and safety. Food Chemistry, 278, 144–162.

Meijboom, F. L., Visak, T., & Brom, F. W. (2006). From trust to trustworthiness: Why information is not enough in the food sector. Journal of Agricultural and Environmental Ethics, 19(5), 427-442.

Meulenberg, M. T. G., & Viaene, J. (2005). Changing agri-food systems in Western countries: a marketing approach. Innovation in Agri-Food Systems. Wageningen Academic Publishers, Wageningen, 17-51.

Motoshita, M., Sakagami, M., Kudoh, Y., Tahara, K., & Inaba, A. (2015). Potential impacts of information disclosure designed to motivate Japanese consumers to reduce carbon dioxide emissions on choice sof shopping method for daily foods and drinks. Journal of Cleaner Production, 101, 205–214.

Nalca, A., Boyaci, T., & Ray, S., (2018). Brand positioning and consumer taste information. European Journal of Operational Research, 268(12), 555-578.

Newman, C.L., Turri, A.M., Howlett, E. & Stokes, A. (2014), "Twenty years of country-of origin food labeling research: a review of the literature and implications for food marketing systems", Journal of Macromarketing, 34, 505-519.

NFC Forum. (2014). NFC forum what is NFC. Consulted 10/10/2014 http://nfc-forum. org/what-is-nfc/about-the-technology/.

Nielsen, K. E. (2016). Health beneficial consumer products—status and trends. In Developing food products for consumers with specific dietary needs (15-42). Woodhead Publishing.

Ning, Z., Dequan, Z., Shurong, L., & Qingpeng, L. (2008). Preliminary study on origin traceability of mutton by near infrared reflectance spectroscopy coupled with SIMCA method. Transactions of the Chinese Society of Agricultural Engineering, 12.

Nsibande, S. A., & Forbes, P. B. C. (2016). Fluorescence detection of pesticides using quantum dot materials–a review. Analytica Chimica Acta, 945, 9-22.

Ogba, I., E. & Tan, Z. (2009). Exploring the impact of brand image on customer loyalty and commitment in China. Journal of Technology Management in China, 4(2), 132-144.

Pappu, R., & Quester, P. (2006). A consumer-based method for the retailer equity measurement: results of an empirical study. Journal of Retailing and Consumer Services, 13(5), 317-339.

Patel, H. K. (2014). The electronic nose: Artificial olfaction technology. New Delhi: Springer.

Pigini, D., & Conti, M. (2017). NFC-based traceability in the food chain. Sustainability, 9(10), 1910.

Pitta, D. A., & Katsanis, L. P. (1995). Understanding brand equity for successful brand extension. Journal of Consumer Marketing, 12(4), 51-74.

Posudin Y.I., Peiris K.S., & Kays S.J, (2015). Non-destructive detection of food adulteration to guarantee human health and safety, Ukrainian Food Journal 4 (2) 207–260.

Poulli, K. I., Mousdis, G. A., & Georgiou, C. A. (2007). Rapid synchronous fluorescence method for virgin olive oil adulteration assessment. Food Chemistry, 105(1), 369-375.

Qi, L., Xu, M., Fu, Z., Mira, T., & Zhang, X. (2014). C2SLDS: a WSN-based perishable food shelf-life prediction and LSFO. Food Control, 38(2014), 19-29.

Qian, J., Yang, X., Wu, X., Zhao, L., Fan, B., & Xing, B. (2012). A traceability system incorporating 2D barcode and RFID technology for wheat flour mills. Computers and Electronics in Agriculture, 89(0), 76-85.

Ramasamy, K, & Kalaivanan, G (2005), "Consumer Behaviour towards Instant Food Products", International Journal of Marketing, 35(6), 24-58.

Ramsey, B. (2003), "Whither global branding? The case of food manufacturing", Journal of Brand Management, 11 (1), 9-21.

Regattieri, A., Gamberi, M., & Manzini, R. (2007). Traceability of food products: General framework and experimental evidence. Journal of food engineering, 81(2), 347-356.

Ritota, M., Casciani, L., & Valentini, M. (2013). PGI chicory (Cichorium intybus L.) traceability by means of HRMAS-NMR spectroscopy: a preliminary study. Journal of Science and Food Agriculture, 93(7), 1665–1672.

Rozin, P. (2005). The meaning of "natural" process more important than content. Psychological Science, 16, 652-658.

Ruiz-Garcia, L., & Lunadei, L. (2011). The role of RFID in agriculture: applications, limitations and challenges. Computers and Electronics in Agriculture, 79(2011), 42-50.

Sakai, Y., Ishihata, K., Nakano, S., Yamada, T., Yano, T., & Uchida, K. (2010). Specific detection of banana residue in processed foods using polymerase chain reaction. Journal of Agricultural and Food Chemistry, 58, 8145-8151.

Scarano, D., & Rao, R. (2014). DNA markers for food products authentication. Diversity, 6(3), 579-596.

Schaefer, D., & Cheung, W. M. (2018). Smart packaging: Opportunities and challenges. Procedia CIRP, 72:1022–1027.

Schmidt, J. B. (2005). What we still need to learn about developing successful new products: a commentary on Van Kleef, Van Trijp & Luning. Food Quality and Preference, 16, 213–216.

Shafiee, M., Gheidi, S. & Khorrami, M.S., (2020). Proposing a new framework for personal brand positioning. European Research on Management and Business Economics, 26(1), 45-64.

Shaked, N. T., Katz, B., & Rosen, J. (2009). Review of three-dimensional holographic imaging by multiple-viewpoint-projection based methods. Applied Optics, 48(34), H120-H136.

Siegrist, M., Visschers, V. H. M., & Hartmann, C. (2015). Factors influencing changes in sustainability perception of various food behaviors: Results of a longitudinal study. Food Quality and Preference, 46, 33–39.

Silva, P., Freitas, J., Nunes, F. M., & Câmara, J. S. (2021). A Predictive Strategy Based on Volatile Profile and Chemometric Analysis for Traceability and Authenticity of Sugarcane Honey on the Global Market. Foods, 10(7), 1559.

Singh, G., & Pandey, N. (2018). The determinants of green packaging that influence buyers' willingness to pay a price premium. Australasian Marketing Journal, 26, 221–230.

Spence, M., Stancu, V., Elliott, C. T., & Dean, M. (2018). Exploring consumer purchase intentions towards traceable minced beef and beef steak using the theory of planned behavior. Food Control, 91, 138-147.

Spieth, P., Roeth, T., & Meissner, S. (2019). Reinventing a business model in industrial networks: Implications for customers' brand perceptions. Industrial Marketing Management, 83(7), 275–287.

Srinivasan, K. & Nirmala, R. (2014). A Study on Consumer Behaviour towards Instant Food Products (With Special References to Kanchipurm Town). Journal of Business and Management, 16(11), 17-35.

Sun, J., Zhang, G. L., Li, S., Ivanov, A. R., Fenyo, D., Lisacek, F., ... & Brusic, V. (2014). Pathway analysis and transcriptomics improve protein identification by shotgun proteomics from samples comprising small number of cells-a benchmarking study. BMC genomics, 15(9), 1-10.

Sung-Kyoung, K., & Sang Bok, L. (2009). Highly encoded one-dimensional nanostructures for rapid sensing. Journal of Materials Chemistry, 19, 1381–1389.

"UK Food Law News (02-35)". www.reading.ac.uk.

Vandeginste, B. (2013). 7-Chemometrics in studies of food origin. In P. Brereton (Ed.), New analytical approaches for verifying the origin of food (pp. 117e145). Woodhead Publishing.

Versari, A., Laurie, V.F., Ricci, A., Laghi, L., & Parpinello, G.P. (2014). Progress in authentication, typifcation and traceability of grapes and wines by chemometric approaches. Food Research International, 60, 2–18.

Videbæk, P. N., & Grunert, K. G. (2017). Forbrugerinteresse i at spise insekter. Litteraturstudie od.

Vyshnavi, P. V. S., & Rao, P. V. (2016). Factors influencing consumer behaviour: brand preferences, usage patterns, and satisfaction levels of using mobile phones. International Journal of Innovative Research and Development, 5(9), 9-31.

Wang, S., Zheng L., Cai, G., Liu, N., Liao, M., Li, Y., Zhang, X., & Lin, J. (2019). A microfluidic biosensor for online and sensitive detection of Salmonella typhimurium using fluorescence labeling and smartphone video processing. Biosensors & Bioelectronics, 140, 111333

Winger, R. and Wall, G. (2006). Food product innovation: a background paper.

Xu, M., Zhu S., & Yu, Y. (2017). Quality assessment of Chinese liquor with different ages and prediction analysis based on gas chromatography and electronic nose. Scientific Reports, 7 (1), 1–8

Yadav, S., Carvalho, J., Trujillo, I., & Prado, M. (2021). Microsatellite Markers in Olives (Olea europaea L.): Utility in the Cataloging of Germplasm, Food Authenticity and Traceability Studies. Foods, 10(8), 1907.

Zepeda, L., & Deal, D. (2009). Organic and local food consumer behaviour: Alphabet theory. International Journal of Consumer Studies, 33(6), 697-705.

Zhao, Y., Zhang, B., Chen, G., Chen, A., Yang, S., & Ye, Z. (2014). Recent developments in application of stable isotope analysis on agro-product authenticity and traceability. Food Chemistry, 145, 300-305.

Zhang, M., & Li, P. (2012). RFID application strategy in agri-food supply chain based on safety and benefit analysis. Physics Procedia, 25, 636-642.

20

Emerging Issues in Preventing Food Fraud

Neha Chaubey, Narender Raju Panjagari

Dairy Technology Division, ICAR-National Dairy Research Institute Karnal, Haryana, India

Introduction

Food fraud is not new it can be traced back to ancient Rome and Athens, where regulations prohibiting the adulteration of beverages with flavours and colours were in place. It exists since the food trade has been started. Food fraud affects all food supply chains, thus the entire food sector. Due to globalisation and the increasing complexity of supply lines, food supply chains are becoming more vulnerable to food fraud (Robson *et al.*, 2021a). A fraud is "any act or omission relating to the use or presentation of false, incorrect, or incomplete statements or documents, or non-disclosure of information in violation of a specific obligation, and any other illegal activity affecting the financial interests" (EU No. 2017/1371). Food related frauds can not only harm food enterprises and industrial associations, but also lead the competent authorities in charge of ensuring food safety to lose public faith due to serious threat involved to consumer health (Visciano & Schirone, 2021). They refer to the deliberate substitution, manipluation, tampering, or misrepresentation of food ingredients or food or food packaging for commercial advantage (Silvis *et al.*, 2017). Any individual or group involved in the entire supplychain can commit food fraud (Johnson, 2014). Food fraud is gaining a lot of attention these days resulting in an increase in the publications. Frederick Carl Accum, who made the first serious attempt to reveal both the magnitude and risks of food adulteration in 1820, was the first researcher to use scientific procedures to detect food adulteration (Valand *et al.*, 2020).

Food fraud is incredibly expensive to combat, the cost of food fraud to the global food sector was estimated to be around EUR 30 billion per year (European Commission, 2018). It is not only about the global economic concern but also about the serious health consequences that a consumer faces such as nausea, paralysis, allergic reactions, mental retardness, and even sometimes death (Valand *et al.*, 2020).

Renewed public health programs including food grew and evolved, resulting in diverse definition sets and scope breadths. There are numerous terminologies with ambiguous, even contradictory definitions (MSU, 2018). Given below are some of the definitions related to food fraud for a better understanding of the topic.

Contaminant: Any substance not intentionally added to food, which is present in such food as a result of the production, manufacture, processing, preparation, treatment, packing, packaging, transport or holding of such food or as a result of environmental contamination. Owever, the term exclude insect fragments, rodent hairs and other extraneous matter (Codex Alimentarius, 2014).

Economically Motivated Adulteration (EMA): It is to increase the apparent value of the product or reduce the cost of its production, i.e., for economic gain. EMA includes dilution of products with increased quantities of an already-present substance to the extent that such dilution poses a known or public health concern, as well as the inclusion or removal of substances to mask dilution (FDA, 2009).

Food Authenticity: It is about ensuring that food offered for sale or sold is of the nature, substance, and quality expected by the purchaser. Authenticity can be a particular issue for consumers or religious groups with particular food preferences who do not want to purchase products prepared not according to their faith or containing certain ingredients not allowed in their religion (DEFRA, 2014).

Food Integrity: It referes to the state of being whole, entire, or undiminished or in perfect condition.

Food Defense- "The process to ensure the security of food and drink and their supply chains from all forms of intentional malicious attack including ideologically motivated attack leading to contamination or supply failure" (GFSI, 2017)

Food Crime: Dishonesty relates to the production or supply of food that is either complex or likely to be seriously detrimental to consumers, businesses, or the overall public interest. Food fraud becomes a food crime when the scale and potential impact of the activity is considered to be serious (NFCU, 2017).

International and National Food Fraud Scenario

The majority of occurrences go unnoticed because they rarely pose a food safety danger, and consumers are generally unaware of a quality issue. Although the exact extent of food fraud is unknown, the number of documented occurrences may be a small percentage of the true number (Johnson, 2014). In 1981 in Spain toxic olive oil incident occurred in which industrial fuel oil was sold as olive oil resulting in about tweleve hundred deaths and twenty thousand hospitalizations (Visciano & Schirone, 2021). In the USA, melamine-laced pet food killed a huge number of dogs and cats in 2007. In an another incident, Peanut Corporation of America's actions related to Salmonella epidemic, which killed nine individuals and sickened 700 in the year in 2009, was also motivated by fraud (Johnson, 2014). In China in 2008, the addition of melamine into the baby formula has created a lot of questions regarding the safety of food (Pei *et al.*, 2011). A recent one is the China gutter oil scandal, recycled oil from grease traps, slaughterhouse and restaurant fryers or extracted from discarded animal internal organs, fat, and skin was sold out as cooking oil. Reports revealed that the contaminated products were found in Taiwan, Singapore and Hong Kong also (Visciano & Schirone, 2021). Almost 20% of honey sold in Australia is contaminated with ingredients like cane sugar or corn syrup (Zhou *et al.*, 2018). In another study, DNA testing on 153 samples from multiple brands (30) of roasted cod fillet (Xue Yu) conducted by Chinese and Italian scientists revealed that 58 percent of the samples were substituted with other fish species (Xiong *et al.*, 2017). In Taiwan, a food processing company was found guilty of mislabelling high-grade olive oil with low-grade palm oil and other low-cost oils and the mixture also contained artificial colorants that were dangerous to human health (Logan, 2016). In Bangladesh, authorities shut down a synthetic fruit juice production plant because it was producing a synthetic beverage that didn't contain any fruit, instead made with harmful chemical substances. The horse meat scandal (2013) in Europe is one of the food scandals where cheap horse meat was substituted with high price beef meat. In 2017 fipronil an insecticide was found in eggs in Europe which resulted in a massive recall of eggs from shelves (Fengou *et al.*, 2021).

In India, the Food Safety and Standards Authority of India (FSSAI) is the central regulatory body which looks after the food-related matters across the country, was established in 2008, however, its functioning started in 2011. Before FSSA, the Prevention of Food Adulteration (PFA) 1954, an act was there which now is all covered under Food Safety and Standard Act (FSSA). However, scenario of the situation is no good, according to a report by the Authentic solution provider association (ASPA), in 2018-19 when around 11000 samples were analyzed, it was found that the percentage of substandard food is 15.8%, mislabelled food 9% and unsafe food 3.7%. the counterfeiting incidences have risen. In 2019, around 249 (129 in dairy, 103 in alcohol, and 17 in agriculture) incidences were reported which were 21% more than what was reported in 2018 (Authentication Solution Providers' Association, 2020). Maggi was prohibited by India's food authorities in 2015 after tests revealed it had an excessive lead and was mislabelled with the flavor enhancer MSG. Since then, Nestle has withdrawn the claim "No added MSG" (ET Bureau, 2019). A recent one of the highlighted case was adulteration found in honey, NMR test was conducted by the Centre for Science and Environment (CSE) laboratory on 13 popular brands and found out that 10 i.e.,77 % of the market samples were adulterated with sugar syrup (Jacob Koshy, 2020). In 2021, police have seized around 4000 packets of tea, mislabelled as a popular brand i.e. Tata tea (JRC Food Fraud Monthly report,2021).

Intentional and Unintentional Contaminants

Contaminants present in food may be intentional or unintentional. Sometimes, it happens that the contaminants are present in food accidentally, then this is more of a food safety issue than a case of food fraud. Ignorance, lack of facilities, lack of knowledge, and inappropriate food and packaging handling results in unintentional contamination (Aswal, 2016). These unintentional contaminants can be divided into two categories environmental contaminants and processed toxins. Environmental contaminants include- Pesticides, Dioxins, furans (sources are municipal and industrial waste, fire, and wood), and heavy metals such as As, Hg, Cd, and Pb (Heavy metals are present in food due to consequence of human activity, contamination during processing or storage). Processed toxins include acrylamide (it is a carcinogenic compound formed in carbohydrate-containing food when processed at a high temperature like more than 120 °C), Maillard reaction derivatives (in the last stage of reaction advanced glycation end products are formed and responsible for diabetes and kidney disorder), biogenic amines (produced by microbial decarboxylation of aa, mainly in the processes involving LAB like fermentation and cured foods like fish sauce and fermented meat), fat peroxidation products, polycyclic aromatic hydrocarbons and heterocyclic amines (these compounds are formed at a very high temperature above 300°C having toxicity, carcinogenicity and even teratogenicity), nitrosamines (these are created when secondary or tertiary amines react with a nitrosating agent such as nitrous anhydride, which is made from nitrite in an acidic aqueous solution. To prevent the production of nitrosamine, ascorbic acid and sulphur dioxide are utilized. Unintentional food adulteration or contamination includes quality and safety-related aspects while intentional contamination covers food fraud and food defense (Spink & Moyer, 2011).

Table 1: Risk cause and effects for the food disciplines

Discipline Risk type	**Example**	**Cause and motivation**	**Effect**	**Public health risk type**	**Secondary effect**
Food quality	Bruising of fruits accidentally	Mishandling	Unsalable product or possible additional contamination with public health concern microbes	None or Food Safety	Reduced brand equity or incidence of food safety incident
Food fraud	Adulteration of milk with urea or melamine	Increased margin	Toxic poisonings	Food Safety	Public fear and possible lower prices
Food safety	Unintentional contamination of raw vegetables with public helath concern microbes	Limited field protection and control during post-harvest processing	Illnesses and/ or deaths	Food Safety	Foo recall and expenses related to that, and fear among people
Food defense	Intentional contamination of ground beef with nicotine	Revenge intent against the store/manager by way of injury to consumers	Nonlethal poisonings	Food Defence	Adulterated and inferior product, food recall expenses, and fear among the public

Source: (Spink & Moyer, 2011)

Databases to Track the Food Fraud

Around the globe, there are different databases to get information regarding food fraud (Fig.1). To effectively combat food fraud, stakeholders must have access to resources that provide reliable information (Ulberth, 2020). The Center for Food Protection and Defense in the United States (US) maintains the Economically Motivated Adulteration (EMA) database, which includes information about product, origin, year, type of fraud, effect on health, and their discovery. The United States Pharmacopoeia (USP) is a standards-setting organization that has been followed by countries since the late 1800s, including the United States. For product quality, the USP specifies specifications, test methodologies, and reference samples. The Rapid Alert System for Food and Feed (RASFF) is a network that connects the EU Member States, their competent authorities, the European Commission, the European Food Safety Authority (EFSA), the European Free Trade Association Surveillance Authority (i.e., Iceland, Liechtenstein, and Norway), and Switzerland to share details on the actions taken in the event of serious health problems caused by food or feed imported into the EU. RASFF site is an open-access platform, the other databases either require a subscription or have only a few free components (Manning & Soon, 2019). "HorizonScan" is UK based and it monitors over 536 commodities in 202 nations using official websites from over 65 countries and over 100 sources examined daily. On an average, more than two dozen new data records are added to the database daily (Bouzembrak *et al.*, 2018). The Food Adulteration Incidents Registry (FAIR) database is maintained by Food Protection and Defense Institute (FPDI) at–the University of Minnesota, which is a collection of current and historical food fraud events (Ulberth, 2020).

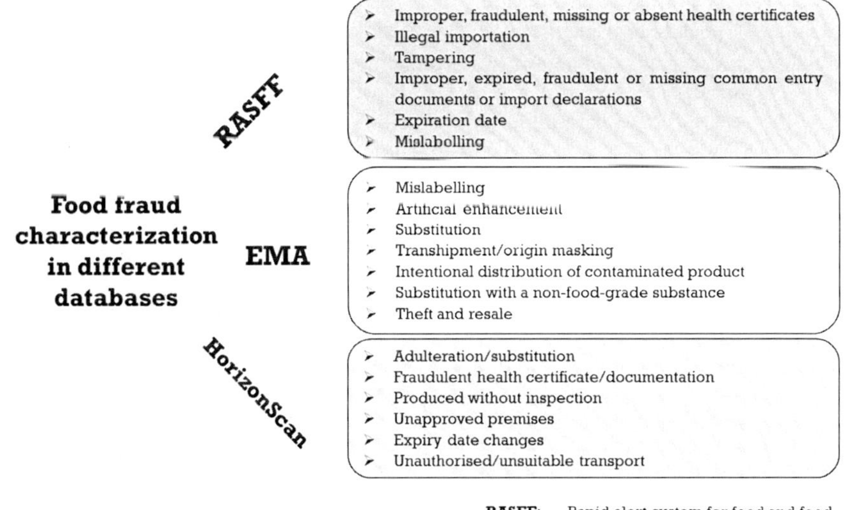

Fig.1. Some selected global databases to identify food fraud

Types of Food Fraud

Food fraud can be achieved in many ways, so can be categorized into different types these types are discussed below:

Table 2: Different types of Food Fraud

Term	Definition	Example	Potential public health threat
Adulteration	A component of the finished product is fraudulent	Melamine added to milk	Fake component
Tampering and mislabelling	Legitimate products and packaging are used in a fraudulent way	Changed expiry information; fraudulent description of production method or country of origin	Fake packaging information
Over-run	The legitimate product is made more than production agreements	Under-reporting of production	Fake product is distributed outside of regulated or controlled supply chain
Theft	Legitimate product is stolen and passed off as legitimately procured	Stolen products are mixed with legitimate products	Fake product is distributed outside of regulated or controlled supply chain
Diversion	The sale or distribution of legitimate products outside of intended markets	Relief food redirected to markets where aid is not required	Shortages or delays of relief food to needy populations
Simulation	An illegitimate product is designed to look like but not exactly copy the legitimate product	Imitation of popular foods not produced with the same food safety guarantees	Fake product of lesser quality
Counterfeit	All aspects of the fake product and packaging are fully replicated	Copies of popular foods not produced with the same food safety guarantees	Fake product

Source: (Spink & Moyer, 2011)

Adulteration

Adulteration is an umbrella term including substitution, dilution, and addition (Visciano & Schirone, 2021), and it is the most common type of food fraud (Fiorino *et al.*, 2018). In literature in many places food fraud and food adulteration are used interchangeably, however food adulteration is a subset of food fraud. FSSAI defined “adulterant” as any material which is or could be employed for making the food unsafe or sub-standard or misbranded or containing extraneous matter. The main motive behind adulteration is economic gain. It is harmful because

(a) it can be toxic and create health problems

(b) it can deprive people of essential nutrients

(c) it can cause intoxication or allergies in people who are allergic to certain foods.

Adulteration can be of two types intentional adulteration and unintentional adulteration, In the former adulteration material of cheap quality is added to food, which shows similar properties as that of food, e.g. addition of water in milk while in the latter one occurs when undesired substances are introduced into food due to ignorance, negligence, or a lack of sufficient facilities e.g. presence of a toxic variety of pulse or mushroom, presence of pesticide (Bansal *et al.*, 2017). Goyal *et al.*, 2022, stated that among food products dairy products account for 31% of the adulteration, followed by fruits at 30%, vegetables at 24%, and spices at 12%.

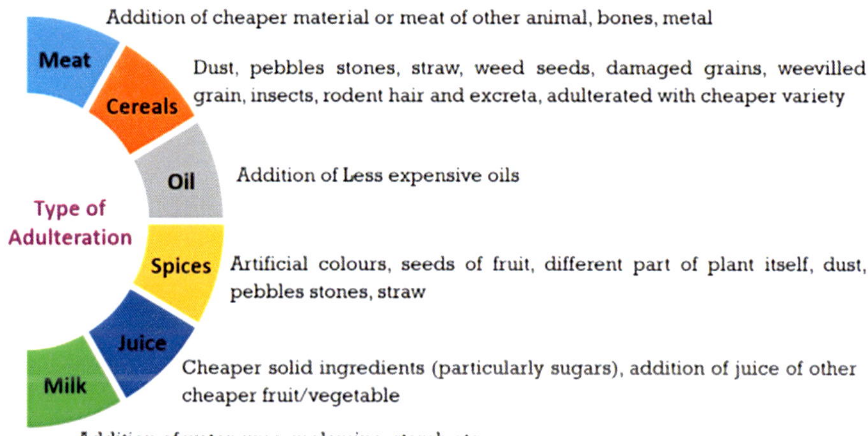

Fig. 2: Different food types and their common adulterants
(Modified from Bansal *et al.*, 2017)

Mislabelling

Food labelling is an indirect way of communication between manufacturer and consumer (Moreira *et al.*, 2021). People are now more aware of the package labels. Moreira *et al.*, 2021 mentioned that even reducing the font size decreases consumer trust in the quality of the product. The significance of food package labelling and the information offered to consumers has expanded greatly in the recent times due to the rise in consumer demands for healthier and safer products (Aung & Chang, 2014). Food labelling must be precise to ensure consumer safety along with providing the right choice. Mislabelling of food can include the following -

a) Ingredients and allergens (gluten, milk, egg, hazelnut, peanut, etc.)

b) Unmentioned false country of origin, low fat mentioned while giving more than 3g fat per serving)

c) Insufficient labelling (missing of warnings on the packet)

Research done in Itlay revealed that around 57% of the meat was mislabelled (Pinto *et al.*, 2014). Another research done at the EU for seafood, reveals similar data, Overall mislabeled samples were 31.1%. The most frequently mislabelled fish varieties are Cod (13.1%) and sole (11.1%). Bluefin tuna was 95% mislabelled (substituted by other species of tuna) (Christiansen *et al.*, 2018).

Counterfeiting

It is imitating of almost identical or identical to the original. They have an impact on food authenticity. One cause for the rapid expansion of food counterfeiting could be the increase in international trade and growing novel markets, as well as the fact that food prices are steadily rising globally (Huck *et al.*, 2016). Counterfeit alcohol is a common occurrence, particularly in colder climates. Several incidences of methanol poisoning caused by counterfeit spirits have happened in various parts of Russia, killing several individuals (Fiorino *et al.*, 2018).

Food Fraud in Different Food Products

In this section food frauds in different food products such as spices, fats, oils, cereal, pulses, etc. are discussed:

Spices

The spice business is considered one of the most vulnerable in terms of food fraud (Morling & McNaughton, 2016) along with the olive oil, fish and meat. Spices are frequently related to difficulties of authenticity due to the insertion of lower-value foreign and/or product-owned material. It is an alluring sector for possible fraud because the products have a high value by weight and customers are unable to find the adulteration. Also, in the case of ground spices, the original morphology is changed, so it becomes more difficult to identify the extraneous matter. E.g., the presence of pulverized peanut shells in powdered cumin was identified in 2014. Because of the allergenic qualities of the peanut substance, this resulted in a massive recall, posing a serious risk to the person with a peanut allergy (Silvis *et al.*, 2017). In 2003, red chili powder was found contaminated with Sudan I, which is banned as a food additive as it possesses carcinogenic and mutagenic properties (Visciano & Schirone, 2021). Incidences are also reported where lead chromate is added to turmeric to increase the intensity of color, which is detrimental to human health (Erasmus *et al.*, 2021). The most extensively used spice in the world is black pepper, which is commonly adulterated with papaya seeds, husk, and pinheads (Wilde *et al.*, 2019).

Vegetable Oils and Fat

Vegetable oils and fats play an important role in our diet as cooking or frying oil, salad oil, or used as an ingredient in the different food matrices. Also, from an economic standpoint, they are critical. Because some oils and fat have a high price, fraudsters may be tempted to adulterate them with less expensive oils and fats to make more money. In general, there are two types of fraud in oil, the first one is mixing less expensive oil with the final product and the second is admixing cold press oil (more expensive) with refined one (Azadmard-Damirchi & Torbati, 2015). According to Valand *et al.*, 2020, the most common food susceptible to food fraud is olive oil. It is adulterated with rapeseed and corn oil (De Lima *et al.*, 2020), also with the cheaper variety of olive oil from Turkey, vegetable oil, palm oil, sunflower, safflower e (Johnson, 2014), the hazelnut has been used to adulterate olive oil due to its similar composition(Azadmard-Damirchi & Torbati, 2015). Palm oil is another popular food, because of the high level of β-carotene, but it is frequently contaminated with Sudan IV, which is used as a color enhancer (MacArthur *et al.*, 2020). Edible oil such as soybean and sunflower are added to avocado oil (Quiñones-Islas *et al.*, 2013). Similarly, due to the premium price of cocoa butter and milk fat they are prone to fraudulent activity. Every vegetable oil and fat has a specific profile of fatty acids which is used as a specific marker for adulteration (Azadmard-Damirchi & Torbati, 2015).

Meat and Seafood Market

Meat has high nutritional significance it is a good source of high-quality protein, iron, vitamin B12, and other B vitamins, as well as zinc, selenium, and phosphorus (Pereira & Vicente, 2013). The horse meat scandal is a popular food fraud that happened in 2013. The presence of horse meat in processed food products labeled as beef was first discovered during routine control analyses in the United Kingdom and Ireland, but the real concern was the possibility of finding phenylbutazone, a nonsteroidal veterinary drug used in horses suffering from musculoskeletal disorders and soft or hard tissue injuries (Annunziata *et al.*, 2018). This compound is banned by FDA as it causes bone marrow toxicity (Dodman *et al.*, 2010). Some other scandals include the distribution of 150 tons of rotten meat in German restaurants in 2006. Wang *et al.*, 2020 determined goat meat was adulterated with pork. The injection of gel into shrimp to raise their weight and make them more palatable has become a significant problem in various Asian and Pacific countries. The health effects of such procedures are unknown, but they are likely to be dangerous because the items used to inject shrimp are not intended for human consumption. (FAO, 2021). The growing awareness of the health benefits of fish consumption, particularly the high level of PUFA, has increased by the manipulation and adulteration of seafood products, both fresh and processed (Fiorino *et al.*, 2018).

Milk and Milk Products

Milk and milk products are consumed by people of all age groups due to their taste, nutritional value, and palatability, hence there is a great scope of profit from fraudulent activity in this sector. Different examples of fraud will be discussed in this section. Water is added to milk for economic gain. For increasing the nitrogen content of milk which gives a false indication of high protein content different compounds are being added such as melamine, 3-amino-1, 2, 4-triazine (ATZ), and Urea (Sharma & Kaushal, 2021). Several cases have been found in which bovine milk is adulterated with milk of other species, rennet, reconstituted milk powder, detergents, etc. (Johnson, 2014). Melamine levels were found to be excessive in the products of 22 Chinese dairy enterprises in 2008, resulting in the deaths of six infants and the hospitalization of over 52,000 small children (Pei *et al.*, 2011). In ghee, vanaspati is added as an adulterant.

Fruits and Vegetables

Fruits and vegetables are widely regarded as being beneficial to one's health. Fruits and vegetables are good providers of vitamins and minerals, as well as phytochemicals, which act as antioxidants, phytoestrogens, anti-inflammatory agents, and other protective mechanisms, and also reduce the risk of cardiovascular diseases and obesity because of good fiber content (Slavin & Lloyd, 2012). Different processed products are made from them such as juices, concentrate, pickles, etc., which have a good market value and are so more susceptible to fraud. Juices are generally adulterated with water or cheaper juices are mixed with an expensive one. Sometimes unlisted juices such as lemon juice, mandarin juice, or grapefruit juice are added to orange juice, similar is the case with apple juice (Johnson, 2014).

Cereals and Pulses

Cereals and pulses are the staple diet of the Indian population. It also undergoes many types of food fraud such as toor dhal and Bengal gram dhal is adulterated with khesari dal, which contains B-Oxalyl-Amino L-Alanine (BOAA) and causes health risk called lathyrism. Cases of color enhancement of arhar dhal using artificial colorant metanil yellow were also seen(Sharma & Kaushal, 2021).

Others

Tea is a widely consumed beverage in the world and cases of added carmine a harmful color was also reported (Wei *et al.*, 2020). Honey was reported to be diluted by molasses sugar, fructose, glucose, and high fructose corn syrup (Johnson, 2014). Chalk powder substitution in salt (K. *et al.*, 2017). Coffee is generally adulterated with chicory powder.

The established methods to identify the adulteration of a food product are using either microbiological or chemical signals, and they can be classed as targeted or untargeted. In targeted ones, specific and selected molecular markers are monitored to identify the presence or absence of substances potentially linked to food frauds. While in untargeted approaches, a holistic approach and aim to provide information on the entire food item, often in the form of a spectral fingerprint, providing a simpler and overall image of the food under investigation (Fiorino *et al.*, 2018). As the metabolites profile is specific to a food product, it can also be used as a biomarker for the detection of fraud (Medina *et al.*, 2019). Along with the conventional detection methods, artificial intelligence is also now being used for the detection of adulteration (Goyal *et al.*, 2022).

Table 3: Overview of Analytical techniques used for detecting food frauds

Product Category	Fraud Type	Analytical approach	Predictors
Fish and fish products	Authentication	Mitochondrial DNA sequences	Primers for *CYTB* and *COI* genes Primer for *NAD*2 gene
			Primers for *NAD*2 and *NAD*5 genes
		DNA barcoding and multiplex PCR	Primers for *COI* gene
		qPCR	FASTFISH-ID
		DNA microarray	16S rDNA and *CYTB* target gene
	Authentication/ Traceability	Chemometric	NIR spectroscopy
	Substitution	Hyperspectral imaging	Reflectance/NIR/Fluorescence/Raman
	Substitution or mixing	Mid Infrared Spectroscopy	Spectral signature between 4000 and 700 cm^{-1}
	Identification	Loop-mediated isothermal amplification (LAMP)	Real-time fluorescence LAMP and visual LAMP
	Adulteration	Chemometric	Headspace-gas chromatography-ion mobility spectrometry
	Mislabelling	Peptide biomarkers	UPLC-Q/Exactive-HRMS
		DNA barcoding	Primers for *COI* gene
		DNA amplification	Primers of 16S rDNA
		Chemometric	$\delta^{13}C$ (carbon) and $\delta^{15}N$ (nitrogen) isotope
	Freezing-thawing	Nuclear Magnetic Resonance Spectroscopy (NMR)	H NMR
		Histological	Morphological/morphometrical parameters
		Proteomic	Two-dimensional electrophoresis and high-resolution LC-MS/MS
Meat and meat products	Animal species detection	Qualitative duplex real-time PCR	*CYTB* barcoding gene
	Adulteration	Chemometric	Raman spectra of 1800-700 cm^{-1}
		Chemometric	NIR spectroscopy
		Chemometric	Laser-induced plasma spectroscopy

Product Category	Fraud Type	Analytical approach	Predictors
	Authentication	Mitochondrial DNA	Real-time LAMP
		Mitochondrial DNA	Primers for *CYTB* gene
		Peptide markers	LC-QTOF-MS
		Vis/NIR reflectance spectroscopy	Reflectance spectra at 350–2500 nm
	Mislabelling	Colorimetric	Polydopamine coated magnetic nanoparticles
		Peptide markers	LC-MS
		Spectrometry	Laser ablation ICP-MS
		DNA-based	Real-time PCR and ELISA
		DNA sequences	Multiplex PCR
Milk and dairy products	Adulteration	Real-time PCR	High-resolution melting/rhAmp® SNP
		Proteomic-based	Protein marker β-lactoglobulin
		Spectroscopy and chemometric	Laser-induced breakdown spectroscopy
		Chemometric	Differential scanning calorimetry
		DNA-based	Quadruplex PCR
	Authentication	DNA-based	Real-time PCR
		Mitochondrial DNA	Triplex real-time PCR
		Spectroscopy and chemometric	FT-NIR spectra of 10000-4000 cm^{-1}
	Authentication/ Identification	Spectroscopy	ATR/FTIR
	Dilution	Photometric	Digital image colorimetry
	Substitution	Synchronized fluorescence spectroscopy	Fluorescence spectra 250–550 nm
	Identification	Direct-triplex PCR and capillary electrophoresis	Primers for *COI* and *CYTB* genes
	Detection of maltodextrin	Protein extraction	MALDI-TOF MS
		Immunological and DNA-based	ELISA and RT-PCR
		Chemometric	FT-Raman spectroscopy
	Authentication	Spectroscopy and chemometric	^{13}C High Resolution NMR spectroscopy

Product Category	Fraud Type	Analytical approach	Predictors
Cereals and bakery products	Characterization	Spectroscopy	Pulse-echo ultrasonic system
	Authentication	Target metabolomic	HPLC/MS
		DNA-based	Multiplex real-time PCR
		Convolutional neural networks	Images
		Mid-infrared spectroscopy	Spectral range at 4000-700 cm^{-1}
		Spectroscopy	NIR hyperspectral imaging
	Authentication/ traceability	Spectroscopy and chemometric	Laser induced breakdown spectroscopy
		NIR	HS GC-MS
	Adulteration	Gas chromatography	HS solid-phase microextraction couple with GC-MS-MS
Fats and oils	Authentication	Spectroscopy and chemometric	^{13}C High-Resolution NMR spectroscopy
		Spectroscopy and chemometric	Dielectric spectroscopy range 0.04–20 MHz
		FT-IR spectroscopy	Spectra interval of 450–4000 cm^{-1}
	Adulteration	Mid-Infrared (MIR) spectroscopy	FT-MIR spectra of 4000-600 cm^{-1}
	Adulteration/ geographical origin	VIS/NIR spectroscopy	Spectral range of 400–1100 nm
		Spectroscopy	1H NMR
	Addition	Mass spectrometry	Direct infusion by electrospray

Source: Visciano & Schirone (2021)

Emerging Issues in Preventing Food Fraud

The points given below are the compiled information on food fraud mitigation measures from different organizations such as British Standards Institution (BSI), Nestle, FSSC 22000, Global Food Safety Initiative, National Food Crime Unit, Food and Drink Federation, Chartered Institute of Environmental Health, etc. These are the general guidelines for all supply chains to be followed in a large-scale industry that carries out activities such as processing, manufacturing, storage packaging, and transportation (Robson *et al.*, 2021b).

- Supply chain mapping
- Assessment of contributing factors:
  - Supply chain
  - Audit strategy
  - Supplier relationship
  - History of suppliers
  - Testing frequency
  - Geopolitical considerations
  - Fraud history
  - Economic anomalies
- Identification of impacts, risks, and opportunities
- Assessing and prioritizing findings
- Creating a plan of action
- Implementation, Tracking, Reviewing, and Communication.
- Potential impact assessment of fraud
  - Public health impact
  - Economic impact
- Focusing on fraud reliance
- Establish the nature and scale of the problem
- Designing and implementing fraud prevention, fraud detection and fraud investigation measures
- Mitigation strategies development
  - Raw Material Specifications
  - Analytical surveillance
  - Supplier relationship
  - Supplier audit
  - A science-based framework for understanding food fraud risk
  - Opportunities
  - Motivations

- Threats to the food supply chain
    - Malicious tampering
    - Theft
    - Food Safety and temperature control
- Understanding the attacker
- Establishing Threat Assessment Critical Control Point (TACCP)
- Critical Controls
- Response to an incident
- Review of food protection arrangements
- FSSC 22000 scheme Requirements
- Food fraud Mitigation Team Training
- The GFSI requirements on food fraud
- The implementation
- Auditing a vulnerability assessment and a food fraud mitigation plan

These mitigation and preventive measures are still lacking the guidelines, many documents give the same advice just presented in a different way. The core advice will include i) A system to monitor suppliers, ii) supplier audits, iii) creating open and transport relationships with suppliers, iv) analyzing threats and opportunities in all aspects of political, geographical, economical, and climate, v) finding vulnerabilities and risk assessment (Robson *et al.*, 2021b).

Conclusion

Food fraud is an existing problem, but now-a-days food supply chains are becoming more vulnerable to fraud due to globalization and increasing complexity. It has a negative impact on both economy as well as public health. Food fraud is done intentionally and possesses a food safety issue. Adulteration, counterfeiting, mislabelling, theft and tempering are different types of food frauds, adulteration being the most common type. Several incidences of fraud occurred globally in the past. Several databases are being maintained by different organizations or universities which track and record the information related to food fraud such as EMA, RASFF, HorizonScan, FAIR, etc. If relevant information is provided these databases can also be used as a key for fighting food fraud. There are several analytical methods are available for identifying the type of fraud type, and artificial intelligence is also now being used for the purpose. Food fraud mitigation guidelines are given by several organizations.

References

Annunziata, L., Visciano, P., Stramenga, A., Colagrande, M. N., Campana, G., Scortichini, G., Migliorati, G., & Compagnone, D. (2018). Investigation of phenylbutazone and its metabolite oxyphenbutazone in horse meat products during years 2013–2017. Drug Testing and Analysis, 10(8), 1251–1257. https://doi.org/10.1002/dta.2386

Aswal, M (2016). Food Adulteration and Consumer Awareness. Review Journal Philosophy & Social Science, Vol.41 pp 90-95

Aung, M. M., & Chang, Y. S. (2014). Traceability in a food supply chain: Safety and quality perspectives. Food Control, 39(1), 172–184. https://doi.org/10.1016/J.FOODCONT.2013.11.007

Authentication Solution Providers Association (ASPA) (2020). Importance of Authentication & Traceability in Indian Food Value Chain report for nation. www.aspaglobal.com

Azadmard-Damirchi, S., & Torbati, M. (2015). Adulterations in Some Edible Oils and Fats and Their Detection Methods. Journal of Food Quality and Hazards Control, 2, 38–44. http://www.jfqhc.com

Bansal, S., Singh, A., Mangal, M., Mangal, A. K., & Kumar, S. (2017). Food adulteration: Sources, health risks, and detection methods. Critical Reviews in Food Science and Nutrition, 57(6), 1174–1189. https://doi.org/10.1080/10408398.2014.967834

Bouzembrak, Y., Steen, B., Neslo, R., Linge, J., Mojtahed, V., & Marvin, H. J. P. (2018). Development of food fraud media monitoring system based on text mining. Food Control, 93, 283–296. https://doi.org/10.1016/J.FOODCONT.2018.06.003

Christiansen, H., Fournier, N., Hellemans, B., & Volckaert, F. A. M. (2018). Seafood substitution and mislabeling in Brussels' restaurants and canteens. Food Control, 85, 66–75. https://doi.org/10.1016/J.FOODCONT.2017.09.005

CODEX, Codex Alimentarius (2014). Procedural Manual, Twenty-second edition. World Health Organization /Food and Agriculture Organization of the United Nations, Geneva/Rome

DA, Food and Drug Administration (2009) "Economically Motivated Adulteration; Public Meeting; Request for Comment, [Docket No. FDA-2009-N-0166]." Federal Register74,15497

DEFRA, United Kingdom Department for Environment, Food & Rural Affairs (2014). Elliott review into the integrity and assurance of food supply networks, Independent report, Ref: PB14089, PDF, 539KB, 84 pages, URL: Retrieved fromhttps://www.gov.uk/government/uploads/system/uploads/attachment_data/file/350726/elliot-review-final-report-july2014.pdf

Dodman, N., Blondeau, N., & Marini, A. M. (2010). Association of phenylbutazone usage with horses bought for slaughter: A public health risk. Food and Chemical Toxicology, 48, 1270–1274.

Erasmus, S. W., van Hasselt, L., Ebbinge, L. M., & van Ruth, S. M. (2021). Real or fake yellow in the vibrant colour craze: Rapid detection of lead chromate in turmeric. Food Control, 121, 107714. https://doi.org/10.1016/J.FOODCONT.2020.107714

ET Bureau (2019,). Nestle India to contest fine in Maggi noodles case. Economic Times. September 11; https://economictimes.indiatimes.com/industry/cons-products/fmcg/nestle-india-to-contest-fine-in-maggi-noodles-case/articleshow/71073426.cms?from=mdr

EU FIP, European Union Food Integrity Project (2017). URL: Retrieved from https://secure.fera.defra.gov.uk/foodintegrity/index.cfm

European Commission. 2018. The EU Food Fraud Network and the System for Administrative

Assistance - Food Fraud. Annual Report 2018. [online]. [Cited 7 October 2020]. https://ec.europa.eu/food/sites/food/files/safety/docs/ff_ffn_annual-report_2018.pdf

Fengou, L. C., Lianou, A., Tsakanikas, P., Mohareb, F., & Nychas, G. J. E. (2021). Detection of meat adulteration using spectroscopy-based sensors. Foods, 10(4). https://doi.org/10.3390/FOODS10040861/S1

Fiorino, G. M., Garino, C., Arlorio, M., Logrieco, A. F., Losito, I., & Monaci, L. (2018). Overview on Untargeted Methods to Combat Food Frauds: A Focus on Fishery Products. Journal of Food Quality, 2018. https://doi.org/10.1155/2018/1581746

GFSI, Global Food Safety Initiative (2017). "Guidance Document, Benchmarking Document, Version 7, URL: Retrieved from http://www.theconsumergoodsforum.com/files/Publications/GFSI_Guidance_Document_Intro.pdf

Goyal, K., Kumar, P., & Verma, K. (2022). Food Adulteration Detection using Artificial Intelligence: A Systematic Review. In Archives of Computational Methods in Engineering (Vol. 29, Issue 1, pp. 397–426). Springer Science and Business Media B.V. https://doi.org/10.1007/s11831-021-09600-y

Huck, C. W., Pezzei, C. K., & Huck-Pezzei, V. A. (2016). An industry perspective of food fraud. Current Opinion in Food Science, 10, 32–37. https://doi.org/10.1016/J.COFS.2016.07.004

Jacob Koshy (2020). 10 out of 13 honey brands fail 'purity test', finds CSE investigation. The Hindu. December 3; https://www.thehindu.com/news/national/10-out-of-13-honey-brands-fail-purity-test-finds-cse-investigation/article33230094.ece

Johnson, R. (2014). Food fraud and "economically motivated adulteration" of food and food ingredients. www.crs.govR43358

JRC Food Fraud Monthly Report (2021). Summary of Articles on Food Fraud and Adulteration Retrieved mainly from the JRC tool Medisys (http://medisys.newsbrief.eu). September; http://medisys.newsbrief.eu

K., S. M., P, N., & K., R. (2017). Analysis of Food Adulterants in Selected Food Items Purchased From Local Grocery Stores. International Journal of Advances in Scientific Research, 3(7), 82. https://doi.org/10.7439/IJASR.V3I7.4299

Logan, W. 2016. Taiwanese businessman sentenced to four years in prison for olive oil fraud. In: Olive Oil Times. Rhode Island, USA.. https://www.oliveoiltimes.com/business/asia/taiwanese-businessman-sentenced-to-four-years-in-prison-for-olive-oil-fraud/51023

MacArthur, R. L., Teye, E., & Darkwa, S. (2020). Predicting adulteration of Palm oil with Sudan IV dye using shortwave handheld spectroscopy and comparative analysis of models. Vibrational Spectroscopy, 110, 103129

Manning, L., & Soon, J. M. (2019). Food fraud vulnerability assessment: Reliable data sources and effective assessment approaches. Trends in Food Science & Technology, 91, 159–168

Medina, S., Pereira, J. A., Silva, P., Perestrelo, R., & Câmara, J. S. (2019). Food fingerprints – A valuable tool to monitor food authenticity and safety. Food Chemistry, 278, 144–162. https://doi.org/10.1016/J.FOODCHEM.2018.11.046

Moreira, M. J., García-Díez, J., de Almeida, J. M. M. M., & Saraiva, C. (2021). Consumer knowledge about food labeling and fraud. Foods, 10(5). https://doi.org/10.3390/foods10051095

Morling, A., & McNaughton, R. (2016). A 2016 baseline food crime annual strategic assessment. Retrieved from https://www.food.gov.uk/sites/default/files/fsa-food-crime-assessment-2016.pdf.

MSU FFI, Michigan State University Food Fraud Initiative (2018). Glossary of Food Fraud Related Terms, Edited by John Spink, URL: Retrieved fromwww.foodfraud.msu.edu/resources/publications/

NFCU, UK National Food Crime Unit. (2017). "What is food crime and how does it differ from food fraud? URL: Retrieved fromhttps://www.food.gov.uk/enforcement/the-national-food-crime-unit/what-is-food-crime-and-food-fraud

Pei, X., Tandon, A., Alldrick, A., Giorgi, L., Huang, W., & Yang, R. (2011). The China melamine milk scandal and its implications for food safety regulation. Food Policy, 36, 412–420. https://doi.org/10.1016/j.foodpol.2011.03.008

Pereira, P. M. de C. C., & Vicente, A. F. dos R. B. (2013). Meat nutritional composition and nutritive role in the human diet. Meat Science, 93(3), 586–592. https://doi.org/10.1016/J.MEATSCI.2012.09.018

Pinto, A. di, Bottaro, M., Bonerba, E., Bozzo, G., Ceci, E., Marchetti, P., Mottola, A., & Tantillo, G. (2014). Occurrence of mislabeling in meat products using DNA-based assay. https://doi.org/10.1007/s13197-014-1552-y

Quiñones-Islas N, Meza-Márquez OG, Osorio-Revilla G, Gallardo-Velazquez T. Detection of adulterants in avocado oil by Mid-FTIR spectroscopy and multivariate analysis. Food Research International. 2013 Apr 1;51(1):148-54

Robson, K., Dean, M., Brooks, S., Haughey, S., & Elliott, C. (2020). A 20-year analysis of reported food fraud in the global beef supply chain. Food Control, 116, 107310

Robson, K., Dean, M., Haughey, S., & Elliott, C. (2021b). A comprehensive review of food fraud terminologies and food fraud mitigation guides. Food Control, 120, 107516. https://doi.org/10.1016/J.FOODCONT.2020.107516

Schmutzler M, Huck C W :Automatic sample rotation for simultaneous determination of geographical origin and quality characteristics of apples based on near infrared spectroscopy.VibrSpectrosc2014,72:97-104.

Sharma, S., & Kaushal, P. (2021). Food fraud: A menace to public health. Journal of Advanced Research and Reviews, 2021(03), 466–475. https://doi.org/10.30574/wjarr.2021.12.3.0671

Silvis, I. C. J., van Ruth, S. M., van der Fels-Klerx, H. J., & Luning, P. A. (2017). Assessment of food fraud vulnerability in the spices chain: An explorative study. Food Control, 81, 80–87. https://doi.org/10.1016/J.FOODCONT.2017.05.019

Slavin, J. L., & Lloyd, B. (2012). Health Benefits of Fruits and Vegetables. Advances in Nutrition, 3(4), 506. https://doi.org/10.3945/AN.112.002154

Spink, J., & Moyer, D. C. (2011). Defining the Public Health Threat of Food Fraud. Journal of Food Science, 76(9), R157–R163. https://doi.org/10.1111/J.1750-3841.2011.02417.X

Ulberth, F. (2020). Tools to combat food fraud – A gap analysis. Food Chemistry, 330, 127044. https://doi.org/10.1016/J.FOODCHEM.2020.127044

Valand, R., Tanna, S., Lawson, G., & Bengtström, L. (2020). A review of Fourier Transform Infrared (FTIR) spectroscopy used in food adulteration and authenticity investigations. Food Additives and Contaminants - Part A Chemistry, Analysis, Control, Exposure and Risk Assessment, 37(1), 19–38. https://doi.org/10.1080/19440049.2019.1675909

Visciano, P., & Schirone, M. (2021). Food frauds: Global incidents and misleading situations. In Trends in Food Science and Technology (Vol. 114, pp. 424–442). Elsevier Ltd. https://doi.org/10.1016/j.tifs.2021.06.010

Von Bargen, C., Dojahn, J., Waidelich, D., Humpf, H. U., & Brockmeyer, J. (2013). New sensitive high-performance liquid chromatography-tandem mass spectrometry method for the detection of horse and pork in halal beef. Journal of Agricultural and Food Chemistry, 61(49), 11986–11994. https://doi.org/10.1021/jf404121b

Wang Z, Wang Z, Li T, Qiao L, Liu R, Zhao Y, Xu Z, Chen G, Yang S, Chen A. Real-time PCR based on single-copy housekeeping genes for quantitative detection of goat meat adulteration with pork. International Journal of Food Science & Technology. 2020 Feb;55 (2):553-8.

Wei, L., Yang, Y., & Sun, D. (2020). Rapid detection of carmine in black tea with spectrophotometry coupled predictive modelling. Food Chemistry, 329. https://doi.org/10.1016/J.FOODCHEM.2020.127177

Wilde AS, Haughey SA, Galvin-King P, Elliott CT. The feasibility of applying NIR and FT-IR fingerprinting to detect adulteration in black pepper. Food Control. 2019 Jun 1;100:1-7

Xiong, X., Yao, L., Ying, X., Lu, X., Guardone, L., Armani, A., Guidi, A. & Xiong, X. 2017. Multiple fish species identified from China's roasted Xue Yu fillet products using DNA and mini-DNA barcoding: Implications on human health and marine sustainability. Science Direct Volume 88:123–130.

Index